PERIODIC TABLE OF THE ELEMENTS

Key:

6	Atomic number
C	Symbol
12.0	Atomic weight

IA	IIA	IIIB	IVB	VB	VIB	VIIB	VIIIB	VIIIB	VIIIB	IB	IIB	IIIA	IVA	VA	VIA	VIIA	Noble gases
1 H 1.0																	2 He 4.0
3 Li 6.9	4 Be 9.0											5 B 10.8	6 C 12.0	7 N 14.0	8 O 16.0	9 F 19.0	10 Ne 20.2
11 Na 23.0	12 Mg 24.3											13 Al 27.0	14 Si 28.1	15 P 31.0	16 S 32.1	17 Cl 35.5	18 Ar 39.9
19 K 39.1	20 Ca 40.1	21 Sc 45.0	22 Ti 47.9	23 V 50.9	24 Cr 52.0	25 Mn 54.9	26 Fe 55.8	27 Co 58.9	28 Ni 58.7	29 Cu 63.5	30 Zn 65.4	31 Ga 69.7	32 Ge 72.6	33 As 74.9	34 Se 79.0	35 Br 79.9	36 Kr 83.8
37 Rb 85.5	38 Sr 87.6	39 Y 88.9	40 Zr 91.2	41 Nb 92.9	42 Mo 95.9	43 Tc (98) ‡	44 Ru 101.1	45 Rh 102.9	46 Pd 106.4	47 Ag 107.9	48 Cd 112.4	49 In 114.8	50 Sn 118.7	51 Sb 121.8	52 Te 127.6	53 I 126.9	54 Xe 131.3
55 Cs 132.9	56 Ba 137.3	57 La* 138.9	72 Hf 178.5	73 Ta 180.9	74 W 183.9	75 Re 186.2	76 Os 190.2	77 Ir 192.2	78 Pt 195.1	79 Au 197.0	80 Hg 200.6	81 Tl 204.4	82 Pb 207.2	83 Bi 209.0	84 Po (210)	85 At (210)	86 Rn (222)
87 Fr (223)	88 Ra (226)	89 Ac† (227)	104 Unq (261)	105 Unp (262)	106 Unh (263)	107 Uns (262)	108 — 	109 Une (266)									

* Lanthanide series:

58 Ce 140.1	59 Pr 140.9	60 Nd 144.2	61 Pm (147)	62 Sm 150.4	63 Eu 152.0	64 Gd 157.3	65 Tb 158.9	66 Dy 162.5	67 Ho 164.9	68 Er 167.3	69 Tm 168.9	70 Yb 173.0	71 Lu 175.0

† Actinide series:

90 Th 232.0	91 Pa (231)	92 U 238.0	93 Np (237)	94 Pu (242)	95 Am (243)	96 Cm (247)	97 Bk (247)	98 Cf (249)	99 Es (254)	100 Fm (253)	101 Md (256)	102 No (254)	103 Lr (256)

† Parentheses around atomic weight indicate that weight given is that of the most stable known isotope.

‡ Parentheses around atomic weight indicate that weight given is that of the most stable known isotope.

Chemistry
for
Changing Times

Fifth Edition

Chemistry
for
Changing Times

John W. Hill
University of Wisconsin—River Falls

With Special Contributions By

Cynthia S. Hill, R.D.
South Dakota Department of Health

Macmillan Publishing Company
New York

Macmillan Publishing Company
866 Third Avenue, New York, New York 10022
Collier Macmillan Canada, Inc.

Library of Congress Cataloging in Publication Data
Hill, John William (date)
    Chemistry for changing times.

    Includes index.
    1. Chemistry. I. Hill, Cynthia S. II. Title.
QD33.H65 1988   540   87-18502
ISBN 0-02-355010-4

Printing: 1 2 3 4 5 6 7 8   Year: 8 9 0 1 2 3 4 5 6 7

Preface

The previous editions of *Chemistry for Changing Times* were written with the firm conviction that chemistry can be presented in an intellectually honest way to students who have little background and no prior interest in chemistry. That assumption also underlies this revision, as does the belief that a course for students not majoring in science must be different from a course taught to science majors. This philosophy is spelled out in some detail in my paper "Chemistry for Citizens: Content and Strategies," *Journal of Chemical Education*, **62**, 765–767 (1985), reprints of which are available from the author.

Our Goals

We can no longer feel we have done our duty to nonscience students and to society by teaching only traditional chemistry courses. A traditional course will not stimulate a typical nonscience student's interest or kindle much of a desire to know more about the world. Most of the world's problems won't be solved without intelligent applications of chemistry, and if we fail to motivate our students or if we allow them to complete our courses without having gained an understanding of or an appreciation for the chemical world, all of society loses. We are living in an age of difficult daily decisions that affect our health and the future of the world, and the stakes are enormously high. Three-fourths of all legislation considered by the United States Congress involves scientific or technological questions. Our government is now committed to improving science education, especially for those who will not become scientists. If we who teach chemistry fail to take advantage of our present opportunities, we will be contributing to the very maladies we hope to correct.

Our Students

Generally, students enrolled in this course are not interested in the austere abstractness and the elegant mathematics of the physical theories that we scientists find so beautiful. If they were, they would most likely be science majors. Because of their temperament and

v

training, these students are not prepared to understand the awesome mathematical theories of quantum mechanics and thermodynamics; even the simplest stoichiometric relationships inspire fear in many of them. The essential quantitative nature of chemistry must be presented to students with care to avoid intimidation.

Our Objectives

The principal objectives in a chemistry course for nonscience students are these.

- Design your chemistry course to attract as many students as possible. If they don't enroll, you won't have the opportunity to teach them.
- Have students study examples of current topics in chemistry so that they will incorporate into their lives a sense for how chemists approach and solve problems.
- Induce students to relate chemical problems to their own lives so that they will be better able to appreciate the significance of such problems.
- Instill in students an appreciation for chemistry as an open-ended learning experience that will continue throughout their lives. Chemistry should not be a subject that is memorized before the final exam and then quickly forgotten.
- Acquaint students with scientific methods so that they will be able to distinguish between science and technology.

These objectives have been met to a most gratifying extent in my own course. Course enrollment rose sharply during the early years of the course, and it has continued to rise more rapidly than overall university enrollment.

Major Changes in This Revision

The entire book has been updated. New sections have been added on nucleic acids, protein synthesis, and genetic engineering (Chapter 11). Sections on electrochemical cells and batteries are now included in the chapter on oxidation and reduction (Chapter 8). For those who wish to discuss electron configuration using energy sublevels, a section on the quantum mechanical description of atomic structure has been added (Chapter 3). Sections on bonding forces and the states of matter now appear at the end of Chapter 5.

Many other parts of the book have been changed a bit, but this revision still retains the spirit and philosophy of previous editions.

Readability

A chemistry text need not be stodgy in order to present its subject well. Learning chemistry is not always easy, but it can be rewarding if the information is conveyed in a clear and understandable way. With this thought in mind, we have tried to create a chemistry textbook that is both clearly readable and thoroughly enjoyable—and we believe we have succeeded in doing so.

Units of Measurement

The metric system has been adopted by most of the world, and the modern version, the International System of Units (SI), is used in many countries, especially by scientists. What units of measurement should be used in a chemistry text designed for use by students of the humanities and the social sciences? We feel that students will encounter both the old metric system and the new SI version, in addition to the customary United States system, in their everyday lives. Therefore, we have used the units that they are most likely to encounter—Calories, not joules, when discussing weight loss or gain, for example. Nevertheless, we have used SI units in the presentation of most chemical principles, and we have used the SI symbols (K, not °K) where they do not present a barrier to learning.

Complicated Chemical Structures

Structures of complicated molecules are presented in the text, particularly toward the end, but students should not feel that they are expected to memorize them. They have been presented merely to emphasize the fact that these structures are known and that molecular properties depend upon them. Students may, however, come to recognize familiar functional groups, even with respect to the most complicated molecules.

Glossary

A glossary is provided in Appendix E. Definitions are given there for all terms in **boldface** in the text.

Problems

The number of end-of-chapter problems has been increased more than threefold, from 444 in the fourth edition to 1364 in this edition. In addition, detailed worked-out examples are interspersed in the text, particularly in the earlier, more quantitative chapters (Chapters 1–6). Instructors should assign the problems at their own discretion. Answers to all odd-numbered numerical and short-answer problems are given in Appendix F. Detailed, worked-out solutions to all problems (except those that are projects and those that ask for an opinion) are provided in the Instructor's Manual.

References and Suggested Readings

A list of recommended books and articles appears at the end of each chapter. Students whose interest and enthusiasm have been sparked can delve more deeply into those subjects about which they are curious. Instructors will also find this information helpful.

SUPPLEMENTARY MATERIALS

The most important teaching aid in any course is the teacher. To make your task a bit easier as you teach the course, we have provided a variety of supplementary materials. These will also serve to enrich and round out the education of your students.

Study Guide

A study guide is available to help students learn the material more effectively. For each chapter of the text, this guide provides an overview, a list of objectives, and a series of short-answer questions. The guide also provides additional problems as well as worked-out sample problems where appropriate. Use your own judgment to make additions to and deletions from the lists of objectives.

Instructor's Manual

An instructor's guide has also been prepared in conjunction with the text. This guide suggests course outlines for courses ranging in length from one quarter to one full year. The flexibility of these outlines will enable you to choose the material that best suits the particular needs of your students. The guide also includes a test bank and transparency masters as well as source lists for other printed educational material, audiovisual aids, computer software, and lecture demonstrations. Teaching strategies for presenting difficult material are also provided. Many of the suggestions that have come from you, the users of *Chemistry for Changing Times*, have been incorporated in this guide, and additional suggestions are welcome any time.

The Laboratory Manual

In the fifth edition of the laboratory manual, *Chemical Investigations for Changing Times*, you will find short, cheap, and safe experiments. The large number of experiments enables you to select experiments that are best suited to your style of teaching.

ACKNOWLEDGMENTS

For this edition, the following people provided penetrating, challenging, and greatly helpful prepublication reviews: Professors Marcia Bailey, Central Michigan University; Frank Fazio, Indiana University of Pennsylvania; and Keith J. Harper, North Texas State University.

We are also indebted to the respondents to the users' survey for the fourth edition. We especially want to acknowledge the contributions of Albert Cheh of American University, Jerry Mitchell of Sacramento City College, and Charles Park of Grossmont Community College who went beyond the confines of the survey to provide innumerable helpful suggestions. C. W. Schimelpfenig of Dallas Baptist University also gave significant assistance.

Cynthia S. Hill, who has contributed so much to earlier editions, prepared the new material on nucleic acids, protein synthesis, genetic engineering, and phenylethylamine for the fifth edition. Her contributions are acknowledged on the title page. Lori Bakke and Tom Nagel provided invaluable service in the preparation of the glossary.

Above all, though, I would like to thank the many students who with zest and enthusiasm have gone on to learn for themselves more about chemistry that I ever taught them. Teaching is a joy, but learning is a real celebration. I have learned far more from my students than I can ever teach them; for that I am eternally grateful. Comments, corrections, suggestions, and criticisms are always welcome.

J. W. H.

Contents

To the Student

Welcome to Our Chemical World!

Chemistry is fun. Through this book, I would like to share with you some of the excitement of chemistry and some of the joy in learning about it. I hope to convince you that chemistry does not need to be excluded from your learning experiences. Learning chemistry will enrich your life—now and long after this course is over—through a better understanding of the natural world, the technological questions now confronting us, and the choices we must face as citizens within a scientific and technological society.

Chemistry Directly Affects Our Lives

How does the human body work? How does aspirin cure our headaches? Do steroids enhance athletic ability? Is table salt poisonous? Can scientists cure genetic diseases? Why do most weight-loss diets seem to work in the short run but fail in the long run? Does fasting "cleanse" the body? Why do our moods swing from happy to sad? Can a chemical test on urine predict possible suicide attempts? How does penicillin kill bacteria without harming our healthy body cells? Chemists have found answers to questions like these and continue to seek the knowledge that will unlock still other secrets of our universe. As these mysteries are resolved, the direction of our lives often changes—sometimes dramatically.

We live in a chemical world—a world of drugs, biocides, food additives, fertilizers, detergents, cosmetics, and plastics. We live in a world with toxic wastes, polluted air and water, and dwindling petroleum reserves. Knowledge of chemistry will help you to better understand the benefits and hazards of this world and enable you to make intelligent decisions in the future.

Chemical Dependency

We are all chemically dependent. Even in the womb, we depend on a constant supply of oxygen, water, glucose, and a multitude of other chemicals.

Our bodies are intricate chemical factories. They are durable but delicate systems. A myriad of chemical reactions are constantly taking place within us that allow our bodies to function properly. Thinking, learning, exercising, feeling happy or sad, putting on too much weight or not gaining enough, and virtually all life processes are made possible by these chemical reactions. Everything that we ingest is part of a complex process that determines whether our bodies work effectively or not. The consumption of some substances can initiate chemical reactions that will stop body functions altogether. Other substances, if consumed, can cause permanent handicaps, and others can make living less comfortable. A proper balance of the right foods provides the chemicals and generates the reactions we need in order to function at our best. The knowledge of chemistry that you will soon be gaining will help you to better understand how your body works so that you will be able to take proper care of it.

Changing Times

We live in a world of increasingly rapid change. It has been said that the only constant is change itself. At present, we are facing some of the greatest problems that humans have ever encountered, and the dilemmas with which we are now confronted seem to have no perfect solutions. We are sometimes forced to make a best choice among only bad alternatives, and our decisions often provide only temporary solutions to our problems. Nevertheless, if we are to choose properly, we must understand what our choices are. Mistakes can be costly, and they cannot always be rectified. It is easy to pollute, but cleaning up pollution once it is there is enormously expensive. We can best avoid mistakes by collecting as much information as possible before making critical decisions. Science is a means of gathering and evaluating information, and chemistry is central to all the sciences.

Chemistry and the Human Condition

Above all else, my hope is that you will learn that chemistry need not be dull and difficult. Rather, it can enrich your life in so many ways—through a better understanding of your body, your mind, your environment, and the world in which we live. After all, the search to understand the universe is an essential part of what it means to be human.

1

Chemistry

A Science for All Seasons

Chemists find toxic chemicals in our drinking water. Lethal chemicals are released in industrial accidents. Death-dealing chemicals are stockpiled as weapons. Cancer-causing chemicals are detected in the air of our cities. Lives are diminished and destroyed through chemical abuse.

Chemicals relieve pain. Chemicals kill bacteria and viruses that cause disease and death. Chemicals increase our food supply and improve our nutrition. Chemicals clothe us and provide us with cheaper and better housing. Chemicals fuel the machines that relieve us of back-breaking labor and that transport us rapidly to the far reaches of the world—and even to other worlds. Chemicals increase our wealth and improve our leisure time. Chemicals provide us with luxuries unavailable even to the mightiest kings of ages past.

Chemicals are toxic. Some are lethal. Some cause cancer. Chemicals are also beneficial. Some save lives. Many are useful. Often a particular chemical is both useful and dangerous. Chemicals may be and often are both good and bad.

Perhaps a better understanding of chemistry would enable us to control the uses of chemicals so that we could maximize their benefits and minimize the risk involved in their use.

Figure 1.1 Aristotle (384–322 B.C.), Greek philosopher and tutor of Alexander the Great, believed that we could understand nature through logic. The idea of experimental science did not triumph over Aristotelian logic until about A.D. 1500. [Courtesy of the Smithsonian Institution, Washington, DC.]

Just what is chemistry anyway? Consider the usual definition: **chemistry** is a study of matter and the changes it undergoes. What is **matter**? It is anything that has mass and occupies space. Matter is the stuff of which all material things are made. We change matter to make it more useful. Most changes in matter are accompanied by changes in energy. Some matter we change to extract a part of its energy; for example, we burn gasoline to get energy to propel our automobiles.

More about matter and energy later. Indeed, matter and energy are what this book—and all of chemistry—is about. But chemistry isn't just something you hear or read about. You practice chemistry every day.

You practice chemistry in the kitchen when you cook. You practice chemistry when you clean your house, wash your car, or paint a fence. You practice chemistry in the bathroom when you bathe and apply cosmetics. You practice chemistry when you take medicine or treat an injury. Indeed, some remarkable chemistry occurs while you eat or breathe and even while you sleep. Your body is the most miraculous of all the chemical factories on Earth. It takes the food you eat and turns it into muscle and blood and skin and bones and brains and a myriad of other marvelous things. Your body takes oxygen from the air and combines it with part of the food you eat to provide you with energy for every activity you undertake.

What is chemistry? It is a science that touches your life every moment. It deals with matter from the tiniest parts of atoms to the minutest materials of the complex human body. It goes beyond the individual to affect society as a whole, and it shapes our civilization.

Science and Technology: The Roots of Knowledge

Chemistry is a *science*, but what is a science? Let's examine the roots of science. Our study of the material universe has two facets: the *technological*, or *factual*, and the *philosophical*, or *theoretical*.

Technology arose long before science, having its origins in antiquity. The ancients used fire to bring about chemical changes. For example, they cooked food, baked pottery, and smelted ores to produce metals such as copper. They made beer and wine by fermentation, and obtained dyes and drugs from plant materials. These things—and many others—were accomplished without an understanding of the scientific principles involved.

The Greek philosophers, about 2500 years ago, were perhaps the first to formulate theories explaining the behavior of matter. They generally did not test their theories by experimentation, however. Nevertheless, their view of nature—attributed mainly to Aristotle—was consistent internally, and it dominated natural philosophy for 2000 years.

The experimental roots of chemistry are planted in **alchemy**, a mystical chemistry that flourished in Europe during the Middle Ages (about A.D. 500 to 1500). Modern chemists inherited from the alchemists an abiding interest in human health and the quality of life. Consider, for example, that alchemists not only searched for a philosophers' stone that would turn cheaper metals into gold but also sought an elixir that would confer immortality on those exposed to it. Alchemists never achieved these goals, but they discovered many new chemical substances and perfected techniques such as distillation and extraction that are still used today.

Technology also developed rapidly during the Middle Ages in Europe, in spite of the generally nonproductive Aristotelian philosophy that prevailed. The beginnings of modern science were more recent, however, coming with the emergence of the experimental method. What we now call science grew out of **natural philosophy**, that is, out of philosophical speculation about nature. Science had its true beginnings in the seventeenth century, when the work of astronomers, physicists, and physiologists was characterized by a reliance on experimentation.

Figure 1.2 "The Alchemist," a painting done by the Dutch artist Cornelis Bega around 1660, depicts a laboratory of the seventeenth century. [Courtesy of Aldrich Chemical Company, Milwaukee, WI.]

Figure 1.3 Sir Francis Bacon (1561–1626), English philosopher and Lord Chancellor to James I. [Courtesy of the Smithsonian Institution, Washington, DC.]

The Baconian Dream

It was a philosopher, Sir Francis Bacon (1561–1626), who first dreamed about how science could enrich human life with new inventions and increased prosperity. By the middle of the twentieth century, science and its application in technology appeared to have made the Baconian dream come true. Many dread diseases—smallpox, polio, plague—had been virtually eliminated. Fertilizers, pesticides, and scientific animal breeding had increased and enriched our food supply. Transportation was swift, communication nearly instantaneous. New power sources had been discovered. Nuclear energy seemed to promise an unlimited quantity of power for our every need. New materials—plastics, fibers, metals, ceramics—were developed to improve our clothing and shelter.

Much of twentieth-century technology has grown out of scientific discoveries, and technological developments are used by scientists as tools for even more discoveries. These developments in science and technology are, to a considerable extent, the base of what we mean by the "modern" world.

The Carsonian Nightmare

The Baconian dream has lost much of its lustre in recent decades. People have learned that the products of science are not an unmitigated good. Some people have predicted that science might bring not wealth and happiness but death and destruction.

Perhaps most noteworthy among these critics of modern technology was Rachel Carson, a biologist. Her poetic and polemic book *Silent Spring* was published in 1962. The book's main theme is that, through our use of chemicals to control insects, we are threatening the destruction of all life, including ourselves. People in the pesticide industry (and their allies) roundly denounced Carson as a "propagandist," while other scientists rallied to her support. By the late 1960s, though, we had experienced massive fish kills, the threatened extinction of several species of birds, and the disappearance of fish from rivers, lakes, and areas of the ocean that had long been productive. The majority of scientists had moved into Carson's camp. Popular support for Carson's views was overwhelming.

Carson was not the first prophet of doom. As early as 1798, Thomas Malthus, in his "Essay upon the Principles of Population," had predicted that an increase in population more rapid than the increase in food supply would lead to great famine. During the nineteenth century and for more than half of the twentieth, science and technology seemed to make a fool of Malthus. Food was abundant, at least in developed countries, and scientific discoveries and technological developments enabled us to increase food production as rapidly as the population grew.

The last few decades have brought changes, however. Population growth does threaten to outpace even the most optimistic projections of food production. Some scientists project a dismal future; others confidently predict that science and technology, properly applied, will save us from disaster.

Science: Testable, Explanatory, and Tentative

What *is* science if scientists dispute what is and what will be? Is science merely a guessing game in which one guess is as good as another? We cannot *define* science precisely. Rather, we must resort to *describing* it.

One essential characteristic of science is that its tenets are *testable*. Scientists make **hypotheses** (guesses) that can be tested by **experiment**. This is the main characteristic that distinguishes science from the arts and humanities. We can learn from individual experience, and we can learn about historical events, but the knowledge gained through science is different: it depends upon phenomena that can be verified through repeated testing. Even educated guesses are of little value to scientists unless they can devise experiments to test their guesses. You may be elated over a good grade in chemistry, but that experience is uniquely yours; others might not be at all pleased with the same grade. Scientific **facts**—such as the boiling point of water and the speed of light—remain the same, however, no matter who does the measuring. These facts are verified by repeated testing.

Scientists must make careful observations and accurate measurements. They record facts based upon their observations, but nothing really counts as science until those observations have been verified by others. If something is false, a scientist can't get away for long with saying that it is true.

Experimental observations are only a bare (but necessary) beginning to the intellectual processes of science. Science is not a straightforward and logical process for cranking out discoveries. It is a way of *explaining* nature, but the explanations must be tested against a sometimes less-than-agreeable reality. The most beautiful hypothesis can be destroyed by one ugly fact. Our ideas about the universe must correspond to our observations. Scientists test ideas by predicting what

Figure 1.4 Rachel Carson (1907–1964) at Woods Hole, Massachusetts, in 1951. [Edwin Gray Studio, copyright © 1951.]

they should observe if the ideas are true. Their understanding of nature is refined constantly by the interplay of ideas and observations.

Science is a body of knowledge, but that knowledge is always *tentative*. Detailed explanations, called **theories**, are quite useful as a framework for the organization of scientific knowledge. Sometimes, though, a theory has to be modified or discarded in the light of new observations. The body of knowledge we call science is alive, ever changing, and rapidly growing.

Science is a way to cope with the environment. It involves the experimental establishment of cause and effect. For example, scientists have learned that water vapor condenses on small dust particles, called nuclei, to form raindrops. Therefore, scientists try to induce rainfall by seeding clouds with artificial nuclei. That such seeding is only somewhat successful in producing rain and that it raises innumerable economic, political, and ethical questions also serve to point out some of the limitations of science.

Scientists often use **models** to help explain complicated phenomena. The word *model* has a somewhat different meaning in science than in everyday life. A scientific model can be used to visualize the invisible.

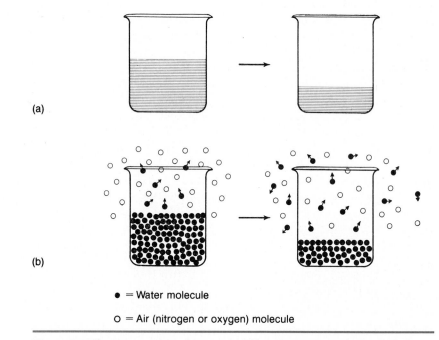

(a)

(b)

● = Water molecule

○ = Air (nitrogen or oxygen) molecule

Figure 1.5 The evaporation of water. (a) When a container of water is left standing open to the air, the water slowly disappears. (b) Scientists explain evaporation in terms of the motion of molecules.

For example, when a glass of water stands for a period of time, the water disappears (Figure 1.5). The process is called evaporation. Scientists explain this phenomenon by the kinetic-molecular theory. According to the kinetic-molecular model, the liquid (water, in this case) is made up of small, invisible particles called *molecules*. The molecules are in constant motion and, in the bulk of the liquid, are held together by forces of attraction. Some molecules near the surface of the liquid gain sufficient energy (through collisions with other molecules) to break the attraction of their neighbors, escape from the liquid, and disperse among the widely spaced air molecules. Thus, the water in the glass disappears. For the scientist, understanding evaporation is much more satisfying than merely having a name for it.

What *is* science, then? We can only state some of its characteristics: it is *testable*, *explanatory*, and *tentative*. Contrary to a popular notion, scientific knowledge is not absolute. Science is cumulative, but the body of knowledge is growing, changing, and never final. New facts and new concepts are always being added. Old concepts, or even old "facts," are discarded when new tools, new questions, and new techniques reveal new data and generate new concepts. To understand what science is, we have to observe what the worldwide community of scientists has done over several years; we cannot just look over the shoulder of one scientist for a few days.

What Science Cannot Do

We sometimes hear scientists and nonscientists alike state that we could solve all our problems if we would only attack them using the scientific method. We have seen already that there is no single scientific method. But why can't the procedures of the scientist be applied to social, political, ethical, and economic problems? Why do scientists disagree when they try to predict the future?

The answer usually lies in the ability to control *variables*. If, for example, we wanted to study in the laboratory how the volume of gases varies with changes in pressure, we would hold constant such factors as temperature and the amount and kind of matter. If, on the other hand, an economist wished to determine the effect of increased interest rates on the rate of unemployment, he or she would find it difficult, if not impossible, to control such variables as the level of governmental expenditures, the rate of business expansion, the number of high-school and college graduates (and dropouts) entering the job market, and so on. Imagine, then, the difficulties encountered by a sociologist trying to predict the effect of a technological innovation, such as a communications satellite, upon whole populations.

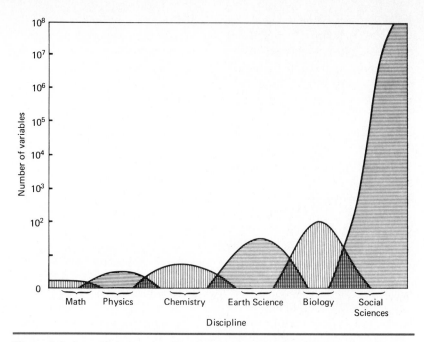

Figure 1.6 A rough estimate of the number of variables involved in scientific disciplines.

We cannot control variables in social "experiments" (for example, public-school desegregation) as we can in laboratory experiments. Therefore, a scientist would not be in any better position than any other citizen to decide whether desegregation is good or evil.

Figure 1.6 is a rough graph showing how the number of variables increases as we go from an exact science such as physics to the complex social sciences. Notice that there is overlap between disciplines. The boundaries in the graph are crude approximations at best. A more accurate representation (but one much harder to draw) would show overlap between *all* the disciplines—even between physics and the social sciences.

Social scientists have become more productive by using some of the methods of science. We can make observations, formulate hypotheses, and conduct experiments even if most of the variables are not subject to control. Interpretation of the results, however, is much more difficult and much more subject to disagreement. Other nonscientists use some of the methods and language of scientists. Artists experiment with new techniques and new materials. Playwrights and novelists observe life as it is before trying to express its essence in their writings. It is easy to see some of the methods of science, as well as the influence of science, in nearly all of the endeavors of modern men and women.

Science and Technology:
Risks and Benefits

Science and technology are interrelated. In everyday life, people often fail to distinguish between the two. A distinction can be useful, however. In an earlier section some of the distinguishing characteristics of science were noted, but no simple definition was given. **Technology**, on the other hand, can be defined as the sum total of the processes by which humans modify the materials of nature to better satisfy their needs and wants. These processes need not be based on scientific principles. For example, ancient peoples were able to smelt ores to produce metals, such as copper and iron, without having any understanding of the chemistry involved.

Few people question that modern Western society has benefited from science and technology, but there are risks associated with technological advances. How can we determine when the benefits outweigh the risks? One approach, called **risk–benefit analysis**, involves the calculation of a desirability quotient (DQ).

$$DQ = \frac{\text{Benefits}}{\text{Risks}}$$

A **benefit** can be described as anything that promotes well-being or has a positive effect. Benefits may be economic, social, or psychological. A **risk** can be defined as any hazard that leads to loss or injury. Some of the risks in modern technology have led to disease, death, economic loss, and environmental deterioration. Risks may involve one individual, a group, or society as a whole, and they may be local, regional, or worldwide.

Individual risk can be defined as the product of the probability that an incident will occur and the degree of severity of the incident.

$$\text{Individual risk} = \text{Probability} \times \text{Severity}$$

For example, the automobile is a technological development that benefits us by making it easy to travel from one place to another. Most of us know the risks involved in driving a car. A severe risk is the possibility of death in a traffic accident. The probability that this will happen is remote if an individual drives carefully and within the speed limit. Chances are, if we want to go to a place 50 km away, we will find that the benefits of driving outweigh the chance that we might be injured or killed in an accident.

Societal risk involves more people. We must therefore add another factor—the number of people affected—to our equation.

$$\text{Societal risk} = \text{Probability} \times \text{Severity} \times \text{Population affected}$$

Weighing the benefits and risks associated with a product of modern technology is decidedly more difficult when one considers a group of people. For example, for most people the benefit of pasteurized milk is that it is a safe, clean beverage that provides needed nutrition. For some people, however, milk poses a risk. Some people can't tolerate lactose, the sugar in milk. People with cardiovascular disease may be harmed by the saturated fats in whole milk. To these people, and to others who are allergic to milk protein, milk is a harmful substance. Fortunately, milk allergies and lactose intolerance are relatively rare in North America and northern Europe. In these societies, milk is generally beneficial.

In a risk–benefit analysis for milk, we find large benefits and small risks, resulting in a large DQ.

$$\frac{\text{Large benefits}}{\text{Small risks}} = \text{Large DQ}$$

Thus, most of us readily accept the risks associated with milk. Interestingly, however, this analysis generally fits only people of northern European descent. Adults in much of the rest of the world are lactose intolerant and would find that milk has a small DQ. This explains the lack of gratitude on the part of those who receive dried milk under United States foreign aid programs and the failure of soft-ice-cream stands in areas with predominantly black populations.

An important technological contribution is the production of natural and synthetic chemicals used as drugs. One example—among many—was the introduction of the drug thalidomide to prevent morning sickness in pregnant women. The drug was introduced in the 1950s, but was soon shown to be of little benefit. Several years later, it was shown to present a large risk: many malformed infants were born to women who took the drug during pregnancy. For thalidomide, then, we have small benefits, large risks, and a small DQ.

$$\frac{\text{Small benefits}}{\text{Large risks}} = \text{Small DQ}$$

Thalidomide was eventually judged to present unacceptable risks and was banned. (Other drugs are discussed in Chapters 22 and 23. Many have high DQs, some have low DQs, and others are more difficult to evaluate.)

It is easy to judge the desirability of milk and thalidomide, but other products present difficult choices. Consider a product whose benefits and risks both are small: the artificial sweetener aspartame. Studies have shown that artificial sweeteners generally are of little benefit to those who use them to replace sugar in an effort to lose weight. Aspartame is the most studied of all food additives. There are anecdotal reports of problems with the sweetener, but these have not been confirmed in controlled studies. To most people, the risk involved in using aspartame is small. This leads to an uncertain DQ—and, it seems, to endless debate over the safety of aspartame.

$$\frac{\text{Small benefits}}{\text{Small risks}} = \text{Uncertain DQ}$$

(Aspartame may provide some benefit to diabetics, however, because sugar consumption presents a large risk to them.)

Other technologies provide large benefits and present large risks. For these technologies, too, the DQ is uncertain.

$$\frac{\text{Large benefits}}{\text{Large risks}} = \text{Uncertain DQ}$$

An example is the conversion of coal to liquid fuels. Liquid fuels provide large benefits to society in the areas of transportation, home heating, and industry. The risks associated with coal conversion are also large, however. These include air and water pollution and the exposure of workers to toxic chemicals. The result again is an uncertain DQ and political controversy.

There are yet other problems in risk–benefit analysis. Some technologies benefit one group of people (population A), while presenting a risk to another (population B). For example, it may be economically advantageous to a community to spend as little as possible on a sewage treatment plant and to dump raw wastes into a nearby stream. These wastes might present a hazard to downstream communities, however. Difficult political decisions are needed in such a case.

Other technologies provide benefits now but present risks later. For example, although nuclear power now provides useful electricity, wastes from nuclear power plants, if improperly stored, might present hazards for centuries. Thus, the use of nuclear power is highly controversial.

It is important to remember that science and technology involve *both* risks and benefits. The determination of benefits is almost entirely a social judgment; risk assessment also involves social decisions, but scientific investigation can help considerably in risk evaluation.

Chemistry: A Study of Matter and Energy

At the beginning of this chapter, we defined chemistry as a study of matter and the changes it undergoes. Now let's look at matter and energy a little more closely.

Figure 1.7 Astronaut John W. Young leaps from the lunar surface, where gravity pulls at him with only one sixth the force on Earth. [Courtesy of National Aeronautics and Space Administration.]

Since the entire physical universe is made up of nothing more than matter and energy, the field of chemistry extends from atoms to stars, from rocks to living organisms. Matter and energy are such fundamental concepts that definitions are difficult. **Matter** is the stuff that makes up all material things. It occupies space and has *mass*. Wood, sand, water, air, and people have mass and occupy space. These are material things. **Mass** is a measure of the *quantity* of matter. **Weight**, on the other hand, measures a force. On Earth, it measures the force of attraction between the Earth and the mass in question.

For most of its history, the human race was restricted to the surface of the Earth. The terms *mass* and *weight* can be used interchangeably in such restricted circumstances. When the exploration of space began, however, it became apparent to most people that mass and weight are not the same thing. An astronaut has the same mass on the moon and on Earth (Figure 1.7). The matter that makes up the astronaut does not change. John Young's weight on the moon, however, was one sixth his weight on Earth because the moon's pull is only one sixth as strong as the Earth's. Weight varies with gravity; mass does not.

Matter is characterized by its *properties*. This means that we know we have water and not, say gasoline, because water has certain characteristics or properties that distinguish it from gasoline. For example, it is hard to start a fire by dousing an object with water and then lighting the object with a match. **Chemical properties** describe how one substance reacts with other substances. To demonstrate a chemical property, a substance must undergo a change in composition. In exhibiting its **physical properties**, a substance undergoes no change in composition. Characteristics such as color, hardness, density, and melting point are physical properties. Sulfur is a brittle yellow solid that is more dense than water. Each of these physical properties of sulfur can be evaluated without changing the composition of the sulfur. We can also cite some chemical properties of sulfur: it reacts with oxygen, with carbon, and with iron. These reactions yield, in turn, sulfur dioxide, carbon disulfide, and iron sulfide—all new substances.

The burning of a candle demonstrates both a physical and a chemical change. After the candle is lighted, the solid wax near the burning wick melts (a physical change, for in both solid form and liquid form the composition of the wax is the same). Some of the melted wax is drawn into the burning wick, where a chemical change occurs. The wax combines with oxygen in the air to form carbon dioxide gas and water vapor. As the candle burns and the wax undergoes this chemical change, the candle becomes smaller and smaller. However, the apparent disappearance of something is not necessarily a sign that we are observing a chemical change. If a glass of water is allowed to stand for many days, the water seems to disappear. In this instance, however, we are observing a physical change. The water is evaporating, changing from a liquid to a gas, but in both forms it is still water. Thus, it is not always easy to identify a physical or a chemical change. The critical question is: Has the fundamental composition of the substance been changed? In a chemical change it has; in a physical change it has not.

Energy is the capacity for doing work. **Work** must be done to make something happen that wouldn't happen by itself. Getting out of bed, building a house, or mining coal all require energy. Eating requires energy; that forkful of spaghetti would never make it to your mouth by itself. Energy is the basis for change in the material world.

Energy appears in several forms. The source of nearly all of the energy on the Earth is the sun. Solar energy radiates through space. A small portion of this radiant energy reaches the Earth, where some of it is converted to heat energy. This heat causes water to evaporate and rise to form clouds. The water in the clouds has **potential energy** (energy by virtue of its position). As the water falls through the air and then flows into rivers, the potential energy is converted to **kinetic energy** (energy of motion). The kinetic energy of a flowing stream can be used to turn a turbine, which converts part of the kinetic energy to electrical energy. The electricity thus produced can be carried by wires to homes and factories, where it can be converted to light energy or to heat or to mechanical energy.

Some of the solar energy striking the Earth's surface is absorbed by green plants, which use a complicated chemical process called **photosynthesis** to convert solar energy into chemical energy. The chemical energy stored by plants—today and in ages past—is used by animals and humans for food and fuel (Figure 1.8). Nearly all of the vast quantities of energy being used in our modern civilization came originally from green plants. Plants of the current age are harvested by foresters and farmers. Those of ancient ages are reaped as fossil fuels—coal, oil, and gas.

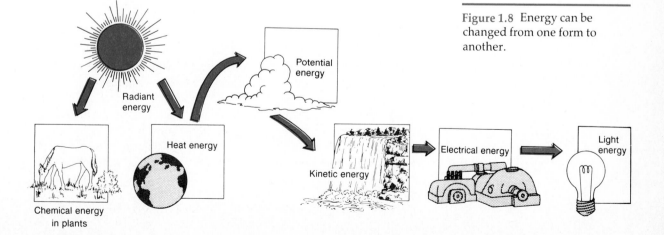

Figure 1.8 Energy can be changed from one form to another.

Radiant energy

Potential energy

Heat energy

Kinetic energy

Electrical energy

Light energy

Chemical energy in plants

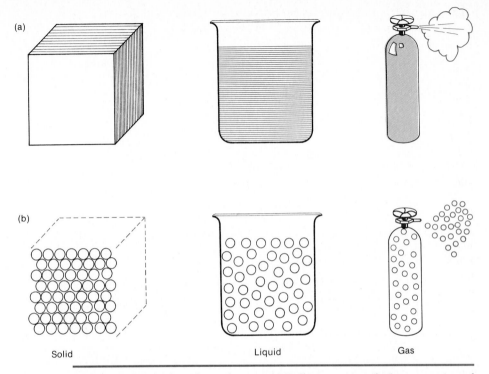

Figure 1.9 Solids, liquids, and gases. (a) Bulk properties. (b) Interpretation of bulk properties in terms of the kinetic molecular theory.

Chemistry is concerned not only with matter and its changes but with the energy transformations that accompany those changes.

The States of Matter

There are three familiar **states of matter**: solid, liquid, and gas. **Solid** objects ordinarily maintain their shape and volume regardless of their location. A **liquid** occupies a definite volume, but assumes the shape of the occupied portion of its container. If you have a 12-oz soft drink, you have 12 oz whether the soft drink is in a can, a bottle, or, through some slight mishap, on the floor—which demonstrates another property of liquids. Unlike solids, liquids flow readily. **Gases** maintain neither shape nor volume. They expand to fill completely whatever container one puts them in. Gases can be easily compressed. For example, enough air for many minutes of breathing can be compressed into a steel tank for underwater diving. We shall take up the topic of the states of matter in more detail in Chapter 5.

Matter: Pure Substances and Mixtures

Matter can be classified in yet other ways. For example, we can subdivide matter into pure substances and mixtures (Figure 1.10). **Pure substances** have a definite, or fixed, composition. The composition of **mixtures** may vary. Water is a pure substance; it always contains 11% hydrogen and 89% oxygen by weight. Similarly, pure gold is pure gold, that is, 100% gold. A milk shake, on the other hand, is a mixture. The proportions of milk, ice cream, and flavorings change depending on who is preparing the shake. Mixed nuts are a mixture; the ratio of peanuts to pecans depends on how much you are willing to pay per pound.

Pure substances may be either elements or compounds. **Elements** are those fundamental substances from which all material things are constructed. **Compounds** are pure substances that are made up of two or more elements combined in fixed proportions. Our ideas about elements have changed during historical times. Fire was once considered an element, but it is now regarded as nonmaterial, that is, as a form of energy. Water was once thought to be an element, but we now know it to be a compound composed of two elements, hydrogen and oxygen. We presently regard as elements a few more than 100 pure substances that cannot be broken down by chemical means into simpler substances. Sulfur, oxygen, carbon, and iron are elements. Sulfur dioxide, carbon disulfide, and iron sulfide are compounds.

Because elements are so fundamental to our study of chemistry, we find it useful to refer to them in a shorthand form. Each element can be represented by a symbol made up of one or two letters derived from the

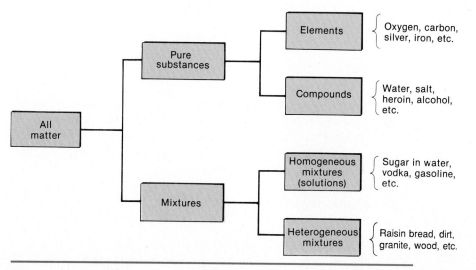

Figure 1.10 A scheme for classifying matter.

Table 1.1 Some Elements and Their Symbols

Element	Symbol	Element	Symbol	Element	Symbol
Hydrogen	H	Phosphorus	P	Silver (argentum)	Ag
Helium	He	Sulfur	S	Tin (stannum)	Sn
Carbon	C	Chlorine	Cl	Iodine	I
Nitrogen	N	Potassium (kalium)	K	Barium	Ba
Oxygen	O	Calcium	Ca	Mercury (hydrargyrum)	Hg
Fluorine	F	Iron (ferrum)	Fe	Gold (aurum)	Au
Sodium (natrium)	Na	Copper (cuprum)	Cu	Lead (plumbum)	Pb
Magnesium	Mg	Zinc	Zn	Uranium	U
Aluminum	Al	Bromine	Br	Plutonium	Pu
Silicon	Si				

name of the element (or, sometimes, from the Latin name of the element). The first letter of the symbol is always capitalized; the second is always lower case. (It does make a difference. For example, Hf is the symbol for hafnium, an element; HF is the formula for hydrogen fluoride, a compound. Similarly, Co is cobalt, an element; CO is carbon monoxide, a compound.)

Symbols for some of the more important elements are given in Table 1.1. Memorization of the names and symbols of the elements shown in that table is well worth your time. The structure of compounds will be examined in detail in Chapters 5 and 6. You will find that discussion much easier to follow if you are familiar with the common elemental symbols.

Example 1.1 Which of the following represent an element and which represent a compound?

$$Hg \quad HI \quad BN \quad In$$

Hg and In represent elements (each is a single symbol). HI and BN are composed of two symbols each and represent compounds.

The Measurement of Matter

Accurate measurement of such quantities as mass (or weight), volume, time, and temperature are essential to the compilation of dependable scientific "facts." Such facts may be used by a chemist interested in basic research, but similar information is of critical importance in every science-related field. Certainly we are all aware that measurements of both temperature and blood pressure are routinely made in medicine. It is also true that modern medical diagnosis depends on a whole battery of other measurements, including careful chemical analyses of blood and urine.

The system of measurement used by most scientists is an updated metric plan called the International System of Measurements, or SI (from the French Système International). Indeed, SI has been adopted worldwide for everyday use. Even the United States is committed to change to SI, but conversion is voluntary and has proceeded rather slowly so far.

The beauty of SI is that it is based on the decimal system. This makes conversion from one unit to another rather simple. The SI has only a few basic units. For example, the basic unit of length is the **meter** (m), a distance only slightly greater than a yard. The SI unit of mass is the **kilogram** (kg), a quantity slightly greater than two pounds. All other units for length, mass, and volume can be derived from these basic units. For example, area can be measured in square meters (m^2) and volume in cubic meters (m^3).

A disadvantage of the basic SI units is that they are often of awkward magnitude. We seldom work with kilogram quantities in the laboratory. A cubic meter of liquid would fill a very large test tube. More convenient units can be derived by the use of (or, in the case of the kilogram, the deletion of) prefixes (Table 1.2). For example, in the

Table 1.2 Approved Numerical Prefixes
The most commonly used units are shown in color.

Exponential Expression	Decimal Equivalent	Prefix	Phonic	Symbol
10^{12}	1 000 000 000 000.	Tera-	ter′ a	T
10^9	1 000 000 000	Giga-	ji′ ga	G
10^6	1 000 000	Mega-	meg′ a	M
10^3	1 000	Kilo-	kil′ o	k
10^2	100	Hecto-	hek′ to	h
10	10	Deka-	dek′ a	da
10^{-1}	0.1	Deci-	des′ i	d
10^{-2}	0.01	Centi-	sen′ ti	c
10^{-3}	0.001	Milli-	mil′ i	m
10^{-6}	0.000 001	Micro-	mi′ kro	μ
10^{-9}	0.000 000 001	Nano-	nan′o	n
10^{-12}	0.000 000 000 001	Pico-	pe′ ko	p
10^{-15}	0.000 000 000 000 001	Femto-	fem′ to	f
10^{-18}	0.000 000 000 000 000 001	Atto-	at′ to	a

laboratory we may work with grams (g) or milligrams (mg) of material. The prefix **milli-** means 1/1000 or 0.001. Thus, one milligram equals 1/1000 gram or 0.001 gram. The relationship between milligrams and grams is given by

$$1 \text{ mg} = 0.001 \text{ g}$$

or

$$1000 \text{ mg} = 1 \text{ g}$$

You should learn the more common prefixes (printed in color in Table 1.2) right away.

For volume we may choose cubic centimeters (cm^3) or cubic decimeters (dm^3) as more convenient units. The cubic decimeter has a special name, the **liter** (L). From that is derived the milliliter (mL), a unit that is the same as the cubic centimeter.

$$1 \text{ dm}^3 = 1 \text{ L}$$
$$1 \text{ L} = 1000 \text{ mL}$$
$$1 \text{ mL} = 1 \text{ cm}^3$$

Conversions within the SI system are much easier than those using the more familiar units of pounds, feet, and pints. This is best shown by examples.

Example 1.2   Convert 0.742 kg to grams.

$$0.742 \text{ kg} \times \frac{1000 \text{ g}}{1 \text{ kg}} = 742 \text{ g}$$

Note: If you are not familiar with the use of conversion factors or would like to review the mathematics of conversions, see Appendix C.

Example 1.3   Convert 0.742 lb to ounces.

$$0.742 \text{ lb} \times \frac{16 \text{ oz}}{1 \text{ lb}} = 11.9 \text{ oz}$$

Example 1.4   Convert 1247 mm to meters.

$$1247 \text{ mm} = \frac{1 \text{ m}}{1000 \text{ mm}} = 1.247 \text{ m}$$

Example 1.5   Convert 1247 in. to yards.

$$1247 \cancel{\text{in.}} \times \frac{1 \text{ yd}}{36 \cancel{\text{in.}}} = 34.64 \text{ yd}$$

In conversions involving pounds, ounces, inches, and yards, you multiply and divide by numbers such as 16 or 36. In metric conversions you multiply by 10 or 100 or 1000, and so on; you need only to shift the decimal point.

Conversions between systems are seldom necessary. (If you need to do this, however, such conversions are discussed in Appendix C.) What you *do* need is some idea of the relative sizes of comparable units. Some comparisons are shown in Figure 1.11. Other comparisons easily remembered are that the United States 10-cent coin, the dime, is about 1 mm thick and that the 5-cent piece, the nickel, weighs about 5 g.

Measuring Energy: Temperature and Heat

The SI unit for temperature is the kelvin (K), but for much of their work, scientists still use the Celsius scale. On this scale, the freezing temperature of water is 0 °C and the boiling point is 100 °C. The scale

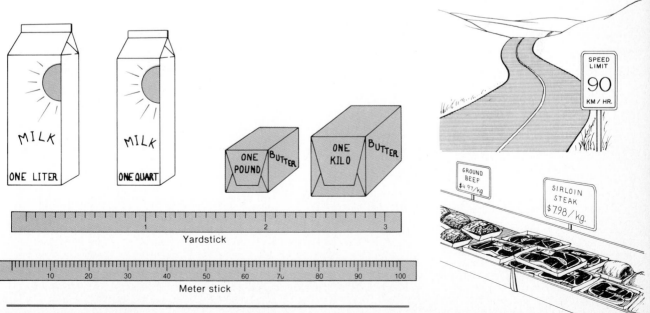

Figure 1.11 A metric America!

between these two reference points is divided into 100 equal divisions, each a Celsius degree.

Degrees on the Kelvin and Celsius scales are the same size. The Kelvin scale is called an **absolute scale** because its zero point is the coldest temperature possible, or absolute zero. This fact was determined by theoretical considerations and has been confirmed by experiment. The zero point on the Kelvin scale is equal to -273 °C. Note that there are no negative temperatures on the absolute scale. To convert from Celsius to Kelvin, you merely add 273 to the Celsius temperature.

$$K = °C + 273$$

Example 1.6   What is the boiling point of water in kelvins? The boiling point of water is 100 °C.

$$100 °C + 273 = 373 K$$

In the United States, weather reports and food recipes still use the Fahrenheit scale. This scale defines the freezing temperature of water as 32 °F and the boiling point as 212 °F. Exact conversions are seldom necessary. Figure 1.12 compares the Fahrenheit and Celsius scales.

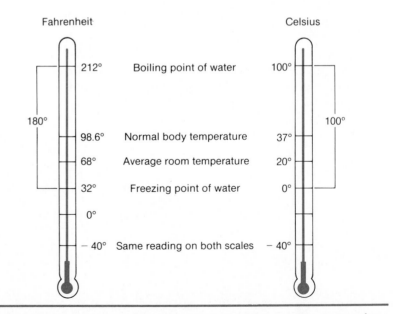

Figure 1.12 A comparison of the Fahrenheit and Celsius temperature scales.

Some additional temperature equivalents and formulas for converting one scale to the other are shown in Appendix A.

Scientists often need to measure amounts of heat energy. You should not confuse heat with temperature. **Heat** is a measure of quantity, that is, of *how much energy* a sample contains. **Temperature** is a measure of intensity, that is, of *how energetic* each particle of the sample is. A glass of water at 70 °C contains less heat than a bathtub of water at 60 °C. The particles of water in the glass are more energetic, on the average, than those in the tub, but there is far more water in the tub and its total heat content is greater.

The SI unit of heat is the **joule** (J), but we will use the more familiar **calorie** (cal).

$$1 \text{ cal} = 4.18 \text{ J}$$
$$1000 \text{ cal} = 1 \text{ kcal} = 4184 \text{ J}$$

A calorie is the amount of heat required to raise the temperature of 1 g of water 1 °C. There is a more precise definition, but this version will do for our purposes.

For measuring the energy content of foods, the *large* **Calorie** (note the capital *C*), or **kilocalorie** (kcal), is sometimes used. A dieter might be aware that a banana split contains 1500 Calories. If the same dieter realized that that meant 1 500 000 calories, giving up the banana split might be easier.

Density

An important property of matter, particularly in scientific work, is **density**. When one speaks of lead as "heavy" or aluminum as "light," one is referring to the density of these metals. This term is defined as the amount of mass (or weight) per unit of volume.

$$d = \frac{m}{V}$$

Rearrangement of this equation gives

$$m = d \times V \quad \text{and} \quad V = \frac{m}{d}$$

These equations are useful for calculations. Densities are usually reported in grams per milliliter (g/mL) or grams per cubic centimeter (g/cm^3).

Example 1.7 What is the density of iron if 156 g of iron occupy a volume of 20 cm^3?

$$\frac{156 \text{ g}}{20 \text{ cm}^3} = 7.8 \text{ g/cm}^3$$

Example 1.8 How much will a liter of gasoline weigh if its density is 0.66 g/mL?

$$\frac{0.66 \text{ g}}{1 \text{ mL}} \times 1\text{ L} \times \frac{1000 \text{ mL}}{1 \text{ L}} = 660 \text{ g}$$

The density of water is 1 g/mL, or 1 g/cm^3 (remember that 1 mL = 1 cm^3). This nice round number for the density of water is no accident. The metric system was originally set up to ensure that this was the case.

Measurement is discussed further in Appendix A, and a variety of conversion tables are collected there for convenient reference.

Chemistry: Its Central Role

Chemistry is not only useful in itself but also fundamental to other scientific disciplines. Biology has been revolutionized by the application of chemical principles. Psychology, too, has been profoundly influenced by chemistry and stands to be even more radically altered as the chemistry of the nervous system is unraveled. The social goals of better health and more and better food, housing, and clothing are dependent to a large extent on the knowledge and techniques of chemists. The recycling of basic materials—paper, glass, and metals—is primarily a matter of chemical processes. Devising new, more specific pesticides that entail less risk to useful organisms will require the application of chemical principles and skills. Chemistry is indeed a central science. There is scarcely a single area of our daily lives that is not affected by chemistry.

Chemistry is also important to the *economy* of industrial nations. In the United States, the chemical industry employs over a million people in more than 10 000 plants. It is the nation's fifth largest industry with sales approaching $200 billion per year. Incidentally, according to the

National Safety Council, the chemical industry ranks first in worker safety among 42 basic industries. It has the fewest incidents of occupational illness and injury involving death or days away from work. The rate of such incidents in the chemical industry is only one third of the average rate for all United States industries. This is in sharp contrast to the popular belief that chemicals are so dangerous. (Some are, of course, but even those can be used safely with proper precautions.)

An understanding of chemistry is essential to an understanding of many of the problems facing society. If you want to understand how a drug or a pesticide acts, how to clean up a pollution problem, or how to avoid consumer rip-offs in cosmetics, you must know some chemistry. Much of the remainder of this book deals with such subjects. Before going on, though, let's look briefly at some of the things chemists do.

Solving Society's Problems: Applied Research

Most chemists work in the area of *applied research*. They analyze polluted soil, air, and water. They synthesize new chemical compounds for use as drugs or pesticides. They formulate plastics for new applications. They analyze foods, fuels, cosmetics, detergents, and drugs. These are examples of **applied research**—work oriented toward the solution of a particular problem in industry or the environment.

As an example, let's consider the synthesis of an antifungal drug. Many drugs are effective against bacteria, but few are useful and safe against fungal infections in humans. Seeking an effective fungal antibiotic, Rachel Brown and Elizabeth Hazen discovered nystatin, a compound effective against such fungal infections as thrush and vaginitis. Nystatin also has been used in veterinary medicine, in agriculture (to protect fruit), and in the art world (to prevent the growth of mold on art treasures in Florence, Italy, after the objects were damaged by floods). Brown and Hazen set out to find a fungicide and did so—an excellent example of applied research.

In Search of Knowledge: Basic Research

Many chemists are involved in basic research, the search for knowledge for its own sake. Some chemists work out the fine points of atomic and molecular structure. Others measure the intricate energy changes that accompany complex chemical reactions. Chemists synthesize new compounds and determine their properties. This type

Figure 1.13 Rachel Brown (left) and Elizabeth Hazen, codiscoverers of nystatin, a safe fungal antibiotic. [Courtesy of Rachel Fuller Brown, Research Corporation, New York.]

Figure 1.14 Joel Hildebrand (1881–1983), noted chemist and teacher, whose theoretical studies on solubility led to the use of helium in deep-sea diving. [Courtesy of Joel Hildebrand.]

of investigation is called **basic research**. Done for the sheer joy of unraveling the secrets of nature and discovering order in our universe, basic research is characterized by the absence of any predictable, usable product.

Findings from basic research often *are* applied at some point, but that is not the primary goal of the researcher. In fact, most of our modern technology is based on results obtained in basic research. Without this base of factual information, technological innovation would be haphazard and slow. Warnings have been sounded that the United States is falling behind Germany, Japan, and other nations because it has failed to provide adequate support for basic research. How costly is research? All research done since the time of Aristotle has cost less than 2 weeks' output in the modern industrial world.

Applied research is carried on mainly by industries seeking a competitive edge with a novel, better, or more salable product. Its ultimate aim is usually profit for the stockholders. Basic research is conducted mainly at universities and research institutes. Most of its support comes from federal and state governments and foundations, although some of the larger industries also support it.

As an example of basic research that was later applied for human welfare, consider the work of Joel Hildebrand, a chemist at the University of California at Berkeley. In 1916 Hildebrand published a theoretical paper on solubility. He deduced that in any liquid, helium would be found less soluble than any other gas. He then proposed that a mixture of helium and oxygen (rather than air, a mixture of nitrogen and oxygen) be used for deep-sea diving. If air is used, nitrogen dissolves in the blood. As the diver rises toward the surface, bubbles of nitrogen separate from the blood, causing a painful affliction called the *bends*. Helium, which is less soluble in blood than nitrogen, prevents the bends. Helium–oxygen mixtures are now commonplace in deep-sea diving. Because Hildebrand was curious about the solubility of gases in liquids, divers need not worry as much about the bends as they once did.

"There are two compelling reasons why society must support basic science. One is substantial: The theoretical physics of yesterday is the nuclear defense of today; the obscure synthetic chemistry of yesterday is curing disease today. The other reason is cultural. The essence of our civilization is to explore and analyze the nature of man and his surroundings. As proclaimed in the Bible in the Book of Proverbs: 'Where there is no vision, the people perish.'"

Arthur Kornberg (1918——)
American Biochemist
Nobel Prize in Physiology and Medicine, 1959

Problems

1. Define chemistry.

2. What is matter?

3. Which of the following are examples of matter?
   a. iron
   b. air
   c. love
   d. the human body
   e. gasoline
   f. an idea

4. List five chemical activities that you have engaged in today.

5. State three distinguishing characteristics of science. Which characteristic best serves to distinguish science from other disciplines?

6. Why were the ancient Greek philosophers, such as Aristotle, not successful as scientists?

7. What is alchemy?

8. What is natural philosophy?

9. What did Francis Bacon envision for us as a result of science?

10. What was the main theme of Rachel Carson's *Silent Spring*?

11. Why have Thomas Malthus's predictions not been fulfilled in developed countries?

12. What is a hypothesis? How are hypotheses tested?

13. What is a theory?

14. Why can't scientific methods always be used to solve social, political, ethical, and economic problems?

15. How does technology differ from science?

16. What is risk–benefit analysis?

17. What sort of judgments go into the evaluation of benefits?

18. What sort of judgments go into the evaluation of risks?

19. What is a desirability quotient?

20. Why is it often difficult to estimate desirability quotients?

21. Synthetic food colors make food more attractive and increase sales. Some such dyes are suspected carcinogens (cancer inducers). Who derives most of the benefits from the use of food colors? Who assumes most of the risk associated with use of these dyes?

22. Penicillin kills bacteria, thus saving the lives of thousands of people who otherwise might die of infectious diseases. Penicillin causes allergic reactions in some people; in extreme cases it can cause death if the resulting condition is not treated. Do a risk–benefit analysis of the use of penicillin for society as a whole.

23. Do a risk–benefit analysis of the use of penicillin for a person who is allergic to it. (See Problem 22.)

24. An artificial sweetener is 4000 times as sweet as table sugar, but exhibits possible toxic side effects. Do a risk–benefit analysis of the sweetener.

25. Nitrogen mustard is extremely toxic, but is an effective anticancer drug. Do a risk–benefit analysis of nitrogen mustard.

26. The generation of electricity from coal results in acid rain. Do a risk–benefit analysis of this process.

27. Explain the difference between mass and weight.

28. Two samples are weighed under identical conditions in a laboratory. Sample A weighs 1 lb and Sample B weighs 2 lb. Does Sample B have twice the mass of Sample A?

29. Which has changed when a person completes a successful diet: the person's weight or the person's mass?

30. Sample A, which is on the moon, has exactly the same mass as Sample B, which is on Earth. Do the two samples weigh the same?

31. Distinguish between chemical and physical properties.

32. Which of the following describes a physical change, and which describes a chemical change?
   a. Sheep are sheared, and the wool is spun into yarn.
   b. Silkworms feed on mulberry leaves and produce silk.

33. Which describes a physical change, and which describes a chemical change?
   a. Because a lawn is watered and fertilized, it grows more thickly.
   b. An overgrown lawn is manicured by mowing it with a lawn mower.

34. Which describes a physical change, and which describes a chemical change?
   a. Ice cubes form when a tray filled with water is placed in a freezer.
   b. Milk, which has been left outside a refrigerator for many hours, turns sour.

35. What is energy?

36. What is the difference between kinetic energy and potential energy?

37. How do pure substances and mixtures differ?

38. All samples of the sugar glucose consist of 8 parts (by weight) oxygen, 6 parts carbon, and 1 part hydrogen. Is glucose a pure substance or a mixture?

39. Identify each of the following as a pure substance or a mixture.
  a. carbon dioxide   b. oxygen
  c. smog

40. Identify each of the following as a pure substance or a mixture.
  a. gasoline   b. mercury   c. soup

41. Identify each of the following as a pure substance or a mixture.
  a. a carrot   b. 24-karat gold

42. How do gases, liquids, and solids differ in their properties?

43. Which of the following represent elements, and which represent compounds?
  a. H   b. He   c. HF

44. Which of the following represent elements, and which represent compounds?
  a. C   b. CO   c. Ca

45. Which of the following represent elements, and which represent compounds?
  a. CO_2 b. Cl_2 c. $CaCl_2$

46. Without consulting Table 1.1, write the symbol for each of the following.
  a. carbon   b. calcium
  c. chlorine   d. potassium
  e. phosphorus   f. plutonium

47. Without consulting Table 1.1, name each of these elements.
  a. H   b. N   c. O
  d. Fe   e. Na   f. U

48. What are the basic units of length, volume, and mass in the SI system? What derived units are more often employed in the laboratory?

49. How many millimeters and how many centimeters are there in 1 m?

50. For each of the following, indicate which is the larger unit.
  a. mm or cm   b. kg or g   c. dL or μL

51. For each of the following, indicate which is the larger unit.
  a. L or cm^3 b. cm^3 or mL

52. How many meters are there in each of the following?
  a. 50 km   b. 25 cm

53. How many millimeters are there in each of the following?
  a. 1.5 m   b. 16 cm

54. How many liters are there in each of the following?
  a. 2056 mL   b. 47 kL

55. Make the following conversions.
  a. 15 000 mg to g   b. 0.086 g to mg

56. Make the following conversions.
  a. 0.149 L to mL   b. 47 mL to L

57. How many milliliters are there in 1 cm^3? In 15 cm^3?

58. Convert the following to kelvins.
  a. 37 °C   b. 273 °C

59. Convert the following to degrees Celsius.
  a. 298 K   b. 373 K

60. How many calories are there in each of the following?
  a. 2.75 kcal   b. 0.74 Cal

61. What is the density of a salt solution if 50 cm^3 of the solution weighs 57 g?

62. What is the density of a liquid that has a mass of 60.2 g and a volume of 25.0 mL?

63. A 1.0-L container of carbon tetrachloride weighs 1.6 kg. What is the density of the carbon tetrachloride?

64. A piece of tin has a volume of 16.4 cm^3. The density of tin is 5.75 g/cm^3. What is the mass of the tin?

65. What is the mass of 50.0 mL of mercury? The density of mercury is 13.6 g/mL.

66. What is the density of a urine sample if 150 mL has a mass of 157 g?

67. A 59.0-g piece of lead has a volume of 5.20 cm^3. Calculate the density of lead.

68. The density of ethyl alcohol at 20 °C is 0.789 g/mL. What is the mass of a 50.0-mL sample?

69. What volume is occupied by 253 g of bromoform? The density of bromoform is 2.90 g/mL.

70. What is applied research?

71. What is basic research?

References and Readings

1. Boulding, Kenneth E. "Science: Our Common Heritage." *Science*, 22 February 1980, pp. 831–836.

2. Bronowski, J. *The Common Sense of Science*. New York: Vantage Books (Random House), 1960. Chapter 1, "Science and Sensibility." Written for the nonscientist, shows the sciences and the arts to be complementary.

3. Carson, Rachel. *Silent Spring*. Boston: Houghton Mifflin, 1962. The classic book on dangers to the environment.

4. Committee on Chemistry and Public Affairs. *Chemistry in the Economy*. Washington, DC: American Chemical Society, 1973. Explains the vital role of chemistry in the United States economy. Plastics, textiles, rubbers, drugs, detergents, cosmetics, fertilizers, pesticides, and many other chemicals play a part.

5. Curtis, Harry A. "Alchemy." *The Hexagon*, June 1984, pp. 24–26.

6. Handler, Philip. "Science, Technology, and Social Achievement." *EPRI Journal*, September 1979, pp. 15–19.

7. Harrison, Anna J. "Common Elements and Interconnections." *Science*, 1 June 1984, pp. 939–942. Discusses science, technology, and engineering.

8. Ihde, A. J. *The Development of Modern Chemistry*. New York: Harper and Row, 1964. Chapter 1.

9. National Science Foundation. "How Basic Research Reaps Unexpected Rewards." Washington, DC: U.S. Government Printing Office, February 1980.

10. Pauling, Linus C. "Chemistry and the World of Tomorrow." *Chemical and Engineering News*, 16 April 1984, pp. 54–56.

11. Porter, Sir George. "Chemistry's Cornucopia." *Chemistry International*, Vol. 8, No. 1, 1986.

12. Wilson, Richard, and E. A. C. Crouch. "Risk Assessment and Comparisons: An Introduction." *Science*, 17 April 1987, pp. 267–270. Risk assessment is the focus of several articles in this issue.

2

Atoms
Are They for Real?

Figure 2.1 Democritus. [Courtesy of the Smithsonian Institution, Washington, DC.]

An **atom** is defined as the smallest characteristic particle of an element. This concept has been around for thousands of years. For most of that time, however, it was not a popular idea. The development of chemistry as a science was largely a matter of accepting the existence of atoms and then working to define the properties of the atom more and more precisely.

About 90 kinds of atoms occur in nature—one kind for each element. From these, though, millions of different compounds are possible. Over 7 million chemical compounds have been characterized by chemists around the world. As far as we can tell, the entire universe is made up of less than 100 kinds of atoms.

In this chapter, we see how the concept of the atom developed and how it has changed forever the way we look at things. Who cares about atoms? You do, that's who. For all practical purposes, atoms are eternal. We can't really get rid of those we don't want. We can bury them under the ground or throw them into the sea, and sometimes we can disperse them into the air. We can combine them in different ways, and occasionally, by using a lot of energy, we can even change one kind of atom into another. But they still won't go away. Keep that in mind when you think about disposing of wastes.

Learn all you can about atoms. Remember that you and the

universe are made of atoms. So, to start our study, let's go back for a moment to ancient Greece.

Atoms: The Greek Idea

In the fifth century B.C., Leucippus and his pupil Democritus strolled along an Aegean beach. Leucippus is said to have wondered aloud whether the water of the sea was continuous, as it appeared, or whether it might not be composed of tiny, separate particles like the grains of sand on the beach. From a distance, the sand appeared continuous, but closer inspection revealed that it was made up of separate grains. Leucippus could divide the water into drops and each of those into smaller ones. Was there any reason that this process could not be continued indefinitely, yielding ever smaller drops of water? This idea of endless divisibility was the prevailing view of the Greek philosophers of that time, but Leucippus, on the basis of intuition alone, concluded that there must be a limit to divisibility—that there must be ultimate particles that could not be subdivided.

Democritus, who lived from about 470 to 380 B.C., gave these ultimate particles a name. He called them *atomos*, which meant "indivisible." Democritus also expanded the theory of matter. He believed that atoms of each element* were distinct in shape and size (Figure 2.2). Real substances were considered to be mixtures of atoms of different elements in different proportions. One substance could be changed into another by altering the proportions.

This atomic theory was expanded five centuries later by the Roman Lucretius. In a long poem, *On the Nature of Things*, Lucretius gave strong arguments for the atomic nature of matter.

These ancient theories seem remarkably modern in many respects. Neither Greeks nor Romans, however, had the means to determine which view of matter—atomistic or continuous—was correct. In fact, the ancients almost never experimented, preferring to reason from what they called "first principles." Therefore, the atomistic theory remained a minority view for 2000 years. It just didn't seem reasonable to the ancient theorists that a piece of matter could be so small that it could not be split into still smaller pieces.

Figure 2.2 Democritus imagined that "atoms" of water might be smooth round balls and that "atoms" of fire could have sharp edges.

---

*Elements were considered basic substances that were not composed of some combination of more basic substances. The Greeks believed that there were only four elements: earth, air, fire, and water. The relationship among the four elements and the four "principles"—hot, moist, dry, and cold—is shown in Figure 2.3.

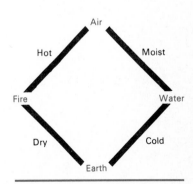

Figure 2.3 The Greek view of matter.

Figure 2.4 Antoine Lavoisier. [Courtesy of the Smithsonian Institution, Washington, DC.]

Lavoisier: The Law of Conservation of Mass

By about A.D. 1700, scientists were observing more carefully and measuring more accurately. Antoine Laurent Lavoisier, a Frenchman (1743–1794), perhaps did more than anyone else to establish chemistry as a quantitative science. He found that when a chemical reaction was carried out in a closed system, the total mass of the system was not changed. Perhaps the most important chemical reaction that Lavoisier performed was the decomposition of the red oxide of mercury to form metallic mercury and a gas he named oxygen. The reaction had been carried out before—by Karl Wilhelm Scheele, a Swedish apothecary (1742–1786), and by Joseph Priestley, a Unitarian minister who later fled England and settled in America—but Lavoisier was the first to weigh all the substances present before and after the reaction. He was also the first to interpret the reaction correctly.

Lavoisier carried out many quantitative experiments. He found that when coal was burned it united with oxygen to form carbon dioxide. He also experimented with animals. When a guinea pig breathed, oxygen was consumed and carbon dioxide was formed. Lavoisier therefore concluded that respiration was related to combustion. In each of these reactions he found that matter was conserved.

Lavoisier summarized his findings in these and other experiments by formulating a **scientific law**.* His **law of conservation of mass** holds that matter is neither created nor destroyed during a chemical change. In other words, if one weighed all the products of a reaction—solids, liquids and gases—the total would be the same as the mass of all the original substances (called **reactants**, or starting materials).

Scientists had by this time abandoned the Greek idea of the four elements and were almost universally using Robert Boyle's operational definition, put forth over a century before. Boyle, an Englishman, in his book *The Sceptical Chymist* (published in 1661), said that supposed *elements* must be tested to see if they really were simple. If a substance could be broken down into simpler substances, it was not an element. The simpler substances might be elements and would be so regarded until such time (if it ever came) as they in turn could be broken down into still simpler substances. On the other hand, two or more elements might combine to form a complex substance, called a *compound*.

Using Boyle's definition, Lavoisier included a table of elements in

---

* Scientific laws merely summarize experimental data. For example, Lavoisier found that in each of the reactions he carried out, the total weight of products was equal to the total weight of reactants. He summarized these findings as the law of conservation of mass. This law has been verified repeatedly through the years.

his book *Elementary Treatise on Chemistry*. The table included some substances we now know to be compounds. Lavoisier was the first to use modern and somewhat systematic names for the chemical elements. He is often called the "father of chemistry," and his book is regarded as the first chemistry textbook. Incidentally, Lavoisier lost his head (on the guillotine) during the French Revolution—but not because of his chemical research. In those days no one was a full-time chemist. Lavoisier had another job on the side: he was a tax collector for Louis XVI, and it was in this capacity that he incurred the wrath of the French revolutionaries.

The law of conservation of mass is of more than academic interest. The law states that we cannot create materials from nothing; we make new materials only by changing the way atoms are combined. Nor can we get rid of wastes by the destruction of matter. We must put the wastes somewhere. But chemistry does offer an alternative: we can change potentially harmful wastes to less dangerous forms. Indeed, such transformations of matter from one form to another are what chemistry is all about.

Figure 2.5 Robert Boyle. [Courtesy of the Smithsonian Institution, Washington, DC.]

Proust: The Law of Definite Proportions

In 1799, Joseph Louis Proust showed that a substance called copper carbonate, whether prepared in the laboratory or obtained from natural sources, contained the same three elements—copper, carbon, and oxygen—and always in the same proportion by weight—5.3 parts of copper to 4 parts of oxygen to 1 part of carbon. To summarize this and numerous other experiments, Proust formulated a new law. A compound, he said, always contains elements in certain definite proportions, and in no other combinations. This generalization he called the **law of definite proportions** (sometimes referred to as the **law of constant composition**).

Proust, like Lavoisier, was a member of the French nobility, but he was working in Spain, temporarily safe from the ravages of the French Revolution. His laboratory was destroyed, however, and he was re-

Figure 2.6 Whether synthesized in the laboratory or obtained from various natural sources, copper carbonate always has the same composition. Analysis of this compound led Proust to formulate the law of definite proportions.

103 g of
copper carbonate

53 g of copper   40 g of oxygen   10 g of carbon

duced to poverty when the French troops of Napoleon Bonaparte occupied Madrid in 1808.

One of the earliest illustrations of the law of definite proportions is found in the work of a Swedish chemist, J. J. Berzelius (1779–1848). Berzelius heated 10 g of lead with various amounts of sulfur to form lead sulfide. Since lead is a soft, grayish metal, sulfur a yellow solid, and lead sulfide a shiny, black solid, it was easy to tell when all the lead had reacted. Excess sulfur was easily washed away by carbon disulfide, a liquid that dissolves sulfur but not lead sulfide. As long as he used at least 1.56 g of sulfur, Berzelius got exactly 11.56 g of lead sulfide. Any sulfur in excess of 1.56 g was left over, unreacted. If he used more than 10 g of lead with 1.56 g of sulfur, he got 11.56 g of lead sulfide, with lead left over. These reactions are illustrated in Figure 2.7 (and explained in Figure 2.10).

In 1783, Henry Cavendish, a wealthy and eccentric Englishman, found that water was produced when hydrogen was burned in oxygen. (It was Lavoisier, however, who correctly interpreted Cavendish's experiment and first used the modern names for hydrogen and oxygen.) In 1800, the reverse reaction was accomplished by two English chemists, William Nicholson and Anthony Carlisle. They passed electric current through water and decomposed it into hydrogen and oxygen. This they did only 6 weeks after the electric battery was invented by an Italian, Alessandro Volta. This scientific breakthrough led to very rapid de-

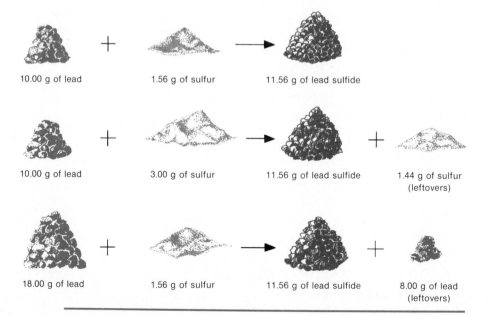

10.00 g of lead	+	1.56 g of sulfur	→	11.56 g of lead sulfide
10.00 g of lead	+	3.00 g of sulfur	→	11.56 g of lead sulfide	+	1.44 g of sulfur (leftovers)
18.00 g of lead	+	1.56 g of sulfur	→	11.56 g of lead sulfide	+	8.00 g of lead (leftovers)

Figure 2.7 Berzelius's experiment illustrating the law of definite proportions.

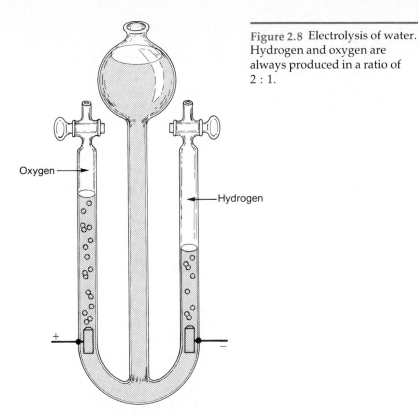

Figure 2.8 Electrolysis of water. Hydrogen and oxygen are always produced in a ratio of 2 : 1.

Oxygen

Hydrogen

+

−

velopments in chemistry and provided the death blow to the old Greek idea of water as an element— if indeed the idea was not already dead.

The law of definite proportions is the basis for chemical formulas (Chapter 6)—but it also has a wider meaning. In electrolysis, for example, water always yields (by volume) 2 parts of hydrogen and 1 part of oxygen (Figure 2.8). Yet it doesn't matter *where* the water comes from or *how* it is made. Furthermore, pure water always has the same *properties*: it is wet, it dissolves salt and sugar, and it freezes at 0 °C. The properties of water do not depend on our wants or wishes.

Dalton's Atomic Theory

Lavoisier's law of conservation of mass and Proust's law of definite proportions were repeatedly confirmed by experiment. This led to attempts to formulate theories that would account for these laws. In science, a *theory* is a model that consistently explains observations.

John Dalton, an English schoolteacher, was one of those who proposed a model to explain the accumulating experimental data. As he developed the details of his model, he uncovered a third "law" that his

Figure 2.9 John Dalton. [Courtesy of the Smithsonian Institution, Washington, DC.]

theory would have to explain. Proust had previously observed that a compound contains elements in certain definite proportions and only those proportions. Dalton's new law stated that elements might combine in *more* than one set of proportions. If they did, though, each different combination would produce a different compound. For example, carbon combined with oxygen in a mass proportion of 3 parts of the former to 8 parts of the latter to form carbon dioxide—a gas familiar as a product of respiration and of the burning of wood or coal. But Dalton found that 3 parts of carbon would also combine with 4 parts of oxygen by weight to form a poisonous gas that we know today as carbon monoxide. After observing similar multiple combinations for other sets of elements, Dalton put forth his **law of multiple proportions**.

In the same year (1803), he set down the details of his **atomic theory**, a model that offered a logical explanation for the laws we have mentioned. The most important points of Dalton's atomic theory are

1. All elements are made up of small, indestructible, and indivisible particles called **atoms**.
2. All atoms of a given element are identical, but the atoms of one element differ from the atoms of any other element.
3. Atoms of different elements can form combinations to give compounds.
4. A chemical reaction involves a change not in the atoms themselves, but in the way atoms are combined to form compounds.

Dalton's reasoning went something like this. If matter is continuous, why should 1.0 g of hydrogen always combine with 19.0 g of fluorine? Why shouldn't 1.0 g of hydrogen also combine with 18.9 g of fluorine? Or 19.1 g of fluorine? Or any other mass of fluorine? On the other hand, if matter is made of atoms, only whole atoms should combine. If an atom of fluorine has a mass 19 times that of a hydrogen atom, then the compound formed by the union of one of each element would have to consist of 1 part by mass of hydrogen and 19 parts by mass of fluorine. An analogy may clarify this point. If you have mashed potatoes and gravy (call them continuous foods), you can have a lot of mashed potatoes and a little bit of gravy, or vice versa. There are no set amounts that go together. On the other hand, if you have a hot-dog sandwich (an atomistic food), you need a bun for every hot dog. If the hot dogs weigh twice as much as the buns, then the law of definite proportions simply notes that hot-dog sandwiches always consist of 2 parts by mass of meat and 1 part by mass of bread. If you have 27 hot dogs and only 25 buns, you can only make 25 hot-dog sandwiches. The other 2 hot dogs are simply left over (see Berzelius's experiment in Figure 2.7, then consider Figure 2.10). Dalton concluded that matter must be atomistic for the law of definite proportions to hold.

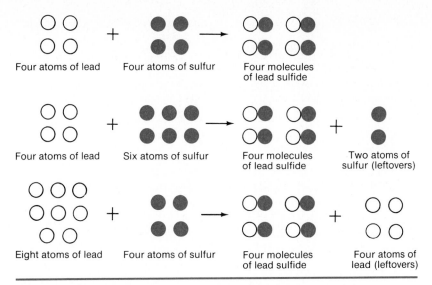

Figure 2.10 The law of definite proportions: Berzelius's experiment interpreted in terms of Dalton's atomic theory.

The atomic theory also explained the law of multiple proportions. The difference between carbon dioxide and carbon monoxide is that, in the former compound, one atom of carbon always combines with two atoms of oxygen and, in the latter, one atom of carbon always combines with only one atom of oxygen. Figure 2.11 shows how oxygen and nitrogen combine.

Finally, the law of conservation of mass can also be understood in terms of the atomic theory. When a reaction occurs and the reactants change to products, this change simply involves a reordering of atoms.

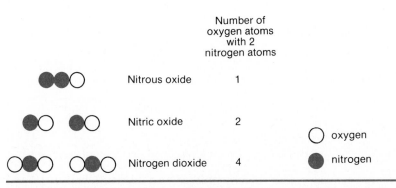

Figure 2.11 The law of multiple proportions. The amount of nitrogen is the same in each sample (two atoms). The oxygen ratio of the three compounds is 1:2:4.

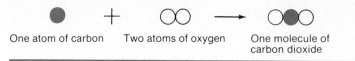

One atom of carbon   Two atoms of oxygen   One molecule of
carbon dioxide

Figure 2.12 The law of conservation of mass. In a chemical reaction atoms are merely rearranged (not created or destroyed); thus matter is conserved.

Matter is neither lost nor gained; it is simply rearranged (Figure 2.12).

Dalton set up a table of relative (atomic) masses as a part of his theory. In this table—based on hydrogen as 1—oxygen had a mass of 8, carbon 6. Thus, carbon monoxide was made up of one atom of carbon combined with one atom of oxygen to give a mass ratio of 6 parts of carbon to 8 of oxygen (or 3 to 4). Carbon dioxide was made up of one atom of carbon combined with two atoms of oxygen to give a mass ratio of 6 parts of carbon to 16 parts of oxygen (or 3 to 8). Many of Dalton's atomic masses were incorrect, however, as we might expect with the equipment available at that time. Dalton also invented a set of symbols to represent the different kinds of atoms. In fact, we used Dalton's symbols in Figures 2.11 and 2.12. Those symbols, too, have been replaced by modern symbols of one or two letters (see Table 1.1, page 16).

We can use proportions, such as those determined by Dalton, to calculate the amount of one substance needed to combine with a given amount of another substance. To learn how to do this, let's look at some examples.

Example 2.1   Carbon combines with hydrogen in a ratio of 3 parts by mass of carbon to 1 part by mass of hydrogen to form a gas called methane. How much hydrogen is needed to combine with 900 g of carbon to form methane?

$$900 \text{ g carbon} \times \frac{1 \text{ g hydrogen}}{3 \text{ g carbon}} = 300 \text{ g hydrogen}$$

Example 2.2   Nitrous oxide ("laughing gas") can be decomposed to give 7 parts by mass of nitrogen and 4 parts by mass of oxygen. How much nitrogen (by mass) would be obtained if enough nitrous oxide was decomposed to give 36 g of oxygen?

$$36 \text{ g oxygen} \times \frac{7 \text{ g nitrogen}}{4 \text{ g oxygen}} = 63 \text{ g nitrogen}$$

> *Example 2.3*   Hydrogen sulfide gas can be decomposed to give sulfur and hydrogen in a mass ratio of 16:1. If the relative mass of sulfur is 32 when hydrogen is taken to be 1, how many hydrogen atoms are combined with each sulfur atom in the gas?
>
> $$\frac{32 \text{ units of sulfur}}{1 \text{ atom of sulfur}} \times \frac{1 \text{ unit of hydrogen}}{16 \text{ units of sulfur}} \times \frac{1 \text{ atom of hydrogen}}{1 \text{ unit of hydrogen}} = \frac{2 \text{ atoms of hydrogen}}{1 \text{ atom of sulfur}}$$

Despite its inaccuracies, Dalton's atomic theory was a great success. Why? Because it served—and still serves—to explain a large body of experimental data. It also successfully predicts how matter will behave under a wide variety of circumstances. Dalton arrived at his atomic theory purely on the basis of reasoning power, and with modest modification it has stood the test of time and the assault of modern, highly sophisticated instrumentation. Formulation of so successful a theory was quite a triumph for a Quaker schoolteacher in the year 1803!

Atoms: Real and Relevant

Are atoms real? Certainly they are real as a concept, and a highly useful concept at that. Still, no one has ever *seen* an atom. Perhaps no one ever will.

Are atoms relevant? Much of modern science and technology—including the production of new materials and the technology of pollution control—is ultimately based on the concept of atoms. We have seen that atoms are conserved in chemical reactions. Thus, material things—things made of atoms—can be recycled, for the atoms are not destroyed no matter how we use them. The one way we might lose a material from a practical standpoint is to spread the atoms so thinly that it would take too much time and energy to put them back together again.

Consider iron atoms in a sample of hematite, an iron ore. The hematite might be converted into pig iron and then into steel and made into an automobile. After the automobile is worn out, the steel could be recovered and used again in a new automobile. Thus, the atoms could be changed from one combination to another, but in each conversion they would be conserved.

But suppose at one stage or another the iron was dissolved in sulfuric acid. The iron sulfate formed would be soluble in water. If someone poured the iron sulfate down the drain and it eventually wound up in the ocean, the iron atoms would still be there, but they would spread so thinly in the vast waters of the ocean that there would be no practical way to recover them. We may conclude, then, that atoms can be recycled—provided we do not spread them too thinly.

Leucippus Revisited

Now back to Leucippus' musings by the seashore. We now know that if we keep dividing those drops of water into smaller drops, we will ultimately obtain a small particle—called a **molecule**—that is still water. If we divide that particle still further, we will obtain two *atoms* of hydrogen and one *atom* of oxygen. And if we divide those . . . but that is a story for another time.

Dalton regarded the atom as indivisible, as did his successors up until the discovery of radioactivity in 1895. We examine the changing concept of the atom in the next chapter.

Problems

1. What is the distinction between the atomistic view and the continuous view of matter?

2. Why did the theory that matter was continuous (rather than atomic) prevail for so long?

3. What discoveries finally refuted the theory that matter was continuous?

4. What is Democritus's contribution to atomic theory?

5. If foods were described as atomistic or continuous, which designation would you use for each of the following?
   a. peas   b. mashed potatoes
   c. milk   d. hot dogs
   e. hard-boiled eggs   f. scrambled eggs

6. Describe Lavoisier's contribution to the development of modern chemistry.

7. State the law of conservation of mass.

8. How does the ancient Greek definition of an element differ from the modern one?

9. How did Robert Boyle define an element?

10. State the law of definite proportions.

11. What is a scientific law? How does a scientific law differ from a governmental law?

12. When we burn a 10-kg piece of wood, only 0.05 kg of ash is left. Explain this apparent contradiction of the law of conservation of mass.

13. When 3 g of carbon is burned in 8 g of oxygen, 11 g of carbon dioxide is formed. What mass of carbon dioxide would be formed if 3 g of carbon was burned in 50 g of oxygen? What law does this illustrate?

14. Outline the main points of Dalton's atomic theory.

15. Heptane *always* contains 84% carbon and 16% hydrogen. What law does this illustrate?

16. Sulfur and oxygen form two compounds. One is 50% S and 50% O; the other is 40% S and 60% O. What law does this illustrate?

17. A photographic flash bulb weighing 0.750 g contains magnesium and air. The flash converts the magnesium to magnesium oxide. After cooling, the bulb weighs 0.750 g. What law does this illustrate?

18. An atom of calcium has a mass of 40 amu and an atom of cobalt has a mass of 59 amu. Are these findings in agreement with Dalton's atomic theory?

19. An atom of calcium has a mass of 40 amu and an atom of potassium has a mass of 40 amu. Are these findings in agreement with Dalton's atomic theory?

20. An atom of calcium has a mass of 40 amu and another atom of calcium has a mass of 44 amu. Are these findings in agreement with Dalton's atomic theory?

21. Use Dalton's atomic theory to explain the law of conservation of mass. Give an example.

22. Use Dalton's atomic theory to explain the law of definite proportions. Give an example.

23. Use Dalton's atomic theory to explain the law of multiple proportions. Give an example.

24. What did Nicholson and Carlisle prove when they decomposed water to hydrogen and oxygen by passing electricity through the water?

25. What did each of the following contribute to the development of modern chemistry?

   a. J. J. Berzelius   b. Henry Cavendish
   c. Joseph Proust

26. When 18.0 g of water is decomposed by electrolysis, 16.0 g of oxygen and 2.0 g of hydrogen are formed. According to the law of definite proportions, how much hydrogen is formed by the electrolysis of 180 g of water?

27. Hydrogen, from the decomposition of water, has been promoted as the fuel of the future (Chapter 14). How much water would have to be electrolyzed to produce 1000 kg of hydrogen? (See Problem 26.)

28. Use the mass ratios on page 34 to calculate how much carbon is required to produce 2200 g of carbon dioxide.

29. In limited air, carbon burns to produce carbon monoxide. Use the mass ratios on page 34 to calculate how much carbon monoxide can be formed from 24 g of carbon.

30. The gas silane can be decomposed to yield silicon and hydrogen in a ratio of 7 parts by mass of silicon to 1 part by mass of hydrogen. If the relative mass of silicon atoms is 28 when the mass of hydrogen atoms is taken to be 1, how many hydrogen atoms are combined with each sillicon atom?

31. Jan Baptista van Helmont, a Flemish alchemist (1579–1644), performed an experiment in which he planted a young willow tree in a weighed bucket of soil. After 5 years, he found that the tree had gained 75 kg, yet the soil had lost only 57 g (0.057 kg). Van Helmont had added only water to the system; hence, he concluded that the substance of the tree had come from water. Criticize this conclusion.

References and Readings

1. Asimov, Isaac. *A Short History of Chemistry*. Garden City, NY: Doubleday, 1965. Chapters 1, 5, 12, and 13.
2. Cole, K. C. "On Imagining the Unseeable." *Discover*, December 1982, pp. 70–72.
3. Jaffe, Bernard. *Crucibles: The Story of Chemistry*. New York: Fawcett World Library, 1957.
4. Kolb, Doris. "Chemical Principles Revisited: But If Atoms Are So Tiny . . . ." *Journal of Chemical Education*, September 1977, pp. 543–547.
5. Kolb, Doris. "Chemical Principles Revisited: What Is an Element?" *Journal of Chemical Education*, November 1977, pp. 696–700.
6. Patterson, Elizabeth C. *John Dalton and the Atomic Theory*. New York: Doubleday, 1970. The story of Dalton and his formulation of the atomic theory.
7. Ritchie-Calder, Lord. "The Lunar Society of Birmingham." *Scientific American*, June 1982, pp. 136–145. John Watt, Ben Franklin, Erasmus Darwin, Josiah Wedgwood, and Joseph Priestley were members of the Lunar Society.
8. Stillman, John Maxon. *The Story of Alchemy and Early Chemistry*. New York: Dover Publications, 1960. Chapter 14, "Lavoisier and the Chemical Revolution."
9. Young, Louise B. (Ed.). *The Mystery of Matter*. New York: Oxford University Press, 1965. Contains excerpts from the original works of Lucretius, Dalton, and others.

3

Atomic Structure

Images of the Invisible

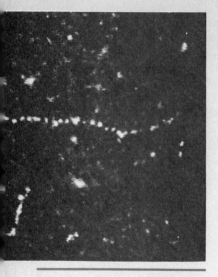

Figure 3.1 The bright spots in this photomicrograph are images of thorium atoms complexed with an organic chemical called benzenetetracar-boxylic acid. The images were made by Albert Crewe with a scanning electron microscope. [Courtesy of the Enrico Fermi Institute, Chicago.]

Dalton's concept of hard, indivisible atoms was accepted for nearly all the nineteenth century. By the beginning of the present century, however, it had become obvious that atoms are more complex. The concept of the atom changed rapidly: atoms were no longer seen as hard spheres differing only in mass; they were conceived of as being made up of component parts and of having different sizes and shapes.

Atoms are so small they are invisible. It is impossible to see an atom in ordinary light, even with the most powerful optical microscopes. There are techniques for recording images of atoms, however. In 1970, Professor Albert Crewe of the University of Chicago announced that he had photographed images of single uranium and thorium atoms (Figure 3.1). Color Plate A shows images of uranium atoms made with an electron microscope. In 1976, a group of scientists led by George W. Stroke of the State University of New York at Stony Brook photographed images showing the locations and relative sizes of tiny carbon,

40

magnesium, and oxygen atoms in a section of crystal (Figure 3.2). These photographs were remarkable achievements, but the images in them were only light spots against dark backgrounds. They did little to reveal the structure of atoms.

Why do we care about the structure of atoms? Because it is the arrangement of the parts of atoms that determines the properties of different kinds of matter. Only by understanding atomic structure can we know how atoms combine to make the many different substances in nature and, even more important, how we can modify materials to meet our needs more precisely. A knowledge of atomic structure is even essential to your health. Many medical diagnoses are based on the analysis of blood and urine. Many such analyses depend on knowledge of how the structure of atoms is changed when energy is absorbed.

Perhaps of greater immediate interest to you is the fact that your success in the study of chemistry (as well as much of biology and other sciences) will depend at least in part on your knowledge of atomic structure. Let's start our study of atomic structure by going back to the time of John Dalton.

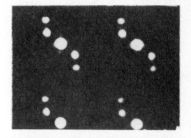

Figure 3.2 Images of magnesium, oxygen, and carbon atoms in a section of a crystal. George Stroke combined X rays, computer techniques, and holography to obtain this remarkable picture. [Courtesy of Prof. Dr. George W. Stroke, Technische Universität, Munich.]

The Atom in the Nineteenth Century

Dalton, who set forth his atomic theory in 1803, regarded the atom as hard and indivisible. It wasn't long, however, before evidence accumulated to show that matter is electrical in nature. Indeed, the electrolytic decomposition of water by Nicholson and Carlisle in 1800 (Chapter 2) had already indicated as much.

Humphry Davy, an English chemist, built a powerful battery. With it, he discovered a number of new elements. In 1807, he liberated a bright, silvery metal from molten potash (now known to chemists as potassium hydroxide). This very reactive metal he named potassium. A short time later, he produced sodium metal from molten soda (sodium hydroxide). Within a year, Davy also had produced the metals magnesium, strontium, barium, and calcium.

Sir Humphry's protégé, Michael Faraday (1791–1867), greatly extended the new science of electrochemistry. He termed the process of splitting compounds by means of electricity **electrolysis** (Figure 3.3). A compound that, when melted or taken into solution, could conduct an electric current was named an **electrolyte**. The carbon rods or metal strips inserted into a melt or a solution were named **electrodes**. One electrode, called the **anode**, bears a positive charge. The other, which carries a negative charge, Faraday called the **cathode**.

Faraday hypothesized that the electric current was carried through the melted compound or the solution by charged atoms, which he called **ions**. One type of ion, which bears a negative charge and travels toward

Figure 3.3 Electrolysis apparatus.

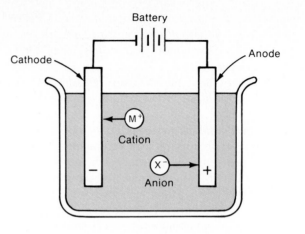

the anode, he called the **anion**. The other, which has a positive charge and travels toward the cathode, he called the **cation**.

Faraday's work with solutions and melts established that atoms are electrical in nature. Further detail in the structure of atoms had to wait for the development of gas discharge tubes and of still more powerful sources of electrical voltage. In fact, Faraday himself tried and failed to pass electricity through a tube that had part of the air pumped out. His vacuum just wasn't good enough.

By 1875, tubes with better evacuation were available. William Crookes passed an electric current through such a tube (Figure 3.4 and Color Plate B). The beam of current traveled in straight lines from cathode to anode. This beam was said to be composed of **cathode rays**.

But just what are these cathode rays? Streams of particles, the British scientists stoutly maintained. No, they were much more likely to be a form of light made up of waves, the Germans insisted. Who was right? The answer in such cases comes—at least it should come—from

Figure 3.4 A simple gas discharge tube.

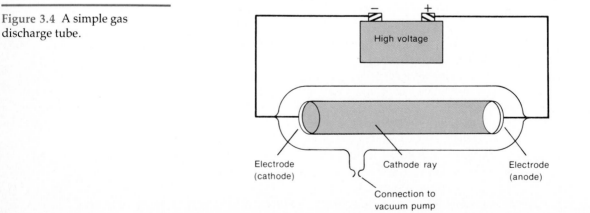

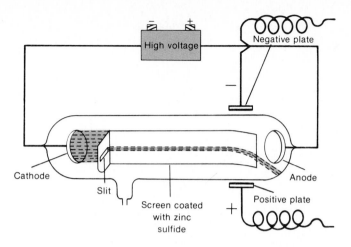

Figure 3.5 Thomson's apparatus showing deflection of an electron beam in an electric field. The screen is coated with zinc sulfide, a substance that glows when struck by electrons.

experimentation, not politics. An Englishman, Joseph John Thomson, provided the answer in 1897. He showed that the cathode rays were deflected in an electric field (Figure 3.5). Since they were deflected toward the positive plate, the rays must be composed of negatively charged particles. Each particle carried an identical charge. The name **electron** was given to those units of negative charge. Thomson also was able to measure the mass-to-charge ratio of the cathode ray particle by determining the amount of deflection in a magnetic field of known strength. He couldn't measure either the mass or the charge separately, though. That is like determining that an apple weighs 1000 times as much as an apple seed, but not knowing the mass of the apple or its seed. Thomson won the Nobel Prize in physics in 1906.

The same type of ray was obtained regardless of the material from which the cathode was made and regardless of the gas in the tube. Thus, electrons came to be regarded as constituents of all matter.

In 1886, a German scientist named Eugen Goldstein experimented with gas discharge tubes that had perforated cathodes (Figure 3.6). He found that while electrons were formed and sped off toward the anode, positive rays were formed and shot in the opposite direction through the

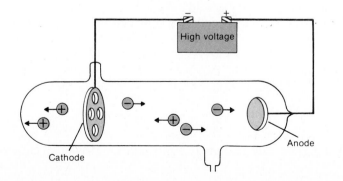

Figure 3.6 Goldstein's apparatus for study of positive particles.

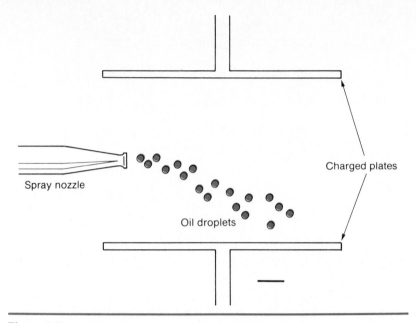

Figure 3.7 Oil-drop method for determining the charge of an electron.

holes in the cathode. It was not until 1907, however, that a study of the deflection of these particles in a magnetic field revealed that they were of varying mass. The lightest, formed when there was little hydrogen gas in the tube, was found later to be 1837 times as massive as an electron.

The charge of an electron was determined by an American, Robert A. Millikan, in 1909. A crude diagram of Millikan's apparatus is shown in Figure 3.7. An atomizer is used to spray tiny droplets of oil between two charged plates. Gravity tends to cause the droplets to settle. If the droplets can be given a negative charge, however, they will be attracted upward (toward the positive plate). Movement in either direction is opposed by air resistance. X rays can be used to knock electrons loose from atoms. Some of these will adhere to oil droplets, giving the latter the desired negative charge. Through a telescope, Millikan observed the speed of droplets as they moved toward the positive plate. From the rate of rise and the magnitude of the electric field, he was able to calculate the charge on an individual droplet. The smallest possible charge on a droplet was taken to be the charge of an individual electron, a quantity now considered to be the basic unit of charge and often referred to as a charge of 1− (one minus). Millikan was awarded the Nobel Prize in physics in 1923.

From Millikan's value for the charge and Thomson's value for the mass-to-charge ratio, the mass of the electron was readily calculated. Electrons are extremely light particles, having a mass of only

9.1×10^{-28} g. (If you do not understand numbers like 9.1×10^{-28}, you can find out about them in Appendix B.)

Out of Chaos: The Periodic Table

Before moving on, let's take a look at a remarkable parallel development. New elements were being discovered with surprising frequency. By 1830, there were 55 known elements, all with different properties and with no apparent order in these properties. John Dalton had set up a table of relative* atomic masses in his book *A New System of Chemical Philosophy*, published in 1808. Many of Dalton's atomic masses were wrong. These were improved in subsequent years, notably by Berzelius, who published a table of atomic masses in 1828 that contained 54 elements. Except for three cases, Berzelius's atomic masses are in quite good agreement with modern values.

Several attempts were made to arrange the elements in some sort of systematic fashion. The first really successful arrangement was that of Dmitri Ivanovich Mendeleev (1834–1907), a Russian chemist. He published a **periodic table** of the elements in 1869. His table arranged the elements primarily in order of increasing atomic mass, although in a few cases he placed a slightly heavier element before a lighter one. He did this to get elements with similar *chemical properties* in the same column. For example, he placed tellurium (with an atomic mass of 127.6) ahead of iodine (with an atomic mass of 126.9) because the former resembled sulfur and selenium in its properties whereas the latter was similar to chlorine and bromine.

Mendeleev left a number of gaps in his table. Instead of looking upon these blank spaces as defects, he boldly predicted the existence of elements yet undiscovered. Furthermore, he even predicted the properties of some of these missing elements.

In succeeding years, many of the gaps were filled in by the discovery of new elements. The properties were often quite close to those

Figure 3.8 Dmitri Mendeleev, the Russian chemist who invented the periodic table of the elements. [Reprinted with permission from Mary E. Weeks, *Discovery of the Elements* (Easton, PA: Chemical Education Publishing, 1968). Copyright © 1968 by Chemical Education Publishing Company.]

---

* Atoms are extremely minute. During the nineteenth century, it was impossible to determine actual masses of atoms. Indirect measurements, however, could indicate relative atomic masses. Dalton's atomic masses were based on an atomic mass of 1 for hydrogen. As more accurate atomic masses were determined, this standard was replaced by one in which oxygen was assigned a value of 16.0000. This latter standard survived until the middle of the present century, when it was replaced by a slightly more logical one based on an isotope (Chapter 4) of carbon. Adoption of this new standard caused little change in the relative atomic masses.

These relative atomic masses usually are expressed in terms of **atomic mass units** (amu).

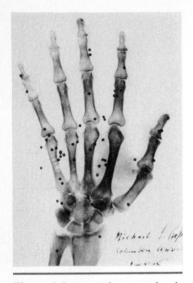

Figure 3.9 An early example of the use of X rays in medicine. Professor Michael Pupin of Columbia University made this X ray in 1896 to aid in the removal by surgery of gunshot pellets (the dark spots) from the hand of a patient. [Courtesy of Burndy Library, Norwalk, CT.]

Mendeleev had predicted. It was the predictive value of this great innovation that led to the wide acceptance of Mendeleev's table.

The modern periodic table (inside front cover) contains over 100 elements.

Serendipity in Science: X Rays and Radioactivity

Let's return now to the structure of the atom and look at a little scientific serendipity. Many scientific discoveries often are described as happy accidents. Have you ever wondered why these accidents always seem to happen to scientists? It is probably because scientists are trained observers. The same accident could happen right before the eyes of an untrained person and go unnoticed. Or, if it was noticed, its significance might not be grasped.

Two such happy accidents occurred in the last years of the nineteenth century. In 1895, a German scientist, Wilhelm Konrad Röntgen, was working in a dark room, studying the glow produced in certain substances by cathode rays. To his surprise, he noted this glow on a chemically treated piece of paper some distance from the cathode ray tube. The paper even glowed when taken to the next room. Röntgen had discovered a new type of ray that could travel through walls! The rays were given off from the anode whenever the cathode ray tube was operating. With seeming lack of imagination, Röntgen called the mysterious, penetrating rays **X rays**.

As is often the case when an exciting new discovery is made, many other scientists began to study the new phenomenon. One, Antoine Henri Becquerel, a Frenchman, was studying **fluorescence**. Certain chemicals, when exposed to strong sunlight, continued to glow even when taken into a dark room. Was this phenomenon related to X rays? Becquerel wrapped photographic film in black paper, placed a few crystals of the fluorescing chemical on top of the paper, and placed the paper in strong sunlight. If the glow was ordinary light, it would not pass through the black paper. If, on the other hand, the glow was similar to X rays, it would pass through the black paper and fog the film.

Before Becquerel found out very much about fluorescence, he accidentally made an important discovery. He was testing a crystal of a uranium compound. When placed in sunlight, the compound fogged the photographic film. On several cloudy days when work in sunlight was impossible, Becquerel prepared samples and placed them in a drawer. To his great surprise, the photographic film was exposed even though the uranium compound had not been exposed to sunlight. Further experiments showed that this radiation had no connection with fluorescence but was a characteristic of the element uranium.

Right away many other scientists began to study this newly discovered type of radiation. One, Marie Sklodowska Curie, gave this new phenomenon the name **radioactivity**. Marie Sklodowska was born in Poland in 1867. She went to Paris to work for her doctor's degree in mathematics and physics. There she met and married Pierre Curie, a French physicist of some note. With the help of her husband, she discovered a number of new radioactive elements, including radium. Pierre Curie was killed in a traffic accident in 1906, 3 years after he, his wife, and Becquerel were awarded the Nobel Prize in physics. Marie Curie continued to work with radioactivity, winning the Nobel Prize for chemistry in 1911. She died in 1934 of pernicious anemia, perhaps brought on by deprivation, hard work, and long exposure to radiation from the materials with which she had worked.

Figure 3.10 Marie Sklodowska Curie. [Courtesy of Culver Pictures.]

A New Atom for a New Century

It was soon realized that three types of radiation emanated from uranium and thorium. When passed through a strong magnetic field, the second type was deflected mildly in a direction opposite to the first; the third type was not deflected at all (Figure 3.11). These types of radiation were named **alpha** (α), **beta** (β), and **gamma** (γ) rays, respectively, by Ernest Rutherford, a New Zealander working at McGill University in Montreal. The alpha rays were shown to have a mass four times that of the hydrogen atom and a charge twice that of the electron and opposite in sign. Beta rays were shown to be identical to cathode rays; that is,

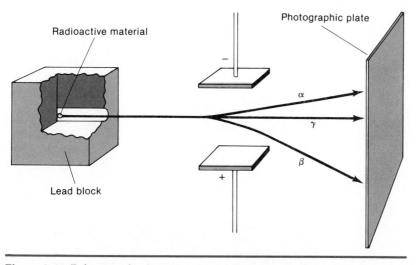

Figure 3.11 Behavior of radioactive rays in an electric field.

Table 3.1 Types of Radioactivity

Name	Greek Letter	Mass (atomic mass units)	Charge
Alpha	α	4	2+
Beta	β	$\frac{1}{1837}$	1−
Gamma	γ	4	0

they were shown to be streams of electrons. Electronic charge is usually assigned a value of 1−. Thus, alpha particles have charges of 2+. Gamma rays were shown to be very much like X rays, only more penetrating. These properties are summarized in Table 3.1.

Rutherford soon used the positively charged alpha particles to make an important discovery. He placed some highly radioactive material in a lead-lined box with a tiny hole. Most of the alpha particles were absorbed by the lead, but those escaping through the hole made up a narrow stream of very high energy particles. This apparatus, then, could be aimed like a gun at some target (Figure 3.12).

One target Rutherford selected was a thin piece of gold foil. He expected most of the positively charged alpha particles to be deflected only slightly by the positive charges in the atoms of the gold foil. What he found was that most of the alpha particles went through the foil—which was about 2000 atoms thick—without being deflected at all. This result is understandable if one assumes that the positive charges in the atom are rather thinly spread and no match for a high-energy alpha bullet. The tiny electrons in the atom would not be expected to

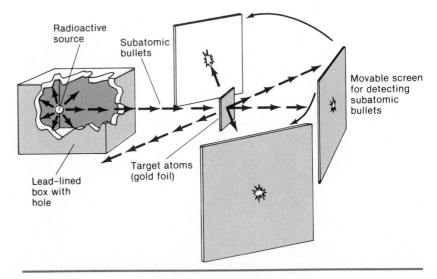

Figure 3.12 Rutherford's gold-foil experiment.

Alpha
particles   Gold foil

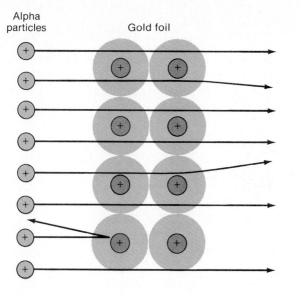

Figure 3.13 Model explaining
the results of Rutherford's
gold-foil experiment.

stop the more massive alpha particle either. What really amazed Rutherford was the fact that a few of the particles were deflected very sharply, some even bouncing back in the direction from which they came (Figure 3.13).

On the basis of these results, Rutherford proposed an arrangement of the positive and negative parts of the atom. To account for the strong deflections of some of the alpha particles, he pictured all the positive charge and almost all of the mass concentrated in an extremely small space. When an alpha particle approached this concentrated, positively charged matter (called the **nucleus** of the atom), it was strongly repelled and, thus, strongly deflected. Since few alpha particles were deflected, Rutherford concluded that the nucleus must occupy only a fraction of the total volume of the atom. Most of the particles passed right through the atom because most of the atom was empty space. But this space was not completely empty: it was here that Rutherford placed the negatively charged electrons. He concluded that electrons were so small that their presence in no way interfered with the passage of the alpha particles through the atom, a situation analogous to that of a mouse trying to stop the charge of an enraged bull elephant.

Rutherford's nuclear theory of the atom, set forth in 1911, was revolutionary indeed. He postulated that all the positive charge and virtually all the mass were concentrated in the tiny, tiny nucleus. The negatively charged electrons, he said, had virtually no mass yet occupied nearly all the volume of the atom. Rutherford's model can be pictured as follows: a BB (the nucleus) in the middle of a volume of space equivalent to that inside the New Orleans Superdome (the atom) surrounded by some mosquitoes (electrons) that flit here and there throughout the stadium.

The Structure of the Nucleus

In 1914, Rutherford suggested that the smallest positive-ray particle (i.e., that which is formed when there is hydrogen gas in the Goldstein apparatus) is the unit of positive charge in the nucleus. This particle, called a **proton**, has a charge equal in magnitude to that of the electron and has nearly the same mass as the hydrogen atom. Rutherford's suggestion was that protons constitute the positively charged matter in all atoms. The hydrogen atom's nucleus consisted of one proton, and the nuclei of larger atoms contained a nmber of protons. Except for hydrogen atoms, though, atomic nuclei were found to be heavier than would be indicated by the number of positive charges (number of protons). For example, the helium nucleus was found to have a charge of 2+ (and, therefore, it contained two protons, according to Rutherford's theory), but its mass was four times that of hydrogen. This excess mass puzzled scientists until 1932, when the English physicist James Chadwick discovered a particle with about the same mass as a proton, but with *no* electrical charge. This particle was called the **neutron**, and its existence made possible an explanation of the unexpectedly high mass of the helium nucleus. Whereas the hydrogen nucleus contained only a proton of 1 amu, the helium nucleus contained not only two protons (2 amu) but also two neutrons (2 amu), giving a total mass to the nucleus of 4 amu.*

With the discovery of the neutron, the list of "building blocks" we will need for "constructing" atoms is complete. The properties of these particles are summarized in Table 3.2.

The number of protons in the nucleus of an atom of any element is equal to the **atomic number** of that element. This number determines the kind of atom, that is, the identity of the element. Dalton had said that the mass of an atom determines the element. We now say it is not the mass but the number of protons that determines the element. For example, an atom with 26 protons (one whose atomic number is 26) is an atom of iron (Fe). An atom with 92 protons is an atom of uranium (U).

---

* The alpha particle, which has a mass of 4 and a charge of 2+, is identical to the nucleus of a helium atom.

Table 3.2 Subatomic Particles

Particle	Symbol	Mass (atomic mass units)	Charge	Location in Atom
Proton	p^+	1	1+	Nucleus
Neutron	n	1	0	Nucleus
Electron	e^-	$\frac{1}{1837}$	1−	Outside nucleus

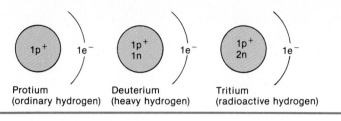

Figure 3.14 The isotopes of hydrogen.

The number of neutrons in the nuclei of atoms of a given element may vary. For example, most hydrogen (H) atoms have a nucleus consisting of a single proton and no neutrons (and therefore a mass of 1 amu.) About 1 hydrogen atom in 5000, however, does have a neutron as well as a proton in the nucleus. This heavier hydrogen is called **deuterium** and has a mass of 2 amu. Both kinds are hydrogen atoms (any atom with atomic number 1, that is, with one proton, is a hydrogen atom). Atoms that have this sort of relationship—the same number of protons but differing numbers of neutrons—are called **isotopes** (Figure 3.14). A third, very rare isotope of hydrogen is **tritium**, which has two neutrons and one proton in the nucleus (and thus a mass of 3 amu). Most, but not all, elements exist in nature in isotopic forms. This fact also requires a major modification of Dalton's original theory. He said that all atoms of the same element have the same mass. We now say that most elements have several isotopes, that is, atoms with different numbers of neutrons and, therefore, different masses.

Electron Structure: The Bohr Model

First, an aside. Light, such as that emitted by the sun or an incandescent bulb, is a form of pure energy. When white light from an incandescent lamp is passed through a prism, it is separated into a **continuous spectrum**, or rainbow, of colors (Figures 3.15 and 3.16 and Color Plate C). When sunlight passes through a raindrop, the same phenomenon occurs. The different colors of light represent different amounts of energy. Blue light packs more energy than red light of the same intensity. White light is simply a combination of all the various colors of different energies.

Now back to atoms. If the light from a flame in which a certain element is being heated is passed through a prism, narrow colored lines are obtained (Figure 3.17 and Color Plate D). Each line corresponds to light of definite energy. The pattern of colored lines emitted by each element (called its **line spectrum**) is characteristic of that element and

Figure 3.15 The rainbow is an example of a continuous spectrum.

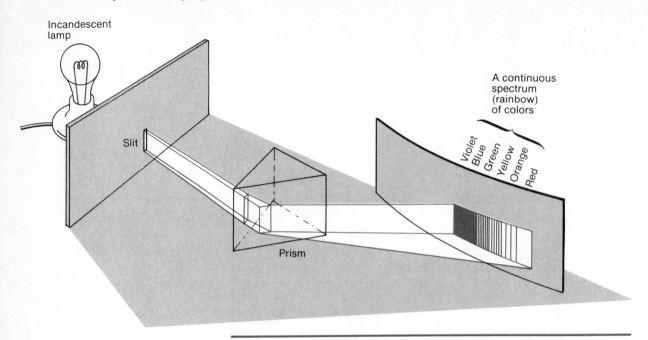

Figure 3.16 White light is separated into a continuous spectrum.

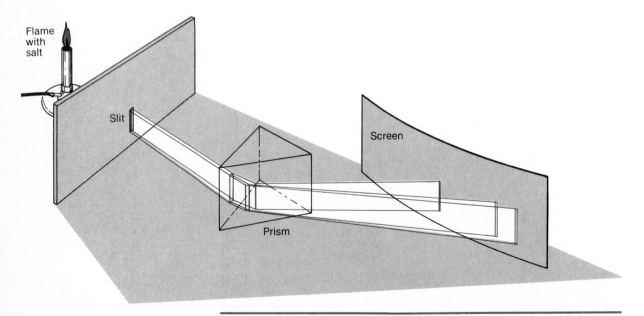

Figure 3.17 A line spectrum.

Red   Violet

Figure 3.18 Characteristic line spectra of some elements.

can be used to identify it (Figure 3.18). Consider, for example, that the only thing reaching the Earth from the stars or the planets is light. Until recently, everything we knew about these heavenly bodies had to be deduced from our examination of this light. Scientists have used line spectra to determine the chemical makeup of the stars and the atmospheres of the planets.

> In a crude way, the color of a flame can be used to identify an element (Color Plate E). Sodium salts give a persistent yellow flame; potassium salts, a fleeting lavender; lithium salts, a brilliant red. The colors in fireworks displays are due to the characteristic energies of the light given off by specific elements (Color Plate F). Sodium compounds are used for yellow fireworks, strontium salts for red, and barium compounds for green. Copper salts produce an ordinary blue. No substance produces a brilliant blue.

When last we saw the atom, it included electrons flitting about the nucleus like mosquitoes. The motion of the electrons kept them from falling into the nucleus. Remember that the nucleus is positively charged and should attract the negatively charged electrons. The motion of the electrons opposes the attraction, so the electrons remain outside the nucleus. (The moon resists the pull of the Earth, and the Earth does not fall into the sun for the same reason.) Like anything in motion, electrons have kinetic energy. And because they are outside the nucleus, they have a potential energy. Electrons, in this respect, are like rocks on a cliff. The rocks can fall and give up energy; if electrons were to fall into the nucleus, they too would give up energy.

How does all this concern the line spectra of the elements? Well, the first satisfactory explanation of line spectra was set forth by Niels Bohr, a Danish physicist, in 1913. He made the revolutionary suggestion that electrons cannot have just *any* amount of energy but can have only certain specified values; that is to say, the total energy of an electron is *quantized*. An electron, by absorbing a **quantum** (a packet of specific size) of energy (for example, when atoms of the element are heated in a

Figure 3.19 Possible electron shifts between energy levels in atoms to produce the lines found in spectra. Not all the lines are in the visible portion of the spectrum.

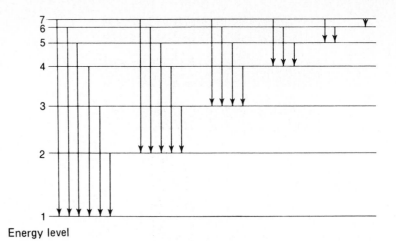

Energy level

flame), is elevated to a higher energy level. By giving up a quantum of energy, the electron can return to a lower energy level. The energy given off, having only certain specified values, shows up as a line spectrum (Figure 3.19). An electron moves instantaneously from one energy level to another.

An analogy would be a person on a ladder. The person can stand on the first rung, the second rung, the third rung, and so on, but is unable to stand between rungs. As the person goes from one rung to another, the potential energy (energy due to position) changes by definite amounts, or quanta. For an electron, the total energy (both potential and kinetic) changes as the electron moves from one energy level to another.

Bohr's model of the atom was based on the laws of planetary motion that had been set down by the Danish astronomer Johannes Kepler three centuries before. Bohr imagined the electrons to be orbiting about the nucleus much as planets orbit the sun (Figure 3.20). Different energy levels can be pictured as different orbits. The electron in the hydrogen atom is usually in the first energy level (the innermost orbit). Given the choice, electrons usually remain in their lowest possible energy levels; atoms whose electrons are in this level are said to be in their **ground state**. When a flame supplies energy to an atom of hydrogen, for example, and the electron jumps from the first level to the second or a higher level, the atom is said to be **excited**. An atom in an excited state eventually emits energy (light) as the electron jumps back down to one of the lower levels and, ultimately, reaches the ground state.

Atoms larger than hydrogen have more than one electron, and Bohr also deduced that the various energy levels of an atom could only handle a certain number of electrons at one time. We shall simply state Bohr's findings in this regard. The maximum number of electrons that can be in a given level is indicated by the formula $2n^2$, where n is equal to the

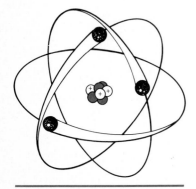

Figure 3.20 The nuclear atom, as envisioned by Bohr, has most of its mass in an extremely small nucleus. Electrons orbit about the nucleus, occupying most of the volume of the atom but contributing little to its mass.

energy level being considered. For the first energy ($n = 1$), the maximum population is $2(1^2)$ or 2. For the second energy level ($n = 2$), the maximum number of electrons is $2(2^2)$ or 8. For the third level, the maximum is $2(3^2)$ or 18.

Example 3.1 What is the maximum number of electrons in the fifth energy level?

$$2n^2 = 2 \times 5^2 = 2 \times 25 = 50$$

Imagine building up atoms by adding one electron to the proper energy level as protons are added to the nucleus. Electrons will go to the lowest energy level available. For hydrogen (H), with a nucleus of only one proton, the one electron goes into the first energy level. For helium (He), with a nucleus of two protons (and two neutrons), two electrons go into the first energy level. According to Bohr, the maximum population of the first energy level is two electrons; thus, that level is filled in the helium atom. With lithium (Li), which has three electrons, the extra electron goes into the second energy level. This process of adding electrons is continued until the second energy level is filled with eight electrons, as in the neon (Ne) atom (which has a total of 10 electrons, two in the first level and eight in the second). This buildup is diagrammed in Figure 3.21. One could continue to build atoms (in one's imagination) in this manner and, with a few modifications, build up the entire periodic chart of elements.

Example 3.2 Draw a Bohr diagram for fluorine (F).
  Fluorine is element number nine; it has nine electrons. Two of these go into the first shell, the remaining seven into the second.

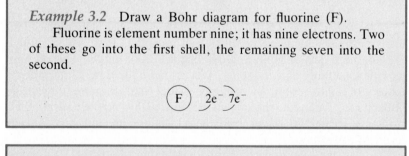

Example 3.3 Draw a Bohr diagram for sodium (Na).
  Sodium is element number eleven; it has 11 electrons. Two of these go into the first shell. Of the nine remaining, eight go into the second shell. That leaves one electron for the third shell.

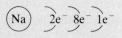

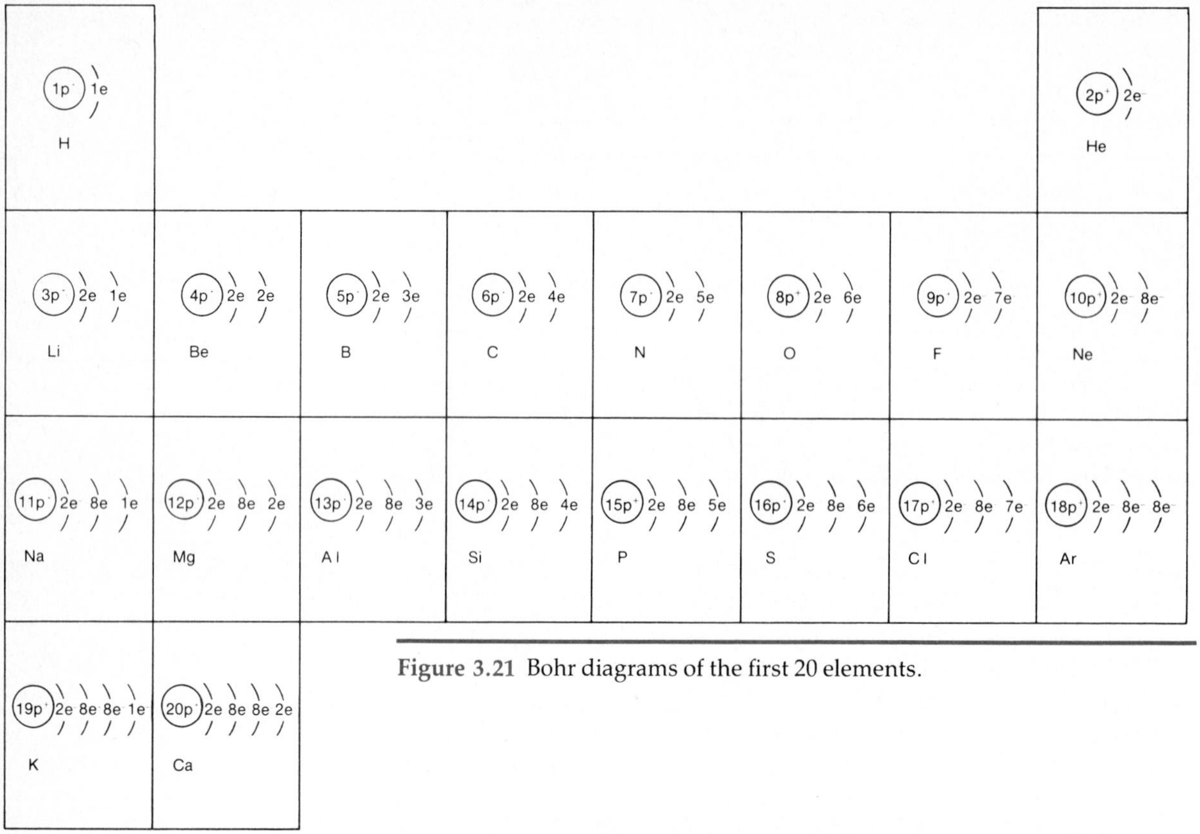

Figure 3.21 Bohr diagrams of the first 20 elements.

The Quantum Mechanical Atom

The simple planetary Bohr model of the atom has been replaced for many purposes by more sophisticated models. These newer models explain more data in greater detail than the Bohr model. This presents us with something of a quandary. We can more accurately interpret the nature of matter only by using a model that is more difficult to understand. Quantum theory has become highly mathematical. Electrons are treated as waves, and their locations are indicated only in terms of probabilities.

It was the young French physicist Louis de Broglie who first suggested (in 1924) that the electron should have wavelike properties. In other words, de Broglie said that a beam of electrons should behave very much like a beam of light. Since Thomson had proved (in 1897) that electrons were particles, this suggestion that they be treated as waves was hard to accept. Nevertheless, de Broglie's theory was experimentally verified within a few years. Electron microscopes, which make use of the wave nature of electrons, now are found in many scientific laboratories.

Erwin Schrödinger, an Austrian physicist, used highly mathematical, quantum mechanics in the 1920s to develop elaborate equations that describe the properties of electrons in atoms. (Fortunately, we need not understand the elaborate mathematics to make use of some of the results.) The solutions to these equations are called **eigen functions**. These functions can be manipulated to provide a measure of the probability of finding an electron in a given volume of space. The definite planetary orbits of the Bohr model are replaced by shaped volumes of space in which electrons move. The term **orbital** (replacing Bohr's *orbits*) was used for this new description of the location of electrons.

Suppose you had a camera that could photograph electrons (there is no such thing, but we are just supposing) and you left the shutter open while an electron zipped about the nucleus. When you developed the picture, you would have a record of where the electron had been. Doing the same thing with an electric fan that was turned on would give you a picture in which the blades of the fan look like a disk of material. The blades move so rapidly that their photographic image is blurred. Similarly, electrons in the first energy level of an atom would appear in our imaginary photograph as a fuzzy ball (Figure 3.23). The fuzzy ball (frequently referred to as a *charge cloud* or *electron cloud*) is the rough equivalent of an orbital.

Like Bohr, Schrödinger concluded that only two electrons could occupy the first energy level in an atom. In the quantum mechanical atom, this level is referred to as the 1*s* orbital (which is spherical). Also like Bohr, Schrödinger stated that the second energy level of an atom could contain a maximum of eight electrons. However, Schrödinger also concluded that these eight electrons must be located in four different orbitals. Each individual orbital could contain a maximum of two electrons. One of the orbitals of the second energy level is spherical in shape and is referred to as the 2*s* orbital. The remaining three orbitals of the second level have identical dumbbell shapes (Figure 3.23). They differ in their orientation in space (i.e., the direction in which they

Figure 3.22 Erwin Schrödinger, the Austrian physicist who developed mathematical equations for the structure of atoms. The equations treated electrons as waves. [Courtesy of the Nobel Foundation, Stockholm.]

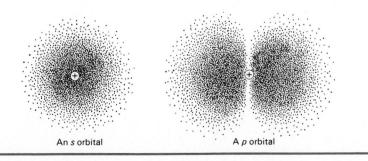

An *s* orbital   A *p* orbital

Figure 3.23 Charge-cloud representations of atomic orbitals.

Figure 3.24 Electron orbitals. In these drawings the nucleus of the atom is located at the intersection of the axes. The eight electrons that would be placed in the second energy level of Bohr's model are distributed among these four orbitals in the current model of the atom, two electrons per orbital.

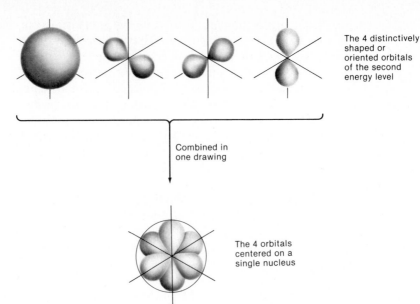

The 4 distinctively shaped or oriented orbitals of the second energy level

Combined in one drawing

The 4 orbitals centered on a single nucleus

point; see Figure 3.24). As a group these orbitals are called the $2p$ orbitals and, to distinguish them from one another, they are individually referred to as the $2p_x$, $2p_y$, and $2p_z$ orbitals. Electrons in these three orbitals all lie at the same energy level and possess slightly more energy than the electrons in the $2s$ orbital. Thus, the quantum mechanical atom distinguishes between main energy levels (1, 2, 3, and so on) and sublevels ($2s$ and $2p$, for example).

The third energy level of Bohr's model, and of Schrödinger's, could hold 18 electrons. In the quantum mechanical atom, however, the third main energy level is divided into three sublevels totaling nine orbitals: one $3s$ orbital, three $3p$ orbitals, and five $3d$ orbitals. Each higher main energy level adds another sublevel. The orbitals in each new sublevel have their own special shapes.

Just as one could build up atoms for all the elements by fitting electrons into the orbits of the Bohr model of the atom, so it is possible to construct a table of the elements by placing electrons into the orbitals of the quantum mechanical model. The Bohr diagram for nitrogen (which has seven electrons) is

The quantum mechanical description of the nitrogen atom focuses on a more detailed description of its **electron configuration** (arrangement): $1s^2 2s^2 2p^3$. The superscripts indicate the total number of electrons contained in a particular energy sublevel. For example, in nitrogen

Table 3.3 Electron Structures for Atoms of the First 20 Elements

Name	Atomic Number	Electron Structure
Hydrogen	1	$1s^1$
Helium	2	$1s^2$
Lithium	3	$1s^2 2s^1$
Beryllium	4	$1s^2 2s^2$
Boron	5	$1s^2 2s^2 2p^1$
Carbon	6	$1s^2 2s^2 2p^2$
Nitrogen	7	$1s^2 2s^2 2p^3$
Oxygen	8	$1s^2 2s^2 2p^4$
Fluorine	9	$1s^2 2s^2 2p^5$
Neon	10	$1s^2 2s^2 2p^6$
Sodium	11	$1s^2 2s^2 2p^6 3s^1$
Magnesium	12	$1s^2 2s^2 2p^6 3s^2$
Aluminum	13	$1s^2 2s^2 2p^6 3s^2 3p^1$
Silicon	14	$1s^2 2s^2 2p^6 3s^2 3p^2$
Phosphorus	15	$1s^2 2s^2 2p^6 3s^2 3p^3$
Sulfur	16	$1s^2 2s^2 2p^6 3s^2 3p^4$
Chlorine	17	$1s^2 2s^2 2p^6 3s^2 3p^5$
Argon	18	$1s^2 2s^2 2p^6 3s^2 3p^6$
Potassium	19	$1s^2 2s^2 2p^6 3s^2 3p^6 4s^1$
Calcium	20	$1s^2 2s^2 2p^6 3s^2 3p^6 4s^2$

there are a total of three electrons in the $2p$ orbitals. Table 3.3 lists the electron configurations of the first 20 elements. Notice that the orbitals fill in order of increasing energy: first the $1s$, then the $2s$, $2p$, $3s$, and so on. The d orbitals introduce a minor complication because the $3d$ orbitals turn out to be at a higher energy level than are the $4s$ orbitals. Figure 3.25 illustrates the order in which orbitals are filled.

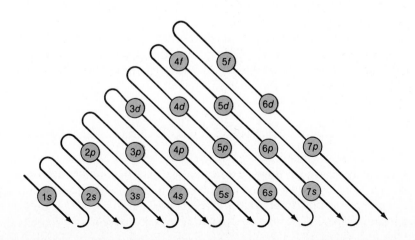

Figure 3.25 An order-of-filling chart for determining the electron configurations of atoms.

To check whether you understand the fundamentals of orbitals and electron configurations, work through the following examples.

Example 3.4   Write the electron configuration for carbon, atomic number 6.

The atomic number indicates that there are six electrons in the carbon atom. The first energy level can accommodate only two of these electrons in its $1s$ orbital.

$$1s^2 \dots$$

The remaining four electrons must occupy orbitals on the second main energy level. The lowest energy orbital on the second level is the $2s$ orbital, which can accommodate two electrons.

$$1s^2 2s^2 \dots$$

The remaining two electrons must go to the $2p$ orbitals, which can handle up to six electrons if necessary.

$$1s^2 2s^2 2p^2$$

Example 3.5   What is the electron configuration of hydrogen (atomic number 1)?

The single electron in hydrogen resides in the lowest energy orbital. Therefore, the electron configuration for hydrogen is $1s^1$.

Example 3.6   What is the electron configuration for phosphorus (atomic number 15)?

We have 15 electrons to distribute. Two go into the first main energy level ($1s^2$), eight go into the second ($2s^2 2p^6$), and five go into the third ($3s^2 3p^3$). In summary, the configuration of phosphorus is

$$1s^2 2s^2 2p^6 3s^2 3p^3$$

Quantum mechanics offers to scientists detailed descriptions of the electron structure of atoms. However, we shall make most use of the picture it paints of electrons as clouds of negative charge. Sometimes the shape of the cloud (that is, the orbital) is presented simply in outline. (Figure 3.24 uses shaded drawings to present the combined orbitals of the second energy level.) Later we shall see how the shape and orientation of the electron cloud can determine the shapes of molecules.

Electron Configuration and the Periodic Table

It is now known that properties of an element depend mainly on the number of electrons in the outermost energy level of the atoms of the element (Chapter 5). Sodium atoms have one electron in their outermost energy level (the third). Lithium atoms have a single electron in their outermost level (the second). The chemical properties of sodium and lithium are similar. The atoms of helium and neon have filled outer electron energy levels, and both elements are inert, that is, they do not undergo chemical reactions readily. Apparently, not only are similar chemical properties shared by elements whose atoms have similar electron **configurations** (arrangements), but also certain configurations appear to be more stable (less reactive) than others. This is a point we shall explore in Chapter 5.

In Mendeleev's table (page 45), the elements were arranged by atomic weights for the most part, and this arrangement revealed the periodicity of chemical properties. Because the number of electrons determines chemical properties, that number should (and now does) determine the order of the periodic table. In the modern periodic table, the elements are arranged according to atomic number. Remember, this number indicates both how many protons and how many electrons there are in a neutral atom of the element. The modern table, arranged in order of increasing atomic number, and Mendeleev's table, arranged in order of increasing atomic weight, parallel one another because an increase in atomic number is generally accompanied by an increase in atomic weight. In only a few cases (noted by Mendeleev) do the weights fall out of order. Atomic weights do not increase in precisely the same order as atomic numbers because *both protons and neutrons* contribute to the mass of an atom. It is possible for an atom of lower atomic number to have more neutrons than one with a higher atomic number. Thus, it is possible for an atom with a lower atomic number to have a greater mass than an atom with a higher atomic number.

The modern periodic table (inside front cover) has vertical columns called **groups** or **families**. Each group includes elements with the same number of electrons in their outermost energy levels and, therefore, with similar chemical properties (Figure 3.26). The horizontal rows of the table are called **periods**. Each new period indicates the opening of the next main electron energy level. For example, sodium starts row

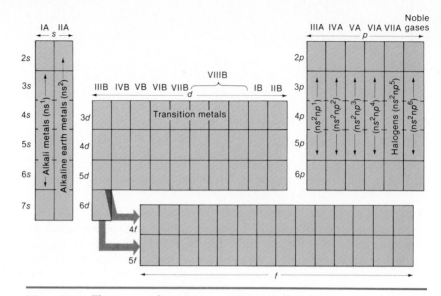

Figure 3.26 Electron configurations and the periodic table.

three and the outermost electron in sodium is the first electron to be placed in the third energy level.

Certain groupings of elements in the periodic table are designated by special names. The heavy, stepped, diagonal line on the table divides the elements into two major classes. Those to the left of the line are called **metals** and those to the right, **nonmetals**. Group IA elements are known as **alkali metals**; Group IIA are **alkaline earth metals**; Group VIIA, **halogens**. The group at the extreme right of the table contains the **noble gases**. All the B group elements are called **transition metals**.

You will find the periodic table an invaluable aid in your study of chemistry.

Which Model to Choose

For some purposes in this text, Bohr diagrams of atoms are used to picture the distribution of electrons among the main energy levels. At other times, the electron clouds of the quantum mechanical model are more useful. Even Dalton's model will sometimes prove to be the best way to describe certain phenomena (the behavior of gases, for example). The choice of model will always be based on which one is most helpful for understanding a particular concept. That, after all, is the whole purpose of scientific models.

Problems

1. What did each of the following scientists contribute to our knowledge of the atom?
   a. Crookes
   b. J. J. Thomson
   c. Goldstein
   d. Millikan
   e. Röntgen
   f. Becquerel

2. What evidence is there that electrons are particles?

3. How did the discovery of radioactivity contradict Dalton's atomic theory?

4. Define or identify each of the following.
   a. alpha particle
   b. beta particle
   c. gamma ray
   d. deuterium
   e. tritium

5. What are isotopes?

6. How are X rays and gamma rays similar? How are they different?

7. The table below describes four atoms.

	Atom A	Atom B	Atom C	Atom D
Number of protons	10	11	11	10
Number of neutrons	11	10	11	10
Number of electrons	10	11	11	10

   Are atoms A and B isotopes? A and C? A and D? B and C?

8. What are the masses of the atoms in Problem 7?

9. Discuss Rutherford's gold-foil experiment. What did it tell us about the structure of the atom?

10. Give the distinguishing characteristics of the proton, the neutron, and the electron.

11. Should the proton and electron attract or repel one another?

12. Should the neutron and proton attract or repel one another?

13. Which subatomic particles are found in the nuclei of atoms?

14. In an atom, what are the extranuclear (outside the nucleus) subatomic particles?

15. Compare Dalton's model of the atom with the nuclear model of the atom.

16. If the nucleus of an atom contains 10 protons, how many electrons are there in the neutral atom?

17. How did Bohr refine the nuclear model of the atom?

18. What particles travel in the "orbits" of the Bohr model of the atom?

19. According to Bohr, what is the maximum number of electrons in the fourth energy level ($n = 4$)?

20. If the third energy level of electrons contains two electrons, what is the total number of electrons in the atom?

21. Define the following terms.
    a. ground state
    b. excited state

22. When light is emitted by an atom, what change has occurred within the atom?

23. Which atom absorbed more energy: one in which an electron moved from the second energy level to the third energy level, or an otherwise identical atom in which an electron moved from the first to the third energy level?

24. Use the periodic table to determine the number of protons in atoms of the following elements.
    a. helium (He)
    b. sodium (Na)
    c. chlorine (Cl)
    d. oxygen (O)
    e. magnesium (Mg)
    f. sulfur (S)

25. How many electrons are there in the neutral atoms of the elements listed in Problem 24?

26. Draw Bohr diagrams for the elements listed in Problem 24.

27. The following Bohr diagram is supposed to represent the neutral atoms of an element. The diagram is incorrectly drawn. Identify the error.

28. In the quantum mechanical notation $2p^6$ how many electrons are described? What is the general shape of the orbitals described in the notation? How many orbitals are included in the notation?

29. Use quantum mechanical notation to describe the electron configuration of the atom represented in the following Bohr diagram.

30. Give the electron configurations (using quantum mechanical notation) for the elements in Problem 24. You may refer to the periodic table.

31. Identify the elements from their electron configurations. You may refer to the periodic table.

    a. $1s^2 2s^2$ b. $1s^2 2s^2 2p^3$
    c. $1s^2 2s^2 2p^6 3s^2 3p^1$

32. Without referring to the periodic table, give the atomic numbers of the elements described in Problem 31.

33. None of the following electron configurations is reasonable. In each case explain why.

    a. $1s^2 2s^2 3s^2$ b. $1s^2 2s^2 2p^2 3s^1$
    c. $1s^2 2s^2 2p^6 2d^5$

34. Referring only to the periodic table, indicate what similarity in electron structure is shared by fluorine (F) and chlorine (Cl). What is the difference in their electron structures?

35. What is the difference in the electron configurations of oxygen (O) and fluorine (F)?

36. Referring only to the periodic table, indicate what similarity in electron structure is shared by oxygen (O) and sulfur (S). What is the difference in their electron structures?

37. What is the difference in the electron configurations of sulfur (S) and chlorine (Cl)?

38. What is the difference in the electron configuration of fluorine (F) and sulfur (S)?

39. If three electrons were added to the outermost energy level of a phosphorus atom, the new electron configuration would resemble that of what element?

40. If two electrons were removed from the outermost energy level of a magnesium atom, the new electron configuration would resemble that of what element?

41. Elements are defined on a theoretical basis as being composed of atoms that share the same atomic number. On the basis of this theory would you think it possible for someone to discover a new element that would fit between magnesium (atomic number 12) and aluminum (atomic number 13)?

42. What is the difference between a Bohr orbit and the orbital of wave mechanics?

43. Where are the metals and nonmetals located on the periodic table?

44. Identify the following elements as metals or nonmetals. You may refer to the periodic table. The numbers in parentheses are the atomic numbers of the elements.

    a. sulfur (16)   b. chromium (24)
    c. iodine (53)

45. Indicate the group number of the following families.

    a. alkali metals   b. halogens
    c. alkaline earth metals

46. Identify the period of the following elements. You may refer to the periodic table. The numbers in parentheses are the atomic numbers of the elements.

    a. chlorine (17)   b. osmium (76)
    c. hydrogen (1)

47. Which of the following elements are halogens?

    a. Ag   b. At   c. As

48. Which of the following elements are alkali metals?

    a. K   b. Y   c. W

49. Which of the following elements are noble gases?

    a. Fe   b. Ne   c. Ge   d. He   e. Xe

50. Which of the following elements are transition metals?

    a. Ti (22)   b. Tc (43)   c. Te (52)

51. Which of the following elements are alkaline earth metals?

    a. Bi   b. Ba   c. Be   d. Br

52. How many electrons are in the outermost energy level of the halogens?

53. How many electrons are in the outermost energy level of Group IIA elements?

54. In what period of elements are electrons first introduced into the fourth energy level?

References and Readings

1. Andrade, E. N. da C. *Rutherford and the Nature of the Atom.* Garden City, NY: Doubleday, 1964.
2. Angier, Natalie. "The Hot New Science of Fireworks." *Discover*, July 1982, pp. 25–28.
3. Asimov, Isaac. *A Short History of Chemistry.* Garden City, NY: Doubleday, 1965. Chapters 12 and 13.
4. Hill, John W., and Dorothy M. Feigl. *Chemistry*

and Life, 3rd ed. New York: Macmillan, 1987. "Special Topic B," pp. 61–69, is recommended for those who wish to pursue a more detailed study of atomic structure.

5. Houwink, R. *Data: Mirrors of Science*. New York: American Elsevier, 1970. Chapter 3 includes marvelous illustrations showing the minute size of atoms.

6. Jaffe, Bernard. *Moseley and the Numbering of the Elements*. London: Heinemann Educational Books, 1971. Tells how atomic numbers were determined.

7. Radtke, Neil. "Atomic 'Cities.'" *The Science Teacher*, January 1978, p. 35. Relates the filling of electron shells to the placement of people in houses and neighborhoods.

8. Wolff, Peter. *Breakthroughs in Chemistry*. New York: Signet Science Library, 1967. Chapters 6, 7, 8, and 9 relate the research of Faraday, Mendeleev, Marie Curie, and Bohr, respectively.

9. Walker, Jearl. "The Amateur Scientist: The Spectra of Street Lights Illuminate Basic Principles of Quantum Mechanics." *Scientific American*, January 1984, pp. 138–143.

4

Nuclear Chemistry
The Heart of Matter

In Chapter 3, we took a brief look at the atomic nucleus and then turned our attention to electrons. For a major portion of this book, we take special note of the outermost electrons. It is the interaction of these particles that holds atoms together to make up the materials of the world. For the moment, however, let us turn to that tiny, tiny speck of matter called the *nucleus*.

Whole atoms are 10 000 times larger than atomic nuclei, yet from the nucleus come radioactive particles and even the enormous power of nuclear bombs. In this chapter, we discuss radioactivity and some of its applications in medicine, agriculture, and archaeology. We also talk about nuclear bombs, fallout, and the effects of radiation on living things. The important topic of nuclear power plants is covered in Chapter 14.

It seems that much of what we hear today about nuclear processes is negative. We hear about bombs and about problems at nuclear power plants. But there is a positive side, too. Nuclear medicine saves lives. Diseases once regarded as incurable can be diagnosed and treated effectively with radioactive isotopes. Applications of nuclear chemistry to biology, industry, and agriculture have improved the human condition significantly. The use of radioisotopes in biological and agricultural research has led to increased crop production, which provides more food for a hungry world. Although the treat of annihilation from nuclear warfare still hangs over our heads like a dark cloud, it appears more likely that you will be saved by nuclear medicine than that you will be killed by a nuclear explosion.

A Partial Parts List for the Atomic Nucleus

We saw in Chapter 3 that atomic nuclei are made up of protons and neutrons. Although nuclear physicists have extended this parts list to include at least a hundred particles, most of these particles have a transitory existence and are of little interest to a chemist. In this discussion, we take the oversimplified, but useful, view that atomic nuclei are made up of protons and neutrons—and some sort of force that holds them together.

A proton and neutron have virtually the same mass, 1.007 276 amu and 1.008 665 amu, respectively. That's equivalent to saying that two different people weigh 100.7 kg and 100.9 kg. The difference is so small that it can be ignored. Thus, for many purposes, we assume the masses of the proton and the neutron to be the same, 1 amu. The proton has a charge equal in magnitude but opposite in sign to that of an electron. This charge on a proton is written as 1+. The electron has a charge of 1− and a mass of 0.000 549 amu. The electrons in an atom contribute so little of its total mass that they are usually disregarded and are treated as if their mass was 0. The subatomic particles of greatest interest to us are summarized in Table 4.1.

Recall that the number of protons in the nucleus of an atom of any element is the atomic number of that element. Atoms of an element have differing numbers of neutrons, that is, they exist as isotopes (Chapter 3). Most, but not all, elements exist in nature in isotopic forms. An interesting and easy-to-remember example is the element tin (Sn), which exists in 10 isotopic forms. (Tin...ten isotopes.)

Table 4.1   Subatomic Particles

Particle	Symbol	Approximate Mass (atomic mass units)	Charge
Proton	p^+	1	1+
Neutron	n	1	0
Electron	e^-	0	1−

Isotopes usually are of little importance in ordinary chemical reactions. All three hydrogen isotopes (Chapter 3) react with oxygen to form water. Since the isotopes differ in mass, compounds formed with different hydrogen isotopes have different physical properties, but such differences are usually slight. In nuclear reactions, isotopes are of utmost importance.

Nuclear Arithmetic

To represent the different isotopes, symbols with subscript and superscript numbers are used. In the general symbol

$$_Z^A X$$

the Z represents the *atomic number*, that is, the number of protons. The A represents the **mass number**, the number of protons plus the number of neutrons. In the symbol

$$_{17}^{35} Cl$$

the number of protons is 17. The number of neutrons is $35 - 17 = 18$.

The Z number is not necessary to identify the element—the symbol does that—but it is convenient for chemists to use when they write nuclear reactions.

Example 4.1   How many neutrons are there in the $_{92}^{235}U$ nucleus?

235 (mass number) − 92 (atomic number)

$= 143$ (number of neutrons)

Example 4.2   Which of the following are isotopes of the same element? We are using the letter X as the symbol for all so that the symbol will not identify the elements.

$$_8^{16}X \qquad _7^{16}X \qquad _7^{14}X \qquad _6^{14}X \qquad _6^{12}X$$

$_7^{16}X$ and $_7^{14}X$ are isotopes of the element nitrogen (N). $_6^{14}X$ and $_6^{12}X$ are isotopes of the element carbon (C).

Which of the original five isotopes have identical mass numbers?

$_8^{16}X$ and $_7^{16}X$ have the same mass number. The first is an isotope of oxygen, the second an isotope of nitrogen. $_7^{14}X$ and

$^{14}_6$X have the same mass number. The first is an isotope of nitrogen, the second an isotope of carbon.

Which of the original five isotopes have the same number of neutrons?

$^{16}_8$X ($16 - 8 = 8$ neutrons) and $^{14}_6$X ($14 - 6 = 8$ neutrons) have the same number of neutrons.

Natural Radioactivity

Some nuclei are unstable as they occur in nature. Radium atoms with a mass number of 226, for example, break down spontaneously, giving off alpha particles. Since alpha particles are identical with helium nuclei, this process can be summarized by the equation

$$^{226}_{88}Ra \longrightarrow \, ^4_2He \, + \, ^{222}_{86}Rn$$

The new element, with two fewer protons, is identified by its atomic number (86) as radon (Rn). A second example of alpha decay is shown in Figure 4.1(a).

Tritium nuclei are also unstable. Tritium is one of the heavy isotopes of hydrogen first mentioned in Chapter 3. Like all hydrogen nuclei, the tritium nucleus contains one proton. Unlike the most common isotope of hydrogen, however, the tritium nucleus contains two neutrons, and its mass is therefore 3 amu ($^3_1$H). Tritium decomposes

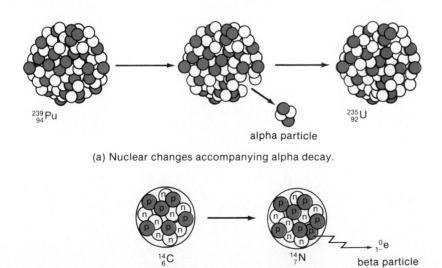

$^{239}_{94}$Pu $^{235}_{92}$U

alpha particle

(a) Nuclear changes accompanying alpha decay.

$^{14}_6$C $^{14}_7$N $^{0}_{-1}$e

beta particle

(b) Nuclear changes accompanying beta decay.

Figure 4.1 Nuclear emission of (a) an alpha particle and (b) a beta particle.

Table 4.2 Radioactive Decay and Nuclear Change

Type of Radiation	Greek Letter	Mass Number	Charge	Change in Mass Number	Change in Atomic Number
Alpha	α	4	2+	Decreases by 4	Decreases by 2
Beta	β	0	1−	No change	Increases by 1
Gamma	γ	0	0	No change	No change

by what is termed beta decay. Since a beta particle is identical to an electron, this process may be written

$$^3_1\text{H} \longrightarrow \ _{-1}^{0}\text{e} + \ ^3_2\text{He} \quad \text{or} \quad ^3_1\text{H} \longrightarrow \beta + \ ^3_2\text{He}$$

The product isotope is identified by its atomic number as helium (He). How can the original nucleus, which contains only a proton and two neutrons, emit an electron?

We can envision one of the neutrons in the original nucleus changing into a proton and an electron.

$$^1_0\text{n} \longrightarrow \ ^1_1\text{p} + \ _{-1}^{0}\text{e}$$

The new proton is retained by the nucleus (therefore, the atomic number of the product increased by one), and the almost massless electron or beta particle is kicked out (the product nucleus has the same mass as the original). A second example of beta decay is pictured in Figure 4.1 (b).

The third type of radioactivity is called gamma decay. Since this type of emission involves no particle, no equation is needed. Gamma emission usually accompanies the emission of alpha or beta particles.

The major types of radioactive decay and the ensuing nuclear changes are summarized in Table 4.2.

Example 4.3 Plutonium-239 emits an alpha particle when it decays. What new element is formed?

$$^{239}_{94}\text{Pu} \longrightarrow \ ^4_2\text{He} + \ ?$$

Mass and charge are conserved. The new element must have a mass of $239 - 4 = 235$ and a charge of $94 - 2 = 92$. The nuclear charge (atomic number) of 92 identifies the element as uranium (U).

$$^{239}_{94}\text{Pu} \longrightarrow \ ^4_2\text{He} + \ ^{235}_{92}\text{U}$$

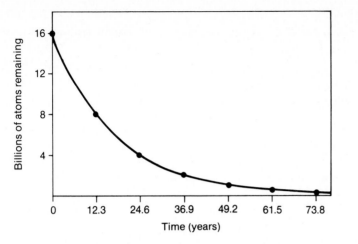

Figure 4.2 The radioactive decay of tritium, $_1^3$H.

Half-life

Thus far we have discussed radioactivity as applied to single atoms. In the laboratory, we generally deal with great numbers of atoms—numbers far larger than the number of people on all the Earth. If we could see the nucleus of an individual atom, we could tell *whether or not* it would undergo radioactive decay by noting its composition. Certain combinations of protons and neutrons are unstable. We could not, however, determine *when* the atom would undergo a change. Radioactivity is a random process, generally independent of outside influences.

With large numbers of atoms, the process of radioactive decay becomes more predictable. A measurable quantity, called a **half-life**, is a characteristic property of each radioactive isotope. This half-life can be very long, as with uranium-238 (with a half-life of 4.5 billion years), or very short, as with boron-9 (with a half-life of 8×10^{-19} second).

The half-life of a radioactive isotope is that period of time in which 50% of the original number of atoms undergo radioactive decay. Suppose, for example, that we had 16 billion atoms of the radioactive hydrogen isotope tritium. The half-life of this isotope is 12.3 years. This means that in 12.3 years 8 billion of the atoms would have undergone radioactive decay. In another 12.3 years, half the remaining atoms would have decayed. That is, after two half-lives, 25% of the original atoms would remain unchanged. Two half-lives do not make a whole! This concept of half-life is shown graphically in Figure 4.2.

It is impossible to say exactly when all of the $_1^3$H atoms would have decayed. For practical purposes, we can assume that nearly all the radioactivity would be gone after about 10 half-lives. For the tritium sample considered here, 10 half-lives would be 123 years.

Example 4.4 What fraction of radioactive atoms remains after five half-lives?

After one half-life, $\frac{1}{2}$ of the original atoms remain. After two half-lives, $\frac{1}{2} \times \frac{1}{2}(\frac{1}{2}$ of $\frac{1}{2}) = \frac{1}{4}$ of the original atoms remain. After three half-lives, $\frac{1}{2} \times \frac{1}{2} \times \frac{1}{2} = 1/(2^3) = \frac{1}{8}$ of the original atoms remain. After four half-lives, $\frac{1}{2} \times \frac{1}{2} \times \frac{1}{2} \times \frac{1}{2} = 1/(2^4) = \frac{1}{16}$ of the original atoms remain. After five half-lives, $\frac{1}{2} \times \frac{1}{2} \times \frac{1}{2} \times \frac{1}{2} \times \frac{1}{2} = 1/(2^5) = \frac{1}{32}$ of the original atoms remain. To determine what fraction of the original sample remains after a certain number (n) of half-lives, calculate the following value.

$$\frac{1}{2^n}$$

Example 4.5 If you started with 16 million radioactive atoms, how many would you have left after four half-lives?

Since $n = 4$, the fraction remaining would be $1/(2^4)$, which is $\frac{1}{16} \times 16$ million atoms = 1 million. A million of the original atoms would be left.

Artificial Transmutation

The forms of radioactivity encountered thus far occur in nature. Other nuclear reactions may be brought about by bombardment of stable nuclei with alpha particles, neutrons, or other subatomic particles. These particles, given sufficient energy, penetrate the formerly stable nucleus and bring about some form of radioactive emission. Like natural radioactivity, this sort of nuclear change brings about a **transmutation**: one element is changed into another. Because the change would not have occurred naturally, the process is called **artificial transmutation**.

Ernest Rutherford, a few years after his famous gold-foil experiment (Chapter 3), studied the bombardment of a variety of light elements with alpha particles. One such experiment, in which he bombarded nitrogen, resulted in the production of protons.

$$^{14}_{7}N + {}^{4}_{2}He \longrightarrow {}^{17}_{8}O + {}^{1}_{1}H$$

(Recall that the hydrogen nucleus is a proton; hence the alternative symbol $^{1}_{1}H$ for the proton.) Notice that the sum of the mass numbers on the left equals the sum of the mass numbers on the right. The atomic numbers are also balanced.

Rutherford had postulated the existence of protons in nuclei in 1914. An experiment published in 1919 gave the first empirical verification of the existence of these fundamental particles. Eugen Goldstein had earlier produced protons in his discharge tube experiments (Chapter 3). He obtained these particles from hydrogen gas in the tube by knocking an electron away from the hydrogen atom. The significance of Rutherford's experiment lay in the fact that he obtained protons from the nucleus of an atom other than hydrogen, thus establishing their nature as fundamental particles. By **fundamental particles** we mean basic units from which more complicated structures (such as the nitrogen nucleus) can be fashioned. Rutherford's experiment was the first induced nuclear reaction.

A great many transmutations were carried out during the 1920s. In the 1930s one such reaction led to the discovery of another fundamental particle. James Chadwick, in 1932, bombarded beryllium with alpha particles.

$$\ce{^9_4Be} + \ce{^4_2He} \longrightarrow \ce{^{12}_6C} + \ce{^1_0n}$$

Among the products was the neutron.

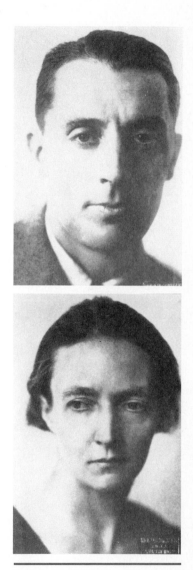

Figure 4.3 Frédéric and Irène Joliot-Curie discovered artificially induced radioactivity in 1934. [Reprinted with permission from Mary E. Weeks, *Discovery of the Elements* (Easton, PA: Chemical Education Publishing, 1968). Copyright © 1968 by Chemical Education Publishing Company.]

Example 4.6   When potassium-39 is bombarded with neutrons, chlorine-36 is produced. What other particle is emitted?

$$\ce{^{39}_{19}K} + \ce{^1_0n} \longrightarrow \ce{^{36}_{17}Cl} + ?$$

To balance the equation, a particle with a mass of 4 and an atomic number of 2 is required. That's an alpha particle.

$$\ce{^{39}_{19}K} + \ce{^1_0n} \longrightarrow \ce{^{36}_{17}Cl} + \ce{^4_2He}$$

Induced Radioactivity

The first artificial nuclear reactions produced isotopes already known to occur in nature. This was perhaps fortuitous, for it was inevitable that an unstable nucleus would be produced sooner or later. Irène Curie (daughter of the 1903 Nobel Prize winners) and her husband, Frédéric Joliot, were studying the bombardment of aluminum with alpha particles. Neutrons were produced, leaving behind an isotope of phosphorus.

$$\ce{^{27}_{13}Al} + \ce{^4_2He} \longrightarrow \ce{^{30}_{15}P} + \ce{^1_0n}$$

Much to their surprise, the target continued to emit particles after the bombardment was halted. The isotope of phosphorus was radioactive, emitting particles equal in mass to the electron but opposite in charge. These particles are called **positrons**. The reaction they were observing is written

$$^{30}_{15}P \longrightarrow \ _{1+}^{0}e \ + \ ^{30}_{14}Si$$

Once again the question arises: Where does this particle comes from if the nucleus contains only protons and neutrons? Previously we accounted for a beta particle (an electron) popping out of a nucleus by saying a neutron changed into a proton and an electron. Perhaps a similar happening can account for the appearance of a positron. Imagine a proton changing to a neutron and a positron (a proton is the same as a hydrogen nucleus).

$$^{1}_{1}H \longrightarrow \ _{1+}^{0}e \ + \ ^{1}_{0}n$$

Everything balances rather nicely in this equation. When the positron is emitted, the original radioactive nucleus suddenly has one less proton and one more neutron than before. Therefore, the mass of the product nucleus is the same, but its atomic number is one less than that of the original nucleus.

This work gave the Joliot-Curies a Nobel Prize of their own in 1935. (The Joliot-Curies adopted the combined surname to perpetuate the Curie name. Marie and Pierre Curie had two daughters, but no son.)

Figure 4.4 presents another example of nuclear decay involving positron emission.

Example 4.7   Carbon-10, a radioactive isotope, emits a positron when it decays. Write an equation for this process.

$$^{10}_{6}C \longrightarrow \ _{1+}^{0}e \ + \ ?$$

To balance the equation, a particle with a mass of 10 and an atomic number of 5 (boron) is required.

$$^{10}_{6}C \longrightarrow \ _{1+}^{0}e \ + \ ^{10}_{5}B$$

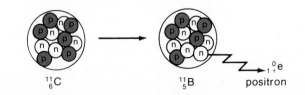

Figure 4.4 Nuclear change accompanying positron emission.

Penetrating Power

Radioactive materials are dangerous because the radiation emitted by decaying nuclei can damage living tissue. The ability of radiation to inflict damage depends, in part, on its penetrating power.

All other things equal, the more massive the particle, the less its penetrating power. Of alpha, beta, and gamma rays, alpha rays are the least penetrating. These are streams of helium nuclei, each particle with a mass number of 4. Beta rays are more penetrating than alpha rays. The electrons that make up the stream of beta particles are assigned a mass number 0. These particles are not really massless, but they are very much lighter than alpha particles. Gamma rays are high-energy radiation and, like light, truly have no mass. This is the most penetrating form of nuclear radiation.

It may seem contrary to common sense that the biggest particles make the least headway. Consider that penetrating power reflects the ability of the radiation to make its way through a sample of matter. It is as if you were trying to roll some rocks through a field of boulders. The alpha particle acts as if it were a boulder itself. Because of its size, it cannot get very far before it bumps into and is stopped by other boulders. The beta particle acts like a small stone. It can sneak between boulders and perhaps ricochet off one or another until it has made its way farther into the field (Figure 4.5). The gamma ray can be compared to an insect that can get through the smallest openings, and although it may brush against some of the boulders, it can, in general, make its way through most of the field without being stopped.

At the beginning of this discussion we said that, all other things being equal, this is how things worked. But all other things are not always equal. The faster a particle moves or the more energetic the radiation is, the more penetrating power it has.

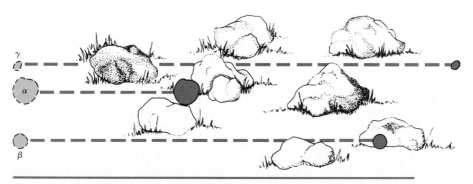

Figure 4.5 Shooting radioactive particles through matter is like rolling rocks through a field of boulders—the larger rocks are more quickly stopped.

If a radioactive substance is *outside* the body, alpha particles of low penetrating power are least dangerous; they are stopped by the outer layer of skin. Beta rays also usually are stopped before reaching vital organs. Gamma rays readily pass through tissues; an external gamma source can be quite dangerous. When the radioactive source is *inside* the body, the situation is reversed. The nonpenetrating alpha particles can do great damage. All such particles are trapped within the body, which must then absorb all the energy released by the particle. Alpha rays inflict all their damage in a very small area because they do not travel far. Beta rays distribute the damage over a somewhat larger area because they travel farther. Tissue may recover from limited damage spread over a large area: it is less likely to survive concentrated damage.

People working with radioactive materials can do several things to protect themselves. The simplest is to move away from the source, for intensity of radiation decreases with distance from the source. Workers can also be protected by shielding. A sheet of paper will stop most alpha particles. A block of wood or a thin sheet of aluminum will stop beta particles. But it takes several meters of concrete or several centimeters of lead to stop gamma rays.

Uses of Radioisotopes

Scientists in a wide variety of fields make use of radioactive isotopes (**radioisotopes**) as **tracers** in physical, chemical, and biological systems. Isotopes of a given element, whether radioactive or not, behave nearly identically in chemical and physical processes. Since radioactive isotopes are easily detected, it is relatively easy to trace their movements, even through a complicated system.

As a simple example, let's consider the flow of a liquid through a pipe. Suppose there is a leak in a pipe buried beneath a concrete floor. We could locate the leak by digging up extensive areas of the floor, or we could add a small amount of radioactive material to liquid going into the drain and trace the flow of liquid with a Geiger counter (an instrument that detects radioactivity). Once the leak was located, only a small area of the floor would have to be dug up to repair the leak. Short-lived isotopes—which disappear soon after doing their job—usually are employed for such purposes.

In a similar manner, we could trace the uptake of phosphorus by a green plant. The plant is fed some fertilizer containing radioactive phosphorus. A simple method of detection involves placing the plant on a photographic film. Radiation from the phosphorus isotopes exposes the film, much as light does. This type of exposure, called an **autoradiograph**, shows the distribution of phosphorus in the plant (Figure 4.6).

Radioactive tracers also are put to good use in agricultural research.

Figure 4.6 Autoradiograph showing the uptake of phosphorus in a green plant. [Courtesy of the U.S. Department of Energy.]

They are used to study the effectiveness of fertilizers and weed killers, to compare the nutritional value of various feeds, and to determine the best methods for controlling insects. The purposeful mutation of plants by irradiation has produced new and improved strains of commercially valuable crop plants ranging from tobacco to peanuts.

Radioisotopes are also used as sources for the irradiation of foodstuffs as a method of preservation (Figure 4.7). The radiation destroys microorganisms that cause food spoilage. Irradiated food shows little change in taste or appearance. Some people are concerned about possible harmful effects of chemical substances produced by the radiation, but there is no good evidence of harm to laboratory animals fed irradiated food, nor are there any known adverse effects in humans in countries where irradiation has been used for several years. There is no residual radiation in the food after the sterilization process.

Radioisotopes have been used extensively in basic scientific research. The mechanism of photosynthesis was worked out in large part by using carbon-14 ($^{14}_{6}C$) as a tracer. Metabolic pathways in plants, animals, and humans are being studied by radioactive tracers. The potential for the use of this knowledge for human good is as enormous as the potential for the use of nuclear bombs for evil.

Figure 4.7 Gamma radiation prevents sprouting in potatoes. Both the potatoes were stored for 16 months. The bottom one was irradiated; the top one was not. [Reprinted from Vernon Pizer, *Preserving Food with Atomic Energy* (Oak Ridge, TN: U.S. Department of Energy, 1970).]

Nuclear Medicine

Nuclear medicine involves two distinct uses of radioisotopes: therapeutic and diagnostic. In radiation therapy, an attempt is made to treat or cure disease with radiation. The diagnostic use of radioisotopes is aimed at obtaining information about the state of an individual's health.

Cancer is not one disease but many. Some forms are particularly susceptible to radiation therapy. Radiation is carefully aimed at the cancerous tissue, and exposure of normal cells is minimized. If the cancer cells are killed by the destructive effects of the radiation, the malignancy is halted. But persons undergoing radiation therapy often get sick from the treatment. Nausea and vomiting are the usual symptoms of radiation sickness. Thus, the aim of radiation therapy is to destroy the cancerous cells before too much damage is done to healthy tissue. Radiation is most lethal to rapidly reproducing cells, and this is precisely the characteristic of cancer cells that allows the therapy to be successfully applied.

The therapeutic use of a radioisotope is intended to treat or cure a disease. Radioisotopes are also used for diagnostic purposes to provide information about the type or extent of illness. Radioactive iodine-131 ($^{131}_{53}I$) is used to determine the size, shape, and activity of the thyroid gland, as well as to treat cancer located in this gland and to control a

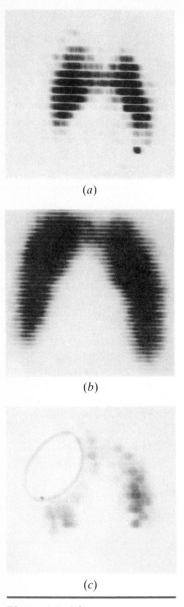

(a)

(b)

(c)

Figure 4.8 A linear photoscanner produced these pictures. (a) A normal thyroid. (b) An enlarged thyroid. (c) A cancerous thyroid. [Courtesy of the U.S. Department of Energy.]

hyperactive thyroid. First, one drinks a solution of potassium iodide incorporating iodine-131. The body concentrates iodide in the thyroid. Large doses are used for treatment of thyroid cancer; the radiation from the isotope concentrates in the thyroid cancer cells even if the cancer has spread to other parts of the body. For diagnostic purposes, however, only a small amount is needed. Again the material is concentrated in the thyroid. A detector is set up so that readings are translated into a permanent visual record showing the differential uptake of the isotope. The "picture" that results is referred to as a **photoscan** (Figure 4.8).

The radioisotope most widely used in medicine is gadolinium-153. This isotope is used to determine bone mineralization. Its popularity is an indication of the large number of people, mostly women, who suffer from **osteoporosis** (reduction in the quantity of bone) as they grow older. Gadolinium-153 gives off two characteristic radiations, a gamma ray and an X ray. A scanning device compares these radiations after they pass through bone. Bone densities are then determined by differences in absorption of the rays.

Technetium-99^m is used in a variety of diagnostic tests. The m stands for *metastable*, which means that this isotope will give up some energy to become a more stable version of the same isotope (same atomic number, same atomic weight). The energy it gives up is the gamma ray needed to detect the isotope.

$$^{99m}_{43}\text{Tc} \longrightarrow\ ^{99}_{43}\text{Tc} + \gamma$$

Notice that the decay of technetium-99^m produces no alpha or beta particles, which could cause unnecessary damage to the body. Technetium-99^m also has a short half-life (about 6 hours), which means that the radioactivity does not linger in the body long after the scan has been completed. With this short a half-life, use of the isotope must be carefully planned. In fact, the isotope itself is not what is purchased. Technetium-99^m is formed by the decay of molybdenum-99:

$$^{99}_{42}\text{Mo} \longrightarrow\ ^{99m}_{43}\text{Tc} +\ ^{0}_{-1}\text{e} + \gamma$$

A container of this molybdenum isotope is obtained, and the decay product, technetium-99^m, is "milked" from the container as needed.

Another scanning technique, positron emission tomography (PET), makes use of modern computer technology. PET can be used to measure dynamic processes occurring in the body, such as blood flow or the rate at which oxygen or glucose is being metabolized. PET scans are being used to pinpoint the area of brain damage which triggers severe epileptic seizures. Compounds incorporating positron-emitting isotopes, such as carbon-11, are inhaled or injected prior to the scan. Before the emitted positron can travel very far in the body, it encoun-

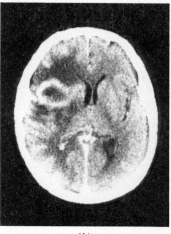

(b)

(a)

Figure 4.9 Modern computer technology used for medical diagnosis. (a) Patient in position for positron emission tomography (PET), a technique that uses radioisotopes to scan internal organs. (b) Image of section of brain created by computer tomography (CT), a scanning technique that uses X rays rather than radioisotopes, shows tumor in left temporoparietal area (white circle). [(a) Courtesy of Brookhaven National Laboratory and New York University Medical Center; (b) courtesy of the National Institute of Neurological and Communicative Disorders and Stroke.]

ters an electron (in any ordinary matter there are numerous electrons) and two gamma rays are produced.

$$^{11}_{6}\text{C} \longrightarrow {}^{11}_{5}\text{B} + {}^{0}_{1+}\text{e}$$
$$^{0}_{1+}\text{e} + {}^{0}_{1-}\text{e} \longrightarrow 2\gamma$$

These exit from the body in exactly opposite directions. Detectors positioned on opposite sides of the patient record the gamma rays. By setting the recorders so that two simultaneous gamma rays must be "seen," gamma rays resulting from natural background radiation are ignored. A computer is then used to calculate the point within the body at which the annihilation of the positron and electron occurred, and an image of that area is produced (Figure 4.9).

Table 4.3 is a list of some radioisotopes in common use in medicine. The list is necessarily incomplete. Even this abbreviated discussion should give you an idea of the importance of radioisotopes in medicine. The claim that nuclear science has saved more lives that nuclear bombs have destroyed is not an idle one.

Table 4.3 Some Radioisotopes and Their Application in Medicine

Isotope	Name	Radiation	Uses
^{51}Cr	Chromium-51	γ	Determination of volume of red blood and total blood volume
^{57}Co	Cobalt-57	γ	Determination of uptake of vitamin B_{12}
^{60}Co	Cobalt-60	β, γ	Radiation treatment of cancer
^{153}Gd	Gadolinium-153	γ	Determination of bone density
^{131}I	Iodine-131	β, γ	Detection of thyroid malfunction; measurement of liver activity and fat metabolism; treatment of thyroid cancer
^{59}Fe	Iron-59	β, γ	Measurement of rate of formation and lifetime of red blood cells
^{32}P	Phosphorus-32	β	Detection of skin cancer or cancer of tissue exposed by surgery
^{226}Ra	Radium-226	α, γ	Radiation therapy for cancer
^{24}Na	Sodium-24	β, γ	Detection of constrictions and obstructions in the circulatory system
^{99m}Tc	Technetium-99^m	γ	Imaging of brain, thyroid, liver, kidney, lung, and cardiovascular system
^{3}H	Tritium	β	Determination of total body water

Radioisotopic Dating

The half-lives of certain isotopes can be used to estimate the age of rocks and archaeological artifacts. Uranium-238 decays with a half-life of 4.5 billion years. The initial products of this decay are also radioactive, and breakdown continues until an isotope of lead ($^{206}_{82}$Pb) is formed. By measuring the relative amounts of uranium-238 and lead-206, chemists can estimate the age of a rock. Some of the older rocks on the Earth have been found to be from 3.0 to 3.5 billion years old. Moon rocks and meteorites have been dated at a maximum age of about 4.5 billion years. Thus, the age of the Earth is generally estimated to be 4.5 to 5.0 billion years.

The dating of artifacts usually involves a radioactive isotope of carbon. Carbon-14 is formed in the upper atmosphere by the bombardment of ordinary nitrogen by neutrons from cosmic rays.

$$^{14}_{7}\text{N} + ^{1}_{0}\text{n} \longrightarrow ^{14}_{6}\text{C} + ^{1}_{1}\text{H}$$

This process leads to a steady-state concentration of carbon-14 on the Earth. Living plants and animals incorporate this isotope as carbon dioxide. When they die, however, the incorporation of carbon-14 ceases, and the carbon-14 in the plants and organisms decays—with a half-life of 5730 years—to nitrogen-14 ($^{14}_{7}\text{N}$). Thus, we merely need to measure the $^{14}_{6}\text{C}$ activity remaining in an artifact of plant or animal origin to determine its age. For example, a sample that has half the $^{14}_{6}\text{C}$ activity of new plant material is 5730 years old; it has been dead for one half-life. Similarly, an artifact with 25% of the $^{14}_{6}\text{C}$ activity of new plant material is 11 460 years old; it has been dead for two half-lives.

Carbon-14 dating, as outlined here, assumes that the formation of the isotope was constant over the years. This is not quite the case. However, for the most recent 7000 years or so, carbon-14 dates have been correlated with those obtained from the annual growth rings of trees. Calibration curves have been constructed from which accurate dates can be determined. Generally, carbon-14 is reasonably accurate for dating objects up to about 50 000 years old. Objects older than that have too little of the isotope left for accurate measurement.

Charcoal from the fires of an ancient people, dated by determining the $_6^{14}C$ activity, is used to estimate the age of other artifacts found at the same archaeological site. Carbon-14 dating also has been used to detect forgeries of supposedly ancient artifacts. Dating procedures based on research in the structure and stability of atomic nuclei have become the routine, yet vital, tools of archaeology.

Tritium, the radioactive isotope of hydrogen, also is useful for dating. Its half-life of 12.3 years makes it useful for dating items up to about 100 years old. An interesting application is the dating of brandies. The alcoholic beverages are quite expensive when aged from 10 to 50 years. Tritium dating can be used to check the veracity of advertising claims about the aging process of the most expensive kinds.

Many other isotopes are useful for estimating the ages of objects and materials. Several of the more important ones are listed in Table 4.4.

Example 4.8 A piece of fossilized wood has a carbon-14 activity that is one eighth that of new wood. How old is the artifact? The half-life of carbon-14 is 5730 years.

The carbon-14 has gone through three half-lives.

$$\tfrac{1}{2} \times \tfrac{1}{2} \times \tfrac{1}{2} = (\tfrac{1}{2})^3 = \tfrac{1}{8}$$

It is therefore about $3 \times 5730 = 17\ 190$ years old.

Table 4.4 Several Isotopes Useful in Radioactive Dating

Isotope	Half-life (years)	Useful Range	Dating Applications
Carbon-14	5730	500 to 50 000 years	Charcoal, organic material
Tritium ($_1^3H$)	12.3	1 to 100 years	Aged wines
Potassium-40	1.3×10^9	10 000 years to the oldest Earth samples	Rocks, the Earth's crust, the moon's crust
Rhenium-187	4.3×10^{10}	4×10^7 years to the oldest samples in universe	Meteorites
Uranium-238	4.5×10^9	10^7 years to the oldest Earth samples	Rocks, the Earth's crust

Einstein and the Equivalence of Mass and Energy

Let's return now to our study of the history of radioactivity. Perhaps the most noted development of our age was a theoretical one, worked out with a pencil and a note pad. These are not the tools we usually associate with a scientist. Albert Einstein may well be the best-known scientist of all time, yet his achievements were those of the mind, not the laboratory (Figure 4.10).

By 1905, Einstein had worked out his special theory of relativity. In doing this, he derived a relationship between matter and energy. The now-famous equation is written

$$E = mc^2$$

where E represents energy, m represents mass, and c is the speed of light. According to Einstein, energy and mass are different aspects of the same thing.

The laws of matter and energy conservation are still useful when applied to changes in chemical reactions. Even there, we presume that the equivalence of mass and energy applies, but changes in mass are much too small to be measured. Conversion of one gram of matter, though, would yield an enormous amount of energy, enough to heat the average home for 1000 years. (We cannot make such complete conversions, however. Even in the explosion of a hydrogen bomb, less than 1% of the matter is converted to energy.)

Figure 4.10 Albert Einstein. [Courtesy of Matheson Gas Products, East Rutherford, NJ.]

The Building of the Bomb

In 1934 the Italian scientists Enrico Fermi and Emilio Segré bombarded uranium atoms with neutrons. They were trying to make elements higher in atomic number than uranium, which then had the highest known number. To their surprise, they found *four* radioactive species among the products. One presumably was element 93, formed by the initial conversion of uranium-238 to uranium-239.

$$^{238}_{92}U + ^{1}_{0}n \longrightarrow ^{239}_{92}U$$

The latter then underwent beta decay.

$$^{239}_{92}U \longrightarrow ^{0}_{-1}e + ^{239}_{93}Np$$

The two scientists were unable to explain the remaining radioactivity.

The German chemists Otto Hahn and Fritz Strassman repeated the Fermi–Segré experiment in 1938. They identified barium (Ba), lanthanum (La), and cerium (Ce) in the reaction products. These elements have atomic weights of a little more than half that of uranium atoms. The uranium atom had been split! In this reaction, it wasn't a matter of a small piece (an alpha or a beta particle, for example) being chipped from the original nucleus. Here the nucleus was cleaved, split into two major fragments—a process called nuclear **fission**.

Hahn was perplexed by these discoveries and relayed them to Lise Meitner, an Austrian Jew who had once worked with him in Berlin. When the Nazis annexed Austria in 1938, Meitner fled to Sweden, where she received the news of the splitting of uranium atoms. She and her nephew, Otto Frisch, an undergraduate student at the University of Copenhagen who was visiting her during the Christmas season, calculated the energy associated with the fission of uranium; they found the energy to be several times greater than that of any previously known nuclear reaction. In addition, the splitting resulted in the release of more neutrons. These could split other uranium atoms, producing enormous amounts of energy (Figure 4.12).

The news of these momentous discoveries was carried to the United States by Niels Bohr, the Danish physicist known for his quantum theory of the electron structure of atoms. Fermi, whose wife was Jewish, received the 1938 Nobel Prize in Physics. He took advantage of his trip to Stockholm to receive the award to flee fascist Italy and take refuge in the United States.

Figure 4.11 Enrico Fermi. [Courtesy of Mrs. Laura Fermi.]

Figure 4.12 The splitting of a uranium atom. The neutrons produced in this fission can split other uranium atoms, thus sustaining a chain reaction. The splitting of one uranium-235 atom yields 8.9×10^{-18} kwh of energy. Fission of a mole of uranium-235 (6.02×10^{23} atoms) produces 5 300 000 kwh of energy.

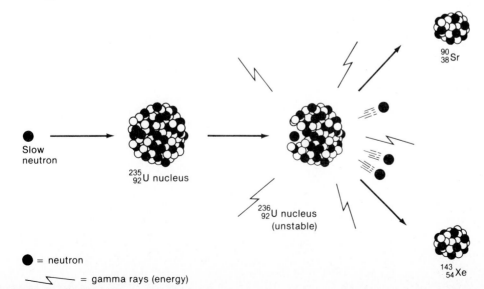

Scientists quickly realized that massive amounts of energy could be obtained from the fission of uranium. They also saw the possibility that neutrons released in the fission of one atom could trigger the fisssion of other uranium atoms, thus setting off a **chain reaction** (Figure 4.13). If rapid enough, this process can produce a bomb with tremendous explosive force.

Aware of the destructive forces that could be produced, and concerned that Germany might develop such a bomb, Fermi prevailed on Einstein to sign a letter to President Franklin D. Roosevelt indicating the importance of the discovery. The United States government launched a massive research project for the study of atomic energy—designating the study the Manhattan Project. Uranium had to be collected and the isotopes separated, for only the relatively rare uranium-235 isotope is fissionable. The neutrons, it was found, had to be slowed by graphite to increase the probability that they would hit a uranium nucleus. Fermi and his group achieved the first sustained nuclear reaction on 2 December 1942 under the bleachers at Stagg Field at the University of Chicago.

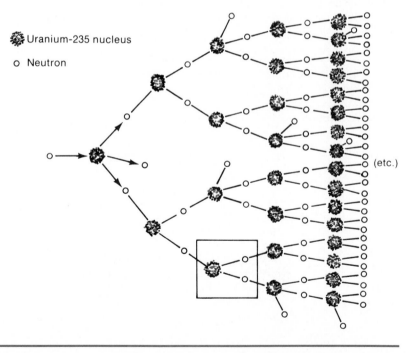

Figure 4.13 Schematic representation of a nuclear chain reaction. Neutrons released in the fission of one uranium-235 nucleus can strike another nucleus, causing it to split and release more neutrons. For simplicity, fission fragments are not shown. An area equivalent to that inside the box is shown in detail in Figure 4.12.

Uranium-235 makes up only 0.7% of natural uranium; the rest is nonfissile uranium-238. To make a bomb, the uranium has to be enriched to about 90% uranium-235. This enrichment proceeded slowly at a top-secret installation in Oak Ridge, Tennessee. Separation could not be accomplished by chemical reactions because the isotopes are nearly identical chemically. Rather, separation was accomplished by conversion of all the uranium to volatile uranium hexafluoride. Molecules of the latter containing the uranium-235 isotope are slightly lighter and move slightly more rapidly than molecules containing the uranium-238 isotope. The vapors of uranium hexafluoride were allowed to pass through thousands of consecutive pinholes, a process in which the molecules that contained uranium-235 gradually outdistanced the others. The scientists eventually obtained 15 kg of the separated uranium-235 isotope, enough to make a small explosive device.

While the tedious work of separating uranium isotopes was under way at Oak Ridge, other workers, led by Glenn T. Seaborg, approached the problem of obtaining fissionable material by another route. It was known that uranium-238 would not fission when bombarded by neutrons. However, it had been determined that when uranium-238 was bombarded by neutrons (and certain other particles) a new element named neptunium (Np) was formed, and this product quickly decayed to another new element, plutonium (Pu).

$$^{238}_{92}\text{U} + {}^{1}_{0}\text{n} \longrightarrow {}^{239}_{92}\text{U}$$

$$^{239}_{92}\text{U} \longrightarrow {}^{239}_{93}\text{Np} + {}^{0}_{-1}\text{e}$$

$$^{239}_{93}\text{Np} \longrightarrow {}^{239}_{94}\text{Pu} + {}^{0}_{-1}\text{e}$$

The isotope plutonium-239 was found to be fissionable and, thus, was suitable material for the making of a bomb. A series of large reactors were built near Hanford, Washington, for the making of plutonium.

Before a fissionable material can sustain a chain reaction, a certain minimum amount, called the **critical mass**, must be brought together. There must be enough fissionable nuclei that the neutrons released in one fission process will have a good chance of finding another fissionable nucleus before escaping from the mass. For uranium-235, this critical mass is about 4 kg, a mass about the size of a baseball. To construct a bomb, separate smaller masses are used. These subcritical masses are then brought together forcefully to trigger the runaway chain reaction of a nuclear explosion.

By July 1945, enough plutonium had been made for a bomb to be assembled. This first atomic bomb was tested in the desert near Alamogordo, New Mexico, on 16 July 1945. The heat from the explosion vaporized the 30-m steel tower on which it was placed and melted the sand for several hectares around the site. The light released was the brightest anyone had ever seen.

Some of the scientists were so awed by the force of the blast that

Figure 4.14 A nuclear bomb of the type exploded over Hiroshima. The bomb is 71 cm in diameter and 305 cm long; it weighs 4000 kg and has an explosive power equivalent to about 18 000 000 kg of high explosive. [Courtesy of the National Air and Space Museum, Smithsonian Institution, Washington, DC.]

Figure 4.15 The now familiar mushroom cloud that follows an atomic explosion. This photograph was taken just after the first atomic bomb destroyed Hiroshima. [Courtesy of the National Air and Space Museum, Smithsonian Institution, Washington, DC.]

they argued against its use on Japan. A few, led by Leo Szilard, argued for a demonstration of its power at an uninhabited site. But fear of a well-publicized "dud" and the desire to avoid millions of casualties in an invasion of Japan led President Harry S Truman to order the dropping of bombs on Japanese cities. A uranium bomb called "Little Boy" was dropped on Hiroshima on 6 August 1945, causing over 100 000 casualties (Figures 4.14 and 4.15). Three days later, a plutonium bomb called "Fat Man" was dropped on Nagasaki with comparable results. World War II ended with the surrender of Japan on 14 August 1945.

Some Chemistry of Nuclear War: Radioactive Fallout

A single explosion involving a nuclear bomb equivalent to 1 megaton (Mt) of conventional explosives would produce a fireball 2 km in diameter. An explosion at ground level would create a crater 300 m across and 100 m deep. Total destruction would occur from the center of the blast out to 10 km. All unprotected people would be killed within a radius of 40 km. Even then, the bomb's destruction would not be over. Radioactive materials may rain upon parts of the Earth thousands of miles away, days and weeks later.

A typical fission reaction might be

$$^{235}_{92}\text{U} + ^{1}_{0}\text{n} \longrightarrow ^{90}_{38}\text{Sr} + ^{143}_{54}\text{Xe} + 3\,^{1}_{0}\text{n}$$

The neutrons may strike additional uranium atoms, carrying on the chain reaction. The strontium (Sr) and xenon (Xe) atoms are radioactive and are a part of the fallout.

The uranium atom can split in 40 or more ways, producing 80 or 90 primary radioactive products. Some of these produce radioactive **daughter isotopes**. For example, xenon-143 undergoes beta decay, with a half-life of 1 second.

$$^{143}_{54}\text{Xe} \longrightarrow ^{0}_{-1}\text{e} + ^{143}_{55}\text{Cs}$$

The cesium isotope is also radioactive—as are its daughters, and their daughters. Still other radioisotopes, such as carbon-14 and tritium, are formed by the impact of neutrons produced in the explosion on molecules of the atmosphere. Thus, fallout is exceedingly complex. We will consider only a few of the more important isotopes.

Of all the isotopes, strontium-90 presents the greatest hazard to people. This isotope has a half-life of 28 years. Strontium-90 reaches us primarily through milk and vegetables (Figure 4.16). Because of its similarity to calcium (both are Group IIA elements), strontium-90 is incorporated into bone. There it remains a source of internal radiation for many years.

Although strontium-90 is a greater long-term hazard, iodine-131 may present a greater threat immediately after a nuclear explosion. The half-life of iodine-131 is only 8 days, but it is produced in relatively large amounts. Iodine-131 is efficiently carried through the food chain. In the

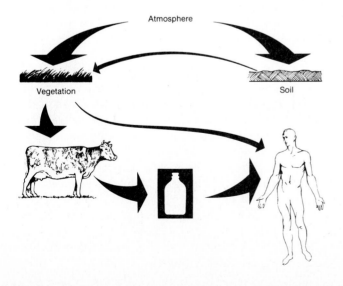

Figure 4.16 Pathways of strontium-90 from fallout. [Reprinted from C. L. Comer, *Fallout* (Oak Ridge, TN: U.S. Department of Energy, 1966).]

body it is concentrated in a small area, the thyroid gland. It is precisely this characteristic that makes this isotope so useful for diagnostic scanning. However, for a healthy individual, the incorporation of radioactive iodine in the thyroid gland offers no useful information, only damaging side effects.

Another important isotope in fallout is cesium-137. Cesium is similar to potassium (both are Group IA elements), and it mimics potassium in the body. Cesium-137 is a gamma emitter and has a half-life of 30 years. It is less of a threat than strontium-90, though, because it is removed from the body more readily. We get cesium-137 through vegetables, milk, and meat.

By the late 1950s, radioactive isotopes from atmospheric testing of nuclear weapons had been detected in the environment. Concern over radiation damage from nuclear fallout led to a movement to ban atmospheric testing. Many scientists were leaders in the movement. Linus Pauling, who won the Nobel Prize in chemistry in 1954 for his work in determining the structure of proteins, was a particularly articulate advocate of banning the bomb. In 1963, a nuclear test ban treaty was signed by the major powers—with the exception of France and the People's Republic of China, who continue above-ground tests. Since the signing of the treaty, India has joined the nuclear club by exploding a bomb above ground. Pauling, who had to endure being called a communist and a traitor for his outspoken position, was awarded the Nobel Prize for peace in 1962.

Nuclear Winter

Radioactive fallout isn't the only concern in the aftermath of nuclear war. The nations of planet Earth have acquired an arsenal of perhaps 50 000 nuclear weapons with an explosive power equal to more than 1 million Hiroshima bombs or to 13 billion tons of TNT. Studies now suggest that explosion of only half these weapons would produce enough soot, smoke, and dust to blanket the Earth, block out the sun, and bring on a "**nuclear winter**" that would threaten the survival of the human race. As much as 90% of the sunlight could be prevented from reaching the surface in the Northern Hemisphere, dropping the temperature to −25 °C for several months. After several weeks, the cloud could blanket the Southern Hemisphere as well. Crops would freeze, farm animals would die of thirst and hunger. Many species would be extinguished, including perhaps our own.

These effects are predicted by mathematical simulations on sophisticated computers. Exact predictions depend upon certain assumptions that are made in the models, but nearly everyone agrees that nuclear war would be the greatest holocaust in human history.

Radiation and Us

Humans have always been exposed to radiation. Some of it, called **cosmic rays**, comes from the sun and outer space. Other radiation reaches us from natural radioactive isotopes in air, water, soil, and rocks. This ever-present radiation is called **background radiation**.

Human activities have added to our exposure to radiation. Figure 4.17 shows that about two thirds of the average radiation exposure comes from background radiation. Most of the remaining third comes from medical irradiation such as X rays. Other sources, such as fallout, releases from the nuclear industry, and occupational exposure account for only a minute fraction of our total exposure. Nevertheless, accidents such as that in 1979 at the Three Mile Island plant near Harrisburg, Pennsylvania, and that in 1986 at Chernobyl in the Ukraine, U.S.S.R., did much to increase public apprehension about nuclear power. Even though no one was hurt at Three Mile Island, some people are still concerned about the long-term effect of exposure to the small amounts of radioactivity released. The accident at Chernobyl was more severe. Several people were killed outright; others died later of severe radiation

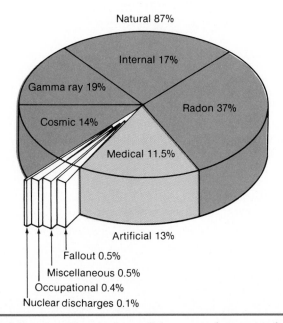

Natural 87%

Internal 17%

Gamma ray 19%

Radon 37%

Cosmic 14%

Medical 11.5%

Artificial 13%

Fallout 0.5%

Miscellaneous 0.5%

Occupational 0.4%

Nuclear discharges 0.1%

Figure 4.17 Most of our exposure to radition comes from natural sources. The next largest contribution is from medical uses of radiation. [Adapted from National Radiological Protection Board publication NRPB-R77. Reprinted with the permission of the American Nuclear Society.]

poisoning. Increased levels of radioactivity throughout Europe brought great concern about possible harmful effects.

Although radiation has always been with us, we didn't know it until the discovery of X rays and radioactivity in the 1890s. Shortly after that, it was noted that such radiation could be both beneficial and harmful. We are all familiar with the use of X rays in medical diagnosis. We have seen how radioisotopes are used in beneficial ways. But radiation also presents a hazard to living things. High-energy particles and rays knock electrons from atoms, forming ions. Such chemical changes, when they occur in living cells, can be highly disruptive. Water in the cells can be transformed to highly reactive hydrogen peroxide (H_2O_2), which can disrupt the delicate chemical balance in the cells. Particularly vulnerable are the white blood cells, the body's first line of defense against bacterial infection. Radiation also affects bone marrow, causing a drop in the production of red blood cells, which results in anemia. Radiation also has been shown to induce leukemia, a cancerlike disease of the blood-forming organs.

Radiation also causes changes in the molecules of heredity (DNA) in reproductive cells. Such changes show up as **mutations** in the offspring of exposed parents. Little is known of the effects of such exposure on humans. However, most of the mutations during the evolution of present species may have been caused by background radiation.

We discuss nuclear power, with its promise and peril, in Chapter 14.

Binding Energy

Nuclear reactions involve tremendous energy changes. Where does all this energy come from? When neutrons and protons are put together to form nuclei, some mass is converted to energy. This is sometimes called **binding energy**. For example, if two protons and two neutrons are put together to form a helium nucleus, the nucleus should have a mass of

2 protons:	$2 \times 1.007276 =$	2.014552 amu
2 neutrons:	$2 \times 1.008665 =$	2.017330 amu
	total mass $=$	4.031882 amu

The actual mass of the helium nucleus is 4.001506 amu. the missing mass, or binding energy, is 0.030376 amu. Remembering Einstein's $E = mc^2$, we see that this small amount of mass is equivalent to a large amount of energy. If we made calculations like this for all nuclei and plotted the data on a graph, a curve like the one in Figure 4.18 would result. The most stable nuclei are those in the vicinity of iron (atomic number 26). Splitting uranium atoms produces daughter nuclei that are more stable than the original uranium atoms; consequently, this splitting releases considerable energy.

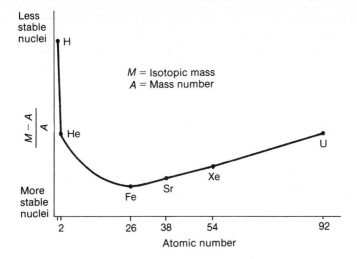

Figure 4.18 Nuclear stability.

We also can see from the graph that we could produce even more energy by combining small nuclei (such as hydrogen) to form larger, more stable nuclei. This sort of **nuclear fusion** occurs in the explosion of a hydrogen bomb.

Thermonuclear Reactions

The source of nearly all our energy on Earth is the thermonuclear reactions taking place in the sun. The intense temperature in the center of the sun causes nuclei to fuse and release tremendous amounts of energy. The principal net reaction is believed to be the fusion of four hydrogen nuclei to produce one helium nucleus.

$$4\,_1^1H \longrightarrow \,_2^4He \; + \; 2\,_{+1}^{\;0}e$$

One gram of hydrogen, upon fusing, releases an amount of energy equivalent to 17 000 kg (nearly 20 tons) of coal.

The hydrogen bomb makes use of a uranium or plutonium (fission) bomb to provide the tremendous heat necessary to start the nuclei fusing. The fusion of ordinary hydrogen ($_1^1H$) occurs much too slowly, so the heavier isotopes deuterium ($_1^2H$) and tritium ($_1^3H$) are employed. The intense heat of the fission explosion starts the fusion of hydrogen nuclei.

$$_1^2H \; + \; _1^3H \longrightarrow \,_2^4He \; + \; _0^1n$$

The neutron released splits lithium atoms, forming more tritium.

$$_3^6Li \; + \; _0^1n \longrightarrow \,_2^4He \; + \; _1^3H$$

To date, the fusion reactions are useful only for making bombs. Research is progressing in the control of nuclear fusion. The prospect for power from controlled fusion is discussed in Chapter 14.

The Nuclear Age

We live in exciting times. The goal of the alchemists, to change one element into another, has been achieved through application of scientific principles. New elements have been formed, and the periodic chart has been extended beyond uranium ($Z = 92$) to element 109. This modern alchemy produces plutonium by the ton; neptunium ($Z = 93$), americium ($Z = 95$), and curium ($Z = 96$) by the kilogram; and berkelium ($Z = 97$) and einsteinium ($Z = 99$) by the milligram. These new elements have been used in medicine and to power spacecraft and build bombs.

We live in an age in which fantastic forces have been unleashed. The threat of nuclear war has been a constant specter for the last four decades. Nuclear bombs have been used to destroy cities—and men, women, children, and other living things. Science and scientists have been deeply involved in it all.

Still, it is hard to believe that the world would be a better place if we had not discovered the secrets of the atomic nucleus. For one thing, more lives have been saved through nuclear medicine than have been destroyed by nuclear bombs. And no nuclear bombs have been used in warefare since 1945. Perhaps the terror of nuclear holocaust, more than anything else, has prevented World War III. But nuclear war hangs over our head like a monstrous sword. There is a chance that it will come in your lifetime. If it does, it could lead to the extinction of human life, especially in the Northern Hemisphere.

Problems

1. Define or identify each of the following.

   a. half-life
   b. positron
   c. background radiation
   d. radioisotope
   e. fission
   f. fusion
   g. artificial transmutation
   h. binding energy

2. How does the size of the nucleus compare to that of the atom as a whole?

3. Why are isotopes important in nuclear reactions?

4. What is meant by the mass number of an isotope?

5. Give the nuclear symbols for protium, deuterium, and tritium (which are hydrogen-1, hydrogen-2, and hydrogen-3, respectively).

6. Give the nuclear symbol for an isotope with a mass number of 8 and an atomic number of 5.

7. Give the nuclear symbol for an isotope with $Z = 35$ and $A = 83$.

8. Give the nuclear symbol for an isotope with 53 protons and 72 neutrons.

9. Give the nuclear symbols for the following isotopes. You may refer to the periodic table.
   a. gallium-69   b. molybdenum-98
   c. molybdenum-99   d. technetium-98

10. Indicate the number of protons and the number of neutrons in atoms of the following isotopes.
    a. $^{62}_{30}\text{Zn}$ b. $^{241}_{94}\text{Pu}$ c. $^{99m}_{43}\text{Tc}$ d. $^{81m}_{36}\text{Kr}$

11. Which of the following pairs represent isotopes?
    a. $^{70}_{34}\text{X}$ and $^{70}_{33}\text{X}$ b. $^{57}_{28}\text{X}$ and $^{66}_{28}\text{X}$
    c. $^{186}_{74}\text{X}$ and $^{186}_{74}\text{X}$ d. $^{8}_{2}\text{X}$ and $^{6}_{4}\text{X}$
    e. $^{22}_{11}\text{X}$ and $^{44}_{22}\text{X}$

12. The two principal isotopes of lithium are lithium-6 and lithium-7. The atomic mass of lithium is 6.9 amu. Which is the predominant isotope of lithium?

13. Out of every five atoms of boron, one has a mass of 10 amu and four have a mass of 11 amu. What is the atomic mass of boron? Use the periodic table only to check your answer.

14. Give the nuclear symbols for the following subatomic particles.
    a. the alpha particle   b. the beta particle
    c. the neutron   d. the positron

15. When a nucleus emits a beta particle, what changes occur in the mass number and atomic number of the nucleus?

16. Lead-209 undergoes beta decay. Write a balanced equation for this reaction.

17. When a nucleus emits an alpha particle, what changes occur in the mass number and atomic number of the nucleus?

18. Thorium-225 undergoes alpha decay. Write a balanced equation for this reaction.

19. When a nucleus emits a gamma ray, what changes occur in the mass number and the atomic number of the nucleus?

20. Write an equation representing the emission of a gamma ray by gold-186.

21. When a nucleus emits a positron, what changes occur in the mass number and the atomic number of the nucleus?

22. Write a balanced equation for the emission of a positron by sulfur-31.

23. When a nucleus emits a neutron, what changes occur in the mass number and the atomic number of the nucleus?

24. Write a balanced equation for the emission of a neutron by bromine-87.

25. When a nucleus emits a proton, what changes occur in the mass number and the atomic number of the nucleus?

26. Write a balanced equation for the emission of a proton by magnesium-21.

27. Complete the following equations.

    a. $^{179}_{79}\text{Au} \rightarrow ^{175}_{77}\text{Ir} + ?$
    b. $^{23}_{10}\text{Ne} \rightarrow ^{23}_{11}\text{Na} + ?$

28. Complete the following equations.

    a. $^{10}_{5}\text{B} + ^{1}_{0}\text{n} \rightarrow ^{4}_{2}\text{He} + ?$
    b. $^{12}_{6}\text{C} + ^{2}_{1}\text{H} \rightarrow ^{13}_{6}\text{C} + ?$
    c. $^{121}_{51}\text{Sb} + ? \rightarrow ^{121}_{52}\text{Te} + ^{1}_{0}\text{n}$
    d. $^{154}_{62}\text{Sm} + ^{1}_{0}\text{n} \rightarrow 2\,^{1}_{0}\text{n} + ?$

29. When magnesium-24 is bombarded with a neutron, a proton is ejected. What new element is formed? (Hint: Write a balanced nuclear equation.)

30. A radioactive isotope decays to give an alpha particle and bismuth-211. What was the original element?

31. C. E. Bemis and colleagues at Oak Ridge National Laboratory confirmed the synthesis of element 104, the half-life of which was only 4.5 seconds. Only 3000 atoms of the element were created in the tests. How many atoms were left after 4.5 seconds? After a total of 9.0 seconds?

32. Krypton-81^m is used for lung ventilation studies. Its half-life is 13 seconds. How long will it take the activity of this isotope to reach one quarter of its original value?

33. Explain how radioisotopes can be used for therapeutic purposes.

34. Which radioisotope has been used extensively for treatment of overactive or cancerous thyroid glands?

35. Describe the use of a radioisotope as a diagnostic tool in medicine.

36. What are some of the characteristics that make technetium-99^m such a useful radioisotope for diagnostic purposes?

37. A radioisotope decays to give an alpha particle and protactinium-233 ($^{233}_{91}\text{Pa}$). What was the original element?

38. List two ways in which workers can protect them-

selves from the radioactive materials with which they work.

39. A pair of gloves would be sufficient to shield the hands from which type of radiation: the heavy alpha particles or the massless gamma rays?

40. Heavy lead shielding is necessary as protection from which type of radiation: alpha, beta, or gamma?

41. Plutonium is especially hazardous when inhaled or ingested because it emits alpha particles. Why would alpha particles cause more damage to tissue than beta particles?

42. What form of radiation is detected in PET scans?

43. Living matter has a carbon-14 content that gives 16 counts per minute per gram of carbon. What is the age of an artifact for which the carbon-14 gives 4 counts per minute per gram of carbon?

44. The ratio of carbon-14 to carbon-12 in a piece of charcoal from an archaeological excavation is found to be one half the ratio in a sample of modern wood. Approximately how old is the site? How old would it be if the ratio was 25% of the ratio in a sample of modern wood?

45. How old is a bottle of wine if the tritium activity is 25% that of new wine?

46. Uranium has a density of 19 g/cm^3. What volume is occupied by a critical mass of 8 kg of uranium?

47. Which subatomic particles are responsible for carrying on the chain of reactions that are characteristic of nuclear fission?

48. Compare nuclear fission and nuclear fusion. Why is energy liberated in each case?

49. To make element 106, a 0.25-mg sample of $^{249}_{98}Cf$ was used as the target. Four neutrons were emitted to yield a nucleus with 106 protons and a mass of 263 amu. What was the bombarding particle?

$$^{249}_{98}Cf + ? \rightarrow 4\,^1_0n + ^{263}_{106}X$$

50. One atom of element 109 with a mass number of 266 was produced in 1982 by bombarding a target of bismuth-209 with iron-58 nuclei for 1 week. How many neutrons were released in the process?

$$^{209}_{83}Bi + ^{58}_{26}Fe \rightarrow ^{266}_{109}X + ?\,^1_0n$$

51. Element 109 undergoes alpha emission to form element 107, which in turn also emits an alpha particle. What is the atomic number and mass number of the isotope formed by these two steps? Write balanced nuclear equations for the two reactions.

52. Did President Harry S Truman make the right decision when he decided to drop the nuclear bombs on Japanese cities? Would your answer be the same if you were living in 1945 and had relatives among the troops preparing for the invasion of Japan? If you were an inhabitant of one of the cities bombed?

53. Discuss the impact of nuclear science on the following topics. (See the list of references and readings for this chapter for resource materials.)
a. war and peace
b. industrial progress
c. medicine
d. agriculture
e. human, animal, and plant genetics

54. Radium-223 nuclei usually decay by alpha emission. For every billion alpha decays, one atom emits a carbon-14 nucleus. Write a balanced nuclear equation for each type of emission.

References and Readings

1. Abrams, Herbert L. "Medical Problems of Survivors of a Nuclear War." *ChemTech*, May 1984, pp. 286–289.

2. Chivian, E., et al. (Eds.). *Last Aid: The Medical Dimensions of Nuclear War*. San Francisco: Freeman, 1981.

3. Ehrlich, Paul R., et al. "Long-Term Biological Consequences of Nuclear War." *Science*, 23 December 1983, pp. 1293–1300.

4. Harper, C. T. (Ed). *Geochronology: Radiometric Dating of Rocks and Minerals*. Stroudsburg, PA: Dowden, Hutchinson, and Ross, 1973.

5. Marshall, Eliot. "Recalculating the Cost of Chernobyl," *Science*, 8 May 1987, pp. 658–659. The accident may cause 39 000 extra cancer deaths.

6. Rawls, Rebecca. "Element 109 Made by West German Researchers." *Chemical and Engineering News*, 11 October 1982, pp. 27–28.

7. Sagan, Leonard. "Radiation and Human Health." *EPRI Journal*, September 1979, pp. 6–13.

8. Schecter, Bruce. "Green Light for the Neutron Bomb." *Discover*, October 1981, pp. 28–30. The neutron bomb produces more radiation, less blast, and less heat than typical thermonuclear weapons.

9. Schecter, Bruce. "The Short, Bright Life of Element 109." *Discover*, December 1982, pp. 98–106.

10. Schell, Jonathan. *The Fate of the Earth*. New York: Knopf, 1982. Presents a scenario for nuclear war and its aftermath.

11. Seaborg, Glenn T. "Modern Alchemy." *The Science Teacher*, November 1983, pp. 29–34.

12. Sun, Marjorie. "Renewed Interest in Food Irradiation." *Science*, 17 February 1984, pp. 667–668.

13. Turco, R. P., et al. "Nuclear Winter: Global Consequences of Multiple Nuclear Explosions." *Science*, 23 December 1983, pp. 1283–1292.

14. Wetherill, George W. "Dating Very Old Objects." *Natural History*, September 1982, pp. 14–20.

15. Yalow, Rosalyn S. "Radioactivity in the Service of Man." *Journal of Chemical Education*, September 1982, pp. 735–738.

5

Chemical Bonds

The Ties That Bind

There are approximately 100 chemical elements. There are millions of chemical compounds. To form these compounds, atoms of different elements must be held together in specific combinations. **Chemical bonds** are the forces that maintain these arrangements. Chemical bonding also plays a role in determining the state of matter. At room temperature, water is a liquid, carbon dioxide is a gas, and table salt is a solid—because of differences in chemical bonding.

As scientists developed an understanding of the nature of chemical bonding, they gained the ability to manipulate the structure of compounds. Dynamite, birth-control pills, synthetic fibers, and a thousand other products were fashioned in chemical laboratories and have dramatically changed the way we live. We are now entering an era that promises (some would say threatens) even greater change.

The DNA molecule—the chemical basis of heredity—carries its genetic message in its bonds. Two DNA molecules in turn are joined to each other by hydrogen bonds (see page 233). Whether an organism is fish, fowl, hippopotamus, or human is determined by the arrangement of bonds in DNA. Scientists already have the ability to rearrange these bonds, and this ability has given them limited control over the structure of living matter. As techniques of genetic engineering improve, scientists may literally be able to custom-tailor genes.

Let us begin our consideration of chemical bonding so that we, too, can understand the forces that control the structure of matter, living and nonliving.

The Art of Deduction: Stable
Electron Configurations

In our discussion of the atom and its structure (Chapters 2 and 3), we followed the historical development of some of the more important atomic concepts. Some of the nuclear concepts (Chapter 4) were approached in the same manner. We could continue to look at chemistry in this manner, but that would require several volumes of print—if we got very far—and perhaps more of your time than you care to spend. We won't abandon the historical approach entirely, but we will emphasize that other aspect of scientific enterprise: deduction.

The art of deduction works something like this.

Fact	*Theory*	*Deduction*
The noble gases, such as helium, neon, and argon, are inert (i.e., they undergo few, if any, chemical reactions).	The inertness of the noble gases is due to their electron structures (each has a filled outermost energy level).	If other elements could alter their electron structures to become more like noble gases, they would become less reactive.

To illustrate, let's look at an atom of the element sodium (Na). It has 11 electrons, 2 in the first energy level, 8 in the second, and 1 in the third. If the atom could get rid of an electron, it would have the same electron structure as an atom of the noble gas neon (Ne).

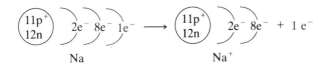

Recall that neon has the structure

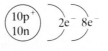

Let's immediately emphasize that the sodium (Na^+) and the neon atom (Ne) are not identical. The electron arrangement is the same, but the nuclei—and resulting charges—are not. As long as sodium keeps its

11 protons, it is still a form of sodium. But it is sodium *ion*, not sodium *atom*.

If a chlorine atom (Cl) could gain an electron, it would have the same structure as argon (Ar).

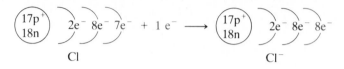

The structure of the argon atom is

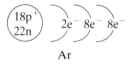

The sodium, having lost an electron, becomes positively charged. It has 11 protons (11+) and only 10 electrons (10−). It is written Na^+ and is called a *sodium ion*. The chlorine atom, having gained an electron, becomes negatively charged. It has 17 protons (17+) and 18 electrons (18−). It is written Cl^- and is called a *chloride ion*. Note that a positive charge, as in Na^+, indicates that one electron has been lost. Similarly, a negative charge, as in Cl^-, indicates that one electron has been gained.

In forming ions, the nuclei of sodium and chlorine and the *inner* shells (i.e., lower energy levels) of electrons do not change. Therefore, it is convenient to let the nucleus *and* the inner shells be represented by the *symbol* alone. Electrons in the outer shell, or **valence shell**, are represented by dots. Thus the shell diagrams for the ionization of sodium and chlorine are reduced to

and

$$Na\cdot \longrightarrow Na^+ + 1\,e^-$$

$$:\overset{..}{\underset{..}{Cl}}\cdot + 1\,e^- \longrightarrow :\overset{..}{\underset{..}{Cl}}:^-$$

Representations of this sort are called **electron-dot** symbols.

Chemical Symbolism

The mystery of chemistry to the nonchemist is probably due in large part to chemists' use of symbolism. Chemists find it *convenient* to represent the sodium atom as Na· rather than the more complex

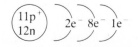

Table 5.1   Electron-Dot Symbols for the First 20 Elements

IA	IIA	IIIA	IVA	VA	VIA	VIIA	Noble Gases
H·							He:
Li·	·Be·	·B·	·C·	:N·	:O·	:F:	:Ne:
Na·	·Mg·	·Al·	·Si·	:P·	:S·	:Cl·	:Ar:
K·	·Ca·						

Therefore, they do just that. And it is easier to write $:\ddot{\underset{..}{Cl}}\cdot$ than

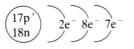

Thus, chemists often use the shorter form.

In practice, it is easy to write electron-dot formulas for elements in the first three periods (horizontal rows) of the periodic chart. The number of electrons in the outer shell is merely equal to the *group number*. Aluminum (Al) is in Group IIIA; therefore, it has three outer electrons. Sulfur (S) is in Group VIA; thus, it has six outer electrons. This generalization works fairly well for elements in A subgroups even beyond the first three periods. Thus, iodine (I) is in Group VIIA and has seven outer electrons, even though it is in the fifth period. Table 5.1 gives the electron-dot formulas for the first 20 elements.

Symbolism is a convenient, shorthand way to convey a lot of information in a compact form. It is a chemist's most efficient and economical form of communication. Learning this symbolism is a good deal like learning a foreign language. Once you have learned a basic "vocabulary," the rest is a lot easier.

Sodium Reacts with Chlorine: The Facts

Sodium is a highly reactive metal. It is soft enough to be cut with a knife. When freshly cut, it is bright and silvery, but it dulls rapidly by reacting with oxygen in the air. In fact, it reacts so readily in air that it is usually stored under oil or kerosene. Sodium reacts violently with water also, getting so hot that it melts. A small piece will form a spherical bead after melting and race around on the surface of the water as it reacts.

Chlorine is a greenish yellow gas. It is familiar as a disinfectant for swimming pools and city water supplies. (The actual substance added may be a compound that reacts with water to form chlorine). Who

Sodium   Chlorine   Sodium chloride

Figure 5.1 Sodium, a soft silvery metal, reacts with chlorine, a greenish gas, to form sodium chloride (ordinary table salt).

hasn't been swimming in a pool that had "so much chlorine in it that you could taste it"? Chlorine is extremely irritating to the respiratory tract. In fact, chlorine was used as a poison gas in World War I.

If a piece of sodium is dropped into a flask containing chlorine gas, a violent reaction ensues. A white solid that is quite unreactive is formed. It is a familiar compound—sodium chloride (table salt) (Figure 5.1).

Sodium Reacts with Chlorine: The Theory

A sodium atom becomes less reactive by *losing* an electron. A chlorine atom becomes less reactive by *gaining* an electron. What happens when sodium atoms come into contact with chlorine atoms?* The obvious: chlorine extracts an electron from a sodium atom.

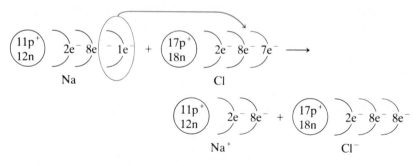

In the abbreviated electron-dot form, this reaction is written

$$Na\cdot \; + \; :\overset{..}{\underset{..}{Cl}}\cdot \; \longrightarrow \; Na^+ \; + \; :\overset{..}{\underset{..}{Cl}}\cdot^-$$

---

* Actually, the greenish yellow gas is composed of chlorine molecules, with each molecule consisting of two atoms. More about that later.

The sodium *ion* (Na$^+$) and the chloride *ion* (Cl$^-$) not only have electron structures like the inert gases (neon and argon, respectively), but they also have opposite charges. Everyone knows that opposites attract. And, though this rule of thumb may not always work when applied to people, it works well for electrically charged particles. Remember that in even a minute amount of salt there are billions and billions of particles. These arrange themselves in an orderly fashion. The arrangements are repeated in all directions—above and below, left and right, top and bottom—to make up a **crystal** of sodium chloride (Figure 5.2). Each sodium ion attracts (and is attracted by) six chloride ions (the ones to the front and back, the top and the bottom, and both sides). Each chloride ion attracts (and is attracted by) the surrounding six sodium ions. The forces holding the crystal together (the attractive forces between positive and negative charges) are called **ionic bonds**.

First layer

Second layer

Scientists sometimes use different models to represent the same system. The model employed in Figure 5.2 is a space-filling model showing the relative sizes of the sodium and chloride ions. Sometimes a ball-and-stick model is employed to better show the geometry of the crystal (Figure 5.3). From this model it is easy to see the cubic arrangement of the ions. In the crystal as a whole, for each sodium ion there is one chloride ion; thus, the ratio of ions is 1:1, and the simplest formula for the compound is NaCl. The symbols Na and Cl, written together, stand for the compound sodium chloride. The formula is also used to represent one sodium ion and one chloride ion.

Figure 5.2 The arrangement of ions in a sodium chloride crystal.

More Ionic Compounds

Potassium (K), a metal similar to sodium, can also react with chlorine to yield a compound called potassium chloride (KCl).

$$\text{K} \cdot \ + \ \cdot \overset{..}{\underset{..}{\text{Cl}}} \colon \ \longrightarrow \ \text{K}^+ \ + \ \colon \overset{..}{\underset{..}{\text{Cl}}} \colon ^-$$

And potassium reacts with bromine, a reddish brown liquid that is chemically similar to chlorine, to form a stable white crystalline solid called potassium bromide (KBr).

$$\text{K} \cdot \ + \ \cdot \overset{..}{\underset{..}{\text{Br}}} \colon \ \longrightarrow \ \text{K}^+ \ + \ \colon \overset{..}{\underset{..}{\text{Br}}} \colon ^-$$

Sodium can also form a compound with bromine: sodium bromide.

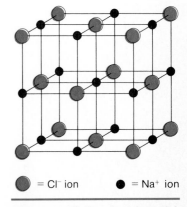

⬤ = Cl$^-$ ion   ● = Na$^+$ ion

Figure 5.3 Ball-and-stick model of a sodium chloride crystal.

Magnesium, a harder and less reactive metal than sodium, reacts with oxygen, a colorless gas, to form another stable white crystalline solid called magnesium oxide (MgO).

$$\dot{Mg}\cdot \ + \ \cdot\ddot{\underset{..}{O}}: \ \longrightarrow \ Mg^{2+} \ + \ :\ddot{\underset{..}{O}}:^{2-}$$

Magnesium must give up two electrons and oxygen must gain two electrons for each to have the same configuration as the noble gas neon.

An atom such as oxygen, which needs two electrons, may react with potassium atoms, which have only one electron to give. In this case, two atoms of potassium are needed for each oxygen atom. The product is potassium oxide (K₂O).

$$\begin{array}{ccccc} K\cdot & & & K^+ & \\ & + \ \cdot\ddot{\underset{.}{O}}: & \longrightarrow & & + \ :\ddot{\underset{..}{O}}:^{2-} \\ K\cdot & & & K^+ & \end{array}$$

By this process, each potassium atom achieves the argon configuration. Oxygen again assumes the neon configuration.

One last example is the reaction of magnesium and nitrogen to give magnesium nitride (Mg₃N₂).

$$\begin{array}{ccccc} \dot{Mg}\cdot & & Mg^{2+} & & \\ & \cdot\ddot{\underset{.}{N}}\cdot & & :\ddot{\underset{..}{N}}:^{3-} & \\ \dot{Mg}\cdot \ + & \longrightarrow & Mg^{2+} \ + & \\ & \cdot\ddot{\underset{.}{N}}\cdot & & :\ddot{\underset{..}{N}}:^{3-} & \\ \dot{Mg}\cdot & & Mg^{2+} & & \end{array}$$

Each of three magnesium atoms gives up two electrons (a total of six) and each of two nitrogen atoms acquires three (a total of six). Notice that the total positive and negative charges on the products are equal (6+ and 6−).

Generally speaking, elements on the left side of the periodic chart (especially those on the far left) react with elements on the far right (excluding the noble gases) to form stable crystalline solids. The theory is that the elements on the left (the *metals*) tend to give up electrons to the elements of the right (*nonmetals*). The crystalline solids are held together by the attraction of oppositely charged ions.

In general atoms of the Group IA elements tend to give up one electron to form 1+ ions and those of Group IIA elements tend to give up two electrons to form 2+ ions. Similarly, atoms of Group VIIA elements tend to take on one electron to form 1− ions, and those of Group VIA tend to pick up two electrons to form 2− ions. Atoms of some Group B metals can give up varying numbers of electrons to form positive ions of various charges. These periodic relationships are summarized in Figure 5.4; some of the ions are discussed further in the next chapter.

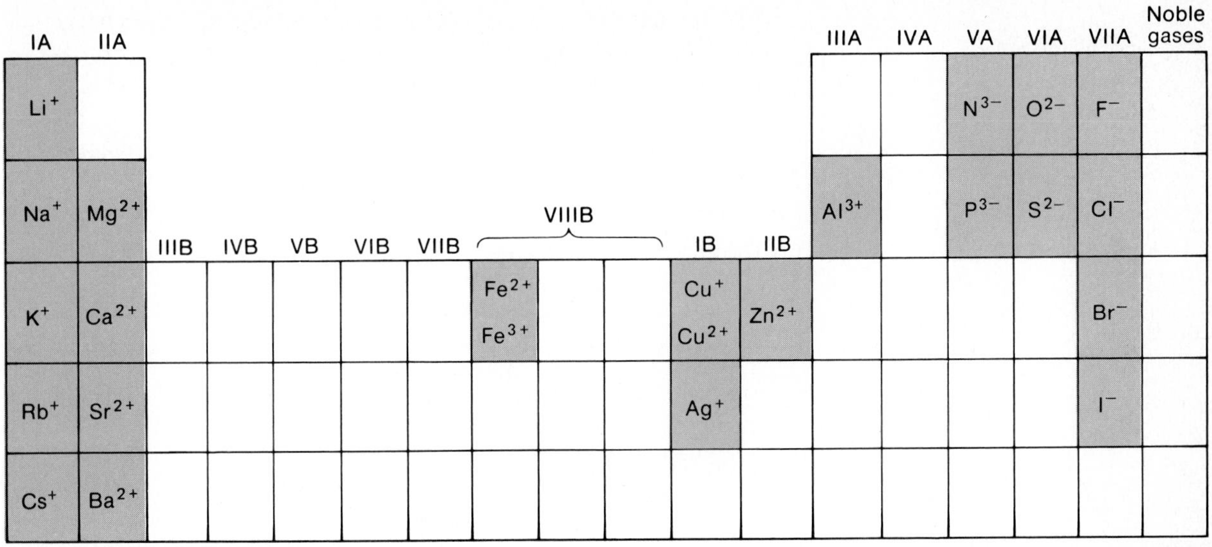

IA	IIA	IIIB	IVB	VB	VIB	VIIB	VIIIB			IB	IIB	IIIA	IVA	VA	VIA	VIIA	Noble gases
Li^+														N^{3-}	O^{2-}	F^-	
Na^+	Mg^{2+}											Al^{3+}		P^{3-}	S^{2-}	Cl^-	
K^+	Ca^{2+}						Fe^{2+} Fe^{3+}			Cu^+ Cu^{2+}	Zn^{2+}					Br^-	
Rb^+	Sr^{2+}									Ag^+						I^-	
Cs^+	Ba^{2+}																

Figure 5.4 The periodic relationship of some simple ions.

Example 5.1 Write the electron-dot formula for the ionic compound formed from potassium and chlorine.

Potassium is in Group IA.

$$K\cdot \longrightarrow K^+ + 1\,e^-$$

Chlorine is in Group VIIA.

$$:\ddot{\underset{..}{C}l}\cdot + 1\,e^- \longrightarrow :\ddot{\underset{..}{C}l}:^-$$

The electron-dot formula for the compound formed from potassium and chlorine is

$$K\cdot + \cdot\ddot{\underset{..}{C}l}: \longrightarrow K^+ + :\ddot{\underset{..}{C}l}:^-$$

Example 5.2 Show the formation of an ionic compound from magnesium atoms and oxygen atoms.

Magnesium is in Group IIA.

$$\dot{M}g\cdot \longrightarrow Mg^{2+} + 2\,e^-$$

Oxygen is in Group VIA.

$$\cdot\dot{\underset{..}{O}}: + 2\,e^- \longrightarrow :\ddot{\underset{..}{O}}:^{2-}$$

Notice that the magnesium ion has a double positive charge because it must give up two electrons to achieve a noble gas configuration. The oxide ion has a double negative charge because it acquires two electrons to fill in its octet. In the compound formed from magnesium and oxygen, one magnesium atom supplies the two electrons required by one oxygen atom.

$$\dot{M}g\cdot \; + \; \cdot \ddot{\underset{\cdot}{O}}: \; \longrightarrow \; Mg^{2+} \; + \; :\ddot{\underset{\cdot\cdot}{O}}:^{2-}$$

Example 5.3   What is the formula for the compound formed from potassium and oxygen?
Potassium is in Group IA.

$$K\cdot \; \longrightarrow \; K^{+} \; + \; 1\,e^{-}$$

Oxygen is in Group VIA.

$$\cdot \dot{\underset{\cdot}{O}}: \; + \; 2\,e^{-} \; \longrightarrow \; :\ddot{\underset{\cdot\cdot}{O}}:^{2-}$$

Oxygen requires two electrons, but a potassium atom has only one to give. The problem can be solved by having each oxygen atom react with two potassium atoms.

$$\begin{matrix} K\cdot \\ \\ K\cdot \end{matrix} \; + \; \cdot \ddot{\underset{\cdot}{O}}: \; \longrightarrow \; \begin{matrix} K^{+} \\ \\ K^{+} \end{matrix} \; + \; :\ddot{\underset{\cdot\cdot}{O}}:^{2-}$$

Covalent Bonds

One might expect a hydrogen atom, with its one electron, to acquire another electron and assume the helium configuration. Indeed, hydrogen atoms do just that in the presence of atoms of a reactive metal such as lithium—that is, a metal that finds it easy to give up an electron.

$$Li\cdot \; + \; H\cdot \; \longrightarrow \; Li^{+} \; + \; H:^{-}$$

But what if there are no other kinds of atoms around? What if there are only hydrogen atoms? One atom can't gain an electron from another, for among hydrogen atoms all have an equal attraction for electrons. They can compromise, however, by *sharing a pair* of electrons.

$$H\cdot + \cdot H \longrightarrow H:H$$

Both electrons occupy one orbital that encompasses both nuclei, with the two electrons spending most of their time somewhere between the two nuclei. The electron-dot representation usually used (H:H) is therefore a fairly accurate representation. This combination of hydrogen atoms is called a hydrogen molecule. The bond formed by a shared pair of electrons is called a **covalent bond**.

A chlorine atom will pick up an extra electron from anything willing to give one up. But, again, what if the only thing around is another chlorine atom? Chlorine atoms, too, can attain a more stable arrangement by sharing a pair of electrons.

$$:\overset{..}{\underset{..}{Cl}}\cdot + \cdot\overset{..}{\underset{..}{Cl}}: \longrightarrow :\overset{..}{\underset{..}{Cl}}:\overset{..}{\underset{..}{Cl}}:$$

Each chlorine atom in the chlorine molecule has eight electrons around it, an arrangement like that of the noble gas argon. This stable *octet* of electrons is the arrangement characteristic of all the noble gases except helium. Covalently bonded atoms that we shall consider, except hydrogen, follow the **octet rule**; that is, they seek an arrangement that will surround them with eight electrons. The shared pair of electrons in the chlorine molecule also creates a covalent bond.

For simplicity, the hydrogen molecule is often represented as H_2 and the chlorine molecule as Cl_2. The subscripts indicate two atoms *per molecule*. In each case, the covalent bond between the atoms is understood. Sometimes the covalent bond is indicated by a dash: H—H and Cl—Cl. The dash means exactly the same thing as two dots: a shared pair of electrons—a covalent bond.

This sharing of electrons is not limited to one pair of electrons. Consider, for example, the nitrogen atom. Its electron-dot symbol is

$$:\overset{..}{\underset{.}{N}}\cdot$$

We know this atom will be reactive, after what we've learned about the octet rule. It has only five electrons in its outermost shell. It could share a pair of electrons with another nitrogen atom and would then look like this:

$$:\overset{..}{\underset{.}{N}}:\overset{..}{\underset{.}{N}}: \qquad \textit{(Incorrect structure)}$$

Each atom in this arrangement has only six electrons around it, and six is not eight. Each has two electrons hanging out there without partners; so, to solve the dilemma, each nitrogen shares two additional pairs of electrons, for a total of three pairs.

$$:N:::N: \quad \text{(or } :N{\equiv}N: \text{ or simply } N{\equiv}N)$$

In drawing the nitrogen molecule (N_2), we have simply drawn all the electrons that are being shared in the space between the two atoms. Each nitrogen has now satisfied the octet rule.

A molecule in which *three pairs* of electrons (a total of six individual electrons) are being shared is said to contain a **triple bond**. Each nitrogen also has an *unshared* pair of electrons. Note that we could have drawn the unshared pair of electrons above or below the atomic symbol. Such a drawing would represent the same molecule.

Polar Covalent Bonds

So far we have seen that atoms combine in two different ways. Some that are quite different in electron structure (from opposite ends of the periodic table) react by the complete transfer of one or more electrons from one atom to another (ionic bond formation). Atoms that are identical combine by sharing one or more pairs of electrons (covalent bond formation). Now let's look at some that are in between.

Hydrogen and chlorine react to form a colorless, toxic gas called hydrogen chloride. This reaction may be represented schematically by

$$H\cdot \ + \ \cdot \ddot{\underset{\cdot\cdot}{C}}l\!: \ \longrightarrow \ H\!:\!\ddot{\underset{\cdot\cdot}{C}}l\!: \quad \text{(or simply} \ \ H\!-\!Cl)$$

Both hydrogen and chlorine want an electron, so they compromise by sharing and forming a covalent bond. Since the substances hydrogen and chlorine actually consist of diatomic molecules rather than single atoms, the reaction is more accurately represented by the scheme

$$H\!:\!H \ + \ :\!\ddot{\underset{\cdot\cdot}{C}}l\!:\!\ddot{\underset{\cdot\cdot}{C}}l\!: \ \longrightarrow \ 2 \ H\!:\!\ddot{\underset{\cdot\cdot}{C}}l\!:$$

This can be more simply written as

$$H_2 \ + \ Cl_2 \ \longrightarrow \ 2 \ HCl$$

One might reasonably ask why the hydrogen molecule and the chlorine molecule react at all. Have we not just explained that they themselves were formed to provide a more stable arrangement of electrons? Yes, indeed, we did say that. But there is stable and there is more stable. The chlorine molecule represents a more stable arrangement than separate chlorine atoms, but given the opportunity, a chlorine atom would rather bond to hydrogen than to another chlorine atom.

In a molecule of hydrogen chloride, a chlorine atom shares a pair of electrons with a hydrogen atom. In this case, however, sharing does not mean sharing equally. Chlorine has a much greater attraction for the electron pair than does hydrogen. Chlorine is said to be more **electronegative** than hydrogen. The shared electrons are held more tightly by the chlorine atom, and this results in the chlorine end of the molecule being more negative than the hydrogen end. When the electrons in a covalent bond are not equally shared, the bond is said to be **polar**. Thus, the bonding in hydrogen chloride is described as **polar covalent**, whereas the bonding in the hydrogen molecule or in the chlorine molecule is **nonpolar covalent**. The polar covalent bond is not an ionic bond. In an ionic bond, one atom completely loses an electron. In a polar covalent bond, the atom at the positive end of the bond (hydrogen in HCl) still has some share in the bonding pair of electrons (Figure 5.6). To distinguish this arrangement from that in an ionic bond, the following notation is used.

$$\overset{\delta+}{H}\text{—}\overset{\delta-}{Cl}$$

The line between the atoms represents the covalent bond, a pair of shared electrons. The $\delta+$ and $\delta-$ (read "delta plus" and "delta minus") signify which end is partially positive and which is partially negative (the word *partially* is used to distinguish this charge from the full charge on an ion).

Figure 5.5 Chlorine hogs the electron blanket, leaving hydrogen partially, but positively, exposed. To hydrogen's pleas for more cover, chlorine's answer is partially negative.

Water: A Bent Molecule

Another polar covalent molecule, unquestionably the most important polar covalent molecule on Earth, is water (H_2O). Oxygen has six outer electrons and needs two more to complete its octet. Hydrogen atoms need only one electron each to complete their duets. Therefore, an oxygen atom bonds with two hydrogen atoms.

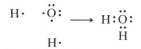

This arrangement completes the outer shell of oxygen, which now has the neon structure, and that of the hydrogens, which now have the helium structure.

We should expect the bonds in water to be polar because oxygen is more electronegative than hydrogen. (Like chlorine, oxygen is to the right in the periodic table, and electronegativity increases as one moves to the right in the table.) Just because a molecule contains polar bonds, however, does not mean that the molecule as a whole is polar. If the atoms in the water molecule were in a straight row (that is, in a linear arrangement), the two polar bonds would cancel one another out.

$$\overset{\delta+}{H}\text{—}\overset{\delta-}{O}\text{—}\overset{\delta+}{H} \qquad \textit{(Incorrect structure)}$$

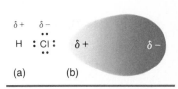

Figure 5.6 Representation of the polar hydrogen chloride molecule. (a) The electron-dot formula, with the shared electron pair shown nearer the chlorine atom. The symbols $\delta+$ and $\delta-$ indicate partial positive and partial negative charges, respectively. (b) A diagram depicting the unequal distribution of electron density in the hydrogen chloride molecule.

Instead of having one end of the molecule positive and the other end negative, the electrons would be pulled toward the right in one bond and toward the left in the other. Overall there would be no net dipole. By a **dipole**, we mean a molecule that has a positive end and a negative end.

But water *does* act like a dipole. If you place a sample of water between two electrically charged plates, the water molecules align themselves, one end attracted toward the positive plate and the other end toward the negative plate. To act like a dipole, the molecule must be bent so that the bonds do not cancel one another out.

Such molecules would align themselves between charged plates as shown in Figure 5.7.

There are molecules with polar bonds that do not act like dipoles. Carbon dioxide (CO_2) is one of these. Its structure is

$$:\overset{..}{O}::C::\overset{..}{O}: \quad \text{or} \quad O{=}C{=}O$$

Since oxygen is more electronegative than carbon, the bonds in this molecule should be polar. Carbon dioxide is a linear molecule, however, and the polar covalent bonds cancel one another to give a non-polar molecule.

Many of the properties of compounds—such as melting point, boiling point, and solubility—depend on the polarity of the molecules of the compound.

Example 5.4   Draw the electron dot structure for H_2S.
The electron-dot symbol for hydrogen is H·, and for sulfur, ·$\overset{..}{\underset{.}{S}}$:. We have two hydrogens and one sulfur to combine.

$$\text{H·} \qquad \text{H·} \qquad \text{·}\overset{..}{\underset{.}{S}}\text{:}$$

If the two hydrogens share electrons, they will be satisfied but the sulfur will be left out. If sulfur shares one of its lone electrons with one hydrogen and its other single electron with the other hydrogen, however, everyone ends up happy.

$$\text{H:}\overset{..}{\underset{\text{H}}{S}}\text{:} \quad \text{or} \quad \text{H}{-}\underset{\overset{|}{H}}{S}$$

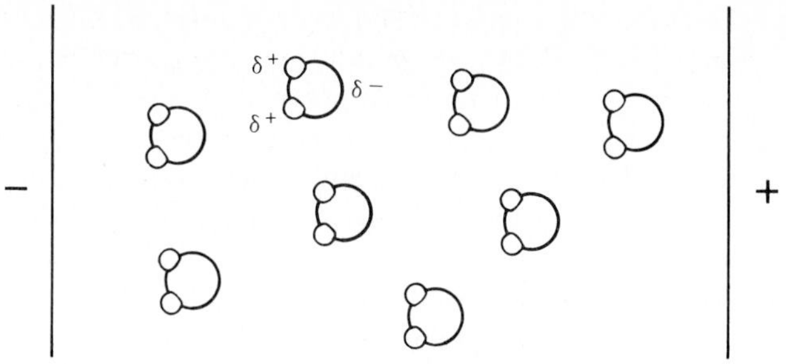

Figure 5.7 Polar molecules are aligned in an electric field, with the positive end of the molecule preferentially pointing toward the negative plate and the negative end of the molecule directed toward the positive plate.

There are several theories of chemical bonding that can be used to account for the shape of the water molecule. One of the simplest, and most satisfying, is the **valence shell electron pair repulsion (VSEPR) theory**. There are four pairs of electrons in the outer (valence) shell of the oxygen atom in the water molecule. Since all electrons bear a like (negative) charge, it is reasonable to expect them to get as far apart as possible. If each pair of electrons is represented as a line extending out from the oxygen atom, the farthest apart these lines can get is 109.5°. Further, if the ends of these lines were all connected, the connecting line would inscribe a regular tetrahedron (Figure 5.8).

The predicted tetrahedral angle of 109.5° is not far off the actual value of 104.5° for water. The disagreement is explained by the fact that only two of the electron pairs in water are shared with hydrogen atoms. The others, called **nonbonding pairs**, occupy a greater volume than do the bonding pairs, thus pushing the latter a little closer together. We take a further look at the unique properties of water in Chapter 16.

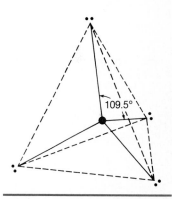

Figure 5.8 The four electron pairs around a central atom point toward the corners of a regular tetrahedron. Each angle is 109.5°.

Ammonia: A Pyramidal Molecule

An atom of the element nitrogen (N) has five electrons in its valence shell. It can assume the neon configuration by sharing pairs of electrons with *three* hydrogen atoms, thereby forming a molecule of ammonia.

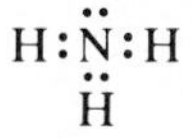

$$H \colon \overset{\displaystyle ..}{\underset{\displaystyle \overset{..}{H}}{N}} \colon H$$

There are four pairs of electrons on the central nitrogen atom in the ammonia molecule. Using the electron pair repulsion theory, we would expect a tetrahedral arrangement of the four pairs and bond angles of 109.5°. The actual bond angles are 107°, close to the theoretical value. Presumably, the unshared pair of electrons occupies a greater volume than does a shared pair, pushing the latter slightly closer together. The arrangement is, therefore, that of a tripod with a hydrogen atom at the

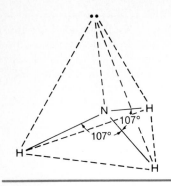

Figure 5.9 The pyramidal ammonia molecule.

end of each leg and the nitrogen atom with its unshared pair sitting at the top (Figure 5.9). Each nitrogen-to-hydrogen bond is somewhat polar, making the ammonia molecule polar.

Ammonia, often written NH_3, is a gas at room temperature. Vast quantities of it are compressed into tanks and used as fertilizer. Ammonia is quite soluble in water. It forms an aqueous solution that is basic; that is, it acts as a base (Chapter 7). Such aqueous preparations are familiar household cleansing solutions.

Methane: A Tetrahedral Molecule

The element carbon (C) has four electrons in the valence shell of each atom. It can assume the neon configuration by sharing electron pairs with four hydrogen atoms, thus forming the compound methane.

$$\text{H} \atop \text{H:C:H} \atop \text{H}$$

There are four pairs of electrons on the central carbon atom in methane. Using the VSEPR theory, we would expect a tetrahedral arrangement and bond angles of 109.5°. The actual bond angles are 109.5°, in perfect agreement with theory (Figure 5.10). All four electron pairs are shared with hydrogen atoms; thus, all four pairs occupy identical volumes.

Each carbon-to-hydrogen bond is slightly polar, but the methane molecule as a whole is symmetrical. The slight bond polarities cancel out, leaving the methane molecule, as a whole, nonpolar. Methane often is represented by the formula CH_4. It is the principal component of natural gas, which is used as a fuel (Chapter 13).

Methane is produced by the decay of plant and animal material. It often is seen bubbling to the surface of swamps—hence, its common name, marsh gas. Bacterial metabolism in the intestinal tract also produces methane, making methane a major component of intestinal gas.

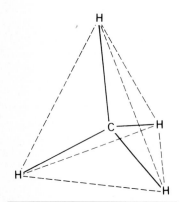

Figure 5.10 The tetrahedral methane molecule. The solid lines indicate covalent bonds; the dotted lines outline the tetrahedron.

Example 5.5   The formula for hydrogen cyanide is HCN. What is the electron-dot structure for this compound?
   The parts are

Hydrogen wants to form one covalent bond (because it has a

single unpaired electron), carbon wants to form four (it has four unpaired electrons), and nitrogen wants to form three (of its five electrons, three are unpaired). The only combination that satisfies everyone is one that incorporates a triple bond.

$$H:C:::\ddot{N} \quad \text{or} \quad H—C≡N$$

Example 5.6 The compound phosgene has the formula $COCl_2$. What is its structure?

Once again we should list the parts.

$$\cdot\dot{C}\cdot \quad \cdot\ddot{O}: \quad :\dot{\ddot{C}l}\cdot \quad :\dot{\ddot{C}l}\cdot$$

Carbon should form four bonds; oxygen, two; the chlorines, one each. The answer is

$$:\ddot{C}l:C:\ddot{C}l: \quad \text{or} \quad Cl—C—Cl$$
$$:\ddot{O}: \qquad\qquad\qquad O$$

Notice that we have an example of a **double bond**, the sharing of two pairs of electrons between carbon and oxygen.

The shapes of molecules are of considerable importance. Rules for predicting these shapes are summarized in Table 5.2.

Example 5.7 What is the shape of the molecules BH_3 and PH_3?

First draw electron-dot structures for each.

$$\begin{array}{cc} H & H \\ \ddot{B}:H & :\ddot{P}:H \\ H & H \end{array}$$

(Boron has only three electrons; BH_3 does not follow the octet rule.) The BH_3 molecule has three bonds and no unshared pairs. The three bonds get as far apart as possible; the molecule is trigonal. The PH_3 molecule has three bonds and one unshared pair; the molecule is pyramidal.

$$\begin{array}{cc} H{\searrow} & \ddot{P} \\ {\quad}B—H & H{\diagup}|{\diagdown}H \\ H{\nearrow} & H \end{array}$$

Table 5.2   Bonding and the Shape of Molecules

Number of Bonds	Number of Unshared Pairs	Shape	Examples
2	0	Linear	$BeCl_2$, $HgCl_2$
3	0	Trigonal	BF_3
4	0	Tetrahedral	CH_4, $SiCl_4$
3	1	Pyramidal	NH_3, PCl_3
2	2	Bent	H_2O, H_2S, SCl_2

Valence Rules

We have seen in the preceding section how different atoms form different numbers of bonds. Hydrogen forms one bond; chlorine, one; oxygen, two; nitrogen, three; and carbon, four. The number of covalent bonds that an atom can form is called its **valence**. Recall that the lowest energy level of an atom can be occupied by at most two electrons. These may be either shared or unshared. The hydrogen atom has one electron. It can form *one* bond by sharing its electron with another atom. The other atom also furnishes one electron to go with the hydrogen electron and form a shared pair. Hydrogen forms one bond and is said to be **univalent**.

The helium (He) atom has a single, filled shell. It cannot share electrons with other atoms; hence, its valence is 0. The second energy level can hold a maximum of eight electrons (i.e., four pairs of electrons). With the element neon (Ne), this shell is filled. Neon has four unshared pairs of electrons and hence forms no bonds. Its valence is 0. The fluorine (F) atom has seven electrons in its outer, or valence, shell. It can, therefore, form one bond by sharing one electron with another element. The other element furnishes one electron to form a shared pair. So fluorine, like hydrogen, is univalent, having three unshared pairs of electrons and the capacity to form one shared pair.

By similar deductions, we can conclude that oxygen is bivalent, nitrogen is trivalent, and carbon is tetravalent. These rules for molecule building are summarized and illustrated in Table 5.3 and Figure 5.11.

Building Molecules

Let's look at a chemical substance called hydrogen peroxide. Chemical analysis shows that each molecule of this substance is composed of two hydrogen atoms and two oxygen atoms. This information

Table 5.3 Valence

Electron Dot Picture	Valence Bond Picture	Number of Bonds (Valence)	Representative Molecules	
H·	H—	1	H—H	H—Cl
He:	—	0	—	
·Ċ·	—Ċ—	4	H—C—H (with H above and H below)	H—C—F (with O double bonded above)
·N̈·	—N—	3	H—N—H (with H below)	N—O—H, H—C—H (C double bonded to N)
·Ö:	—O—	2	H—O—H	H—C—H (with O double bonded above)
·F̈:	—F	1	H—F	F—F
·C̈l:	—Cl	1	Cl—Cl	H—C—Cl (with H above and H below)

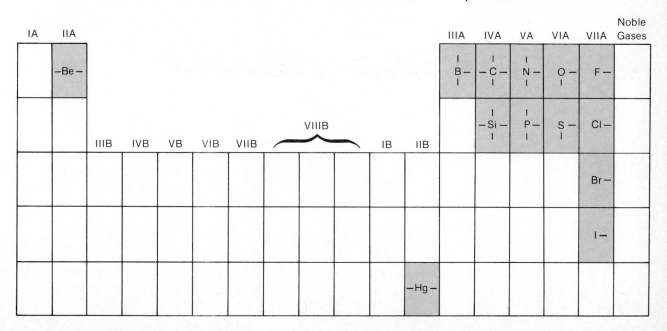

Figure 5.11 Covalent bonding of representative elements of the periodic table.

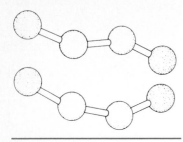

Figure 5.12 Ball-and-stick models of the hydrogen peroxide molecule. One can be made to look like the other by rotation about the bond between the two oxygen atoms.

is indicated in the formula H_2O_2. What is the structure of the hydrogen peroxide molecule? The parts list is

2 each: H—

2 each: —O—

No connections can be made through H, since each H has only one connector. The only kind of molecule we can make, using all connectors, is H—O—O—H. The actual shape of the molecule is shown in Figure 5.12.

Let's build another molecule. Methyl alcohol has the formula CH_4O. The parts list is therefore

4 each: H—

1 each: —C—

1 each: —O—

The only way these parts can be put together according to the rules is

$$\begin{array}{c} \text{H} \\ | \\ \text{H—C—O—H} \\ | \\ \text{H} \end{array}$$

These rules can be used to construct complicated molecules. You should remember them when writing formulas for covalently bonded compounds.

Bonding Forces and the States of Matter

The states of matter—solid, liquid, and gas—are obviously different from one another (Chapter 1). Chemists offer a model to explain these differences. The model is referred to as the **kinetic-molecular theory**. The basic postulates of this theory are

1. All matter is composed of tiny, discrete particles called molecules.
2. Molecules are in constant motion and move in straight lines.
3. The molecules of a gas are very small compared to the distances between them.

4. There is very little attraction between molecules of a gas.

5. Molecules collide with one another, and energy is conserved in these collisions—although one molecule can gain energy at the expense of another.

6. Temperature is a measure of the *average* kinetic energy of the gas molecules.

Right now we need only consider how the model pictures solids, liquids, and gases. *Solids* are viewed as highly ordered assemblies of particles in close contact with one another. *Liquids* are pictured as much more loosely organized collections of particles. In a liquid, the particles are still in close contact with one another, but they are much more free to move about. Finally, in *gases* the particles are no longer in close contact with one another, but are separated by relatively great distances and are moving about at random.

Table salt (sodium chloride) is a typical crystalline solid. In this solid, ionic bonds hold the ions in position and maintain the orderly arrangement. Not all solids are held together by ionic bonds, but some attractive force is necessary to maintain the characteristic orderly array of particles in a solid.

To get a better image of a liquid at the molecular level, think of a box of marbles that is being shaken continuously. The marbles move back and forth, rolling over one another. The particles of a liquid (like the marbles) are not so rigidly held in place as are particles in a solid. The particles in a liquid are being held close together, however, and that means there must be some force attracting them.

Gas particles do not experience significant attractive forces. They move about at great distances from one another and interact only during occasional collisions.

Solids can be changed to liquids; that is, they can be **melted**. The solid is heated, and the heat energy is absorbed by the particles of the solid. The energy causes the particles to vibrate in place with more and more vigor until, finally, the forces holding the particles in a particular arrangement are overcome. The solid has become a liquid. The temperature at which this happens is called the **melting point** of that solid. A high melting point is one indication that the forces holding a solid together are very strong.

A liquid can change to a gas or vapor in a process called **vaporization**. Again, one need only supply sufficient heat to achieve this change. Energy is absorbed by the liquid particles, which move faster and faster as a result. Finally the attractive forces holding the liquid particles in contact are overcome by this increasingly violent motion and the particles fly away from one another. The liquid has become a gas.

The entire sequence of changes can be reversed by removing energy from the sample and slowing down the particles. Vapor changes to

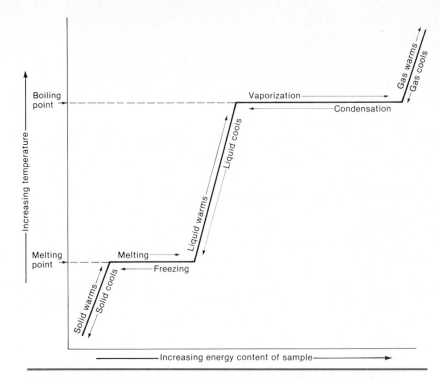

Figure 5.13 Diagram of change in state of matter on heating or cooling.

liquid in a process referred to as **condensation**; liquid changes to solid in a process called **freezing**. Figure 5.13 presents a diagram of the changes in state that occur as energy is added to or removed from a sample.

The amount of energy required to accomplish these changes depends on the type of forces responsible for maintaining the solid or the liquid state. The ionic bonds found in salt crystals are very strong. Sodium chloride must be heated to about 800 °C before it melts. Generally, interionic forces are the strongest of all the forces that hold solids and liquids together. We will now consider some other interactions that hold the particles of solids and liquids together.

Dipole Forces

Hydrogen chloride melts at −112 °C and boils at −85 °C (it is a gas at room temperature). The attractive forces between molecules are not nearly as strong as the interionic forces in salt crystals. We know that covalent bonds hold the hydrogen and chlorine *atoms* together to form the hydrogen chloride *molecule*, but what makes one molecule interact with another in the solid or liquid states?

Remember that the hydrogen chloride molecule is a dipole. It has a positive and a negative end. Two dipoles brought close enough will attract one another. The positive end of one molecule attracts the negative end of another. Such forces may exist throughout the structure of a liquid or solid (Figure 5.14). In general, attractive forces between dipoles are much weaker than the attractive forces between ions. **Dipole interactions** are, however, stronger than the forces between nonpolar molecules of comparable size.

Hydrogen Bonds

Certain polar molecules exhibit stronger attractive forces than expected on the basis of ordinary dipolar interactions. These forces are strong enough to be given a special name, the **hydrogen bond**. Note that "*hydrogen* bond" is a somewhat misleading name, since it emphasizes only one component of the interaction. Not all compounds containing hydrogen exhibit this strong attractive force; the hydrogen *must* be attached to fluorine, oxygen, or nitrogen. It is the presence of these atoms that permits us to offer an explanation for the extra strength of hydrogen bonds as compared with other dipolar forces. Fluorine, oxygen, and nitrogen all have a high electron-attracting power (they are very electronegative), and they are small (they are at the top of the periodic table). A hydrogen–fluorine bond, for example, is very strongly polarized with a negative fluorine end and a positive hydrogen end. Both hydrogen and fluorine are small atoms, so the negative end of one

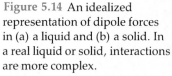

Figure 5.14 An idealized representation of dipole forces in (a) a liquid and (b) a solid. In a real liquid or solid, interactions are more complex.

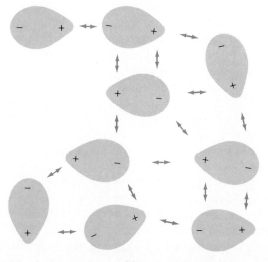

(a)

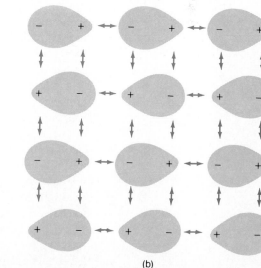

(b)

Figure 5.15 Hydrogen bonding in hydrogen fluoride and in water.

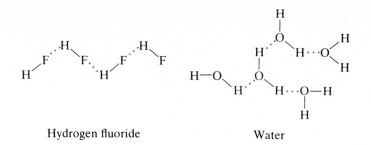

Hydrogen fluoride   Water

Figure 5.15 Hydrogen bonding in hydrogen fluoride and in water.

dipole can approach very closely the positive end of a second dipole. This results in an unusually strong interaction between the two molecules—the hydrogen bond. Hydrogen bonds are often explicitly represented by *dotted* lines to emphasize their unusual strength compared to ordinary dipolar interactions. A *dotted* line is used to distinguish a hydrogen bond from the much stronger covalent bond, which is represented by a *solid* line (Figure 5.15).

Water has both an unusually high melting point and an unusually high boiling point for a compound with molecules of such small size. These abnormal values are attributed to water's ability to form hydrogen bonds. Water is discussed in detail in Chapter 16.

Dispersion Forces

If one understands that positive attracts negative, then it is easy enough to understand how ions or polar molecules maintain contact with one another. But how can we explain the fact that nonpolar compounds can exist in the liquid and solid states? Even hydrogen can exist as a liquid or a solid if the temperature is low enough (its melting point is $-259\,°C$). Some force must be holding these molecules in contact with one another in the liquid and solid states.

Up to this point we have pictured the electrons in a covalent bond as being held in place between the two atoms sharing the bond. But the electrons are *not* really static; they actually move about in the bonds. On the average, the two electrons in the hydrogen molecule (or any nonpolar bond) are between and equidistant from the two nuclei. At any given instant, however, the electrons may be at one end of the molecule. At some other time, a moment later, the electrons may be at the other end of the molecule. At the instant the electrons of one molecule are at one end, the electrons in the next molecule will move away from its adjacent end. Thus, at this instant there will be an attractive force between the electron-rich end of one molecule and the electron-poor end of the next. These momentary, usually weak, attractive forces between molecules are called **dispersion forces**. To a large extent, dispersion forces determine the physical properties of nonpolar compounds.

Forces in Solutions

To complete our look at chemical bonding, we shall briefly examine the interactions that occur in solutions. A **solution** is an intimate, homogeneous mixture of two or more substances. By *intimate* we mean that the mixing occurs down to the level of individual ions and molecules. In a salt–water solution, for example, there are not clumps of ions floating around, but single, randomly distributed ions among the water molecules. **Homogeneous** means that the mixing is thorough. All parts of the solution have the same distribution of components. For the salt solution, the saltiness is the same at the top, bottom, and middle of the solution. The substance being dissolved and usually present in lesser amount (salt in a salt solution) is called the **solute**. The substance doing the dissolving and usually present in greater amount is the **solvent** (water in the salt–water solution) (Figure 5.16).

Ordinarily, solutions form most readily when the substances involved have *similar* bonding characteristics. An old chemical rule is "Like dissolves like." Nonpolar solutes dissolve best in nonpolar solvents. For example, oil and gasoline, both nonpolar, mix (Figure 5.17a), but oil and water do not (Figure 5.17b). Alcohol is a liquid held together

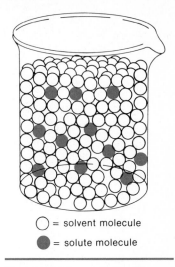

○ = solvent molecule
● = solute molecule

Figure 5.16 In a solution the solute molecules are randomly distributed among the solvent molecules.

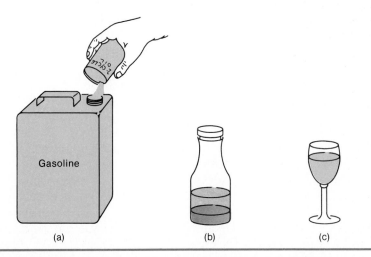

(a) (b) (c)

Figure 5.17 (a) Lawn mowers with two-cycle engines are fueled and lubricated with a solution of nonpolar lubricating oil in nonpolar gasoline. (b) In Italian dressing, polar vinegar and nonpolar olive oil are mixed. However, the two liquids do not form a solution and will separate on standing. (c) Wine is a solution of polar alcohol in polar water.

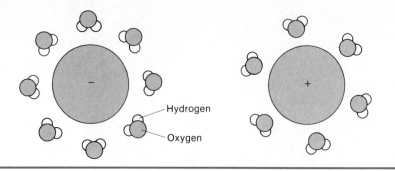

Figure 5.18 The interaction of polar water molecules with ions.

by hydrogen bonds. So is water. Alcohol readily dissolves in water because the two substances can hydrogen bond with one another (Figure 5.17c). In general, a solute dissolves when attractive forces between it and the solvent overcome the attractive forces operating in the pure solute and in the pure solvent.

Why, then, does salt dissolve in water? Ionic solids are held together by strong ionic bonds. We have already indicated that very high temperatures are required to melt ionic solids and break these bonds. Yet, simply by placing sodium chloride in water at room temperature we can dissolve the salt (or, rather, the water can). And when such a solid dissolves, its bonds *are* broken. The difference between the two processes is the difference between brute force and persuasion. In the melting process, we are simply pouring in enough energy (as heat) to pull the crystal apart. In the dissolving process, we offer the ions an attractive alternative to the ionic interactions in the crystal.

It works this way. Water molecules surround the crystal. Those that approach a negative ion align themselves so that the positive ends of their dipoles point toward the ion. With a positive ion the process is reversed, and the negative end of the water dipole points toward the ion. Still, the attraction between a dipole and an ion is not as strong as that between two ions. To compensate for their weaker attractive power, several molecules surround each ion, and in this way the many **ion–dipole** interactions overcome the **ion–ion** interactions (Figure 5.18).

In an ionic solid, the positive and negative ions are strongly bonded together in an orderly crystalline arrangement. In solution, cations and anions move about more or less independently, each surrounded by a cage of solvent molecules. Water—including the water in our bodies— is an excellent solvent for many ionic compounds. Water also dissolves molecules that are polar covalent like itself. These solubility principles explain how nutrients (Chapter 18) get to the cells of our bodies (dissolved in blood, which is mostly water) and how pollutants get into our water supplies (Chapter 16).

Problems

1. What is the structural difference between a sodium atom and a sodium ion?

2. What is the structural difference between a sodium ion and a neon atom? In what way are these two particles similar?

3. What are the structural differences among a chlorine atom, a chlorine molecule, and a chlorine ion?

4. What is wrong with the expression "a molecule of sodium chloride"?

5. Draw electron-dot symbols for each of the following elements. You may use a periodic chart.
   a. sodium   b. fluorine   c. carbon

6. Draw electron-dot symbols for each of the following elements. You may use a periodic chart.
   a. magnesium   b. nitrogen   c. oxygen

7. Draw electron-dot symbols for each of the following elements. You may use a periodic chart.
   a. aluminum   b. potassium   c. chlorine

8. Draw electron-dot formulas for each of the following compounds.
   a. sodium fluoride   b. potassium chloride
   c. potassium fluoride

9. Draw electron-dot formulas for each of the following compounds.
   a. magnesium fluoride   c. calcium chloride
   b. sodium oxide   d. potassium sulfide

10. Draw electron-dot formulas for each of the following compounds.
    a. sodium nitride   b. aluminum chloride

11. Draw electron-dot formulas for each of the following compounds.
    a. magnesium oxide   b. aluminum nitride

12. Draw electron-dot formulas for each of the following compounds.
    a. magnesium nitride   b. aluminum sulfide

13. Classify bonds in each of the following as ionic, polar covalent, or nonpolar covalent.
    a. KF   b. NO   c. F_2

14. Classify bonds in each of the following as ionic, polar covalent, or nonpolar covalent.
    a. IBr   b. I_2 c. KBr

15. Classify bonds in each of the following as ionic, polar covalent, or nonpolar covalent.
    a. MgS   b. NaI   c. HI

16. Consider the hypothetical Elements X, Y, and Z with electron-dot formulas.

$$:\ddot{X}\cdot \qquad :\ddot{Y}\cdot \qquad :\dot{Z}\cdot$$

   a. To which group in the periodic table would each belong?
   b. Write the electron-dot formula for the simplest compound of each with hydrogen.
   c. Write electron-dot formulas for the ions formed when X and Y react with sodium.

17. Solutions of iodine chloride (ICl) are used as disinfectants. Are the molecules of ICl ionic, polar covalent, or nonpolar covalent?

18. Draw electron-dot formulas for the following covalent molecules.
    a. CH_4O b. NOH_3 c. CH_5N d. N_2H_4

19. Draw electron-dot formulas for the following covalent molecules.
    a. NF_3 b. C_2H_4 c. C_2H_2 d. CH_2O

20. Two different covalent molecules are found with the formula C_2H_6O. Draw electron-dot formulas for the two molecules.

21. Shared pairs of electrons can be represented by dashed lines. The electron-dot formula for the hydrogen molecule, H:H can be translated to H—H. Draw dashed-line formulas for the molecules in Problem 18.

22. Draw dashed-line formulas for the molecules in Problem 19.

23. Draw dashed-line formulas for the molecules in Problem 20.

24. Use the valence shell electron pair repulsion (VSEPR) theory to predict the shape of each of the following molecules.
    a. hydrogen sulfide (H_2S)
    b. silane (SiH_4)

25. Use the VSEPR theory to predict the shape of each of the following molecules.
    a. beryllium chloride ($BeCl_2$)
    b. boron fluoride (BF_3)

26. Use the VSEPR theory to predict the shape of each of the following molecules.
    a. arsine (AsH_3)
    b. carbon tetrafluoride (CF_4)

27. Use the valence rules to draw structural formulas for each of the following.
    a. CH_3F b. NH_2Cl c. CCl_4

28. Use the valence rules to draw structural formulas for each of the following.
    a. C_2H_5F b. OF_2 c. NCl_3

29. Use the valence rules to draw structural formulas for each of the following.
    a. HCOF   b. CH_2O c. CH_2O_2

30. Use the valence rules to draw structural formulas for each of the following.
    a. C_2H_3N b. C_3H_4 c. CH_3NO

31. In what ways are liquids and solids similar? In what ways are they different?

32. List four types of interactions between particles in the liquid and in the solid states. Give an example of each type.

33. Define
    a. melting   b. vaporization
    c. condensation   d. freezing

34. In which process is energy absorbed by the material undergoing the change of state?
    a. melting or freezing
    b. condensation or vaporization

35. Label the arrows with the term listed in Problem 33 that correctly identifies the process represented.

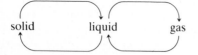

solid   liquid   gas

36. Use the kinetic-molecular theory to explain the properties (Chapter 1) of solids, liquids, and gases.

37. Use the kinetic-molecular theory to explain the process of melting.

38. Use the kinetic-molecular theory to explain the process of vaporization.

39. In which of the following would hydrogen bonding be an important intermolecular force?

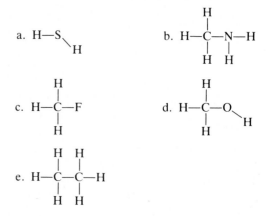

40. Define
    a. solution   b. solute   c. solvent

41. Alcohol that is used as a disinfectant to clean the skin prior to an injection is actually a solution of 3 parts of water and 7 parts of alcohol. Which component is the solvent and which is the solute?

42. Explain why a salt dissolves in water.

43. Benzene (C_6H_6) is a nonpolar solvent. Would you expect NaCl to dissolve in benzene? Explain.

44. Motor oil is nonpolar. Would you expect it to dissolve in water? In benzene? Explain.

References and Readings

1. Asimov, Isaac. *A Short History of Chemistry*. Garden City, NY: Doubleday, 1965. Chapter 7, "Molecular Structure."
2. Benfey, O. T. *Classics in the Theory of Chemical Combination*. New York: Dover Publications, 1963. A history of bonding.
3. Companion, Audrey. *Chemical Bonding*, 2d ed. New York: McGraw-Hill, 1979.
4. Dahl, Peter. "The Valence-Shell Electron-Pair Repulsion Theory." *Chemistry*, March 1973, pp. 17–19.
5. Pauling, L., and R. Hayward. *The Architecture of Molecules*. San Francisco: W. H. Freeman, 1964. Parts 1–14. Easy reading, great artwork.
6. Price, Charles C. *Geometry of Molecules*. New York: McGraw-Hill, 1971. The shapes of molecules from the small to the very large.

Names, Formulas, and Equations

The Language of Chemistry

Have you ever listened to people speaking a language you didn't understand? To you, it probably seemed just a collection of meaningless sounds, but, to the speakers, the language was perfectly intelligible. The language chemists use is much like a foreign language. An equation like

$$4\,Fe \ + \ 3\,O_2 \ \longrightarrow \ 2\,Fe_2O_3$$

probably won't mean much to you unless you already have some knowledge of chemical equations. But, if I translated the chemical symbols into English—"iron combines with oxygen (air) to form iron rust—you'd probably understand my meaning.

There are two ways to find out what is being said in a foreign language: you can hire a translator or you can learn to understand the

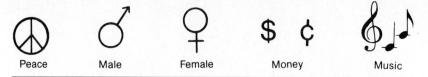

Figure 6.1 The use of symbols is not unique to chemistry. Symbols can be quite helpful—when you know what they mean.

language yourself. To enable you to learn the language of chemistry (so you won't have to depend on a translator), we offer in this chapter a short, intensive course. It won't make you a chemist, but it will make it possible for you to understand a little more of what those mysterious strangers—the chemists—are talking about.

One minor caution before we proceed: the language of chemists often is closely related to that spoken by mathematicians. Don't worry, though; the math that we use is much like the arithmetic you use in everyday life. So hang in there. Before you know it, you, too, will be able to speak the language of chemists.

Names and Symbols for Simple Ions

In Chapter 1, you were introduced to chemical symbols. A symbol of one or two letters is used to represent each of the chemical elements. If you have not yet memorized the list in Table 1.1, you should do so now. You will save time in the long run.

In the last chapter, we saw how certain metals (those from the left side of the periodic chart) react with nonmetals (those from the right side) to form ionic compounds. Recall that in forming compounds each atom of metal tends to give up the electrons in its outer shell, and each atom of nonmetal takes on enough electrons to complete its valence shell. For example, aluminum atoms with three electrons in their outermost energy levels give up three electrons to form triply charged ions. In electron-dot symbols this reaction may be written

$$\cdot \overset{\textstyle \cdot}{Al} \cdot \longrightarrow Al^{3+} + 3\,e^-$$

Oxygen, with six electrons in its outermost energy level, tends to acquire two more.

$$:\overset{\textstyle \cdot \cdot}{O}\cdot\ +\ 2\,e^- \longrightarrow :\overset{\textstyle \cdot \cdot}{\underset{\textstyle \cdot \cdot}{O}}:^{2-}$$

The charged atoms formed by the gain or loss of elctrons are called **ions**. Table 6.1 lists symbols and names for some simple ions formed in this manner. Note that the charge on an ion of a Group IA element is

1+ (usually written simply as +). The charge on an ion of a Group IIA element is 2+, and the charge on an ion of a Group IIIA element is 3+. You can calculate the charge on the negative ions in the table by subtracting 8 from the group number. For example, the charge on the oxide ion (oxygen is in Group VIA) is $6 - 8 = -2$. The charge on a nitride ion (nitrogen is in Group VA) is $5 - 8 = -3$.

There is no simple way to determine the most likely charge on ions formed from group VIII elements and from those in B sub-groups. Indeed, you may have noticed that these can form ions with different charges. In such cases, chemists use roman numerals with the names to indicate the charge. Thus, *iron(II) ion* means Fe^{2+} and *iron(III) ion* means Fe^{3+}. In an older system of terminology, Fe^{2+} was called a *ferrous ion* and Fe^{3+} was called a *ferric ion*. See similar terms for the two copper ions in Table 6.1.

Names of simple positive ions (*cations*) are derived from those of their parent elements by the addition of the word *ion*. A sodium atom (Na), upon losing an electron, becomes a *sodium ion* (Na^+). A magnesium atom (Mg), upon losing two electrons, becomes a *magnesium ion* (Mg^{2+}). Names of simple negative ions (*anions*) are derived from those of their parent elements by changing the usual ending to *-ide* and adding the word *ion*. A chlor*ine* atom (Cl), upon gaining an electron, becomes a chlor*ide ion* (Cl^-). A sulf*ur* atom (S), upon gaining two electrons, becomes a sulf*ide ion* (S^{2-}).

Table 6.1 Symbols and Names for Some Simple Ions

Group	Element	Name of Ion	Symbol for Ion
IA	Hydrogen	Hydrogen ion*	H^+
	Lithium	Lithium ion	Li^+
	Sodium	Sodium ion	Na^+
	Potassium	Potassium ion	K^+
IIA	Magnesium	Magnesium ion	Mg^{2+}
	Calcium	Calcium ion	Ca^{2+}
IIIA	Aluminum	Aluminum ion	Al^{3+}
VA	Nitrogen	Nitride ion	N^{3-}
VIA	Oxygen	Oxide ion	O^{2-}
	Sulfur	Sulfide ion	S^{2-}
VIIA	Chlorine	Chloride ion	Cl^-
	Bromine	Bromide ion	Br^-
	Iodine	Iodide ion	I^-
IB	Copper	Copper(I) ion (cuprous ion)	Cu^+
		Copper(II) ion (cupric ion)	Cu^{2+}
	Silver	Silver ion	Ag^+
IIB	Zinc	Zinc ion	Zn^{2+}
VIIIB	Iron	Iron(II) ion (ferrous ion)	Fe^{2+}
		Iron(III) ion (ferric ion)	Fe^{3+}

* Does not exist independently in aqueous solution.

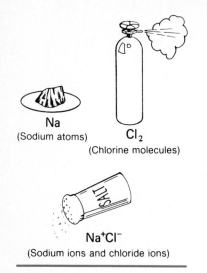

Na
(Sodium atoms)

Cl₂
(Chlorine molecules)

Na⁺Cl⁻
(Sodium ions and chloride ions)

Figure 6.2 Ions differ greatly from the atoms from which they are made. Sodium atoms are the constituents of a soft, highly reactive metal. Chlorine atoms—paired in chlorine molecules—make up a corrosive greenish yellow gas. Sodium ions and chloride ions make up ordinary table salt.

We cannot emphasize too strongly the difference between ions and the atoms from which they are made. They are as different as a whole peach (an atom) and a peach pit (an ion). The names and symbols may look a lot alike, but the substances themselves are quite different. Unfortunately, the situation is confused because people talk about needing "iron" to perk up "tired blood" and "calcium" for healthy teeth and bones. What they really mean is iron(II) *ions* (Fe^{2+}) and calcium *ions* (Ca^{2+}). You wouldn't think of eating iron nails to get "iron." Nor would you eat highly reactive calcium metal. Although careful distinction is not always made by persons who are not chemists, we try to use precise terminology here.

Formulas and Names for Binary Ionic Compounds

Simple ions of opposite charge can be combined to form **binary** (two-component) **compounds**. To get the correct formula for a binary compound, simply write each ion with its charge (positive ion to the left), then cross over the numbers (but not the plus or minus signs) and write them as subscripts. The process is best learned by practice. Work through Examples 6.1–6.5, then do Problems 13–16 at the end of the chapter.

Example 6.1   Write the formula for calcium chloride.
First, write the symbols for the ions.

$$Ca^{2+} \quad Cl^{1-}$$

Then cross over the numbers as subscripts.

$$Ca_1^{2+} \quad\quad Cl_2^{1-}$$

Then rewrite the formula, dropping the charges. The formula for calcium chloride is

$$Ca_1Cl_2 \quad\text{or}\quad CaCl_2$$

Example 6.2 Write the formula for aluminum oxide.
Write the symbols for the ions.

$$Al^{3+} \quad O^{2-}$$

Cross over the numbers as subscripts.

$$Al_2^{3+} \quad O_3^{2-}$$

Then rewrite the formula, dropping the charges. The formula for aluminum oxide is

$$Al_2O_3$$

Note that the cross-over method works because it is based on the transfer of electrons and the conservation of charge. Two aluminum atoms lose three electrons each (that's six electrons lost), and three oxygen atoms gain two electrons each (that's six electrons gained). Electrons lost equal electrons gained and all is well. Similarly, two aluminum ions have six positive charges (three each) and three oxide ions have six negative charges (two each). The net charge on aluminum oxide is 0, just as it should be.

Example 6.3 Determine the formula for magnesium oxide.
The ions are

$$Mg^{2+} \quad O^{2-}$$

Crossing over, we get

$$Mg_2^{2+} \quad O_2^{2-}$$

Dropping the charges, we get

$$Mg_2O_2$$

Such formulas usually are reduced to the lowest ratio, and we write magnesium oxide as MgO.

Now you are able to translate the "English," such as aluminum oxide, into the "chemistry," Al_2O_3. You also can translate in the other direction.

Example 6.4   What is the name for MgS?
Table 6.2 tells us that MgS is made up of Mg^{2+} (magnesium ions) and S^{2-} (sulfide ions). The name is simple magnesium sulfide.

Example 6.5   What is the name for $FeCl_3$?
The ions are

$$Fe^{3+} \quad Cl^-$$

(How do we know the iron is Fe^{3+} and not Fe^{2+}? Since there are three Cl ions, each $1-$, the one Fe ion must be $3+$ because the compound $FeCl_3$ is neutral.) The names of these ions are iron(III) ion (or ferric ion) and chloride ion, Therefore, the compound is iron(III) chloride (or, by the older system, ferric chloride).

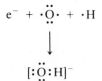

Acetate ion   Ammonium ion

Hydrogen carbonate ion
(Bicarbonate ion)

Carbonate ion   Nitrite ion

Figure 6.3 Polyatomic ions have both covalent bonds (dashes) and ionic charges ($+$ or $-$).

Polyatomic Ions

Many compounds contain both ionic and covalent bonds. Sodium hydroxide, commonly known as lye, consists of sodium ions (Na^+) and hydroxide ions (OH^-). The hydroxide ion contains an oxygen atom covalently bonded to a hydrogen atom, plus an "extra" electron. Whereas the sodium atom becomes a cation by giving up an electron, the hydroxide group becomes an anion by gaining an electron.

$$e^- + \cdot \ddot{O} \cdot + \cdot H$$
$$\downarrow$$
$$[:\ddot{O}:H]^-$$

The formula for sodium hydroxide is NaOH—for each sodium ion there is one hydroxide ion.

Table 6.2 Some Common Polyatomic Ions

Charge	Name	Formula
1+	Ammonium ion	NH_4^+
	Hydronium ion	H_3O^+
1−	Hydrogen carbonate (bicarbonate) ion	HCO_3^-
	Hydrogen sulfate (bisulfate) ion	HSO_4^-
	Acetate ion	$CH_3CO_2^-$ (or $C_2H_3O_2^-$)
	Nitrite ion	NO_2^-
	Nitrate ion	NO_3^-
	Cyanide ion	CN^-
	Hydroxide ion	OH^-
	Dihydrogen phosphate ion	$H_2PO_4^-$
	Permanganate ion	MnO_4^-
2−	Carbonate ion	CO_3^{2-}
	Sulfate ion	SO_4^{2-}
	Monohydrogen phosphate ion	HPO_4^{2-}
	Oxalate ion	$C_2O_4^{2-}$
	Dichromate ion	$Cr_2O_7^{2-}$
3−	Phosphate ion	PO_4^{3-}

There are many groups like the hydroxide ion in nature. They hang together through most chemical reactions. **Polyatomic ions** are charged particles containing two or more covalently bonded atoms. A list of the common polyatomic ions is given in Table 6.2. You can use these ions, in combination with the ions in Table 6.1, to write formulas for compounds containing polyatomic ions.

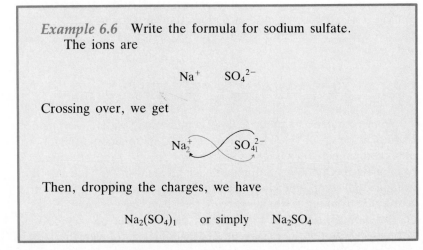

Example 6.6 Write the formula for sodium sulfate.
The ions are

$$Na^+ \qquad SO_4^{2-}$$

Crossing over, we get

$$Na_2^+ \qquad SO_{4}^{2-}$$

Then, dropping the charges, we have

$$Na_2(SO_4)_1 \qquad \text{or simply} \qquad Na_2SO_4$$

Example 6.7 What is the formula for ammonium sulfide? The ions are

$$NH_4^+ \qquad S^{2-}$$

Crossing over, we get

$$NH_4^+ \; \diagup\!\!\!\!\diagdown \; S^{2-}$$

Dropping the charges gives

$$(NH_4)_2S$$

The parentheses with a subscript 2 indicate that the entire ammonium unit is taken twice; there are two nitrogen atoms and eight ($4 \times 2 = 8$) hydrogen atoms.

Example 6.8 What is the name for $CaCO_3$?
The ions are Ca^{2+} (calcium ion) and CO_3^{2-} (carbonate ion). The compound is calcium carbonate.

Example 6.9 What is the name for $Mg(HCO_3)_2$?
The ions are Mg^{2+} (magnesium ion) and HCO_3^- (bicarbonate ion). The compound is magnesium bicarbonate.

Names for Covalent Compounds

Many covalent compounds have common and widely used names. Examples are water (H_2O), methane (CH_4), and ammonia (NH_3). For other compounds, the Greek prefixes *mono-*, *di-*, *tri-*, and *tetra-* are used to indicate the number of atoms of each element in the molecule. For example, the compound N_2O_4 is called *dinitrogen tetroxide*. (The *a* often is dropped from tetra- when it precedes another vowel.) We often leave off the mono- prefix (NO_2 is nitrogen dioxide), but do include it to distinguish between two compounds of the same pair of elements (CO is carbon monoxide, CO_2 is carbon dioxide).

Example 6.10   What is the name for SCl_2? For BF_3?
SCl_2 has one sulfur and two chlorine atoms; it is sulfur dichloride. BF_3 is boron trifluoride.

Example 6.11   Write the formula for carbon tetrachloride.
The name indicates one carbon atom and four chlorine atoms. The formula is CCl_4.

Chemical Equations

Chemistry is a study of matter and the changes it undergoes. More than that, it is a study of the energy that brings about those changes—or the energy that is released when those changes occur. So far we have discussed the symbols and formulas have been invented to represent elements and compounds. Now let's look at a shorthand way of describing chemical changes—the **chemical equation**.

Carbon reacts with oxygen to form carbon dioxide. In the chemical shorthand, this reaction is written

$$C + O_2 \longrightarrow CO_2$$

The plus sign (+) indicates addition of carbon to oxygen (or vice versa) or a mixing of the two in some manner. The arrow ($\rightarrow$) is often read "yields." Substances on the left of the arrow are **reactants**, or **starting materials**. Those on the right are the **products** of the reaction. The conventions here are like those we used in writing nuclear equations (Chapter 4). Now, however, the nucleus will remain untouched. The chemical reactions we are going to look at involve only electronic structures.

Chemical equations have meaning on the atomic and molecular level. The equation

$$C + O_2 \longrightarrow CO_2$$

means that one atom of carbon (C) reacts with one molecule of oxygen (O_2) to produce one molecule of carbon dioxide (CO_2).

Not all chemical reactions are simply represented. Hydrogen reacts with oxygen to form water. We can write this reaction as

$$H_2 + O_2 \longrightarrow H_2O \qquad (\textit{Not balanced})$$

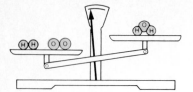

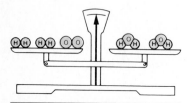

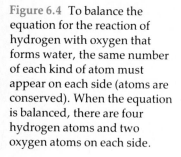

Figure 6.4 To balance the equation for the reaction of hydrogen with oxygen that forms water, the same number of each kind of atom must appear on each side (atoms are conserved). When the equation is balanced, there are four hydrogen atoms and two oxygen atoms on each side.

This representation, however, is not consistent with the law of conservation of matter. There are two oxygen atoms shown among the reactants (as O_2), and only one among the products (in H_2O). For the equation to represent correctly the chemical happening, it must be *balanced*. To balance the oxygen atoms we need only place the coefficient 2 in front of the formula for water.

$$H_2 + O_2 \longrightarrow 2\,H_2O \qquad (\textit{Not balanced})$$

This coefficient means that there are *two* molecules of water involved. As is the case with subscripts, a coefficient of 1 is understood when no other number appears. A coefficient preceding a formula multiplies everything in the formula. In the above equation, the coefficient 2 not only increases the number of oxygen atoms to two, but also increases the number of hydrogen atoms to four.

But the equation is still not balanced. We took care of oxygen at the expense of messing up hydrogen. To balance hydrogen, we place a coefficient 2 in front of the H_2.

$$2\,H_2 + O_2 \longrightarrow 2\,H_2O \qquad (\textit{Balanced})$$

Now there are enough hydrogen atoms on the left. In fact, there are four hydrogens and two oxygens on each side of the equation. Atoms are conserved: the equation is balanced (Figure 6.4).

Note that we could not balance the equation by changing the subscript for oxygen in water.

$$H_2 + O_2 \longrightarrow H_2O_2 \qquad (\textit{Not correct})$$

The equation would be balanced, but it would not mean "hydrogen reacts with oxygen to form *water*." The formula H_2O_2 represents *hydrogen peroxide*, a different compound than water. We cannot change compounds in a chemical equation merely for the convenience of balancing chemical equations.

Example 6.12   Balance the following equation.

$$N_2 + H_2 \longrightarrow NH_3 \qquad (\textit{Not balanced})$$

For this sort of problem we will use the concept of the least common multiple. We will balance the hydrogen first. There are two hydrogens on the left and three and on the right. The least common multiple of 3 and 2 is 6. Six will be the smallest number of hydrogens that can be evenly converted from reactants to products. Thus, we need three molecules of H_2 and two of NH_3.

$$N_2 + 3 H_2 \longrightarrow 2 NH_3 \quad (Balanced)$$

We have balanced the hydrogens and, in the process, the nitrogens. There are two nitrogen atoms on the left and two on the right. The entire equation is balanced (Figure 6.5)!

Example 6.13 Balance the following equation.

$$Fe + O_2 \longrightarrow Fe_2O_3 \quad (Not\ balanced)$$

Begin by balancing the oxygen. The least common multiple is 6. We need three molecules of O_2 and two of Fe_2O_3.

$$Fe + 3 O_2 \longrightarrow 2 Fe_2O_3 \quad (Not\ balanced)$$

We now have four atoms of iron on the right side. We can get four on the left by placing the coefficient 4 in front of Fe.

$$4 Fe + 3 O_2 \longrightarrow 2 Fe_2O_3 \quad (Balanced)$$

The equation is now balanced.

Example 6.14 Balance the following equation.

$$CH_4 + O_2 \longrightarrow CO_2 + H_2O \quad (Not\ balanced)$$

In this equation, oxygen appears in two different products; we will leave the oxygen for last and balance the other two elements first. Carbon is already balanced, with one atom on each side of the equation. The least common multiple of 2 and 4 is 4, so to balance hydrogen we place the coefficient 2 in front of H_2O. Now we have four hydrogens on each side.

$$CH_4 + O_2 \longrightarrow CO_2 + 2 H_2O \quad (Not\ balanced)$$

Now for the oxygen. There are four oxygens on the right. If we place a 2 in front of O_2, the oxygens balance.

$$CH_4 + 2 O_2 \longrightarrow CO_2 + 2 H_2O \quad (Balanced)$$

The equation is balanced.

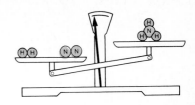
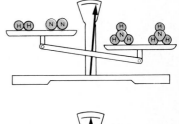
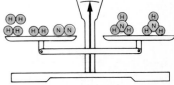

Figure 6.5 To balance the equation for the reaction of nitrogen with hydrogen that forms ammonia, the same number of each kind of atom must appear on each side of the equation. When the equation is balanced, there are two nitrogen atoms and six hydrogen atoms on each side.

Example 6.15 Balance the following equation.

$$H_2SO_4 + NaCN \longrightarrow HCN + Na_2SO_4 \qquad (\textit{Not balanced})$$

Here we have an equation that involves compounds with polyatomic ions. The SO_4 group should be treated as a unit and balanced as a whole. The same is true of the CN group. As the equation is presently written, the SO_4 groups and the CN groups are balanced, but the hydrogens and the sodiums are not. The least common multiple for sodium is 2, so to get two sodiums on the left we place a 2 before NaCN. The least common multiple for hydrogen is 2, so to get two hydrogens on the right we place a 2 before HCN.

$$H_2SO_4 + 2\,NaCN \longrightarrow 2\,HCN + Na_2SO_4 \qquad (\textit{Balanced})$$

The same coefficients balance the CN groups and the SO_4 groups, and the entire equation is balanced.

We have made the task of balancing equations deceptively easy by considering simple reactions. It is more important for you to understand the principle than to be able to balance complicated equations. You should know what is meant by a balanced equation and be able to handle simple systems.

Chemical Arithmetic and the Mole

We said that an equation is meaningful on the atomic and the molecular level. The equation

$$C + O_2 \longrightarrow CO_2$$

says that one carbon atom reacts with the oxygen molecule to produce one carbon dioxide molecule. One carbon atom weighs 12 amu. One oxygen molecule weighs 32 amu (2 × 16 amu per oxygen atom). Thus we also know from the equation that 12 amu of carbon reacts with 32 amu of oxygen. In fact, 12 mass units of carbon react with 32 mass units of oxygen no matter what the mass units are (amu, grams, pounds, anything). If 100 atoms of carbon react with 100 molecules of oxygen, or if 1 million react with 1 million, or if 10^{20} react with 10^{20}, or if 7328 react with 7328, the ratio of the mass of the carbon to the mass of the oxygen will be 12 to 32—because the ratio of the masses of the individual reactants is 12:32.

Carbon and oxygen can react to form a different product, carbon monoxide.

$$2\,C\ +\ O_2\ \longrightarrow\ 2\,CO$$

In this reaction carbon reacts with oxygen in a mass ratio of 24:32 because two carbon atoms, each weighing 12 amu, react with one oxygen molecule.

It should be clear by now that the atomic masses, or **formula masses**, of the species involved in reactions are important, and you should be able to calculate them.

Example 6.16   What is the formula mass of ozone, O_3?

Formula masses are calculated by adding together the masses of the constituent atoms. The ozone molecule contains three oxygen atoms, each of which weighs 16 amu. Therefore, the formula mass of ozone is

$$3 \times 16 = 48 \text{ amu}$$

Example 6.17   What is the formula mass of ammonia, NH_3?

Nitrogen atoms weigh 14 amu, and hydrogen atoms weigh 1 amu. There are three hydrogen atoms in the molecule; therefore, the formula mass is

$$14 + (3 \times 1) = 17 \text{ amu}$$

Example 6.18   What is the formula mass of glucose, $C_6H_{12}O_6$?

Carbon weighs 12 amu; hydrogen 1 amu; oxygen 16 amu. The formula mass of glucose is

$$(6 \times 12) + (12 \times 1) + (6 \times 16) =$$
$$72\ \ +\ \ 12\ \ +\ \ 96\ \ = 180 \text{ amu}$$

Example 6.19 What is the formula mass of ammonium sulfate, $(NH_4)_2SO_4$?

It is imperative that you understand the meaning of parentheses in a formula. $(NH_4)_2$ means that everything within the parentheses is doubled. Ammonium sulfate contains

$$
\begin{array}{llll}
2\,N & = 2 \times 14 & = & 28 \\
8\,H & = 8 \times 1 & = & 8 \\
1\,S & = 1 \times 32 & = & 32 \\
4\,O & = 4 \times 16 & = & 64 \\
\hline
(NH_4)_2SO_4 & & = & 132 \text{ amu}
\end{array}
$$

Because it is not yet possible to measure units as small as individual atoms or molecules, scientists have defined a convenient larger unit, the **mole** (mol). Officially the mole is defined as the amount of a substance containing as many elementary units as there are atoms in exactly 12 g of the carbon-12 isotope. That probably does not sound like a convenient unit to you. You may change your mind, however, when you realize that, in practice, it boils down to the fact that the mass of a mole of any chemical is simply the atomic or formula mass expressed in grams. One mole of carbon weighs 12 g; 1 mol of oxygen molecules weighs 32 g; 1 mol of the compound ammonium sulfate (Example 6.19) weighs 132 g.

How many carbon atoms are there in 12 g of carbon? The number of atoms in 12 g of carbon is unimaginably large. In exponential form it is written as 6.02×10^{23}. Scientists love to find ways to make very large numbers meaningful. Here is one way: if you had a fortune worth 6.02×10^{23} dollars, you could spend a billion dollars each second of your entire life and have used only about 0.001% of your money. Here is another: if carbon atoms were the size of peas, 6.02×10^{23} of them would cover the surface of the Earth to a depth of 15 m, easily burying the Jolly Green Giant (Figure 6.6).

The number 6.02×10^{23} is called **Avogadro's number**. There are 6.02×10^{23} carbon atoms in 1 mol (12 g) of carbon. There are 6.02×10^{23} oxygen molecules in 1 mol (32 g) of oxygen. There are 6.02×10^{23} formula units of ammonium sulfate in 1 mol (132 g) of ammonium sulfate.

The mole used in chemistry is something like the dozen we use every day. You can have a dozen eggs, a dozen strips of bacon, or a dozen breakfasts. In each case, you have 12 of whatever you have specified. A mole simply means of 6.02×10^{23} of whatever you are talking about.

$$1 \text{ mol} = 6.02 \times 10^{23} \text{ units} = 1 \text{ gram formula mass}$$

Back to our favorite equation.

Figure 6.6 The Earth—and the Jolly Green Giant—covered with 6.02×10^{23} peas.

HO HO HOoooooooo

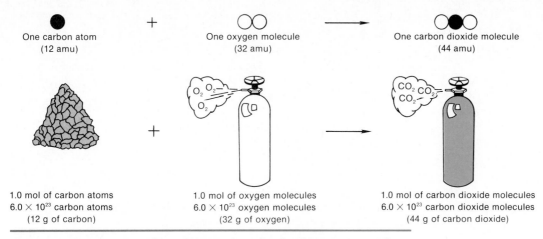

One carbon atom (12 amu) + One oxygen molecule (32 amu) ⟶ One carbon dioxide molecule (44 amu)

1.0 mol of carbon atoms
6.0×10^{23} carbon atoms
(12 g of carbon)

1.0 mol of oxygen molecules
6.0×10^{23} oxygen molecules
(32 g of oxygen)

1.0 mol of carbon dioxide molecules
6.0×10^{23} carbon dioxide molecules
(44 g of carbon dioxide)

Figure 6.7 We cannot weigh single atoms or molecules, but we can weigh equal numbers of these fundamental particles.

$$C + O_2 \longrightarrow CO_2$$

The equation now says that (Figure 6.7)

> 1 atom of carbon reacts with 1 molecule of oxygen to yield 1 molecule of carbon dioxide

or that

> 12 amu of carbon reacts with 32 amu of oxygen to yield 44 amu of carbon dioxide

or that

> 12 g of carbon reacts with 32 g of oxygen to yield 44 g of carbon dioxide

or that

> 1 mol of carbon atoms reacts with 1 mol of oxygen molecules to yield 1 mol of carbon dioxide molecules

We previously referred to a second reaction of carbon with oxygen that produced carbon monoxide.

$$2\,C + O_2 \longrightarrow 2\,CO$$

This equation now tells us that

> 2 atoms of carbon react with 1 molecule of oxygen to yield 2 molecules of carbon monoxide

or that

24 amu of carbon reacts with 32 amu of oxygen to yield 56 amu of carbon monoxide (2 molecules of CO weighing 28 amu each)

or that

24 g of carbon reacts with 32 g of oxygen to yield 56 g of carbon monoxide

or that

2 mol of carbon atoms reacts with 1 mol of oxygen molecules to yield 2 mol of carbon monoxide molecules

Notice that the coefficients in the equation give you directly the combining ratio of atoms and molecules or the combining ratio of moles, but they do not give directly the combining ratio in grams. The formula masses must always be taken into account when dealing with grams.

Another point: although one cannot work with fractions of a molecule or atom, one can work with fractions of a mole. There's no such thing as half an atom of carbon, but half a mole is perfectly reasonable. Half a mole of carbon atoms weighs 6 g and contains 3.01×10^{23} atoms.

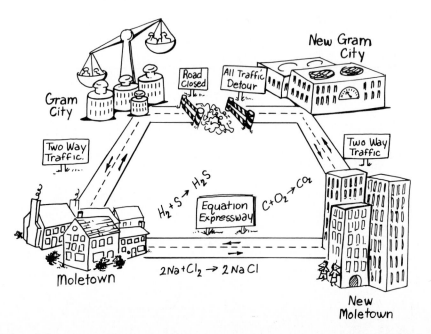

Figure 6.8 The chemical equation relates moles of reactants to moles of products. For mass relationships between reactants and products, we must convert masses to moles, then moles back to masses.

Example 6.20   What is the mass of half a mole of carbon dioxide, CO_2?

$$1 \text{ mol} = 1 \text{ gram formula mass}$$

$$1 \text{ mol } CO_2 = 44 \text{ g } CO_2$$

This equivalence can be used as a conversion factor.

$$0.5 \text{ mol } CO_2 \times \frac{44 \text{ g } CO_2}{1 \text{ mol } CO_2} = 22 \text{ g } CO_2$$

Example 6.21   What is the mass of 3 mol of carbon monoxide?

The formula mass of carbon monoxide, CO, is 28. Therefore,

$$1 \text{ mol } CO = 28 \text{ g } CO$$

and

$$3 \text{ mol } CO \times \frac{28 \text{ g } CO}{1 \text{ mol } CO} = 84 \text{ g } CO$$

Example 6.22   How many moles of carbon monoxide are there in 7 g of carbon monoxide?

You need only invert the conversion factor used in Example 6.20 to convert from grams to moles.

$$7 \text{ g } CO \times \frac{1 \text{ mol } CO}{28 \text{ g } CO} = 0.25 \text{ mol } CO$$

Example 6.23 How many moles of ammonium sulfate are there in 1.32 g of the compound?

From Example 6.19 we know that the formula weight of ammonium sulfate, $(NH_4)_2SO_4$, is 132. Thus,

$$1 \text{ mol } (NH_4)_2SO_4 = 132 \text{ g } (NH_4)_2SO_4$$

$$1.32 \text{ g} \times \frac{1 \text{ mol}}{132 \text{ g}} = 0.01 \text{ mol } (NH_4)_2SO_4$$

Once again let us consider the equation

$$C + O_2 \longrightarrow CO_2$$

Carbon and oxygen combine in a ratio of $1:1$. That means 1 mol to 1 mol, 0.5 mol to 0.5 mol, or 10 mol to 10 mol. A $1:1$ ratio just means that the molar amounts must be equal. For

$$2\,C + O_2 \longrightarrow 2\,CO$$

the mole ratio of reactants is $2:1$—that is, 2 mol of carbon to 1 mol of oxygen, or 0.5 mol of carbon to 0.25 mol of oxygen, or 10 mol of carbon to 5 mole of oxygen.

Example 6.24   According to the equation

$$2\,C + O_2 \longrightarrow 2\,CO$$

how many moles of carbon will react with 0.2 mol of oxygen?
The coefficients in the equation can be used to provide a convenient conversion factor.

$$2 \text{ mol C} = 1 \text{ mol O}_2$$
$$\text{(react with)}$$

$$0.2 \text{ mol O}_2 \times \frac{2 \text{ mol C}}{1 \text{ mol O}_2} = 0.4 \text{ mol C}$$

Example 6.25   According to the equation

$$C_3H_8 + 5\,O_2 \longrightarrow 3\,CO_2 + 4\,H_2O$$

how many moles of CO_2 will be produced in the reaction of 1 mol of O_2? How many grams?

$$5 \text{ mol O}_2 = 3 \text{ mol CO}_2$$
$$\text{(yield)}$$

$$1 \text{ mol O}_2 \times \frac{3 \text{ mol CO}_2}{5 \text{ mol O}_2} = 0.6 \text{ mol CO}_2$$

To find the number of grams, just use the gram formula mass (1 mol CO_2 = 44 g CO_2).

$$0.6 \text{ mol CO}_2 \times \frac{44 \text{ g CO}_2}{1 \text{ mol CO}_2} = 26 \text{ g CO}_2$$

Example 6.26 According to the equation in Example 6.25, how many grams of propane, C_3H_8, will react with 3.2 g of O_2?

Always break up problems like this into workable parts. To use the equation you must deal in moles, but you are both given and asked for amounts in grams. First, the amount given in grams must be converted to moles, using the gram formula mass (Figure 6.8). According to the problem, 3.2 g of O_2 react. The formula mass of O_2 is 32, so

$$3.2 \text{ g } O_2 \times \frac{1 \text{ mol } O_2}{32 \text{ g } O_2} = 0.1 \text{ mol } O_2$$

Now you can use the coefficients from the equation to construct a conversion factor relating moles of O_2 to moles of propane. According to the equation

$$1 \text{ mol } C_3H_8 = 5 \text{ mol } O_2$$
$$\text{(reacts with)}$$

$$0.1 \text{ mol } O_2 \times \frac{1 \text{ mol } C_3H_8}{5 \text{ mol } O_2} = 0.02 \text{ mol } C_3H_8$$

Finally, you can convert the moles of C_3H_8 to grams using the gram formula mass of C_3H_8.

$$1 \text{ mol } C_3H_8 = 44 \text{ g } C_3H_8$$

Therefore,

$$0.02 \text{ mol } C_3H_8 \times \frac{44 \text{ g } C_3H_8}{1 \text{ mol } C_3H_8} = 0.88 \text{ g } C_3H_8$$

To become proficient in problems of this sort, you should work as many of the problems as you can. If you have difficulty with them, go back and work through the examples again. If you still have difficulty, seek help from your instructor, tutor, or teaching assistant.

Problems

1. Name the following ions.
   a. Na^+ b. Mg^{2+} c. Al^{3+}

2. Name the following ions.
   a. Cl^- b. O^{2-} c. N^{3-}

3. Name the following ions.
   a. Fe^{3+} b. Cu^+ c. Ag^+

4. Give symbols (e.g., Cl^-) for the following ions.
   a. lithium ion   b. iodide ion   c. calcium ion

5. Indicate the charges on simple ions formed from the following elements.
   a. Group IIIA   b. Group VIA
   c. Group IA   d. Group VIIA

6. Using electron-dot formulas, draw the formation of an ion from an atom for each of the following.
   a. barium   b. bromine
   c. aluminum   d. sulfur

7. Identify the ions of Problem 6 as cations or anions.

8. Give symbols for the following ions.
   a. iron(II) ion   b. copper(II) ion
   c. zinc ion

9. Give the formulas for the following ion
   a. ammonium ion
   b. hydrogen carbonate ion
   c. phosphate ion

10. Give formulas for the following ions.
    a. nitrite ion   b. cyanide ion
    c. sulfate ion

11. Name the following ions.
    a. CO_3^{2-} b. HPO_4^- c. NO_3^- d. OH^-

12. Name the following ions.
    a. $CH_3CO_2^-$ b. HSO_4^- c. $H_2PO_4^-$

13. Write correct formulas for the following compounds.
    a. lithium fluoride   b. calcium iodide
    c. aluminum bromide   d. aluminum sulfide

14. Write correct formulas for the following compounds.
    a. magnesium sulfate   b. potassium nitrate
    c. sodium cyanide   d. calcium oxalate

15. Write correct formulas for the following compounds.
    a. ferrous sulfate
    b. ammonium phosphate
    c. magnesium phosphate
    d. calcium monohydrogen phosphate

16. Give correct names for the compounds represented by the following formulas.
    a. NaBr   b. $CaCl_2$ c. Al_2O_3

17. Give correct names for the compounds represented by the following formulas.
    a. KNO_2 b. LiCN
    c. NH_4I d. $NaNO_3$

18. Give correct names for the compounds represented by the following formulas.

    a. $CaSO_4$ b. $NaHSO_4$ c. $KHCO_3$
    d. $Al(OH)_3$ e. Na_2CO_3

19. Give correct names for the compounds represented by the following formulas.

    a. $Mg(CH_3CO_2)_2$ b. $Al(C_2H_3O_2)_3$
    c. $(NH_4)_3PO_4$ d. $(NH_4)_2C_2O_4$

20. Give correct names for the compounds represented by the following formulas.
    a. Na_2HPO_4 b. $Ca(H_2PO_4)_2$
    c. $Mg(HCO_3)_2$ d. $Ca(HSO_4)_2$
    e. NH_4NO_2

21. Fill in this table assuming that elements X, Y, and Z are all in A subgroups in the periodic table.

	Element X	Element Y	Element Z
Group number	IA	_____	_____
Electron-dot formula	_____	$\cdot \dot{Y} \cdot$	_____
Charge on ion	_____	_____	2−

22. Name the following covalent compounds.
    a. CS_2 b. CBr_4 c. N_2S_4 d. PBr_3

23. Write formulas for the following covalent compounds.
    a. dinitrogen monoxide
    b. oxygen difluoride
    c. sulfur trioxide
    d. sulfur tetrachloride
    e. tetrasulfur tetranitride

24. Chlorine dioxide is used to bleach flour. Write the formula for chlorine dioxide.

25. Tetraphosphorus trisulfide is used in the tops of "strike anywhere" matches. Write the formula for tetraphosphorus trisulfide.

26. Define or illustrate each of the following.
    a. formula mass   b. mole
    c. Avogadro's number

27. Relate the law of conservation of matter to the need for working with balanced chemical equations.

28. How many hydrogen atoms are there per formula unit in each of the following?
    a. NH_4NO_3 b. $(NH_4)_2HPO_4$

29. How many atoms of each kind (Al, C, H, and O) are included in the following notation: 2 $Al(C_2H_3O_2)_3$?

30. Indicate whether the equations are balanced. You need not balance the equation. Just determine whether it is balanced as written.
    a. $Mg + H_2O \rightarrow MgO + H_2$
    b. $FeCl_2 + Cl_2 \rightarrow FeCl_3$
    c. $F_2 + H_2O \rightarrow 2\ HF + O_2$
    d. $Ca + 2\ H_2O \rightarrow Ca(OH)_2 + H_2$
    e. $2\ LiOH + CO_2 \rightarrow Li_2CO_3 + H_2O$

31. Indicate whether the equations are balanced as written.

a. $2 KNO_3 + 10K \rightarrow 6 K_2O + N_2$
b. $2 NH_3 + O_2 \rightarrow N_2 + 3 H_2O$
c. $4 LiH + AlCl_3 \rightarrow 2 LiAlH_4 + 2 LiCl$
d. $SF_4 + 3 H_2O \rightarrow H_2SO_3 + 4 HF$
e. $4 BF_3 + 3 H_2O \rightarrow H_3BO_3 + 3 HBF_4$

32. Indicate whether the equations are balanced as written.

a. $2 Sn + 2 H_2SO_4 \rightarrow$
$$2 SnSO_4 + SO_2 + 2 H_2O$$
b. $3 Cl_2 + 6 NaOH \rightarrow$
$$5 NaCl + NaClO_3 + 3 H_2O$$

33. Balance the following chemical equations.

a. $Al + O_2 \rightarrow Al_2O_3$
b. $C + O_2 \rightarrow CO$
c. $N_2 + O_2 \rightarrow NO$
d. $SO_2 + O_2 \rightarrow SO_3$
e. $NO + O_2 \rightarrow NO_2$

34. Balance the following chemical equations.

a. $Zn + HCl \rightarrow ZnCl_2 + H_2$
b. $H_2S + O_2 \rightarrow H_2O + S$
c. $Al_2(SO_4)_3 + NaOH \rightarrow Al(OH)_3 + Na_2SO_4$
d. $Zn(OH)_2 + HNO_3 \rightarrow Zn(NO_3)_2 + H_2O$
e. $NH_4OH + H_3PO_4 \rightarrow (NH_4)_3PO_4 + H_2O$

35. Balance the following chemical equations.

a. $Cu + H_2SO_4 \rightarrow SO_2 + CuSO_4 + H_2O$
b. $NH_4Cl + CaO \rightarrow CaCl_2 + H_2O + NH_3$

36. Calculate the formula mass (to the nearest whole number) for each of these compounds.

a. CH_4 b. AlF_3 c. UF_6

37. Calculate the formula mass (to the nearest whole number) for the following.

a. NH_4NO_3 b. $BaSO_4$
c. H_3PO_4 d. $KClO_3$

38. Calculate the formula mass (to the nearest whole number) for the following.

a. $Ca(NO_3)_2$ b. $Mg(OH)_2$ c. $(NH_4)_2SO_4$

39. Calculate the number of moles of compound for each of the following.

a. 32 g of CH_4 b. 336 g of AlF_3
c. 17.6 g of UF_6

40. Calculate the number of moles of compound for each of the following.

a. 8.0 g of NH_4NO_3 b. 0.233 g of $BaSO_4$
c. 980 g of H_3PO_4 d. 3.69 g of $KClO_3$

41. Calculate the number of moles of compound for each of the following.

a. 1.64 g of $Ca(NO_3)_2$ b. 5.8 g of $Mg(OH)_2$
c. 0.0132 g of $(NH_4)_2SO_4$

42. Calculate the number of grams in each of these samples.

a. 0.0010 mol of CH_4 b. 6.00 mol of AlF_3
c. 40.0 mol of UF_6

43. Calculate the number of grams in each of these samples.

a. 0.05 mol of NH_4NO_3 b. 3.00 mol of $BaSO_4$
c. 10 mol of H_3PO_4 d. 0.0200 mol of $KClO_3$

44. Calculate the number of grams in each of these samples.

a. 1.50 mol of $Ca(NO_3)_2$ b. 5.8 mol of $Mg(OH)_2$
c. 0.25 mol of $(NH_4)_2SO_4$

45. Consider the reaction

$$S + O_2 \rightarrow SO_2$$

a. How many moles of SO_2 would be formed by the burning of 4 mol of sulfur?
b. How many moles of SO_2 would be formed in the reaction of 0.6 mol of O_2?
c. How many moles of sulfur would be required to react with 0.0684 mol of O_2?

46. Answer the following questions based on the same equation.

$$S + O_2 \rightarrow SO_2$$

a. How many moles of SO_2 would form in the reaction of 32 g of S?
b. How many moles of S would react with 32 g of O_2?

47. Still using the same equation, answer the following questions.

$$S + O_2 \rightarrow SO_2$$

a. How many grams of SO_2 would form in the reaction of 16 g of O_2?
b. How many grams of SO_2 would be formed by the burning of 8.0 g of sulfur?
c. How many grams of sulfur would be needed to produce 3.2 g of SO_2?

48. Consider the reaction

$$CH_4 + 2 O_2 \rightarrow CO_2 + 2 H_2O$$

a. How many moles of water are produced in the reaction of 0.4 mol of CH_4?

b. How many moles of O_2 are required to produce 25 mol of CO_2?

c. How many moles of CH_4 are required to react with 10 mol of O_2?

d. If 0.86 mol of O_2 reacts, how many moles of H_2O form?

49. Answer the following questions based on the reaction 48.

$$CH_4 + 2 O_2 \rightarrow CO_2 + 2 H_2O$$

a. If 9 g of H_2O forms, how many moles of CH_4 react?

b. If 2.0 mol of O_2 reacts, how many grams of CO_2 form?

50. Again consider the reaction

$$CH_4 + 2 O_2 \rightarrow CO_2 + 2 H_2O$$

a. How many grams of O_2 are required to react completely with 8.0 g of CH_4?

b. How many grams of CO_2 are produced if 180 g of H_2O is formed?

51. Consider the equation

$$C_3H_8 + 5 O_2 \rightarrow 3 CO_2 + 4 H_2O$$

a. How many moles of CO_2 are produced in the reaction of 2 mol of C_3H_8?

b. How many moles of H_2O are produced in the reaction of 20 mol of O_2?

52. Again, use the equation

$$C_3H_8 + 5 O_2 \rightarrow 3 CO_2 + 4 H_2O$$

a. How many moles of O_2 react with 22 g of C_3H_8?

b. If 1.8 g H_2O is produced, how many moles of CO_2 are formed?

53. Answer the following questions based on the same reaction.

$$C_3H_8 + 5 O_2 \rightarrow 3 CO_2 + 4 H_2O$$

a. If 16 g of O_2 reacts, how many grams of C_3H_8 react?

b. If 320 g of O_2 reacts, how many grams of CO_2 are formed?

54. What mass of quicklime (calcium oxide) is formed when 100 g of limestone (calcium carbonate) is decomposed by heating?

$$CaCO_3 \rightarrow CaO + CO_2$$

55. How many grams of hydrogen would be required to produce 68 g of ammonia (NH_3)

$$N_2 + 3 H_2 \rightarrow 2 NH_3$$

56. Some of the ammonia produced by the above process (Problem 55) is converted to nitric acid.

$$NH_3 + 2 O_2 \rightarrow HNO_3 + H_2O$$

How much nitric acid can be made from 68 g of ammonia?

57. Nitric acid is used in the production of trinitrotoluene (TNT), an explosive.

$$C_7H_8 + 3 HNO_3 \rightarrow C_7H_5N_3O_6 + 3 H_2O$$
Toluene   TNT

How much TNT can be made from 46 kg of toluene?

58. The burning of acetylene (C_2H_2) in pure oxygen produces a very hot flame. (This is the reaction in the oxyacetylene torch.) How much oxygen would be required to burn 52 kg of acetylene?

$$2 C_2H_2 + 5 O_2 \rightarrow 4 CO_2 + 2 H_2O$$

59. Joseph Priestley discovered oxygen in 1774 by using heat to decompose "red calx of mercury," known today as mercury (II) oxide.

$$2 HgO \rightarrow 2 Hg + O_2$$

How much oxygen is produced by the decomposition of 108 g of HgO?

60. How much iron can be converted to the magnetic oxide of iron, Fe_3O_4, by 800 g of pure oxygen?

$$3 Fe + 2 O_2 \rightarrow Fe_3O_4$$

61. Dinitrogen monoxide (also called nitrous oxide) can be made by cautiously heating ammonium nitrate.

$$NH_4NO_3 \rightarrow N_2O + 2 H_2O$$

How much nitrous oxide can be made from 40 g of NH_4NO_3?

62. Sometimes a small amount of hydrogen gas is made for laboratory use by allowing calcium metal to react with water.

$$Ca + 2 H_2O \rightarrow Ca(OH)_2 + H_2$$

How many grams of hydrogen would be formed by the action of water on 10 g of calcium?

63. For many years it was thought that "inert" gases such as xenon (Xe) would not form chemical compounds. Following Neil Bartlett's breakthrough in 1962, a number of compounds involving "inert" gases have been synthesized. Xenon hexafluoride is made according to the equation

$$Xe + 3 F_2 \rightarrow XeF_6$$

How many grams of fluorine are required to form 49 g of XeF_6?

64. Black lead sulfide (PbS) can be converted to white lead sulfate ($PbSO_4$) by hydrogen peroxide.

$$PbS + 4 H_2O_2 \rightarrow PbSO_4 + 4 H_2O$$

How much hydrogen peroxide would be required to convert 478 mg of PbS to $PbSO_4$?

References and Readings

1. DeLorenzo, Ronald A. *Problem Solving in General Chemistry*. Lexington, MA: D. C. Heath, 1981. Chapters 4 and 6 discuss moles and equations.
2. Gizara, Jeanne M. "Bridging the Stoichiometry Gap." *The Science Teacher*, April 1981. pp. 36–37. Presents a road-map method for solving problems based on equations.
3. Hill, John W., and Dorothy M. Feigl. *Chemistry and Life*, 3d ed. New York: Macmillan, 1987. See Chapters 4 and 5.
4. Kieffer, William F. *The Mole Concept in Chemistry* (Selected Topics in Modern Chemistry series) New York: Reinhold, 1962.

7

Acids and Bases

Please Pass the Protons

Central to much of the chemistry in our everyday lives is the chemistry of two interrelated classes of compounds called *acids* and *bases*. We use many of these compounds around the house. We clean with ammonia and lye (sodium hydroxide), two familiar bases. We use vinegar (acetic acid) in cooking and as dressing on salads. We take vitamin C (ascorbic acid) and drink beverages made tart by citric acid and phosphoric acid. We treat our excess stomach acid (hydrochloric acid) with a variety of bases, including milk of magnesia (magnesium hydroxide) and baking soda (sodium bicarbonate). Our bodies produce and consume acids and bases, maintaining a delicate balance necessary to our good health and well-being.

Three of the four tastes are related to acids and bases. Acids taste *sour* and bases taste *bitter*. Salts, compounds formed when acids and bases react with one another, taste *salty*. Only a *sweet* taste is not related to acids and bases.

146

Acid rain is an environmental problem in industrialized countries. Bitter, undrinkable alkaline (basic) water is often all that is available in areas with dry climates. Much of our understanding of air and water pollution depends on our knowledge of acids and bases.

In this chapter, we discuss some of the chemistry of familiar acids and bases and some of the properties of these compounds. Principles put forth here are used in later chapters. And perhaps of more importance to you is the fact that you will hear and read about acids and bases for the rest of your life. You will use your knowledge of them every day. Let's hope that what you learn here will enrich your life by giving you a greater understanding of the chemicals that you use and that are a part of your environment.

Acids and Bases: Experimental Definitions

We can define acids and bases in two ways. First, we give an experimental definition. Then we relate the properties of acids and bases to our knowledge of atoms and molecules in a conceptual definition. To simplify things somewhat, we deal only with solutions in water (aqueous solutions).

Experimentally, **acids** are compounds that would

1. Turn the indicator dye litmus from blue to red.
2. React with active metals (such as iron, tin, and zinc) to dissolve the metal and produce hydrogen gas.
3. Taste sour (when diluted enough to be safely tasted).
4. React with bases (or alkalis) to form water and ionic compounds called *salts*.

Similarly, **bases** are defined experimentally as compounds that, when dissolved in water, would

1. Turn the indicator dye litmus from red to blue.
2. Feel slippery or soapy on the skin.
3. Taste bitter.
4. React with acids to form water and salts.

How can you tell when a compound is an acid? You can tell by dissolving some of the compound in water. Since tasting may be hazardous (some acids are poisonous and nearly all are very corrosive unless they have been highly diluted), you can stick a piece of blue litmus paper into a sample of the compound. If the paper turns red, the

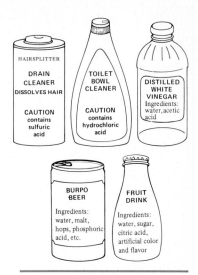

Figure 7.1 Some common acids.

Figure 7.2 Some common bases.

compound is an acid. Similarly, you can identify a base by dissolving it in water and testing it with red litmus paper. If the paper turns blue, the compound is a base.

You can use a taste test on foods. Many foods are acidic. Vinegar contains 4 to 10% acetic acid. Citrus fruits and many fruit-flavored drinks contain citric acid. If a food tastes sour, most likely it contains one or more acids. Lactic acid is formed in sour milk and is responsible for the tart taste of yogurt. Phosphoric acid is used to impart tartness to beer and some forms of soda pop.

Many medicines taste bitter because they contain alkaloid (alkali-like) compounds. Quinine, caffeine, and the antihistamines are familiar examples.

Acids Explained: Hydronium Ions

It is nice to know that certain compounds are acids and others are bases and that each class has characteristic properties. It is more satisfying, however, to know *why* a certain compound is an acid and another a base. The search for meaning is central to science—and to life in general. We will not recount the history of the theory of acids and bases in this discussion, but we will list some of the more important concepts.

A wealth of experimental evidence indicates that, in water solution, the properties of acids are due to H_3O^+, the **hydronium ion**. This ion is a water molecule to which a hydrogen ion (H^+) has been added. Since a hydrogen atom with its electron removed is just a bare nucleus consisting of a single proton, H^+ often is called a **proton**.

Table 7.1 lists a variety of common acids. Every acid contains one or more hydrogen atoms. When dissolved in water, acids transfer hydrogen nuclei to the water molecules. They are proton donors. For hydrogen chloride, the reaction is written

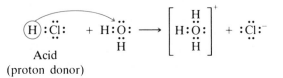

Acid
(proton donor)

A solution of hydronium ions and chloride ions in water is called *hydrochloric acid*. Other acids react similarly with water to produce hydronium ions.

As we mentioned in Chapter 5, oxygen usually forms only two bonds in molecules. Now, in the hydronium ion, it forms three. Why? When you look at the structure of a water molecule, you can see that the

Table 7.1 Some Familiar Acids

Name	Formula	Classification
Sulfuric acid	H_2SO_4	Strong
Nitric acid	HNO_3	Strong
Hydrochloric acid	HCl	Strong
Phosphoric acid	H_3PO_4	Moderate
Hydrogen sulfate ion	HSO_4^-	Moderate
Lactic acid	$CH_3CHOHCOOH$	Weak
Acetic acid	CH_3COOH	Weak
Boric acid	H_3BO_3	Weak
Hydrocyanic acid	HCN	Weak

oxygen atom has two unshared pairs of electrons. The oxygen end of the water molecule carries a partial negative charge (Chapter 5). Isn't it reasonable to expect that a proton (H^+), with its positive charge, would stick to the negative end of a water molecule?

$$H:\overset{..}{\underset{\underset{\delta^+}{..}}{O}}:^{\delta-} \; + \; H^+ \longrightarrow \left[H:\overset{..}{\underset{..}{O}}:H \right]^+$$

$$\qquad\qquad \overset{\delta+}{H} \qquad\qquad\qquad\qquad H$$

Water   Hydronium ion

Notice that in the hydronium ion, the oxygen atom still has eight electrons in its outer shell (the same structure that neon has). The nice thing, from hydrogen's point of view, is that all three hydrogen atoms now have two electrons in their outer shells (the same structure as helium). Oxygen doesn't lose its rare-gas structure, but the proton is able to gain a rare-gas configuration by attaching itself to one of the unshared pairs.

In water, then, the properties of acids are those of the hydronium ion. It is the hydronium ion that turns litmus red, tastes sour, and reacts with active metals and bases. Nowadays, however, chemists and other scientists work with solvents other than water. They have broadened the concept of an acid to operate in these other media. Generally speaking, then, an acid is a **proton donor**—to any receptor, not just to water.

Many acids are made by the reaction of nonmetal oxides with water. For example, sulfur trioxide reacts with water to form sulfuric acid.

$$SO_3 \; + \; H_2O \longrightarrow H_2SO_4$$

Similarly, carbon dioxide reacts with water to form carbonic acid.

$$CO_2 \; + \; H_2O \longrightarrow H_2CO_3$$

This a general reaction of nonmetal oxides.

$$\text{Nonmetal oxide} + H_2O \longrightarrow \text{Acid}$$

Nonmetal oxides are called **acid anhydrides**. *Anhydride* means "without water." These reactions explain why rainwater is acidic, particularly that which forms in air polluted with sulfur oxides (Chapter 15).

Example 7.1 Give the formula for the acid formed when sulfur dioxide reacts with water.
    Simply write the equation

$$SO_2 + H_2O \longrightarrow H_2SO_3$$

Bases Explained: Hydroxide Ions

Plenty of experimental evidence indicates that the properties of a base, in water, are due to OH^-, the **hydroxide ion**. Table 7.2 lists several common bases. Ionic compounds contain two separate and distinct species. In sodium hydroxide, there are sodium ions (Na^+) and hydroxide ions (OH^-). In calcium hydroxide, there are calcium ions (Ca^{2+}) and hydroxide ions (OH^-). When either compound is dissolved in water, a basic solution is formed. The hydroxide ion is the base. Adding sodium hydroxide or calcium hydroxide to water simply supplies an excess of hydroxide ions.

Ammonia seems out of place in Table 7.2, because it contains no hydroxide ions. How can it be a base? The only way to get hydroxide ions from ammonia in water is for the ammonia molecule to accept a proton from water, leaving a hydroxide ion.

$$NH_3 + H_2O \rightleftharpoons NH_4^+ + OH^-$$

The NH_4^+ is the ammonium ion.

Table 7.2 Common Bases

Name	Formula	Classification
Sodium hydroxide	NaOH	Strong
Potassium hydroxide	KOH	Strong
Calcium hydroxide	$Ca(OH)_2$	(see text)
Magnesium hydroxide	$Mg(OH)_2$	(see text)
Ammonia	NH_3	Weak

How can ammonia accept a proton from water? Recall that the nitrogen atom has an unshared pair of electrons. This pair can be used to attach a proton. Removal of a proton from water leaves a hydroxide ion, which is negatively charged because it still has the electrons that bound the "lost" hydrogen.

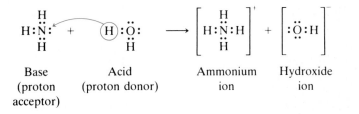

Base	Acid	Ammonium	Hydroxide
(proton	(proton donor)	ion	ion
acceptor)			

In water, then, the properties of bases are those of hydroxide ions. The hydroxide ion is a base in water solutions. It turns litmus blue, makes the skin feel soapy, tastes bitter, and reacts with acids. And, just as the concept of an acid has been expanded to include solvents other than water, the definition of a base has been generalized. A base is a **proton acceptor**. This concept includes not only hydroxide ions but also neutral molecules such as ammonia. It also includes other negative ions such as carbonate (CO_3^{2-}), bicarbonate (HCO_3^-), and phosphate (PO_4^{3-}), which, at least to some extent, accept protons in water solution.

Note also that in reacting with ammonia, water acts as a proton donor—an acid. Thus, the idea of an acid as a proton donor and a base as a proton acceptor greatly expands our concept of acids and bases. No longer do we think of an acid simply as a solution of hydronium ions in water and a base as a solution of hydroxide ions in water. The proton donor–acceptor concept is much broader—but it does *include* the more limited definitions. It's just that the broader concept is useful in a greater variety of situations.

Recall (previous section) that acids can be made from nonmetal oxides. Many common bases can be made from metal oxides. For example, calcium oxide (lime) reacts with water to form calcium hydroxide (slaked lime).

$$CaO + H_2O \longrightarrow Ca(OH)_2$$

Another example is the reaction of lithium oxide with water to form lithium hydroxide.

$$Li_2O + H_2O \longrightarrow 2 LiOH$$

In general, metal oxides react with water to form bases. These metal oxides are called **basic anhydrides**.

> *Example 7.2*   What base is formed by the addition of water to barium oxide (BaO)?
>   Write the equation
>
> $$BaO + H_2O \longrightarrow Ba(OH)_2$$

Strong and Weak Acids and Bases

When gaseous hydrogen chloride (HCl) reacts with water, it reacts completely to give hydronium ions and chloride ions.

$$HCl + H_2O \longrightarrow H_3O^+ + Cl^-$$

Similarly, the poisonous gas hydrogen cyanide (HCN) reacts with water to produce hydronium ions and cyanide ions.

$$HCN + H_2O \longrightarrow H_3O^+ + CN^-$$

The reaction in the latter case takes place only to a slight extent, however, and produces relatively few hydronium ions. Acids that react completely are called **strong acids**. Those that react only slightly are called **weak acids**. Note that Table 7.1 classifies each acid as strong, moderate, or weak. This classification is a measure of the degree of reaction. Strong acids react completely, producing a relatively large number of hydronium ions. Weak acids react only partially, producing fewer hydronium ions.

Strong acids, when moderately concentrated, can cause serious damage to skin and flesh. They eat holes in clothes made of natural fibers such as cotton, silk, and wool. Strong acids also destroy most synthetic fibers such as nylon, polyester, and acrylics. Care always should be taken to prevent spills on skin and clothing. In very dilute solutions, strong acids can be harmless. Hydrochloric acid, for example, is produced in dilute solutions in our stomachs, where it aids in the digestion of certain foodstuffs. Even there, however, under certain conditions, it can cause problems—as anyone with an ulcer can attest. Concentrated weak acids can cause problems, too. Although it is pleasant to eat salads with vinegar as a condiment, concentrated solutions of acetic acid are corrosive to the skin and are especially destructive to the lining of the digestive tract.

Bases also are classified as strong or weak. Perhaps the most familiar **strong base** is sodium hydroxide ($NaOH$), sometimes called *lye*. Even in the solid state, sodium hydroxide is completely ionic; that is, it exists as sodium ions and hydroxide ions. In solution, the hydroxide ions take part in the characteristic reactions we refer to when we say a solution is basic.

Other common bases include potassium hydroxide (KOH), calcium hydroxide [$Ca(OH)_2$], and magnesium hydroxide [$Mg(OH)_2$]. All of these are strong in the sense that they are completely ionic in the solid state; yet only potassium hydroxide gives strongly basic aqueous solutions. Potassium hydroxide, like sodium hydroxide, is quite soluble in water. Therefore, aqueous solutions of the compound are rich in hydroxide ions. Calcium hydroxide is only slightly soluble in water, so the solution contains relatively few dissolved hydroxide ions and is not strongly basic. Magnesium hydroxide is so nearly insoluble that it can be safely taken internally as an antacid (see section on antacids later in this chapter).

Ammonia (NH_3) is a familiar **weak base**. It is a gas at room temperature but readily dissolves in water to yield a basic solution. It reacts with water to a slight extent to produce ammonium ions (NH_4^+) and hydroxide ions.

$$NH_3 \ + \ H_2O \ \longrightarrow \ NH_4^+ \ + \ OH^-$$

Ammonia is classified as a weak base because an ammonia solution contains a relatively low concentration of hydroxide ions. It is not that ammonia is insoluble, however, but that once ammonia is in solution it tends *not* to form ions. The solution so formed sometimes is called *ammonium hydroxide*; but, since most of the nitrogen atoms are still in the form of NH_3, *aqueous ammonia* is perhaps a better name. Many familiar household cleansers contain ammonia, a compound easily detected by its characteristic odor. Brands such as Parsons and Bo Peep are about 3.5% NH_3. Other brands, such as Top Job and Ajax, contain detergents and have less ammonia.

To show a characteristic property of both acids and bases, we listed the reaction of these compounds with one another to form water and a salt. If a solution containing hydronium ions is mixed with one containing the same amount of hydroxide ions, the resulting solution no longer affects litmus paper. Nor does the new solution taste bitter or sour. The solution is neither acidic nor basic. It is neutral. Thus, the reaction of an acid with a base is called **neutralization**.

Sodium hydroxide reacts with hydrochloric acid to produce water and sodium chloride, a salt (in this case, ordinary table salt).

$$NaOH \ + \ HCl \ \longrightarrow \ H_2O \ + \ NaCl$$

If equivalent amounts of acid and base are used, the resulting solution is neutral and tastes salty. The reaction is really that of hydronium ions (acting as an acid) with hydroxide ions (acting as a base) to form water.

$$H_3O^+ + OH^- \longrightarrow 2 H_2O$$

Acid	Base
(proton	(proton
donor)	acceptor)

The pH Scale

It is often convenient in everyday life to count in dozens (a dozen pears, pencils, or people). In chemistry, it is convenient to count in *moles* (see Chapter 6). A mole is an extremely large number of particles: 6.02×10^{23} molecules, atoms, or ions of a substance. The strength of acid solutions often is measured in moles per liter of solution. A 1 molar ($1 M$) solution of hydrochloric acid contains 1 mol of hydronium ions in each liter of solution. That means 6.02×10^{23} H_3O^+ ions in 1 L.

In pure water (or in any neutral aqueous solution), there are 1.0×10^{-7} mol per liter of hydronium ions and an equal number of moles of hydroxide ions. These arise from a slight ionization of water. About 1 molecule in 500 million transfers a proton to another.

$$H_2O + H_2O \longrightarrow H_3O^+ + OH^-$$

Acidic solutions have concentrations of hydronium ions greater than 1.0×10^{-7} mol/L; basic solutions have concentrations less than 1.0×10^{-7} mol/L.

Quite frequently, these concentrations are extremely small numbers (in scientific notation, numbers with negative exponents). It is often inconvenient to work with them. In 1909, the Danish chemist S. P. L. Sorensen proposed that only the number in the exponent be used to express acidity. Sorensen's scale came to be known as the **pH scale**, from the French *pouvoir hydrogene* ("hydrogen power"). On this scale, a concentration of 1×10^{-7} mol of hydronium ions per liter of solution becomes a pH of 7. Similarly, a concentration of 1×10^{-4} mol/L becomes a pH of 4, and so on.

A pH of 7 represents a neutral solution. A pH lower than 7 means the solution is acidic; a pH higher than 7 indicates that the solution is basic. The *lower* the pH, the more acidic the solution; the *higher* the pH, the more basic the solution.

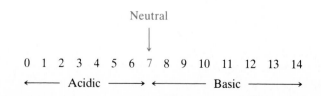

Table 7.3 The Relationship between pH and the Concentration of Hydronium Ions

Concentration of H_3O^+ (moles per liter)	pH
1×10^{0}	0
1×10^{-1}	1
1×10^{-2}	2
1×10^{-3}	3
1×10^{-4}	4
1×10^{-5}	5
1×10^{-6}	6
1×10^{-7}	7
1×10^{-8}	8
1×10^{-9}	9
1×10^{-10}	10
1×10^{-11}	11
1×10^{-12}	12
1×10^{-13}	13
1×10^{-14}	14

Table 7.4 The Approximate pH of Some Common Solutions

Solution	pH
Hydrochloric acid (4%)	0
Gastric juices	1.6–1.8
Lemon juice	2.3
Vinegar	2.4–3.4
Soft drinks	2.0–4.0
Milk	6.3–6.6
Urine	5.5–7.0
Rainwater (unpolluted)	5.6
Saliva	6.2–7.4
Pure water	7.0
Blood	7.35–7.45
Fresh egg white	7.6–8.0
Bile	7.8–8.6
Milk of magnesia	10.5
Washing soda	12.0
Sodium hydroxide (4%)	13.0

Each step on the pH scale corresponds to a tenfold change in the concentration of hydronium ions. For example, a solution with a pH of 3 is *10 times* as acidic as a solution with a pH of 4, *100 times* as acidic as a solution with a pH of 5, and so on. This relationship will be more evident to you if you keep in mind the corresponding concentrations of hydronium ions. A pH of 3 means a hydronium ion concentration of 1×10^{-3}, or 0.001, mol/L. A pH of 4 means a hydronium ion concentration of 1×10^{-4}, or 0.0001, mol/L. Note that 0.001 is 10 times greater than 0.0001.

The relationship between pH and hydronium ion concentration is summarized in Table 7.3. The pHs of a variety of common substances are listed in Table 7.4.

Example 7.3 What is the pH of a solution that has a hydronium ion concentration of 1×10^{-5} *M*?

The exponent is −5; the pH is 5.

Example 7.4 What is the hydronium ion concentration of a solution that has a pH of 4?

The pH value is the negative exponent of 10, so the hydronium ion concentration is 1×10^{-4} *M*.

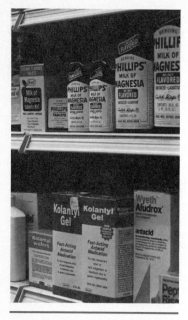

Figure 7.3 All antacids are basic compounds. [Photo by Dennis Tasa.]

Antacids

The stomach secretes hydrochloric acid (HCl) as an aid in the digestion of food. Sometimes overindulgence or emotional stress leads to a condition called **hyperacidity** (too much acid). There are over 8000 brands of antacids available in the United States to treat this condition; many are aggressively advertised (Figure 7.3). Sales are estimated to be about $650 million a year. Despite the variety of brand names, antacids contain only a few different ingredients—such as calcium carbonate, aluminum hydroxide, magnesium hydroxide, and sodium bicarbonate. Let's look at some popular antacids from the standpoint of acid–base chemistry.

Sodium bicarbonate, or baking soda ($NaHCO_3$), is an old standby antacid. It is probably safe and effective for occasional use by most people. Overuse will make the blood too alkaline, a condition called **alkalosis**. Sodium bicarbonate is not recommended for those with hypertension (high blood pressure), because high concentrations of sodium ion tend to aggravate the condition. The antacid in Alka-Seltzer is sodium bicarbonate. This popular remedy also contains citric acid and aspirin. When Alka-Seltzer is placed in water, the familiar fizz occurs, due to the reaction of bicarbonate ions with hydronium ions from the acid.

$$HCO_3^- + H_3O^+ \longrightarrow CO_2 + H_2O$$

Alka-Seltzer may be dangerous because the aspirin in it might be harmful to people with ulcers and other stomach disorders.

Calcium carbonate, commonly called precipitated chalk ($CaCO_3$), is another common antacid ingredient. It is fast acting and safe in small amounts, but regular use can cause constipation. It also appears that calcium carbonate can caused *increased* acid secretion after a few hours. Temporary relief may be achieved at the expense of a worse problem later. Tums is simply flavored calcium carbonate. Alka 2 and Di-Gel liquid suspension also contain calcium carbonate as the antacid ingredient. The carbonate ion neutralizes acid by the following reaction.

$$CO_3^{2-} + 2 H_3O^+ \longrightarrow CO_2 + 3 H_2O$$

Aluminum hydroxide [$Al(OH)_3$] is another popular antacid ingredient. Like calcium carbonate, it can cause constipation in large doses. The hydroxide ions neutralize acid.

$$OH^- + H_3O^+ \longrightarrow 2 H_2O$$

There is concern that antacids containing aluminum ions deplete the body of essential phosphate ions. The aluminum phosphate formed is insoluble and is excreted from the body.

$$Al^{3+} + PO_4^{3-} \longrightarrow AlPO_4$$

Aluminum hydroxide is the only antacid in Amphojel. It occurs in combination in many popular products.

Magnesium compounds constitute the fourth category of antacids. These include magnesium carbonate ($MgCO_3$) and magnesium hydroxide [$Mg(OH)_2$]. Milk of magnesia is a suspension of magnesium hydroxide in water. It is sold under a variety of brand names, the best known of which is probably Phillips. In small doses, magnesium compounds act as antacids. In large doses, they act as laxatives. Magnesium ions are poorly absorbed in the digestive tract. Rather, these small, dipositive ions draw water into the colon (large intestine) causing the laxative effect.

A variety of popular antacids combine aluminum hydroxide, with its tendency to cause constipation, and a magnesium compound, which acts as a laxative. These tend to counteract one another. Maalox and Mylanta are familiar brands.

Rolaids, a highly advertised antacid, contains the complex substance aluminum sodium dihydroxy carbonate [$AlNa(OH)_2CO_3$]. Both the hydroxide ion and the carbonate ion consume acid. New sodium-free Rolaids contains calcium carbonate and magnesium hydroxide.

Antacids interact with other medications. Anyone taking any type of medicine should consult a physician before taking antacids. Generally, antacids are safe and effective for occasional use in small amounts. All antacids are *basic* compounds. If you are otherwise in good health, you can choose a base on the basis of price.

Anyone with severe or repeated attacks of indigestion should consult a physician. Self-medication in such cases might be dangerous.

Acids, Bases, and Human Health

Strong acids and bases cause damage on contact with living cells. Their action is nonspecific. All cells, regardless of type, are damaged more or less equally. These **corrosive** poisons produce what are known as chemical burns. The injuries, once the offending agent is neutralized or removed, are quite similar to burns from heat and they are often treated the same way.

Sulfuric acid (H_2SO_4) is by far the leading chemical product of United States industry. About 40 billion kg are produced annually. Most is used by industries, particularly in the production of fertilizers (Chapter 17) and industrial chemicals. Only small quantities are used in or around the home. Automobile batteries contain sulfuric acid. One type of drain cleaner is sulfuric acid. This strong acid is one of the few substances that dissolves drain-clogging hair. Sulfuric acid is a powerful dehydrating agent. It takes up water from cellular fluid, rapidly killing the cell. The sulfuric acid molecules dehydrate by actually reacting with water in the cells to form hydronium ions and hydrogen sulfate ions.

$$H_2SO_4 + H_2O \longrightarrow H_3O^+ + HSO_4^-$$

Hydration of these ions and other secondary reactions may also be involved in the dehydration process. The burning of sulfur-containing coal produces aerosol mists of sulfuric acid. The effects on the environment and human health of acid air pollutants are discussed in Chapter 15.

Hydrochloric acid (also called muriatic acid) is used in homes to clean calcium carbonate deposits from toilet bowls. It is used in building construction to remove excess mortar from bricks. In industry, it is used to remove scale or rust from metals. Concentrated solutions (about 38% HCl) cause severe burns, but dilute solutions are considered safe enough for use around the house. Even dilute solutions, however, can cause skin irritation and inflammation.

Ingestion of sulfuric, hydrochloric, or any other strong acid causes corrosive damage to the digestive tract. As little as 10 mL of concentrated (98%) sulfuric acid—taken internally—can be fatal.

The most common commercial base is lime (calcium oxide), made by heating limestone (calcium carbonate).

$$CaCO_3 \xrightarrow{\text{heat}} CaO + CO_2$$

Calcium oxide is the third ranked industrial chemical in the United States with an annual production of about 16 billion kg. It is used to make cement (Chapter 9), mortar, and plaster and to neutralize acidic soils.

Sodium hydroxide (lye) is by far the most common of the strong bases used for household purposes. It is used to open clogged drains and as an oven cleaner. It destroys tissue rapidly, causing severe chemical burns.

Both acids and bases, even in dilute solutions, break down the protein molecules in living cells. Generally, the fragments are not able to carry out the functions of the original proteins. In cases of severe exposure, the fragmentation continues until the tissue has been completely destroyed. And, within living cells, proteins function properly only at an optimum pH. If the pH changes much in either direction, the proteins can't carry out their usual functions properly.

When they are misused, acids and bases can be damaging to human health. But acids and bases affect human health in more subtle—and ultimately more important—ways. A delicate balance must be maintained between acids and bases in the blood and body fluids. If the acidity of the blood changes very much, the blood loses its capacity to carry oxygen. Fortunately, the body has a complex but efficient mechanism for maintaining proper acid–base balance. A study of that mechanism is beyond the scope of this book, however. (Consult Reference 1 for an explanation of this mechanism.)

Problems

1. Define and, where possible, illustrate the following terms.
   a. acid
   b. base
   c. salt
   d. hydronium ion
   e. hydroxide ion

2. List four general properties of acidic solutions.

3. List four general properties of basic solutions.

4. What ion is responsible for the properties of acidic solutions (in water)?

5. What ion is responsible for the properties of basic solutions (in water)?

6. Give the formulas and the names of two strong acids and two weak acids.

7. Give the formulas and the names for two strong bases and one weak base.

8. Identify the first compound in each equation as an acid or a base.
   a. $C_5H_5N + H_2O \rightarrow C_5H_5NH^+ + OH^-$
   b. $C_6H_5OH + H_2O \rightarrow C_6H_5O^- + H_3O^+$
   c. $CH_3COCOOH + H_2O \rightarrow$
      $\qquad\qquad CH_3COCOO^- + H_3O^+$
   d. $C_6H_5SH + H_2O \rightarrow C_6H_5S^- + H_3O^+$
   e. $CH_3NH_2 + H_2O \rightarrow CH_3NH_3^+ + OH^-$
   f. $C_6H_5SO_2NH_2 + H_2O \rightarrow$
      $\qquad\qquad C_6H_5SO_2NH^- + H_3O^+$

9. What is meant by the proton as used in acid–base chemistry? How does it differ from the proton of nuclear chemistry (Chapter 4)?

10. What is an acid anhydride?

11. Give the formula for the acid formed when sulfur trioxide reacts with water.

12. What is a basic anhydride?

13. Give the formula for the base formed when potassium oxide reacts with water.

14. Write the equation that shows how ammonia acts as a base in water.

15. What is the effect of strong acids on clothing?

16. What is the effect of strong acids and strong bases on the skin?

17. How does one brand of household ammonia differ from another?

18. What is meant by the neutralization of an acid or base?

19. Write the equation for the reaction of sodium hydroxide with hydrochloric acid.

20. Describe the taste and effect on litmus of a solution that has been neutralized.

21. Magnesium hydroxide is completely ionic, even in the solid state; yet it can be taken internally as an antacid. Explain why it does not cause injury as sodium hydroxide would.

22. How does magnesium hydroxide, in large doses, act as a laxative? Would other magnesium compounds have a similar effect?

23. Name some of the active ingredients in antacids.

24. Should a person who has hypertension be advised to use baking soda or milk of magnesia as an antacid? Why?

25. Strong acids and weak acids both have properties characteristic of hydronium ions. How do strong and weak acids differ?

26. Give a general definition for an acid. Write an equation that illustrates your definition.

27. Indicate whether each of the following pH values represents an acidic, basic, or a neutral solution.
    a. 4   b. 7   c. 3.5   d. 9

28. Lime juice has a pH of about 2. Is lime juice acidic or basic?

29. What is aqueous ammonia? Why is it sometimes called ammonium hydroxide?

30. What is the pH of a solution that has a hydronium ion concentration of 1×10^{-8} M?

31. What is the pH of a solution that has a hydronium ion concentration of 1×10^{-2} M?

32. What is the hydronium ion concentration of a solution that has a pH of 7?

33. What is the hydronium ion concentration of a solution that has a pH of 10?

34. What is the leading chemical product of United States industry?

35. Give two ways in which corrosive acids and alkalis destroy cells.

36. Lime has been used to neutralize water in lakes that have become too acidic from acid precipitation. Assume that the acid is sulfuric acid and write the equation for the reaction. How many metric tons of lime does it take to neutralize 10 t of sulfuric acid?

37. Examine the labels of at least five antacid preparations. Make a list of the ingredients in each. Look up the properties (medical use, side effects, toxicity, etc.) of each ingredient in a reference book such as *The Merck Index* (Reference 7).

38. Examine the labels of at least five toilet-bowl cleaners and five drain cleaners. Make a list of the ingredients in each. Look up the formulas and properties of each ingredient in a reference book such as *The Merck Index* (Reference 7). Which ingredients are acids? Which are bases?

References and Readings

1. Hill, John W., and Dorothy M. Feigl. *Chemistry and Life*, 3rd ed. New York: Macmillan, 1987. See especially Chapters 10 and 11.

2. "How to Choose an Antacid." *Consumer Reports*, August 1983, pp. 412–418.

3. Kohn, Harold W. "The Rolaids Caper or Titration in the Classroom." *Chemistry*, March 1973, pp. 27–28.

4. Kolb, Doris. "Chemical Principles Revisited: Acids and Bases." *Journal of Chemical Education*, July 1978, pp. 459–464.

5. Morris, D. L. "Brønsted-Lowry Acid-Base Theory—A Brief Survey." *Chemistry*, March 1970, pp. 18–19.

6. VanderWerf, Calvin A. *Acids, Bases, and the Chemistry of the Covalent Bond*. New York: Reinhold, 1961.

7. Windholz, Martha (Ed.). *The Merck Index*, 10th ed. Rahway, NJ: Merck and Co., 1983.

8

Oxidation and Reduction

Burn and Unburn

Chemical reactions can be classified in several ways. In this chapter, we consider an important group of reactions called **reduction–oxidation (or redox) reactions**. The two processes—oxidation and reduction—always occur together. You can't have one without the other. One substance is oxidized; another is reduced (Color Plate H). For convenience, however, we may choose to talk about only a part of the process—the oxidation part or the reduction part.

Our cells obtain energy to maintain themselves by oxidizing foods. Green plants, using energy from sunlight, produce food by the reduction of carbon dioxide. We win metals from their ores by reduction, then lose them again to corrosion as they are oxidized. We maintain our technological civilization by oxidizing fossil fuels (coal, natural gas, and petroleum) to obtain the chemical energy that was stored in these materials eons ago by green plants.

Reduced forms of matter—sugars, coal, gasoline—are high in energy. Oxidized forms—carbon dioxide and water—are low in energy. Let's examine the processes of oxidation and reduction in some detail, in order that we might better understand the chemical reactions that keep us alive and enable us to maintain our civilization.

Figure 8.1 Reduced forms of matter, such as foods and fossil fuels, are high in energy. The energy content is released through oxidation-reduction reactions.

161

Oxygen: Abundant and Essential

It would be inaccurate to say that any one element is the most important one, for there are 20 or so elements essential to life. Nevertheless, in any list of important elements, oxygen would be at or near the top. It is an abundant element, making up about half the Earth's crust. Oxygen occurs in each of the three subdivisions of the crust—the atmosphere, the hydrosphere, and the lithosphere. It occurs in the **atmosphere** (the gaseous mass surrounding the Earth) as molecular oxygen (O_2). In the **hydrosphere** (the oceans, seas, rivers, and lakes of the Earth), oxygen occurs in combination with hydrogen in the remarkable compound water (H_2O). In the **lithosphere** (the solid portion of the Earth), oxygen occurs in combination with silicon (pure sand is largely SiO_2) and a variety of metals.

Oxygen is found in most of the compounds that are important to living organisms. Foodstuffs—carbohydrates, fats, and proteins—all contain oxygen. The human body is approximately 65% (by weight) water. Since 89% of the weight of water is due to oxygen—and many other compounds in our bodies also contain oxygen—over 60% of the weight of each one of us is oxygen.

Elemental oxygen (O_2) accounts for about 21% (by volume) of the atmosphere. (The remainder is mainly nitrogen (N_2), which enters into few chemical reactions.) Atmospheric oxygen is said to be free or uncombined, that is, not part of a compound containing another element. Oxygen (in air) is taken into the lungs. From there it passes into our bloodstreams. The blood carries oxygen to the body tissues, and the food we eat combines with this oxygen. This chemical process provides us with energy in the form of heat to maintain our body temperature. It also gives us the energy we need for mental and physical activity.

Oxygen performs many other functions. Fuels such as natural gas, gasoline, and coal need oxygen to burn and release their stored energy. Combustion of fossil fuels currently supplies over 90% of the energy that turns the wheels of civilization.

Pure oxygen is obtained by liquefying purified air, then letting the nitrogen and argon boil off. (Nitrogen boils at −196 °C, argon at −186 °C, and oxygen at −183 °C.) About 13 billion kg of oxygen of 99.5% purity is produced annually in the United States. Most of this oxygen is used directly by industry, much of it in the metals industry. About 1% is compressed into steel tanks for use in welding, in medicine, and many other purposes.

Not all that atmospheric oxygen does is immediately desirable. Oxygen causes iron to rust and copper to corrode, and it aids in the decay of wood. All these—and many other—chemical processes are called oxidation.

Figure 8.2 Fuels burn more rapidly in pure oxygen than in air. Huge quantities of liquid oxygen are used to burn the fuels that blast rockets into orbit. [Courtesy National Aeronautics and Space Administration.]

Chemical Properties of Oxygen: Oxidation

When iron rusts, it combines with oxygen from the atmosphere to form a reddish brown powder.

$$4\,Fe \;+\; 3\,O_2 \;\longrightarrow\; 2\,Fe_2O_3$$

The chemical name for iron rust (Fe_2O_3) is iron(III) oxide. Sometimes it also is called by an older name, ferric oxide. Many other metals react with oxygen to form *oxides*.

Most nonmetals also react with oxygen to form oxides. For example, carbon reacts with oxygen to form carbon dioxide.

$$C \;+\; O_2 \;\longrightarrow\; CO_2$$

Sulfur combines with oxygen to form sulfur dioxide.

$$S \;+\; O_2 \;\longrightarrow\; SO_2$$

And, at high temperatures (such as those that occur in automobile engines), even nitrogen, which is ordinarily quite unreactive, combines with oxygen.

$$N_2 + O_2 \longrightarrow 2\,NO$$

The product is known as nitric oxide.

Example 8.1 Magnesium combines readily with oxygen when ignited in air. Write the equation for the reaction.

Magnesium is an element; its symbol is Mg. Oxygen occurs as O_2 molecules. The reaction is

$$Mg + O_2 \longrightarrow MgO \qquad (\textit{Not balanced})$$

To balance the equation, we need two Mg and two MgO.

$$2\,Mg + O_2 \longrightarrow 2\,MgO$$

Oxygen also reacts with many compounds. Methane, the principal ingredient in natural gas, burns in air to produce carbon dioxide and water.

$$CH_4 + 2\,O_2 \longrightarrow CO_2 + 2\,H_2O$$

Hydrogen sulfide, a gaseous compound with a rotten-egg odor, burns, producing water and sulfur dioxide.

$$2\,H_2S + 3\,O_2 \longrightarrow 2\,H_2O + 2\,SO_2$$

In each example, oxygen combines with each of the elements in the compound.

Example 8.2 Carbon disulfide combines readily with oxygen in the air. What products are formed? Write the equation.

Carbon disulfide is CS_2. The products are CO_2 and SO_2. The balanced equation is

$$CS_2 + 3\,O_2 \longrightarrow CO_2 + 2\,SO_2$$

The combination of elements and compounds with oxygen is called **oxidation**. The substances that combine with oxygen are said to have

Figure 8.3 Cooking, breathing, and burning fuel all involve oxidation.

been **oxidized**. Originally, the term *oxidation* was restricted to reactions involving combination with oxygen. Chemists came to recognize, though, that combination with chlorine (or bromine or other elements in the upper-right of the periodic chart) was not all that different from reaction with oxygen. So they broadened the definition of oxidation. We won't concern ourselves, though, with a formal definition. The three following working definitions should suffice.

1. An element or a compound is oxidized when it gains oxygen atoms. In the reactions we've encountered so far in this section, iron, carbon, sulfur, nitrogen, methane, and hydrogen sulfide all gain oxygen atoms. We say, then, that each of these substances is oxidized. Consider also the "burning" of glucose by living cells. This process, called **respiration**, is the main reaction by which cells obtain energy.

$$C_6H_{12}O_6 \ + \ 6\,O_2 \ \longrightarrow \ 6\,CO_2 \ + \ 6\,H_2O$$

In the reaction, carbon atoms gain oxygen. Each has one oxygen in glucose but two in carbon dioxide. Glucose is oxidized.

2. A compound is oxidized when it loses hydrogen atoms. Methyl alcohol, when passed over hot copper gauze, forms formaldehyde and hydrogen gas.

$$\underset{\text{alcohol}}{\underset{\text{Methyl}}{CH_4O}} \ \longrightarrow \ \underset{\text{Formaldehyde}}{CH_2O} \ + \ H_2$$

Since methyl alcohol loses hydrogen atoms, it is said to have been oxidized.

3. An element is oxidized when it loses electrons. When magnesium metal reacts with chlorine, magnesium ions and chloride ions are formed.

$$Mg + Cl_2 \longrightarrow Mg^{2+} + 2\,Cl^-$$

Since the magnesium atom obviously loses electrons, it is oxidized.

We encounter many reactions that fit these definitions in subsequent sections of this book. For example, it is by oxidation of foodstuffs that organisms derive their life-sustaining energy. Our bodies also oxidize toxic molecules, making them more polar, in order to eliminate them in urine, an aqueous solution.

Why so many different definitions of oxidation? Simply for convenience. Which do we use? Whichever is most convenient. For the reaction

$$C + O_2 \longrightarrow CO_2$$

it is convenient to see that carbon is oxidized because it gains oxygen atoms. Similarly, for the reaction

$$CH_4O \longrightarrow CH_2O + H_2$$

it is easy to see that carbon is oxidized because it loses hydrogen atoms.

Hydrogen: Occurrence, Preparation, and Physical Properties

For every oxidation process, there is a complementary process—reduction. Before discussing reduction in detail, let's look at some of the chemistry of another element—hydrogen.

By weight, hydrogen makes up only about 0.9% of the Earth's crust, so hydrogen is placed far down on the list of abundant elements.* However, hydrogen atoms are quite abundant since hydrogen is the lightest element. If we considered a random sample of 10 000 atoms from the Earth's crust, 5330 would be oxygen atoms, 1590 would

---

*Hydrogen ranks low in abundance on Earth. If we look beyond our home planet, however, hydrogen becomes much more significant. The sun, for example, is made up largely of hydrogen. In fact, in the universe, hydrogen is by far the most abundant element.

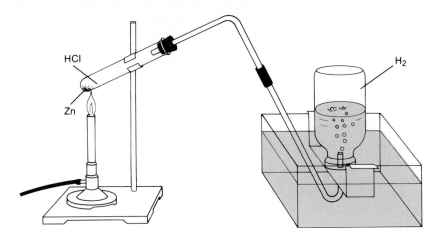

be silicon atoms, and 1510 would be hydrogen atoms. Unlike oxygen, hydrogen is seldom found free in nature on Earth. Most of it is combined with oxygen in water. Nearly all compounds derived from living organisms contain hydrogen. Fats, starches, sugars, and proteins all contain combined hydrogen. Petroleum and natural gas are mixtures composed mainly of *hydrocarbons*, compounds of hydrogen and carbon (Chapter 10).

Since elemental hydrogen does not occur in nature (again, we limit our view to this planet), it is necessary for humans to manufacture it. Small amounts can be made for laboratory use by the reaction of zinc with hydrochloric acid.

$$Zn + 2\,HCl \longrightarrow ZnCl_2 + H_2$$
$$\text{Zinc}$$
$$\text{chloride}$$

or by the reaction of calcium metal with water.

$$Ca + 2\,H_2O \longrightarrow Ca(OH)_2 + H_2$$
$$\text{Calcium}$$
$$\text{hydroxide}$$

Gaseous hydrogen can be easily collected by displacement of water (Figure 8.4). Commercially, hydrogen is usually obtained as a by-product of other processes. Much of it comes from petroleum refineries (Chapter 13). United States production of hydrogen is about 200 million kg per year. About two thirds goes to make ammonia (Chapter 17).

Physically, hydrogen is a colorless and odorless gas. It is essentially insoluble in water. Hydrogen is the least dense of all substances: its density is only 1/14 that of air under comparable conditions. For this

Figure 8.5 Hydrogen is the most effective buoyant gas, but it is highly flammable. Disastrous fires in hydrogen-filled dirigibles such as the Hindenburg (shown here) led to the replacement of hydrogen by nonflammable helium. [Courtesy of National Air and Space Museum, Smithsonian Institution, Washington, DC.]

reason it was once used in dirigibles to give them the necessary buoyancy in air. However, since a spark or flame can trigger the explosion of hydrogen (see next section), the use of this gas represented a considerable danger. When a disastrous explosion occurred aboard a luxury airship called the Hindenburg in 1937, the use of hydrogen was discontinued—and so, indeed, was the commercial use of lighter-than-air ships. Those blimps (small dirigibles) still in use employ the nonreactive gas helium, while hot-air balloons use, as their name indicates, hot air, which is less dense than the atmosphere (Figure 8.5).

Certain metals, such as platinum (Pt), palladium (Pd), and nickel (Ni), collect large volumes of hydrogen (in a condensed form) on their surfaces. This **adsorbed** hydrogen appears to be a great deal more reactive than ordinary molecular hydrogen; thus, these metals often are used as catalysts* for reactions involving H_2 as one of the reactants.

The Chemical Properties of Hydrogen: Reduction

Chemically, a jet of hydrogen (when ignited) burns in air with an almost colorless flame. Hydrogen and oxygen can be mixed at room temperature with no perceptible reaction. When the mixture is ignited by a spark or a flame, however, a tremendous explosion results. The product in both cases is water.

$$2 \ H_2 \ + \ O_2 \ \xrightarrow{\text{spark}} \ 2 \ H_2O$$

It is interesting to note that, when a piece of platinum gauze is inserted into a container of hydrogen and oxygen, the two gases react at room temperature. The platinum acts as a catalyst; it lowers the activation energy for the reaction. The platinum glows from the heat released in the initial reaction and then ignites the mixture, causing an explosion.

Hydrogen reacts with a variety of metal oxides to remove oxygen and yield the free metal. For example, when hydrogen is passed over heated copper oxide, metallic copper and water are formed.

$$CuO \ + \ H_2 \ \longrightarrow \ Cu \ + \ H_2O$$

With lead oxide, the products are metallic lead and water.

---

* A **catalyst** is a substance that increases the *rate* of a chemical reaction without itself being used up. For example, unsaturated fats (Chapter 18) ordinarily react very slowly with hydrogen. In the presence of nickel metal, however, the reaction proceeds readily. The nickel does not increase the *amount* of product formed, though.

$$PbO + H_2 \longrightarrow Pb + H_2O$$

This process, in which a compound is changed or reduced to an element, is labeled reduction.

Like oxidation, which once meant simply the combination of an element or a compound with oxygen, the term **reduction** has a broadened meaning. Instead of a formal definition, however, we use working definitions like those given for oxidation.

1. A compound is reduced when it loses oxygen atoms. In the examples above, CuO and PbO obviously were reduced, since they lost oxygen. When potassium chlorate ($KClO_3$) is heated, it forms potassium chloride and oxygen gas.

$$2\ KClO_3 \xrightarrow{\text{heat}} 2\ KCl + 3\ O_2$$

The potassium chlorate loses oxygen; therefore, it is reduced.

2. A compound is reduced when it gains hydrogen atoms. For example, methyl alcohol is made by reaction of carbon monoxide with hydrogen.

$$CO + 2\ H_2 \xrightarrow{\text{catalyst}} CH_4O$$

The carbon monoxide gains hydrogen; therefore, it is reduced.

3. An atom or an ion is reduced when it gains electrons. When an electric current is passed through a solution containing copper ions (Cu^{2+}), copper metal is plated out at the cathode.

$$Cu^{2+} + 2\ e^- \longrightarrow Cu$$

The copper ion obviously gains electrons; therefore, it is reduced.

Oxidation and reduction go hand in hand. You can't have one

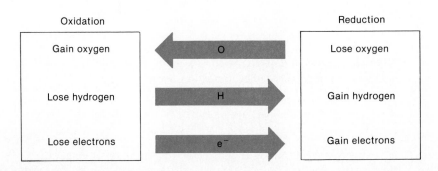

Oxidation		Reduction
Gain oxygen	O	Lose oxygen
Lose hydrogen	H	Gain hydrogen
Lose electrons	e⁻	Gain electrons

Figure 8.6 Three different definitions of oxidation and reduction.

without the other. When one substance is oxidized, another is reduced. For example, in the reaction

$$CuO + H_2 \longrightarrow Cu + H_2O$$

copper oxide is reduced and hydrogen is oxidized. Further, if one substance is being oxidized, the other must be causing it to be oxidized. In the example, CuO is causing H_2 to be oxidized. Therefore, CuO is called the **oxidizing agent**. Conversely, H_2 is causing CuO to be reduced, so H_2 is the **reducing agent**. Each oxidation–reduction reaction has an oxidizing agent and a reducing agent among the reactants. The reducing agent is the substance being oxidized; the oxidizing agent is the substance being reduced.

Reduction:
copper oxide is being reduced;
CuO is the oxidizing agent.

$$CuO + H_2 \longrightarrow Cu + H_2O$$

Oxidation:
hydrogen is being oxidized;
H_2 is the reducing agent.

Example 8.3 Circle the oxidizing agents and underline the reducing agents in the following reactions.

$$2\ C + O_2 \longrightarrow 2\ CO$$
$$N_2 + 3\ H_2 \longrightarrow 2\ NH_3$$
$$SnO + H_2 \longrightarrow Sn + H_2O$$
$$Mg + Cl_2 \longrightarrow Mg^{2+} + 2\ Cl^-$$

The answers can be determined using the definitions previously given.

$$\underline{C} + \boxed{O_2}$$

C gains oxygen and is oxidized, so it must be the reducing agent. Therefore, O_2 is the oxidizing agent.

$$\boxed{N_2} + \underline{H_2}$$

N_2 gains hydrogen and is reduced, so it is the oxidizing agent.

Therefore, H_2 is the reducing agent.

$$\boxed{SnO} \ + \ \underline{H_2}$$

SnO loses oxygen and is reduced, so it is the oxidizing agent. H_2 is the reducing agent.

$$\underline{Mg} \ + \ \boxed{Cl_2}$$

Mg loses electrons and is oxidized, so it is the reducing agent. Cl_2 is the oxidizing agent.

Electrochemical Cells: Batteries

Electricity can cause chemical change (Chapter 3). Conversely, chemical change can produce electricity. Dry cells, storage batteries, and fuel cells all convert chemical energy to electrical energy.

If a strip of zinc metal is placed in a solution of copper(II) sulfate, the following reaction takes place.

$$Zn\colon \ + \ Cu^{2+} \ \longrightarrow \ Zn^{2+} \ + \ Cu\colon$$

The zinc atoms give up their outer electrons to the copper(II) ions. (The sulfate ions are not changed in the reaction.) The zinc metal dissolves, going into solution as Zn^{2+} ions. The copper ions precipitate from the solution as copper metal. The Zn is oxidized; the Cu^{2+} ions are reduced (Color Plate I). Electrons are transferred directly from zinc atoms to copper(II) ions.

Each metal differs in its tendency to give up electrons. Copper is less likely to do so than zinc. If we connect a copper strip to a zinc strip, electrons will flow from the zinc to the copper. This flow of electrons constitutes an **electric current**. If we place copper ions in one compartment and zinc metal in another, physically separating them (Figure 8.7), electrons have to pass through a wire to get from the zinc metal to the copper ions. This flow of electricity can be used to run a motor or light a bulb.

Note in Figure 8.7 that the zinc compartment also contains a solution of zinc sulfate. The compartment with the copper ions also has a strip of copper metal. The metal strips serve as electrodes (Chapter 3), conducting electrons into and out of the solutions. The strips also participate in the chemical reactions. The zinc electrode slowly dissolves

Figure 8.7 A simple electrochemical cell. A battery may consist of one or more such cells.

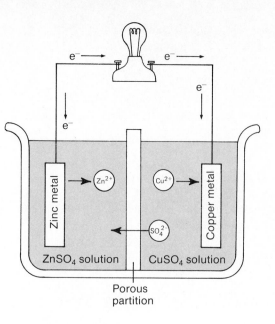

ZnSO$_4$ solution   CuSO$_4$ solution

Porous partition

as zinc atoms give up their electrons. Copper metal is plated out on the copper electrode as the copper(II) ions pick up those electrons after they have traveled through the wire.

Both the compartments in Figure 8.7 contain sulfate ions. These ions don't enter into any chemical reaction, but they do cause complications. Consider the zinc sulfate solution. During the reaction, zinc atoms are converted to zinc ions, so the solution becomes more concentrated in zinc ions. However, nothing happens to the sulfate ions. We are faced with the possibility of the solution containing more positively charged ions (Zn^{2+} ions) than negatively charged ions (SO_4^{2-} ions). This electrical imbalance is an unstable situation, and if something isn't done about it, the reaction will stop. In the other compartment of Figure 8.7, the concentration of copper ions is being reduced while the concentration of sulfate ions remains constant. Thus, there is a tendency for this compartment to develop an excess of negatively charged ions. One way to deal with the problem is to connect the two solutions through a porous plate, made of a material that will retard the mixing of the metal ions in solution and still permit passage of sulfate ions. When a zinc atom is converted to a zinc ion in the left compartment, a copper ion changes to a copper atom in the right, and a sulfate ion moves through the porous plate from the solution in the right compartment to the solution in the left. The charges in the two solutions stay balanced. Why bother with the porous plate? Why not just put both solutions and both electrodes in a single compartment so the sulfate ions don't have to move through pores? If we were to do that, the electrons wouldn't have to make the trip through the wire; they could just be passed from zinc atoms to copper ions, which would now be in contact with one another. And unless the electrons go through the wire, the light bulb won't glow (nor will battery-operated portable radios, hand calculators, or cardiac pacemakers work).

The net reaction for this **electrochemical cell** is

$$Zn + Cu^{2+} \longrightarrow Zn^{2+} + Cu$$

If 1 *M* solutions of $ZnSO_4$ and $CuSO_4$ are used, the system will produce about 1.1 volts (V) at 25 °C. The **volt** is a measure of electrical potential, or of the tendency of the electrons in a system to flow.

A **battery** is a series of electrochemical cells. The familiar 12-V automobile storage battery consists of six cells wired in series (that is, one after the other). Each cell consists of two electrodes. The cells do not incorporate the zinc–copper couple just described but a somewhat more complicated system involving lead, lead oxide, and sulfuric acid. A distinctive feature of the lead storage battery is its capacity for being recharged. It will supply electricity (i.e., it will discharge) when you turn on the ignition to start the car or when the motor is off and the lights are on. But it can be recharged when the car is moving and an electric current is supplied *to* the battery by the mechanical action of the car. The net reaction during discharge is

$$Pb + PbO_2 + 2\ H_2SO_4 \longrightarrow 2\ PbSO_4 + 2\ H_2O$$

That during recharge is just the reverse:

$$2\ PbSO_4 + 2\ H_2O \longrightarrow Pb + PbO_2 + 2\ H_2SO_4$$

These lead storage batteries are durable, but they are heavy and contain corrosive sulfuric acid.

The carbon–zinc dry cell (Figure 8.8) is used in flashlights, tape recorders, camera flash units, and many other small devices that require

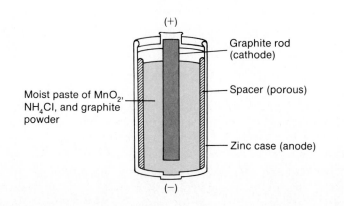

(+)

Graphite rod
(cathode)

Moist paste of MnO_2,
NH_4Cl, and graphite
powder

Spacer (porous)

Zinc case (anode)

(−)

Figure 8.8 Cross section of a zinc–carbon call.

a portable source of power. The case is made of zinc metal and serves as the anode. The carbon rod serves as the cathode. It is surrounded by a moist paste of graphite powder, manganese dioxide (MnO_2), and ammonium chloride. The anode reaction is simple; zinc is oxidized to zinc ions. The reaction at the cathode is more complex, but a major process is the reduction of manganese dioxide. The overall reaction is mainly

$$Zn + 2 MnO_2 + 2 H_2O \longrightarrow Zn^{2+} + Mn_2O_3 + 2 OH^-$$

Alkaline cells are similar, but the zinc case is porous, and the carbon cathode is surrounded by moist manganese dioxide and potassium hydroxide. These cells are more expensive than the usual carbon–zinc cells, but they maintain a high voltage longer.

Corrosion

An oxidation–reduction reaction of particular economic importance is corrosion of metals. It is estimated that in the United States alone corrosion costs $70 billion a year. Perhaps 20% of all the iron and steel production in the United States each year goes to replace corroded items. Let's look first at the corrosion of iron.

In moist air, iron is oxidized, particularly at a nick or scratch.

$$Fe \longrightarrow Fe^{2+} + 2 e^-$$

In order for iron to be oxidized, oxygen must be reduced.

$$O_2 + 2 H_2O + 4 e^- \longrightarrow 4 OH^-$$

The net result, initially, is the formation of insoluble iron(II) hydroxide.

$$2 Fe + O_2 + 2 H_2O \longrightarrow 2 Fe(OH)_2$$

This product is usually further oxidized to iron(III) hydroxide.

$$4 Fe(OH)_2 + O_2 + 2 H_2O \longrightarrow 4 Fe(OH)_3$$

The latter, sometimes written as $Fe_2O_3 \cdot 3H_2O$, is the familiar iron rust.

The corrosion of iron appears to be an electrochemical reaction (Color Plate K). Oxidation and reduction often occur at separate points on the metal surface. Electrons are transferred through the iron metal. The circuit is completed by an electrolyte in aqueous solution. In the snowbelt, this solution is often the slush from road salt and melting

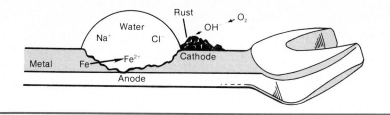

Figure 8.9 The corrosion of iron requires water, oxygen, and an electrolyte.

snow. The metal is pitted in an *anodic* area, where iron is oxidized to Fe^{2+}. These ions migrate to the *cathodic* area, where they react with the hydroxide ions formed by reduction of oxygen.

$$Fe^{2+} + 2 OH^- \longrightarrow Fe(OH)_2$$

As indicated above, this iron(II) hydroxide is then oxidized to $Fe(OH)_3$, or rust. This process is diagrammed in Figure 8.9. Notice that the anodic area is protected from oxygen by the water film, while the cathodic area is exposed to air.

Aluminum is more reactive than iron. We might expect it to corrode more rapidly, but billions of beer cans testify to the fact that it doesn't. How can that be? It so happens that freshly prepared aluminum quickly forms a thin, hard film of aluminum oxide on its surface. This film is impervious to air and protects the underlying metal from further oxidation.

The tarnish on silver is another example of corrosion. Silver is oxidized to silver ions.

$$Ag \longrightarrow Ag^+ + e$$

These silver ions react with sulfur compounds such as hydrogen sulfide to form black silver sulfide.

$$2 Ag^+ + H_2S \longrightarrow Ag_2S + 2 H^+$$

You can use a silver polish to remove the tarnish, but in doing so, you also lose part of the silver. An alternative method involves the use of aluminum metal to reduce the silver ions back to silver metal.

$$3 Ag^+ + Al \longrightarrow 3 Ag + Al^{3+}$$

This reaction also requires an electrolyte. Sodium bicarbonate ($NaHCO_3$) is usually used. The tarnished silver is placed in contact with aluminum foil, covered with a solution of sodium bicarbonate, and heated. A precious metal is conserved at the expense of a cheaper one.

Some Common Oxidizing Agents

Oxygen is undoubtedly the most common oxidizing agent. It oxidizes the wood in our campfires and the gasoline in our automobile engines. It rusts and corrodes the metals that we get by reducing ores. It even "burns" the foods we eat to give us energy to move and to think. It is involved in the rotting of wood and the weathering of rocks. Indeed, we live in an oxidizing atmosphere. Fortunately, though, oxygen is a mild oxidizing agent; otherwise, living things would burn up quickly. As it is, it takes us about 70 years to "burn out."

Oxygen sometimes is used as a laboratory and industrial oxidizing agent. For example, acetylene (used to fuel cutting and welding torches) is made by the partial oxidation of methane.

$$4 \text{ CH}_4 + 3 \text{ O}_2 \longrightarrow 2 \text{ C}_2\text{H}_2 + 6 \text{ H}_2\text{O}$$

Methane   Acetylene

(Complete oxidation of methane gives carbon dioxide and water.) Often, though, it is more convenient to use other oxidizing agents. In the laboratory, sodium dichromate ($\text{Na}_2\text{Cr}_2\text{O}_7$) is used frequently to oxidize alcohols to compounds called *aldehydes* and *ketones*. For the oxidation of ethyl alcohol (found in alcoholic beverages) to acetaldehyde, the reaction is

$$8 \text{ H}^+ + \text{Cr}_2\text{O}_7^{2-} + 3 \text{ C}_2\text{H}_5\text{OH} \longrightarrow 2 \text{ Cr}^{3+} + 3 \text{ C}_2\text{H}_4\text{O} + 7 \text{ H}_2\text{O}$$

Dichromate ion   Chromium(III) ion
(orange-red)   (green)

Some of the tests for intoxication (the breathalyzer used to test drunken drivers, for instance) depend on the oxidation of alcohol by an oxidizing agent that changes color as it is reduced (such as a chromium compound).

Another common oxidizing agent is hydrogen peroxide (H_2O_2). Pure hydrogen peroxide is a syrupy, colorless liquid. It usually is used as an aqueous solution, with concentrations of 30 and 3% generally available. Some interesting uses of hydrogen peroxide are discussed in the next section. For now, let's look at some of its simpler chemistry.

Hydrogen peroxide oxidizes sulfides (S^{2-}) to sulfates (SO_4^{2-}). When lead-based paints are exposed to polluted air containing hydrogen sulfide (H_2S), they turn black because of the formation of lead sulfide (PbS). Hydrogen peroxide oxidizes the black sulfide to white sulfate.

$$\text{PbS} + 4 \text{ H}_2\text{O}_2 \longrightarrow \text{PbSO}_4 + 4 \text{ H}_2\text{O}$$

(black)   (white)

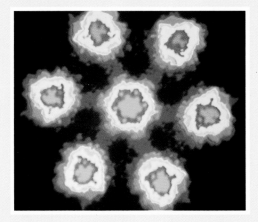

A. Images of Individual Atoms. Shown here are images of seven individual uranium atoms (the colored spots with red-orange centers). The images are made with an electron microscope. Uranyl acetate clusters are pictured on an extremely thin carbon layer that appears black in this photograph. The uranium atoms are 0.34 nm apart. [Courtesy of M. Isaacson (Cornell University) and M. Ohtsuki (University of Chicago).]

B. A Cathode Ray Experiment. In 1879 William Crookes observed the deflection of cathode rays in a magnetic field. Cathode rays are themselves invisible but are observed through the green fluorescence they produce when they strike a zinc sulfide-coated screen. The magnetic field is produced by the large magnet to the right and slightly behind the screen. [Photograph by Carey B. Van Loon.]

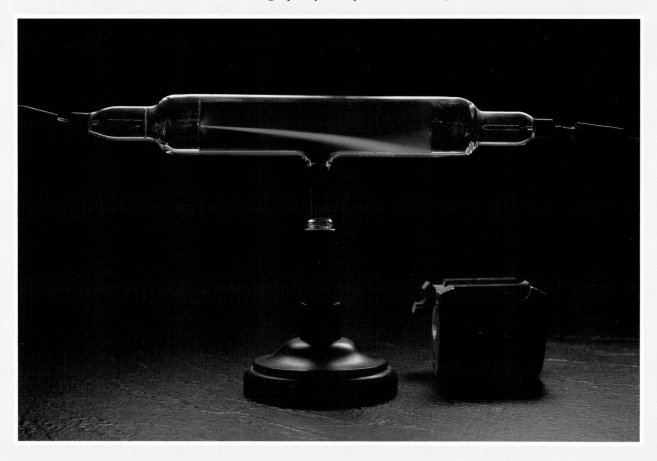

C. The Continuous Spectrum of White Light. A glass prism disperses a beam of white light into a continuous spectrum or rainbow of colors. [Courtesy of Bausch & Lomb.]

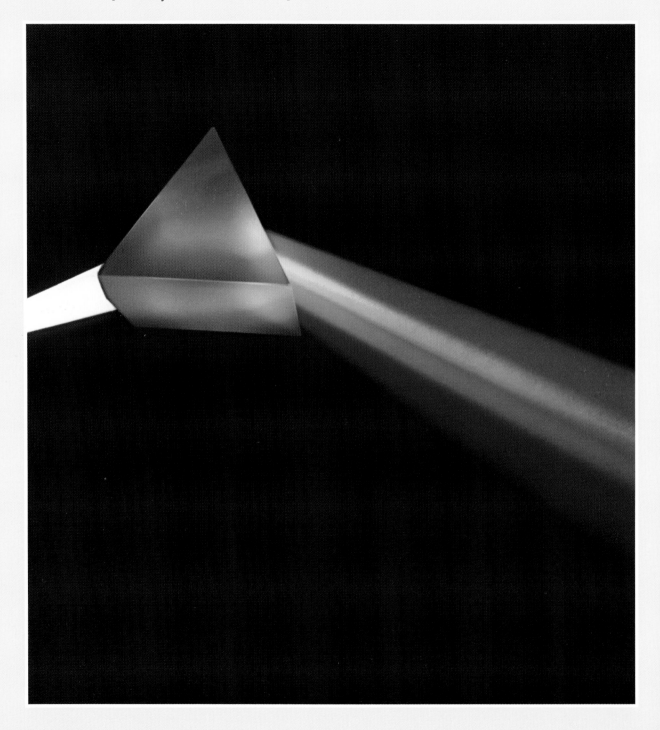

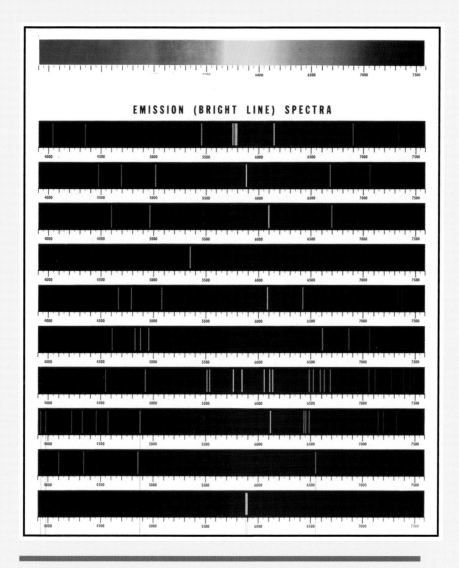

D. Line Spectra of Selected Elements. Some of the components of the light emitted by excited atoms appear as colored lines. Atomic spectra of several elements are shown here, with a continuous spectrum at the top for comparison. Wavelengths are given in angstrom units (1 Å $= 10^{-10}$ m). Each element has its own distinctive spectrum different from all others. In addition to their practical use in analyzing matter, atomic spectra have contributed significantly to the development of ideas concerning atomic structure. [Courtesy of Wabash Instrument Co., Wabash, IN]

Li

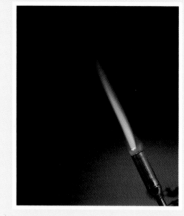

Ca

Na

Sr

K

Ba

E. Flame Tests. Certain chemical elements can be identified through the characteristic colors that their compounds impart to flames. Six examples are shown here. [Photograph by Carey B. Van Loon.]

F. Fireworks. Flame colors (Color Plate E) are also the basis for the brilliant colors of fireworks displays. Red colors are produced by strontium compounds, green by barium compounds, blue by copper compounds, yellow by sodium compounds, and white by powdered magnesium and aluminum metals. [Photograph by Tom Pantages.]

G. Street Lights. A sodium vapor lamp (foreground) produces a yellowish light due to the bright yellow lines in the sodium spectrum (Color Plate D). The spectrum of mercury (Color Plate D) exhibits lines throughout the visible region, but the intense blue lines predominate. The mercury vapor lamp (background) therefore sheds a bluish light. [Photograph by Carey B. Van Loon.]

H. Oxidation–Reduction Reactions. Pictured on the left is ammonium dichromate, which contains both an oxidizing agent ($Cr_2O_7^{2-}$) and a reducing agent (NH_4^+). The equation is

$$(NH_4)_2Cr_2O_7 \longrightarrow Cr_2O_3 + N_2 + 4\ H_2O$$

Considerable heat and light are evolved (center). The water is driven off as vapor, and the nitrogen gas escapes, leaving pure Cr_2O_3 as the visible product (right). [Photographs by Carey B. Van Loon.]

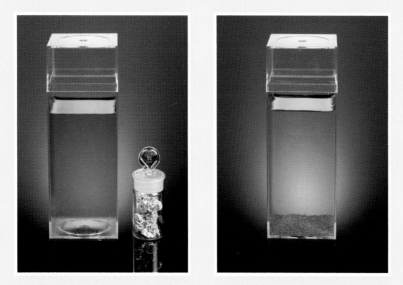

I. Displacement of Copper(II) Ions by Zinc Metal. When zinc metal is placed in a solution of copper(II) ions, the Cu^{2+} ions are displaced from solution as Cu metal. A Cu^{2+} solution and Zn metal are shown in the photograph at the left. The products of the displacement reaction (right) are a red-brown precipitate of Cu and a colorless solution of Zn^{2+}. The equation is

$$Cu^{2+} + Zn \longrightarrow Zn^{2+} + Cu$$

[Photographs by Carey B. Van Loon.]

J. Displacement of Silver Ions by Copper Metal. When a coil of copper wire is placed in a solution of silver nitrate (left), the following reaction begins.

$$2\,Ag^+ + Cu \longrightarrow 2\,Ag + Cu^{2+}$$

After a few minutes, a crystalline deposit of silver can be seen and the solution is colored blue by the Cu^{2+} ions. The photograph at the right was taken when the reaction was essentially complete (after about 2 hours). [Photographs by Carey B. Van Loon.]

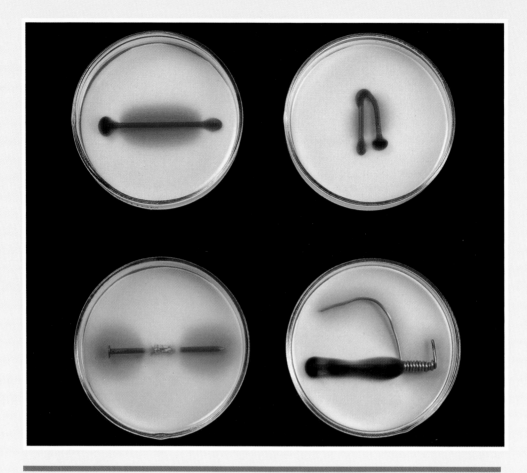

K. The Corrosion of Iron. In the corrosion of an iron nail (top left) oxidation of the iron to Fe^{2+}(aq) occurs at the strained regions of the nail—the head and the tip. The presence of Fe^{2+}(aq) is marked by the formation of a blue precipitate in the presence of $K_3[Fe(CN)_6]$. Reduction of dissolved O_2 to OH^-(aq) occurs on other portions of the nail. The presence of OH^-(aq) is signaled by the pink coloration of phenolphthalein indicator. With a bent nail, oxidation occurs at the additional point of strain, the bend (top right). An iron nail can be protected from corrosion by the more active metal zinc (bottom left). Here, the zinc corrodes and no blue precipitate is seen, indicating the absence of Fe^{2+}(aq). Coating an iron nail with the less active metal copper (bottom right) provides protection only to the regions fully covered by the copper. The exposed portions of the nail corrode even more rapidly than otherwise. [Photograph by Carey B. Van Loon.]

An advantage of hydrogen peroxide as an oxidizing agent is that it is converted to water, an innocuous by-product, in most reactions.

Other common oxidizing agents are the halogens—fluorine (F_2), chlorine (Cl_2), bromine (Br_2), and iodine (I_2). Chlorine, for example, oxidizes magnesium metal to magnesium ions.

$$Mg + Cl_2 \longrightarrow Mg^{2+} + 2\,Cl^-$$

We encounter still other interesting oxidizing agents as we go along.

Some Reducing Agents of Interest

There is no single reducing agent that stands out as oxygen does among oxidizing agents. Hydrogen reduces many compounds, but it would be rather expensive for large-scale processes. Elemental carbon (as coke obtained by driving off the volatile matter from coal) is used frequently to obtain metals from their ores. For example, tin can be obtained from tin oxide by reduction, with carbon used as the reducing agent.

$$SnO_2 + C \longrightarrow Sn + CO_2$$

Hydrogen can be used for the production of expensive metals, such as tungsten (W). First, the tungsten ore is converted to an oxide (WO_3); then it is reduced in a stream of hydrogen gas at 1200 °C.

$$WO_3 + 3\,H_2 \longrightarrow W + 3\,H_2O$$

Perhaps a more familiar reducing agent, by use if not by name, is the developer used in photography. Photographic film is coated with a silver *salt* (Ag^+Br^-). Silver ions that have been exposed to light react with the developer, a reducing agent (such as the organic compound hydroquinone), to form metallic silver.

$$\underset{\text{Hydroquinone}}{C_6H_4(OH)_2} + 2\,Ag^+ \longrightarrow C_6H_4O_2 + \underset{\substack{\text{Silver} \\ \text{metal}}}{2\,Ag} + 2\,H^+$$

Silver ions not exposed to light are not reduced by the developer. The film is then treated with "hypo," a solution of sodium thiosulfate ($Na_2S_2O_3$), which washes out unexposed silver bromide. That leaves the negative dark where the metallic silver has been deposited (where it was originally exposed to light) and transparent where light did not strike it. Light is then shined through the negative onto light-sensitive paper to make the positive print.

Figure 8.10 (a) A photographic negative. (b) A positive print.

Oxidation and Antiseptics

Many common **antiseptics** (compounds that are applied to living tissue to kill microorganisms or to prevent their growth) are mild oxidizing agents. Hydrogen peroxide (H_2O_2), in the form of a 3% aqueous solution, finds use in medicine as a topical antiseptic. It can be used to treat minor cuts and abrasions on certain parts of the body.

Sodium hypochlorite (NaOCl), available in aqueous solution as a laundry bleach (Purex, Clorox, and the like), finds some use in the irrigation of wounds and for bladder infections. It is also used as a disinfectant and deodorizer. Solutions of iodine also are used as antiseptics.

Benzoyl peroxide [$(C_6H_5COO)_2$], a powerful oxidizing agent, has long been used at 5 and 10% concentrations in ointments for treating acne. In addition to its antibacterial action, benzoyl peroxide acts as a skin irritant, causing old skin to slough off and be replaced with newer, fresher looking skin. When used on areas exposed to sunlight, benzoyl peroxide is thought to promote skin cancer; such use is therefore discouraged.

The exact function of these oxidizing agents as antiseptics is unknown. Their action is rather indiscriminate; they attack human cells as well as microorganisms. For many purposes, these simple inorganic compounds have been replaced by organic compounds such as the phenols, which kill by other mechanisms.

Other oxidizing agents are used as disinfectants. Calcium hypochlorite [$Ca(OCl)_2$], a bleaching powder, is used to disinfect clothing and bedding. Chlorine (Cl_2) is used to kill pathogenic (disease-causing) microorganisms in drinking water. Also, wastewater usually is treated with chlorine before it is returned to a stream or a lake (Chapter 16). Swimming pools are usually disinfected by "chlorination." Very large pools are sometimes sanitized by chlorine gas (transported and stored in steel cylinders as liquid chlorine). Pools chlorinated in this way soon become too acidic and must have the pH adjusted by addition of a base (usually sodium carbonate). Smaller pools are often chlorinated with calcium hypochlorite. These pools become too alkaline, and the pH is lowered by adding hydrochloric acid. (Swimming pool water is usually maintained in the pH range of 7.2 to 7.8.)

Oxidation: Bleaching and Stain Removal

Bleaches are compounds that are used to remove unwanted color from fabrics, hair, or other materials. Nearly any oxidizing agent would do the job. However, some also would harm the fabrics, some would be unsafe, some would produce undesirable by-products, and some would simply be too expensive.

The most familiar laundry bleaches (Chapter 20) are aqueous solutions of sodium hypochlorite (NaOCl). Bleaching powder [Ca(OCl)$_2$] generally is preferred for large-scale bleaching operations. The paper industry uses it to make white paper, and the textile industry uses it to make whiter fabrics.

In both aqueous bleaches and bleaching powder, the active agent is the hypochlorite ion (ClO$^-$). Materials appear colored because loosely bound electrons are boosted to higher energy levels by absorption of visible light. Bleaching agents do their work by removing or tying down these mobile electrons. For example, hypochlorite ions (in water) take up electrons to form chloride ions and hydroxide ions.

$$2e^- + ClO^- + H_2O \longrightarrow Cl^- + 2\,OH^-$$

Hypochlorite bleaches are safe and effective for cotton and linen fabrics. They should not be used for wool, silk, or nylon, however.

Other bleaching agents include hydrogen peroxide, sodium perborate (often formulated $NaBO_2 \cdot H_2O_2$ to indicate a loose association of $NaBO_2$ and H_2O_2), and a variety of chlorine-containing organic compounds that release Cl_2 in water. When hydrogen peroxide is used to bleach hair, it acts as an oxidizing agent. In the process, the black or brown pigment in the hair (melanin) is oxidized to colorless products.

Stain removal is not nearly so simple a process as bleaching. A few stain removers are oxidizing agents or reducing agents; others have quite different chemical natures. Nearly all stains require rather specific stain removers. Hydrogen peroxide in cold water removes blood stains from cotton and linen fabrics. Potassium permanganate removes most stains from white fabrics (except for rayon); the permanganate stain then can be removed by treatment with oxalic acid.

$$5\,H_2C_2O_4 + 2\,MnO_4^- + 6\,H^+ \longrightarrow 10\,CO_2 + 2\,Mn^{2+} + 8\,H_2O$$

Oxalic acid   (purple)   (colorless)

(Oxalic acid also removes rust spots, but its action in that case is as a complexing agent, not a reducing agent. The oxalic acid picks up the otherwise insoluble rust and carries it in a solution that is washed away.)

Sodium thiosulfate ($Na_2S_2O_3$) readily removes iodine stains by reducing iodine to a colorless ion (I$^-$).

$$I_2 + 2\,Na_2S_2O_3 \longrightarrow 2\,NaI + Na_2S_4O_6$$

(brown)   (colorless)

Many stain removers are simply adsorbants (for example, cornstarch, which removes grease spots), solvents (for example, amyl acetate, or acetone, which removes ballpoint ink), or detergents (which remove mustard stains).

Oxidation, Reduction, and Living Things

We obtain energy for physical and mental activities by the slow, multistep oxidation of food. These processes will be discussed in detail in later chapters, but for now we will represent them schematically as

$$\text{Carbohydrates (sugars and starches)} + O_2 \longrightarrow CO_2 + H_2O + \text{energy}$$

$$\text{Fats} + O_2 \longrightarrow CO_2 + H_2O + \text{energy}$$

$$\text{Proteins} + O_2 \longrightarrow CO_2 + H_2O + \text{urea} + \text{energy}$$

Green plants make these foods by reducing carbon dioxide. Energy for reaction, called photosynthesis, comes from the sun. The process, shown here for the formation of glucose, a simple sugar, may be written

$$6\,CO_2 + 6\,H_2O + \text{energy} \longrightarrow C_6H_{12}O_6 + 6\,O_2$$

We eat either the plants or the animals that feed on the plants, and we obtain energy from the carbohydrates, fats, and proteins of those organisms. In fact, the photosynthesis reaction is precisely the reverse of that for the oxidation of carbohydrates. If we take the carbohydrate to be glucose, that reaction is

$$C_6H_{12}O_6 + 6\,O_2 \longrightarrow 6\,CO_2 + 6\,H_2O + \text{energy}$$

We can see that green plants carry out the redox reaction that makes possible all life on Earth. Animals can only oxidize the foods that plants provide. We might therefore consider crop farming a process of reduction. Energy captured in cultivated plants, whether the plants are used directly or are fed to animals, is the basis for human life.

Example 8.4   What substance is oxidized in the photosynthesis reaction (above)? What substance is reduced? What substance is the oxidizing agent? What substance is the reducing agent?

Each carbon atom in CO_2 has two oxygen atoms; each carbon in $C_6H_{12}O_6$ has one. Carbon loses oxygen; CO_2 is reduced. Oxygen is combined with hydrogen in water; it loses hydrogen in going to O_2. Thus, H_2O is oxidized. The oxidizing agent is CO_2, and the reducing agent is H_2O.

Problems

1. Give formulas for the products formed in each of the following reactions.
   a. $C + O_2 \rightarrow$ b. $S + O_2 \rightarrow$
   c. $N_2 + O_2 \rightarrow$ d. $H_2 + O_2 \rightarrow$

2. Give formulas for the products formed in each of the following reactions.
   a. $CH_4 + O_2 \rightarrow$ b. $CS_2 + O_2 \rightarrow$
   c. $C_3H_8 + O_2 \rightarrow$ d. $C_6H_{12}O_6 + O_2 \rightarrow$

3. Define oxidation and reduction in terms of the following.
   a. oxygen atoms gained or lost
   b. hydrogen atoms gained or lost
   c. electrons gained or lost

4. In which of the following is the reactant undergoing oxidation? (These are not complete chemical equations.)
   a. $Cl_2 \rightarrow 2\ Cl^-$ b. $WO_3 \rightarrow W$
   c. $2\ H^+ \rightarrow H_2$ d. $CO \rightarrow CO_2$

5. The following "equations" show only part of a chemical reaction. Indicate whether the reactant shown is being oxidized or reduced.
   a. $C_2H_4O \rightarrow C_2H_4O_2$ b. $C_2H_4O \rightarrow C_2H_6O$
   c. $Fe^{3+} \rightarrow Fe^{2+}$

6. Circle the oxidizing agent and underline the reducing agent in these reactions.
   a. $4\ Al + 3\ O_2 \rightarrow 2\ Al_2O_3$
   b. $2\ SO_2 + O_2 \rightarrow 2\ SO_3$
   c. $CS_2 + 3\ O_2 \rightarrow CO_2 + 2\ SO_2$

7. Circle the oxidizing agent and underline the reducing agent in these reactions.
   a. $Cl_2 + 2\ KBr \rightarrow 2\ KCl + Br_2$
   b. $C_2H_4 + H_2 \rightarrow C_2H_6$
   c. $Fe + 2\ HCl \rightarrow FeCl_2 + H_2$

8. Circle the oxidizing agent and underline the reducing agent in these reactions.
   a. $2\ AgNO_3 + Cu \rightarrow Cu(NO_3)_2 + 2\ Ag$
   b. $FeCl_2 + Cu \rightarrow CuCl_2 + Fe$

9. In the following reactions, which element is oxidized and which is reduced?
   a. $2\ HNO_3 + SO_2 \rightarrow H_2SO_4 + 2\ NO_2$
   b. $2\ CrO_3 + 6\ HI \rightarrow Cr_2O_3 + 3\ I_2 + 3\ H_2O$
   c. $5\ C_2H_6O + 4\ MnO_4^- + 12\ H^+ \rightarrow$
   $$5\ C_2H_4O_2 + 4\ Mn^{2+} + 11\ H_2O$$

10. What is an electrochemical cell?

11. What is the purpose of a porous plate between the two electrode compartments in an electrochemical cell?

12. When a strip of copper is placed in a solution of silver nitrate ($AgNO_3$), the copper is plated with silver. Write an equation for the reaction that occurs. (Nitrate ion is not involved in the reaction.)

13. From what material is the case of a carbon–zinc dry cell made? What purpose does the material serve? What happens to it as the cell discharges?

14. How does the alkaline cell differ from a regular carbon–zinc dry cell?

15. What is an electrochemical battery?

16. Describe what happens when the lead storage battery discharges.

17. What happens when the lead storage battery is charged?

18. Describe what happens when iron corrodes. How does road salt speed this process?

19. Why does aluminum corrode more slowly than iron, even though aluminum is more reactive than iron?

20. How does silver tarnish? How can tarnish be removed without the loss of silver?

21. List four common oxidizing agents.

22. List three common reducing agents.

23. Describe how a bleaching agent, such as hypochlorite (ClO^-), works.

24. In the reaction

$$H_2CO + H_2O_2 \rightarrow H_2CO_2 + H_2O$$

which substance is oxidized? Which is the oxidizing agent?

25. Acetylene (C_2H_2) reacts with hydrogen to form ethane (C_2H_6). Is the acetylene oxidized or reduced? Explain your answer.

26. Unsaturated vegetable oils react with hydrogen to form saturated fats. A typical reaction is

$$C_{57}H_{104}O_6 + 3\ H_2 \rightarrow C_{57}H_{110}O_6$$

Is the unsaturated oil oxidized or reduced? Explain.

27. To test for an iodide ion (for example, in iodized salt), a solution is treated with chlorine to liberate iodine. The reaction is

$$2\ I^- + Cl_2 \rightarrow I_2 + 2\ Cl^-$$

Which substance is oxidized? Which is reduced?

28. Molybdenum metal, used in special kinds of steel, can be manufactured by the reaction of its oxide with hydrogen. The reaction is

$$MoO_3 + 3 H_2 \rightarrow Mo + 3 H_2O$$

Which substance is reduced? Which is the reducing agent?

29. Green grapes are exceptionally sour due to a high concentration of tartaric acid. As the grapes ripen, this compound is converted to glucose.

$$C_4H_6O_2 \rightarrow C_6H_{12}O_6$$

Tartaric   Glucose
acid

Is the tartaric acid being oxidized or reduced?

30. The dye indigo (used to color blue jeans) is formed from indoxyl by exposure of the latter to air.

$$2 C_8H_7ON + O_2 \rightarrow C_{16}H_{10}N_2O_2 + 2 H_2O$$

Indoxyl   Indigo

What substance is oxidized? What is the oxidizing agent?

31. Name some oxidizing agents used as antiseptics and disinfectants.

32. Relate the chemistry of photosynthesis to the chemistry that provides energy for your heartbeat.

33. When the water pump failed in the nuclear reactor at Three Mile Island in 1979, zirconium metal reacted with the very hot water to produce hydrogen gas

$$Zr + 2 H_2O \rightarrow ZrO_2 + 2 H_2$$

What substance was oxidized in the reaction? What was the oxidizing agent?

34. To oxidize 1 kg of fat, our bodies require about 2000 L of oxygen. A good diet contains about 80 g of fat per day. What volume of oxygen is required to oxidize that fat?

35. Vitamin C (ascorbic acid) is thought to protect our stomachs from the carcinogenic effect of nitrite ions by converting the ions to NO gas.

$$NO_2^- \rightarrow NO$$

Is the nitrite ion oxidized or reduced? Is ascorbic acid an oxidizing agent or a reducing agent?

36. In the above reaction (Problem 35), ascorbic acid is converted to dehydroascorbic acid.

$$C_6H_8O_6 \rightarrow C_6H_6O_6$$

Is ascorbic acid oxidized or reduced in the reaction?

37. The oxidizing agent we use to obtain energy from food is oxygen (from the air). If you breathe 15 times a minute (at rest), taking in and exhaling 0.5 L of air with each breath, what volume of air do you breathe each day? Air is 21% oxygen by volume. What volume of oxygen do you breathe each day?

References and Readings

1. "Auto Batteries." *Consumer Reports*, February 1985, pp. 94–97.
2. "Chem I Supplement: Everyday Examples of Oxidation–Reduction Processes." *Journal of Chemical Education*, May 1978, pp. 332–333.
3. DeLorenzo, Ronald. "Electrochemical Errors." *Journal of Chemical Education*, May 1985, p. 424.
4. "Fade Creams." *Consumer Reports*, January 1985, p. 12. Hydroquinone is used to get rid of unwanted "age spots."
5. Hill, John W., and James E. Laib. "Acne." *SciQuest*, July–August 1980, pp. 7–10.
6. Kolb, Doris. "The Chemical Equation: Part II: Oxidation–Reduction Reactions." *Journal of Chemical Education*, May 1978, pp. 326–331.
7. Neville, Roy G. "Steps Leading to the Discovery of Oxygen, 1774: A Bicentennial Tribute to Joseph Priestley." *Journal of Chemical Education*, July 1974, pp. 428–431.
8. U.S. Department of Agriculture. "Removing Stains from Fabrics: Home Methods." Washington, DC: U.S. Government Printing Office, 1968. Home and Garden Bulletin no. 62.
9. Worley, John D. "Hydrogen Peroxide in Cleansing Antiseptics." *Journal of Chemical Education*. August 1983, p. 678.

Chemistry of the Earth

Metals and Minerals

This world of ours is indeed a wondrous one—a tiny blue-green gem in the vastness of space. Spaceship Earth has a mass of about 6×10^{21} t, a diameter of 12 600 km, and a surface area of about 500 million km². It is wrapped in a thin blanket of air, and about three fourths of its surface is covered with water.

Our astronauts have walked on the barren surface of Earth's airless moon. We have sent our spaceships to explore the utter desolation of Mars. Spacecraft have measured the crushing pressure and hellish heat at the surface of Venus and have detected that planet's clouds of sulfuric acid driven by winds of greater than hurricane force. They have given us close-up portraits of Jupiter and Saturn—with their turbulent atmospheres of hydrogen and helium rent by horrendous lightning storms—and of arid, airless, pockmarked Mercury. Now it is increasingly clear that the Earth is but a small island in an inhospitable immensity of space—an island uniquely suited to the life that inhabits it.

What sort of materials are we carrying along on Spaceship Earth? Are they sufficient for the 5 billion astronauts aboard? Will they be

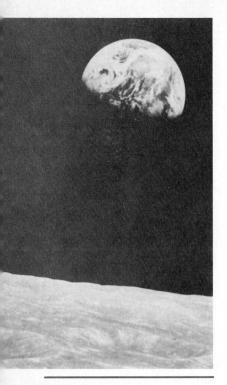

Figure 9.1 Spaceship Earth with the desolate moon in the foreground. [Courtesy of the National Aeronautics and Space Administration.]

sufficient for the 6 billion expected to travel upon it by the end of this century? Let's start to answer these questions by first looking at the elemental composition of the Earth.

Spaceship Earth: What It Is Made of

Structurally, the Earth is divided into three main regions: the *core*, the *mantle*, and the *crust*. The **core** is thought to consist largely of iron (Fe) and nickel (Ni). Since the core is not accessible, and since it does not seem likely that it will become so, we won't consider it as a source of materials.

The **mantle** is believed to be composed mostly of silicates—compounds of silicon and oxygen with a variety of metals. Although it is conceivable that the mantle will be reached eventually, it probably has few useful materials that are not available in the more accessible crust.

The **crust** is the outer shell of the Earth. It has three parts: the **lithosphere** (solid part), the **hydrosphere** (oceans, seas, lakes, rivers, etc.), and the **atmosphere** (air). The lithosphere is about 35 km thick under the continents and only about 10 km thick under the oceans. Extensive sampling of the crust has made it possible for scientists to estimate its elemental composition. Let's consider a random sample of 10 000 atoms from the crust. Of these, more than half (5330 atoms) would be oxygen. This element occurs in the atmosphere as molecular oxygen (O_2), in the hydrosphere in combination with hydrogen as water, and the lithosphere in combination with silicon (pure sand is largely SiO_2) and as various compounds.

The second most abundant element in the Earth's crust is silicon (Si). Of those 10 000 atoms in our sample, 1590 would be silicon. Hydrogen (H) would come in third with 1510 atoms, most of which would be in combination with oxygen in water. Hydrogen is such a light element—the lightest of all—that it makes up only 0.9% of the Earth's crust by weight. Other elements and the number of their atoms in our sample of 10 000 would be: aluminum 480 atoms; sodium, 180 atoms; iron and calcium, 150 atoms each; magnesium, 140 atoms; and potassium, 100 atoms. These nine elements would account for 9630 of the atoms, leaving only 370 atoms of all the other elements (Table 9.1). Some very important elements, however—such as carbon, nitrogen, and phosphorus—are among those elements that are rather scarce on Earth.

So far, our analysis has applied to the crust as a whole. The remainder of this chapter, however, is devoted to one of the divisions of the crust—the lithosphere—and to some of the metals and minerals we use to make a myriad of useful materials. We discuss the atmosphere (Chapter 15) and the hydrosphere (Chapter 16) later.

Table 9.1 Composition of the Earth's Surface: Atoms of Each Element in a Sample of 10 000 Atoms and Percent by Mass

Element	Symbol	Number of Atoms	Percent by Mass
Oxygen	O	5330	49.5
Silicon	Si	1590	25.7
Hydrogen	H	1510	0.9
Aluminum	Al	480	7.5
Sodium	Na	180	2.6
Iron	Fe	150	4.7
Calcium	Ca	150	3.4
Magnesium	Mg	140	1.9
Potassium	K	100	2.4
All others	—	370	1.4
Total		10 000	100.0

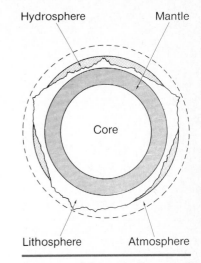

Figure 9.2 Schematic diagram of the Earth (not to scale).

The Lithosphere: Organic and Inorganic

Rocks and minerals make up the greater part of the lithosphere. Prominent among these are **silicates** (compounds of metals with silicon and oxygen), **carbonates** (metals combined with carbon and oxygen), and **sulfides** (metals combined with sulfur only). There are thousands of these mineral compounds making up the **inorganic** portion of the solid crust.

Though much, much smaller in quantity, the **organic*** portion of the lithosphere includes all living creatures, their waste and decomposition products, and fossilized materials (such as coal, natural gas, petroleum, and oil shale) that once were living organisms. This organic material *always* contains the element carbon, nearly always has combined hydrogen, and often contains oxygen, nitrogen, and a variety of other elements.

We devote several later chapters to organic materials, but for the remainder of this chapter, let's concentrate on the inorganic portion of the lithosphere.

---

* Organic chemistry was once simply the study of compounds found in the life-based portion of the lithosphere. With the advent of many synthetic chemicals that have similar properties, however, the definition has been broadened: **organic chemistry** is the study of the compounds of carbon. The chemistry of all the other elements and their compounds is called **inorganic chemistry**. (Simple carbon compounds such as the oxides and ionic carbon compounds such as the carbonates and cyanides often are considered inorganic.)

Meeting Our Needs and Wants: Chemical Change

The needs of primitive people were few. Food was obtained by hunting and gathering. Clothing, when it was needed, was provided by the skins of animals. Shelter was found in a convenient cave or constructed from readily available sticks and stones and mud. At this level of subsistence, though, the Earth could support only a few million people.

After the agricultural revolution (about 10 000 years ago), people no longer were forced to search for food. Domesticated animals and plants supplied their needs for food and clothing. A reasonably assured food supply enabled them to live in cities. City life created a demand for more sophisticated building materials. People learned to convert natural materials to products with superior properties. Clay, a complex mixture of silicates, makes a building material far superior to adobe (sun-dried bricks), after being formed into bricks and heated to high temperature. A similar process was used to make pottery. These ceramic objects made the preparation and storage of food a lot easier.

People also learned quite early to improve the foods provided by nature. Fire was one of the earliest agents of chemical change. Cooking meat and grains improved flavor and probably made digestion easier. People also found ways to convert fruits and starchy grains into alcoholic beverages. The domestication of plants and animals and the ability to manipulate natural materials made it possible for the Earth to support a substantially larger population than it had in the hunting and gathering age.

Silicates and the Shape of Things

There are thousands of different minerals in the lithosphere. We will consider only a few representative ones here. First, let's look at some of the silicates. The basic unit of silicate structure is the SiO_4 tetrahedron.

Silicate tetrahedra can be arranged in a linear manner to form fibers (asbestos), in planar sheets (mica), and in complex three-dimensional arrays (quartz, feldspar).

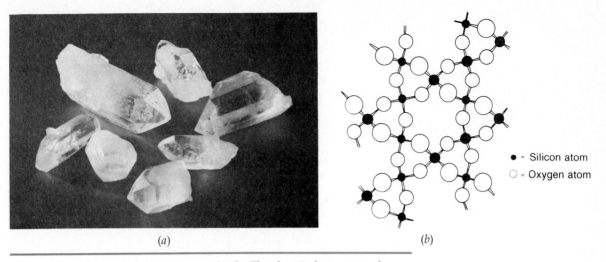

(a)   (b)

Figure 9.3 (a) A sample of quartz crystals. (b) The chemical structure of quartz. [Photo courtesy of Ward's Natural Science Establishment, Inc.]

Pure quartz is composed of silicon and oxygen only. Although the basic unit of **quartz** is the SiO_4 tetrahedron, the ratio of atoms is only $1:2$, since *all* oxygen atoms are shared by adjacent tetrahedra (Figure 9.3). Thus, the simplest formula for quartz is SiO_2.

The **micas** are composed of SiO_4 tetrahedra arranged in a two-dimensional, sheetlike array. Mica is easily cleaved into thin, transparent sheets. Large pieces of mica sometimes were used as windows before glass became generally available. Figure 9.4 shows a photograph of mica and a diagram of its chemical structure.

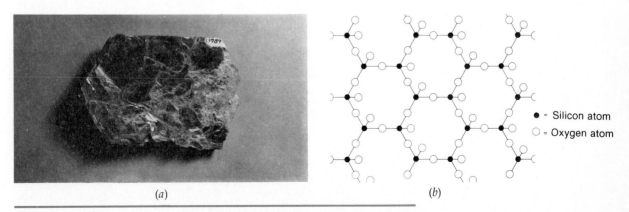

(a)   (b)

Figure 9.4 (a) A sample of mica. (b) The chemical structure of mica. [Photo by Dennis Tasa.]

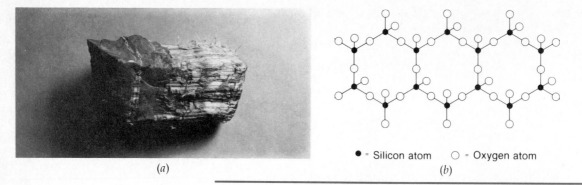

● = Silicon atom   ○ = Oxygen atom

(a)   (b)

Figure 9.5 (a) A sample of chrysotile. (b) The chemical structure of chrysotile. [Photo by Dennis Tasa.]

Asbestos is a generic term for a variety of fibrous silicates. Perhaps the best known of these is **chrysotile**, a magnesium silicate. A sample of this mineral is shown in Figure 9.5. Note that the chrysotile is a double chain of SiO_4 tetrahedra. The oxygen atoms that have only one covalent bond also bear a negative charge. It is with these that the magnesium ions (Mg^{2+}) are associated.

Asbestos has been used widely. It is an excellent thermal insulator. Great quantities of it have been used to insulate furnaces, heating ducts, and steam pipes. It has been used to make protective clothing for firefighters and others who are exposed to flames and high temperatures (Figure 9.6). It is used in brake linings for automobiles.

Perhaps the most notable case of asbestos pollution has occurred in and around Lake Superior. Several cities take their drinking water from the lake. The water was found to contain low levels of asbestos fibers. The pollution was traced to a taconite processing plant at Silver Bay, Minnesota. (Taconite is a low-grade iron ore.) Eventually, the plant was ordered to dispose of its asbestos-contaminated tailings on land, rather than dumping them into the lake. It will probably be many years, though, before the asbestos is gone from these waters.

The health hazards to those who work with asbestos are well known. The tiny, almost indestructible fibers readily penetrate the tissues of the lungs and the digestive tract. Over a period of 10 to 20 years, inhalation of fibers 5 to 50 μm long causes **asbestosis**, a severe respiratory disease. After 30 to 45 years, some workers contract lung cancer—or even mesothelioma, a rare and incurable cancer of the linings of various body cavities.

Exposure to asbestos increases the risk of lung cancer by a factor of five. Cigarette smoking causes a tenfold increase in the risk of lung cancer. People who smoke *and* are exposed to asbestos have a risk 50 to 70 times that of nonsmokers who are not exposed to asbestos. Cigarette smoke and asbestos fibers work together to produce an effect much greater than just the sum of the expected effects. Such an interaction is called a **synergistic effect**.

Occupational exposure to asbestos fibers can be devastating to workers. Indeed, the United States Environmental Protection Agency has banned the use of asbestos in spray-on fireproofing in building construction. But what about the general public? Generally, we are exposed to much lower levels than are those who work with asbestos. The effects of exposure to small quantities of asbestos are unknown. The risk for the general public is probably quite small.

Modified Silicates: Glass, Ceramics, and Cement

Sand, mica, and asbetos are used much as they occur in nature. Some disruption of nature is necessary when the materials are mined, and some environmental hazard is encountered in their use and disposal. Little expenditure of energy is necessary in these processes, however, and no new materials are introduced into the environment.

People long have been able to modify nature's materials to more nearly satisfy their needs and wants. For centuries, the technological developments were based largely on trial and error. Modern science has, by developing an understanding of the structure of materials, greatly increased the human capacity for the modification of natural materials. Thus, technological advances have been greatly accelerated by scientific knowledge.

One of the earliest technologies to develop was that used in the manufacture of pottery. Early potters used natural clays, which they merely hardened by heat. Clays are exceedingly complex and their composition varies widely, but they are basically aluminum silicates formed by the weathering of more complex silicates such as feldspar. When clay is mixed with water, it can be molded into any shape. Firing leaves a hard, resistant (but porous) product. Bricks and tile are made in this manner. When porosity is not desirable—as in a cooking pot or a water jug—the pottery can be glazed by adding various salts to the surface. Heat then converts the entire surface to a glasslike matrix.

A study of the structure and properties of clays and glazes has led to the development of fantastic new ceramic materials—from those that withstand the extreme temperatures of rocket exhaust nozzles to magnetic ceramics that serve as memory elements in computers.

Glass is another technological development of ancient times. The first glass probably was made in ancient Egypt about 5000 years ago by heating a mixture of sand, sodium carbonate (Na_2CO_3), and limestone (calcium carbonate, $CaCO_3$). As the mixture melts, it becomes a homogeneous liquid. When the liquid cools, it becomes a hard, transparent material.

Figure 9.6 Clothing made of asbestos fibers was once widely used to protect workers exposed to high temperatures and hot materials. [Courtesy of Asbestos Information Association of North America.]

This heat-softened glass is shaped by an artist as one step in the production of a Steuben crystal bowl. [Courtesy Corning Glass Works.]

When crystalline materials are heated, they eventually melt over a narrow temperature range. Glass is different; when heated, it gradually softens. While soft, it can be blown (Figure 9.7), rolled, pressed, and molded into almost any shape. The properties of glass are interpreted in terms of an *irregular* arrangement, in three dimensions, of SiO_4 tetrahedra. The chemical bonds in this arrangement are not all equivalent, as they are in a crystalline substance. Thus, when glass is heated, the weaker bonds break first and the glass gradually softens.

The basic ingredients in glass can be used in different proportions. Oxides of various metals can be substituted in whole or in part for the lime, soda, or sand. Thus, many special glasses can be made. Some examples are given in Table 9.2.

No vital raw materials are used in the manufacture of ordinary glass, but a good deal of energy—and a gas or an oil furnace, usually—is required to melt and shape glass. Disposal presents a problem, since glass is one of the most permanent materials known. It doesn't rot when discarded. Glass is easily recycled, however, as long as different kinds are separated. It can be melted and formed into a new object at considerable energy savings compared to the manufacture of new glass.

Cement is still another ancient technological development. The Romans used cement to construct roads, aqueducts, and the famous Roman baths. The raw materials for the production of cement are limestone (calcium carbonate, $CaCO_3$) and clay (aluminum silicates). The materials are finely ground, mixed, and roasted at about 1500 °C (in a kiln heated by burning natural gas or powdered coal) (Figure 9.8). The finished product is mixed with sand and gravel to form concrete.

Our understanding of the complex chemistry of cement is still imperfect. Nevertheless, several varieties of cement with special properties are available. There is a cement that sets quickly, with high early strength; a white cement; a waterproof cement; one that sets at high temperatures; and many other varieties.

The production of cement involves extensive mining, with whole

Table 9.2 Composition and Properties of Glasses

Type	Composition	Special Properties
Ordinary glass (soft glass, soda-lime glass)	Sodium and calcium silicates	Soft, with low melting point
Flint glass (optical glass)	Lead oxide (instead of lime)	Highly refractive
Borosilicate glass (Pyrex, Kimax)	Boron oxide (instead of some of the silica)	Heat resistant
Glass in red signal lights	Selenium and zinc oxide (in addition to silica)	Red, transparent
Glass in ultraviolet lights	Nickel oxide (in addition to silica)	Transparent to ultraviolet rays, opaque to visible light

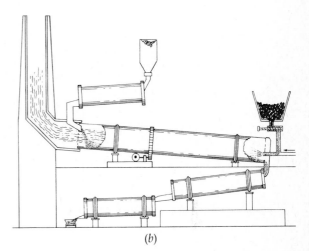

Figure 9.8 (a) A cement kiln. (b) A diagram of a
cement kiln. [Courtesy of National Gypsum Company,
Cement Division.]

(a)

(b)

mountains being torn down for limestone rocks. The rotary kiln process consumes large quantities of irreplaceable fossil fuels. Particulate matter from the crushing operations and smoke and sulfur dioxide from the burning of fossil fuels make air pollution from cement plants especially serious. In use, concrete covers what once were green acres. In cities, the amount of paving with concrete and asphalt is enough to change the climate (temperature and rainfall). Runoff of rainfall from paved areas is especially rapid and contributes to flash flooding. But cement itself doesn't seem to be a bad pollutant. It can be broken up and used as rock fill.

Metals and Ores: The Iron Age Lives!

Human progress through the ages often is described in terms of the materials that were used for making tools. We speak of the Stone Age, the Bronze Age, and Iron Age. Over 2000 years ago, artisans learned to reduce certain metal ores to free metals. For iron, carbon (charcoal) was employed as the reducing agent. We now know the chemistry of the process. First, the carbon is converted to carbon monoxide.

$$2\,C\ +\ O_2\ \longrightarrow\ 2\,CO$$

Then the carbon monoxide reduces the iron oxide to iron metal.

$$Fe_2O_3 + 3\ CO \longrightarrow 2\ Fe + 3\ CO_2$$

The basic raw materials of modern iron production are iron ore, coal, and limestone. The coal is heated to drive off volatile material, leaving behind a relatively pure form of carbon called **coke**. The limestone is added to combine with silicate impurities to form a molten **slag** that is drawn off. Molten iron is drawn off at the bottom of the furnace (Figure 9.9).

The iron can be softened by heating and then shaped into useful items. These items slowly rust; that is, they are oxidized by oxygen in the air.

$$4\ Fe + 3\ O_2 \longrightarrow 2\ Fe_2O_3$$

Thus, a cycle is established by which a metal is obtained from its ore and then slowly is corroded back to the combined state (Figure 9.10). This rust could be reduced back to metal and reused, but this is seldom done. The rust often is scattered in the environment and, for all practical purposes, lost. Many metal objects are melted down, the rust removed or reduced, and new objects molded from the molten metal.

For many purposes, iron has been replaced by other metals and plastics. Our age has been given many names—the Age of Plastics, the

Figure 9.9 A blast furnace for the production of iron.

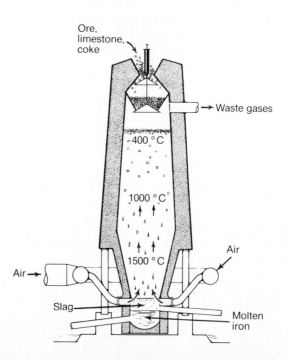

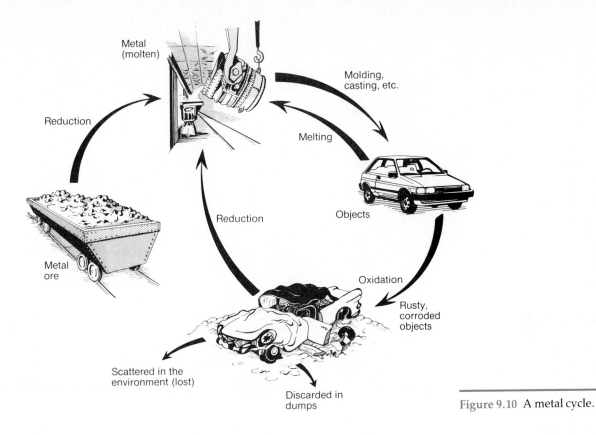

Metal
(molten)

Molding,
casting, etc.

Reduction

Melting

Reduction

Objects

Metal
ore

Oxidation

Rusty,
corroded
objects

Scattered in the
environment (lost)

Discarded in
dumps

Figure 9.10 A metal cycle.

Nuclear Age, and the Computer Age—each appropriate in its own way. As far as the use of materials is concerned, however, we are still in the midst of an iron age. We use far more iron than any other metal for making tools and machines.

Pure iron is too soft and reactive to be useful for most purposes. Most iron is converted to **steel** by adjusting the carbon content. The properties of steel can be adjusted over a wide range, depending on the amount of carbon in it. High-carbon steel is hard and strong. Low-carbon steel is **ductile** (can be drawn into wire) and **malleable**, and it can be forged and welded. Some steel is alloyed with other metals and has still other properties. (An **alloy** is a mixture of two or more elements, at least one of which is a metal. The alloy has metallic properties.) Iron, because it can be formed into steel and so many other alloys, is the most useful of all the metals.

In recent years, iron has been replaced by aluminum for many purposes. Aluminum is the most abundant metal in the Earth's crust, but it is tightly bound in its compounds in nature. Much energy is required to extract aluminum metal from its ores. The principal ore of aluminum is bauxite, a mineral in which aluminum oxide is associated with one, two, or three water molecules per Al_2O_3 unit. The formula for bauxite often is written $Al_2O_3 \cdot xH_2O$. The bauxite also contains iron oxides, silicates, and other impurities. It is extracted with a strong base and then heated to form the oxide, Al_2O_3. The oxide is melted and separated by passing electricity through it.

$$2\ Al_2O_3 \xrightarrow{\text{electricity}} 4\ Al\ +\ 3\ O_2$$

It takes 2 t of aluminum oxide and 17 000 kWh of electricity to produce 1 t of aluminum.

Aluminum is light and strong. It corrodes much more slowly than iron, and, for many purposes, that is a considerable advantage. For one familiar use, though, slow corrosion is a distinct disadvantage. Consider the lowly beverage container. A tin can—really steel with a thin coating of tin on the outside—rusts when thrown away, eventually disintegrating. An aluminum can degrades much more slowly. Scientists at Pennsylvania State University estimate that it would take 500 years for one to degrade completely.

Steel mills and aluminum plants both create considerable pollution. Steel mills discharge lime, acids, grease and oil, and iron salts into the water. They also discharge particulate matter, carbon monoxide, nitrogen oxides, and other air pollutants. The manufacture of aluminum

Table 9.3 Some Important Metals

Metal	Symbol	Important Ore	Selected Properties	Typical Uses
Chromium	Cr	$FeCr_2O_4$	Shiny, resists corrosion	Chrome plating
Copper	Cu	CuS, $CuCO_3$	Good electrical conductor	Electrical wiring
Gold	Au	Au	Yellow metal, soft	Coinage, jewelry, dentures
Lead	Pb	PbS	Low melting, dense, soft	Plumbing, batteries
Magnesium	Mg	$MgCl_2$	Light and strong	Auto wheels, luggage
Mercury	Hg	HgS	Dense liquid	Thermometers, barometers
Nickel	Ni	NiS	Resists corrosion	Coinage, alloy for stainless steel
Platinum	Pt	Pt	Inert, high melting	Catalyst, instruments
Silver	Ag	Ag	Excellent electrical conductor	Electrical contacts, mirrors, jewelry, coins
Sodium	Na	NaCl	Reactive, soft	Heat transfer medium, reducing agent
Tin	Sn	SnO_2	Resists corrosion	Coating for steel cans
Tungsten	W	$CaWO_4$	Very high melting	Light-bulb filaments
Uranium	U	U_3O_8	Fissionable	Energy source
Zinc	Zn	ZnS	Forms protective coating	Galvanizing coating

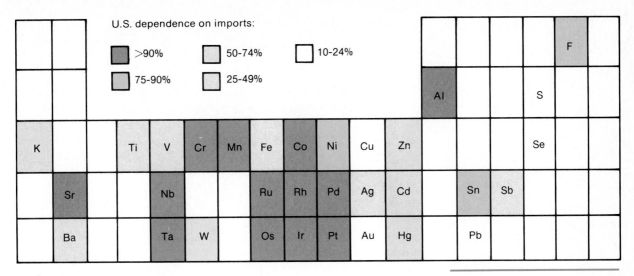

Figure 9.11 The United States imports a variety of strategic metals, minerals, and ores. It depends heavily on imports for some of the elements in the periodic table.

results in the discharge of iron, aluminum, and other metal oxides into waterways and the release of particulate matter, fluorides, and other pollutants into the air.

Steel and aluminum play vital roles in the modern industrial world. Their economic value cannot be overemphasized. But shouldn't we include the *environmental cost* of producing, using, and discarding these materials?

One further comparison of steel and aluminum: Barry Commoner (see Reference 3) estimates that it takes 15 times as much fuel energy to produce aluminum as it does to produce a comparable weight of steel. Aluminum is a lighter (less dense) metal than steel, but making an aluminum can requires 6.3 times as much energy as making a steel can.

There are many other important metals. Several, with typical ores, properties, and uses, are listed in Table 9.3.

Metals and minerals are vital to a vigorous economy. The United States has depleted much of its high-grade ores, and it has become increasingly dependent on other nations for these materials (Figure 9.11). It appears that the United States is as vulnerable to embargoes of these important materials as it was to the oil embargoes of the 1970s.

How can we ever run short of a metal? Aren't atoms conserved? Yes, there is as much iron as there was 100 years ago. But through use we scatter the metal throughout the environment. Gathering it back to a factory to make new objects requires energy, just as obtaining metals from low-grade ores requires more energy than getting them from high-grade ores.

Running Out of Everything: The Materials Crunch

The fuel shortages of the 1970s were quite dramatic. Our potential metals and minerals shortages could be even worse. When the Europeans first came to North America, the native Americans were mining nearly pure copper in Michigan. Now we mine ore with 0.5% or less copper content. For much of the nineteenth century, we mined high-grade iron ore from the Mesabi Range in Minnesota. Now we mine taconite, a hard rock with small amounts of iron oxide dispersed through it. The same is true for gold and silver; the glory holes of western North America are gone. Indeed, there are few high-grade deposits of any metal ores left that are readily accessible to the industrialized nations.

What's wrong with low-grade ores? Considerable energy is required to concentrate them, and energy costs money. Let's consider an analog. A bag of popcorn is useful. You could pop it and eat it. The same popcorn, scattered all over your room, would be less useful. Oh, you could gather it all up and then pop it, but that would take a lot of energy. Perhaps more energy than you would care to put into it. Perhaps more energy than you would get back if you popped the corn and ate it.

Where will we get metals in the future? The sea is one possible source. Nodules rich in manganese cover vast areas of the ocean floor. These nodules also contain copper, nickel, and cobalt. Questions of who owns them and how to mine them without major environmental disruption remain to be resolved.

Land Pollution: Solid Wastes

Productivity and creativity have enabled people in industrialized nations to have a fantastic variety of consumer goods in enormous quantities. As these goods are broken, worn out, or merely become obsolete, their disposal becomes a major problem. The average person in the United States discards 360 kg of waste material each year. About half this waste is paper and other combustible materials. About 7 million automobiles are junked each year in North America, and 170 million tires are thrown away.

In the past, most solid wastes were simply discarded in **open dumps**. This led to infestation by rats, flies, and other pests that often spread to nearby areas. Open burning led to monstrously offensive air pollution.

These open dumps are being rapidly phased out, but they still account for a proportion of the total land pollution.

Another method of solid waste disposal is **incineration**. If this is carried out in properly designed incinerators, air pollution is minimal. A disadvantage is that noncombustibles—metal cans and glass bottles make up a large part of these—must be separated before or after burning. Some cities have used the heat from combustible solid wastes to generate power. The metal and glass components are separated out for recycling.

Another method of disposal is the **sanitary landfill**. Garbage and trash are piled into a trench, compacted, and covered over. This eliminates the problem of rats, flies, and odors. Land for such purposes is scarce where it is most needed, however—near large cities.

Research in the disposal of solid wastes—and in making useful materials from rubbish—is underway and beginning to produce promising results. In Japan, a method for compacting garbage and coating it with asphalt promises to become a source of badly needed building materials. The United States Bureau of Mines has a method for converting garbage to oil, with a yield of about 250 kg from 1000 kg of garbage. Ground-up rubber tires have been found to accelerate plant growth. Waste glass and rubber have been used in road construction. Yet much remains to be done.

We discuss other land-pollution problems, such as pesticides and herbicides, and special problems in the disposal of toxic materials in later chapters.

Recycling: A Way to Save Materials and Energy

One way to postpone the coming shortage of materials is to recycle them. Recycling is not a perfect solution, for in any such process, energy is required and some material is unavoidably lost. But recycling does save materials. Over half of the copper and lead produced each year comes from recycled materials. About one third of the iron and steel and over half of the aluminum also are salvaged. In the future, as raw materials become scarcer, these proportions are bound to increase.

Recycling also saves energy. It takes only 5% as much energy to make new aluminum cans from old ones as it does to make the cans from aluminum ore. Using a ton of scrap to make new iron and steel saves 1.5 t of iron ore and 0.3 t of coal. It also results in a 74% savings in energy, an 86% reduction in air pollution, and a 76% reduction in water pollution.

Let's hear it for scrap!

Reuse: A Better Way

Recycling is obviously a good thing. It isn't necessarily the *best* thing, though. The best thing is **reuse**. Make an item durable enough to withstand repeated use, rather than designing it for a single use after which it is to be discarded. Consider the lowly beverage bottle. It takes about 6300 J (1500 kcal) of energy to make a nonreturnable half-liter bottle. It takes a little more energy, 8300 J (2000 kcal), to make a more durable, returnable bottle of the same capacity. The returnable bottle is used an average of 12.5 times before it is broken. That is about 660 J per use, only about 10% of the energy used to make the nonreturnable bottle for its single use.

We could recycle the nonreturnable bottle. Recall, though, that glass is made from common materials—lime, soda, and sand. No vital material is saved. Recycling does save energy compared to making glass from raw materials, but it is much *less* efficient than reusing returnable bottles. We can save both materials and energy by making durable items and reusing them as much as possible. Where reuse is not possible, materials should be recycled to the fullest extent.

Problems

1. What are the three main regions of the Earth?
2. Name the three parts of the crust of the Earth.
3. What two elements are thought to make up most of the core of the Earth?
4. What is the most abundant element in the crust of the Earth?
5. What are the silicates?
6. What materials make up the organic portion of the lithosphere?
7. What is inorganic chemistry?
8. What is the chemical structure of quartz?
9. Why does mica occur in sheets?
10. Why does asbestos occur as fibers?
11. What is a synergistic effect?
12. How does the structure of glass differ from that of crystalline silicates?
13. How do the properties of glass differ from those of crystalline substances?
14. What are the three principal raw materials for making glass?
15. How is the basic recipe for glass modified to make glasses with special properties?
16. What are the two basic raw materials for making cement?
17. What environmental problems are associated with the manufacture of cement?
18. What is concrete?
19. What are the principal raw materials for modern iron production? What is the purpose of each?
20. What is coke?
21. What is slag?
22. By what kind of chemical process is a metal obtained from its ore?
23. By what chemical process does a metal corrode?
24. What is an alloy?
25. What is steel?
26. What is the most abundant metal in the crust of the Earth?
27. Which metal is the most widely used? Why is the most abundant metal not the most widely used?
28. What environmental problems are associated with steel mills?
29. What environmental problems are associated with aluminum production?

30. If matter is conserved, how can we ever run out of a metal?

31. List three methods of solid waste disposal. Give the advantages and the disadvantages of each.

32. Explain why metals can be recycled fairly easily. What factors limit the recycling of metals?

33. An ore body at Crandon, Wisconsin, is estimated at 75 million t and assays at 5.0% zinc, 0.4% lead, and 1.1% copper. What mass of each metal is contained in the deposit?

34. The Hunter method for the production of titanium uses the reaction

$$TiCl_4 + 4 Na \rightarrow Ti + 4 NaCl$$

What substance is reduced? What is the reducing agent?

35. Close to 81 000 t of gold have been mined throughout history. What size cube of gold would this make if it was all in one piece? The density of gold is 19.3 g/cm^3.

36. The Parc mine at Llanwrst, North Wales, in the United Kingdom, was closed in 1954, leaving behind a mound of about 250 000 t of tailings. The tailings assay at 3.22% Zn, 0.82% Pb, and 0.023% Cd. Erosion has washed away 13 000 t of the tailings. How much of each of the three elements has been carried away, presumably to enter the river ways?

37. Ten million carats of gem-quality rough diamonds are mined each year. What is the mass of these diamonds in kilograms? (1 carat = 200 mg).

38. To obtain 1 carat of gem-quality diamonds, approximately 115 t of earth have to be mined. How much earth is mined to produce 10 million carats of diamonds each year?

39. An aluminum plant produces 65 000 000 kg of aluminum per year. How much aluminum oxide is required? (It takes 2.1 kg of crude bauxite to produce 1 kg of aluminum oxide.) How much bauxite is required? (It takes 17 kWh of electricity to produce 1 kg of aluminum.) How much electricity does the plant use for aluminum production in 1 year?

References and Readings

1. Asimov, Isaac. *A Choice of Catastrophes*. New York: Simon and Schuster, 1980.
2. Chenier, Philip J. "A Summary Chart of the Manufacture of Important Inorganic Chemicals." *Journal of Chemical Education*, May 1983, pp. 411–413.
3. Commoner, Barry. *The Closing Circle*. New York: Bantam, 1972.
4. Dagani, Ron. "The Planets: Chemistry in Exotic Places." *Chemical and Engineering News*, 10 August 1981, pp. 25–36.
5. Denio, Allen A. "Chemistry for Potters." *Journal of Chemical Education*, April 1980, pp. 272–275.
6. Ehrlich, Paul R., Anne H. Ehrlich, and John P. Holdren. *Ecoscience: Population, Resources, Environment*. San Francisco: W. H. Freeman, 1977.
7. Gore, Rick. "The Planets: Between Fire and Ice." *National Geographic*, January 1985, pp. 4–51.
8. Hansen, James. "The Delicate Architecture of Cement." *Science 82*, December 1982, pp. 49–55.
9. Kolb, Doris, and Kenneth E. Kolb. "The Chemistry of Glass." *Journal of Chemical Education*, September 1979, pp. 604–608.
10. Sadoway, Donald R. "The Materials-Energy Symbiosis." *ChemTech*, October 1982, pp. 625–627.
11. Simon Julian L. "Resources, Population, Environment: An Oversupply of Bad News." *Science*, 27 June 1980, pp. 1431–1437. But see also "Letters." *Science*, 19 December 1980, pp. 1296–1308.
12. Turner, Kenneth. "Precious Metals." *Chemistry and Industry*, 19 July 1980, pp. 551–556.
13. White, Peter T. "The Fascinating World of Trash." *National Geographic*, April 1983, pp. 424–427.
14. Zurer, Pamela S. "Asbestos: The Fiber That's Panicking America." *Chemical and Engineering News*, 4 March 1985, pp. 28–41.

10

Hydrocarbons
An Introduction to Organic Chemistry

Figure 10.1 The word *organic* has several different meanings. Organic fertilizer is organic in the original sense that it is derived from a living organism. There is no legal definition of organic foods, but the term generally means foods grown without pesticides or synthetic fertilizers. Organic chemistry is the chemistry of carbon compounds.

In Chapter 9 we defined *organic chemistry* as the chemistry of compounds of the element carbon. In this chapter we consider the compounds of carbon and hydrogen. These two elements combine in many proportions and in almost infinte structural arrangements to form *hydrocarbons*. In the next chapter we consider organic compounds that contain oxygen, nitrogen, and other elements, as well as carbon and hydrogen. In later chapters we consider still more complex compounds of the element carbon—compounds that are an integral part of life processes.

Millions of Hydrocarbons: The Unique Carbon Atom

A unique property of the carbon atom is its ability to bond to another carbon atom, that atom in turn to another carbon atom, and so on. In this way, carbon atoms hook up to form long chains of hundreds or even thousands of atoms. This process is called **catenation**. No other element can form chains longer than four or five atoms. The chains can form branches, and even rings, of carbon atoms. This property makes possible a nearly endless number of **hydrocarbons** (compounds that contain only carbon and hydrogen).

Carbon, near the middle of the periodic chart, also bonds strongly to other elements, such as oxygen and nitrogen. Thus, carbon forms a vast array of compounds with chains of various lengths and rings of various sizes, each with a variety of other constituent and substituent atoms. Some of these, as we see in later chapters, are vital to life processes. Indeed, life itself is a property of some of these compounds.

Alkanes: The Saturated Hydrocarbons

The nearly infinite variety of hydrocarbons is merely hinted at here. Before you can even begin to understand the large, complex molecules on which life is based, it is necessary to learn something about simpler molecules.

The simplest alkane is methane (CH_4). The structure of methane is discussed in some detail in Chapter 5. Methane is the first member of a series of related compounds called **alkanes** or **saturated hydrocarbons**. These are compounds in which there are only single bonds. (Hydrocarbons with double and triple bonds are members of other families and are discussed in subsequent sections.)

The next member of the alkane series is ethane (C_2H_6). We can show how the atoms are attached to one another in this way:

$$
\begin{array}{ccc}
 & H & H \\
 & | & | \\
H- & C-C & -H \\
 & | & | \\
 & H & H
\end{array}
$$

Recall that the methane molecule is tetrahedral in shape. The ethane molecule is also three-dimensional. Figure 10.2 shows ball-and-stick models of methane and ethane. These more accurate pictures are difficult to draw. For that reason we shall ordinarily use **structural formulas** like that shown above for ethane. These formulas show you in what order atoms are attached, but do not attempt to portray accurately the three-dimensional geometry of the molecule. To emphasize the relationship between structural formulas and the more accurate three-dimensional models, we will continue to show both representations while we discuss the smaller alkane molecules.

The three-carbon alkane (C_3H_8) is called propane. A three-

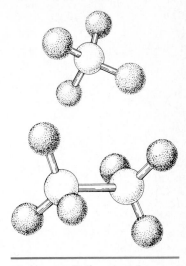

Figure 10.2 Ball-and-stick models of methane and ethane.

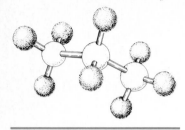

Figure 10.3 Ball-and-stick model of propane.

dimensional model of this compound is shown in Figure 10.3, and the two-dimensional structural formula is shown below.

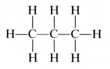

A pattern is now becoming apparent. We can build alkanes of any length simply by tacking carbon atoms together in long chains and adding sufficient hydrogen atoms to give each of the carbon atoms a total of four bonds. Even the naming of these compounds follows a pattern. For compounds of five carbon atoms or more, each stem or root is derived from the Greek or Latin name for the number of carbon atoms in the molecule (Table 10.1). The compound names end in -*ane*, signifying that the compounds are alk*anes*. Table 10.2 gives structural formulas and names for continuous-chain alkanes up to 10 carbon atoms in length. We do not need to stop at 10 carbon atoms. One could hook together 100 or 1000 or 1 million. We can make an infinite number of alkanes simply by lengthening the chain. But lengthening the chain is not the only option. With four carbon atoms, chain branching is also possible. If we extend the chain in the usual fashion, we get

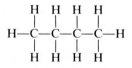

This formula represents butane (C_4H_{10}), a compound that boils at 0 °C. A second compound, whose boiling point is −12 °C, has the same molecular formula: C_4H_{10}. The *structural formula* of the second compound is *not* the same as butane's. Instead of having four carbon atoms connected in a continuous chain, this new compound has a continuous chain of only three carbon atoms. The fourth carbon is branched off the middle carbon of this three-carbon chain.

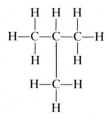

Compounds that have the same molecular formula but different structural formulas are called **isomers**. Because it is an isomer of butane, the

Table 10.1 Prefixes That Indicate the Number of Carbon Atoms in Organic Molecules

Stem	Number
Meth-	1
Eth-	2
Prop-	3
But-	4
Pent-	5
Hex-	6
Hept-	7
Oct-	8
Non-	9
Dec-	10

Table 10.2   The First Ten Continuous-Chain Alkanes

Name	Molecular Formula	Structural Formula	Number of Possible Isomers
Methane	CH_4	H—C—H with H above and H below	1
Ethane	C_2H_6	H—C—C—H (with H above and below each C)	1
Propane	C_3H_8	H—C—C—C—H (with H above and below each C)	1
Butane	C_4H_{10}	H—C—C—C—C—H (with H above and below each C)	2
Pentane	C_5H_{12}	H—C—C—C—C—C—H (with H above and below each C)	3
Hexane	C_6H_{14}	H—C—C—C—C—C—C—H (with H above and below each C)	5
Heptane	C_7H_{16}	H—C—C—C—C—C—C—C—H (with H above and below each C)	9
Octane	C_8H_{18}	H—C—C—C—C—C—C—C—C—H (with H above and below each C)	18
Nonane	C_9H_{20}	H—C—C—C—C—C—C—C—C—C—H (with H above and below each C)	35
Decane	$C_{10}H_{22}$	H—C—C—C—C—C—C—C—C—C—C—H (with H above and below each C)	75

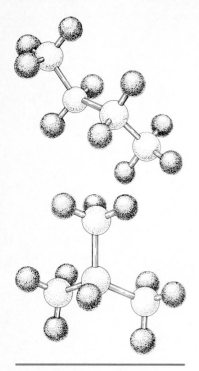

Figure 10.4 Ball-and-stick models of butane and isobutane.

branched four-carbon alkane is called isobutane. Figure 10.4 shows ball-and-stick models of the two isomeric butanes.

The number of isomers increases rapidly with the number of carbon atoms (Table 10.2). There are three pentanes, five hexanes, nine heptanes, and so on. Isomerism is common in the compounds of carbon and provide a third reason for the existence of millions of organic compounds.

Propane and the butanes are familiar fuels. They usually are supplied under pressure in tanks. Although they are gases at ordinary temperatures and under normal atmospheric pressure, they are liquefied under pressure and are sold as liquefied petroleum gas (LP gas). Gasoline (Chapter 13) is a mixture of hydrocarbons, mostly alkanes with 5 to 12 carbon atoms.

Let's return to Table 10.2 for a moment. Notice that the molecular formula of each alkane in the table differs from the one preceding it by precisely one carbon atom and two hydrogens—that is, by a CH_2 group. Such a series of compounds has properties that vary in a regular and predictable manner. This principle, called **homology**, gives meaning to organic chemistry in much the same way that the periodic table gives organization and meaning to the chemistry of the elements. Instead of studying the chemistry of a bewildering array of individual carbon compounds, organic chemists study a few members of a **homologous series** from which they can *deduce* the properties of other compounds in the series.

One other point before leaving Table 10.2: not all the possible isomers of the larger molecules have been isolated. Indeed, the task rapidly becomes even more prohibitive as you proceed up the series. There are, for example, over 4 billion possible isomers with the molecular formula $C_{30}H_{62}$.

Drawing Structural Formulas

So far we have used structural formulas that show all the carbon and hydrogen atoms and how they are attached to one another. These structures convey more information than molecular formulas. For example, the formula C_4H_{10} doesn't even tell us whether we are dealing with butane or with isobutane. The structural formulas (page 202) identify the specific isomers by the order of attachment of the various atoms.

Unfortunately, structural formulas are hard to type and take up a lot of space. Chemists often use **condensed structural formulas** to alleviate these problems. The condensed structures show the hydrogen atoms right next to the carbon atoms to which they are attached. The two butanes become

$$CH_3—CH_2—CH_2—CH_3 \quad \text{and} \quad CH_3—CH—CH_3$$
$$\underset{\displaystyle CH_3}{|}$$

Sometimes these are further simplified by omitting some (or all) of the bond lines.

$$CH_3CH_2CH_2CH_3 \quad \text{and} \quad CH_3CHCH_3$$
$$\underset{\displaystyle CH_3}{|}$$

Example 10.1 Draw the structural formula and the condensed structural formula for heptane.

The name heptane means an alkane with seven carbon atoms. Just write out a string of seven carbon atoms.

$$C—C—C—C—C—C—C$$

Then you attach enough hydrogens to the carbons to give each carbon a valence of 4. This requires three hydrogens on each end carbon and two each on the others.

$$
\begin{array}{c}
\text{H} \quad \text{H} \quad \text{H} \quad \text{H} \quad \text{H} \quad \text{H} \quad \text{H} \\
| \quad | \quad | \quad | \quad | \quad | \quad | \\
\text{H—C—C—C—C—C—C—C—H} \\
| \quad | \quad | \quad | \quad | \quad | \quad | \\
\text{H} \quad \text{H} \quad \text{H} \quad \text{H} \quad \text{H} \quad \text{H} \quad \text{H}
\end{array}
$$

For the condensed form, just write each carbon atom's set of hydrogen atoms next to the carbon. Note that each end carbon has three hydrogen atoms, while the carbons in the middle have only two each.

$$CH_3CH_2CH_2CH_2CH_2CH_2CH_3$$

On Rings and Things: Cyclic Hydrocarbons

The hydrocarbons we have encountered so far (alkanes) have been composed of open-ended chains of carbon atoms. Carbon and hydrogen atoms also can hook up in other arrangements in which closed rings are formed. The simplest possible ring-containing, or **cyclic**, hydrocarbon has the molecular formula C_3H_6.

Figure 10.5 Ball-and-stick model of cyclopropane.

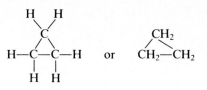

The compound is called cyclopropane (Figure 10.5).

Names of cycloalkanes (cyclic compounds containing only single bonds) are formed by addition of the prefix *cyclo-* to the name of the open-chain compound with the same number of carbon atoms as are in the ring.

Example 10.2 Give the structure for cyclobutane.

Cyclobutane has four carbon atoms arranged in a cyclic fashion.

$$
\begin{array}{c} C-C \\ |\quad| \\ C-C \end{array}
$$

Each carbon atom needs two hydrogen atoms to complete its set of four bonds.

$$
\begin{array}{c} CH_2-CH_2 \\ |\qquad| \\ CH_2-CH_2 \end{array}
$$

There are many such cyclic compounds; they are quite important in organic chemistry and biochemistry—as we shall see.

Unsaturated Hydrocarbons: Alkenes and Alkynes

Two carbon atoms can share more than one pair of electrons. In ethylene (C_2H_4), the two carbon atoms share *two* pairs of electrons and the carbon atoms are joined by a *double bond*.

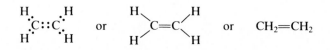

Ethylene is the most important commercial organic chemical. Annual United States production is about 14 billion kg. About 45% of this ethlylene goes into the manufacture of polyethylene, one of the most

Figure 10.6 Ball-and-stick model of ethylene.

Figure 10.7 Ball-and-stick model of acetylene.

familiar plastics. Another 15% or so is converted to ethylene glycol, the major component of most brands of anti-freeze for automobile radiators.

In acetylene (C_2H_2), the two carbon atoms have *three* pairs of electrons; the carbon atoms are joined by a *triple bond*.

$$H\text{:}C\text{:::}C\text{:}H \quad \text{or} \quad H\text{—}C\equiv C\text{—}H$$

About 10% of the acetylene produced in the United States is used in oxyacetylene torches for cutting and welding metals. Such torches can produce very high temperatures. Most acetylene, however, is converted to chemical intermediates that are, in turn, used to make vinyl and acrylic plastics, fibers, resins, and a variety of other chemical products.

Ethylene is the first member of a family of hydrocarbons called **alkenes**. Each alkene contains a carbon-to-carbon double bond. Similarly, acetylene is the first member of the **alkyne** family. Each alkyne contains a triple bond. Collectively, alkenes and alkynes are called **unsaturated hydrocarbons** because they can add more hydrogen atoms to form saturated hydrocarbons (alkanes).

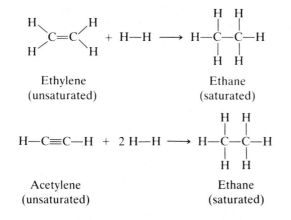

Ethylene
(unsaturated)

Ethane
(saturated)

Acetylene
(unsaturated)

Ethane
(saturated)

We discuss unsaturated fats in Chapter 18. These compounds, like the alkenes, contain carbon-to-carbon double bonds.

Benzene and Relatives: Aromatic Hydrocarbons

Still another type of hydrocarbon is represented by the compound benzene. Benzene was first isolated from a whale oil by-product by Michael Faraday in 1825. (This is the same Michael Faraday whose electrolysis experiments are described in Chapter 3.) The molecular formula of benzene is C_6H_6. One can write many structures that correspond to this formula. Three such structures are

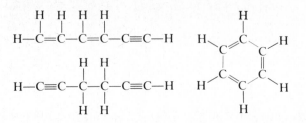

The real substance, benzene, does not have the properties that one would predict from these structures. For example, if benzene really contained double or triple bonds, it would be expected to undergo addition reactions readily. It does not.

Friedrich August Kekulé, a German chemist, proposed the third structure. It is accurate in some respects. Benzene does have a six-member ring structure. It is also an unsaturated molecule; that is, it contains something other than single bonds. To account for benzene's inertness toward the typical reactions of unsaturated compounds, Kekulé's structure has been modified. In order for benzene to behave as it does, the "extra" or double-bond electrons cannot be located between specific carbon atoms as they would be in ordinary double bonds. All six of these electrons are regarded as being equally shared by all six carbon atoms in the ring. Therefore, the structure of benzene is now drawn

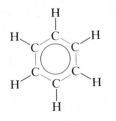

where the circle represents the six "extra" electrons. No real double bonds exist, just a stable ring of electrons that resists being disrupted. Benzene does not enter addition reactions because that would destroy the ring of electrons.

Benzene is a liquid at room temperature. It is used widely as a solvent and as an intermediate product in the production of other aromatic compounds. Annual United States production is about 4.5 billion kg. Benzene is thought to cause leukemia in workers exposed to it over the years, and its use has been restricted.

There are many compounds chemically similar to benzene. Several of those discovered in the early days had pleasant odors. These were called **aromatic compounds**. The label *aromatic* used today simply means a compound related to benzene; it does not imply a pleasing aroma. In fact, some are odorless and others stink.

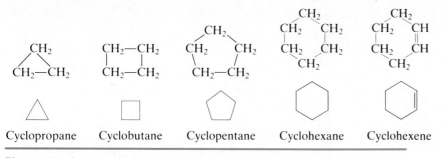

| Cyclopropane | Cyclobutane | Cyclopentane | Cyclohexane | Cyclohexene |

Figure 10.8 Some cyclic hydrocarbons.

Chemical Symbolism

It takes time to draw out structures such as those we have used for cyclopropane and benzene. They are also difficult to type. Therefore, chemists resort to a rather mystical sort of symbolism. For example: a chemist notes that the three carbon atoms of cyclopropane form a triangle; therefore, he or she uses a triangle to represent cyclopropane. This symbolism is illustrated in Figure 10.8 for a variety of cyclic hydrocarbons.

Similarly, we envision the six carbon atoms of benzene as the corners of a hexagon; thus, we represent benzene as a hexagon with an inscribed circle to indicate the six unassigned electrons. This symbolic representation of the gasolinelike liquid we call benzene is shown in Figure 10.9—along with similar structures for other aromatic hydrocar-

Benzene

Figure 10.9 Some aromatic hydrocarbons.

Toluene   Ethylbenzene   Naphthalene

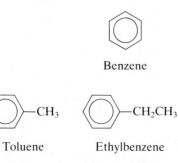

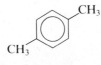

ortho-Xylene
(1,2-Dimethylbenzene)

meta-Xylene
(1,3-Dimethylbenzene)

para-Xylene
(1,4-Dimethylbenzene)

bons. You should recognize these symbols as representations of aromatic hydrocarbons.

Properties of the Hydrocarbons

The alkanes form a homologous series. The properties of the members vary in a regular and predictable manner. For example, their boiling points show a fairly regular increase of from 20 to 30 °C as we go up the series (Table 10.3).

You should note from the table that at room temperature alkanes having from 1 to 4 carbon atoms per molecule are gases. Those alkanes having from 5 to about 16 carbon atoms per molecule are liquids, and alkanes having more than 16 carbon atoms per molecule are solids. Note also that the densities of the liquid alkanes are less than that of water (1.0 g/mL). The alkanes are essentially insoluble in water and hence will float on top of water. Alkanes dissolve many organic substances of low polarity—such as fats, oils, and waxes. Mixtures of alkanes are therefore frequently used as organic solvents.

The physiological properties of alkanes vary in a regular way as we proceed through the homologous series. Methane appears to be totally inert physiologically. We could breathe a mixture of 80% methane and 20% oxygen without ill effect. Such a mixture would be flammable, however, and no fire or spark of any kind could be permitted in an

Table 10.3 Physical Properties of Selected Alkanes

Name	Molecular Formula	Melting Point (degrees Celsius)	Boiling Point (degrees Celsius)	Density at 20°C (grams per milliliter)
Methane	CH_4	−183	−162	(Gas)
Ethane	C_2H_6	−172	−89	(Gas)
Propane	C_3H_8	−187	−42	(Gas)
Butane	C_4H_{10}	−138	0	(Gas)
Pentane	C_5H_{12}	−130	36	0.626
Hexane	C_6H_{14}	−95	69	0.659
Heptane	C_7H_{16}	−91	98	0.684
Octane	C_8H_{18}	−57	126	0.703
Decane	$C_{10}H_{22}$	−30	174	0.730
Dodecane	$C_{12}H_{26}$	−10	216	0.749
Tetradecane	$C_{14}H_{30}$	6	254	0.763
Hexadecane	$C_{16}H_{34}$	18	280	0.775
Octadecane	$C_{18}H_{38}$	28	316	(Solid)
Eicosane	$C_{20}H_{42}$	37	343	(Solid)

atmosphere consisting of methane. Breathing an atmosphere of pure methane (the "gas" of a gas-operated stove) can lead to death—not so much because of the presence of methane but because of the absence of oxygen (asphyxia). The other gaseous alkanes (and vapors of volatile liquid ones) act as anesthetics in high concentrations. They can also produce asphyxiation by excluding oxygen.

Liquid alkanes have varied effects on the body, depending on the part exposed. On the skin, alkanes dissolve body oils. Repeated contact may cause dermatitis. Swallowed, alkanes do little harm in the stomach. However, in the lungs, alkanes cause **chemical pneumonia** by dissolving fatlike molecules from the cell membranes in the alveoli. The cells become less flexible, and the alveoli are no longer able to expel fluids. The buildup of fluids is similar to that which occurs in bacterial or viral pneumonia. People who swallow gasoline, petroleum distillates, or other liquid alkane mixtures should not be made to vomit. That would increase the chance of getting the alkanes into their lungs.

Heavier liquid alkanes, when applied to the skin, act as **emollients** (skin softeners). Such alkane mixtures as mineral oil can be used to replace natural skin oils washed away by frequent bathing or swimming. Petroleum jelly (Vaseline is one brand) is a semisolid mixture of hydrocarbons that can be applied as an emollient or simply as a protective film. Water and water solutions (for example, urine) will not dissolve such a film, which explains why petroleum jelly protects a baby's tender skin from diaper rash.

Physical properties of alkenes are quite similar to those of corresponding alkanes. Alkenes with 1 to 4 carbons are gases at room temperature; those with 5 to 18 carbons are liquids; those with more than 18 are solids. Like the alkanes, the alkenes are insoluble in water and are less dense than water.

The physiological properties of the alkenes are also similar to those of the alkanes. Ethylene has found some use as an inhalation anesthetic. Like the gaseous alkanes, ethylene can cause unconsciousness, and even death, by asphyxiation. Large amounts of liquid and solid (or mixtures of liquid and solid) alkenes are seldom encountered. They would probably act on or in our bodies much as the alkanes do.

Alkenes occur widely in nature. Ripening fruits and vegetables give off ethylene, which triggers further ripening. Food processors artificially introduce ethylene to hasten the normal ripening process: 1 kg of tomatoes can be ripened by exposure to as little as 0.1 mg of ethylene for 24 hours.

Common aromatic hydrocarbons such as benzene, toluene, and the xylenes are liquids at room temperature. Their vapors act as narcotics when inhaled. Naphthalene, a volatile solid, is used as a moth repellant.

The most important property of all the hydrocarbons is that they burn. Their use as fuels is discussed in Chapter 13.

Problems

1. List three characteristics of the carbon atom that make possible the existence of millions of organic compounds.

2. Define, illustrate, or give an example of each of these terms.
   a. hydrocarbons
   b. alkane
   c. saturated
   d. unsaturated
   e. alkene
   f. alkyne
   g. homologous series
   h. isomer
   i. aromatic compounds

3. Classify the following compounds as organic or inorganic.
   a. C_6H_{10}
   b. $CoCl_2$
   c. $C_{12}H_{22}O_{11}$
   d. CH_3NH_2
   e. $NaNH_2$
   f. $Cu(NH_3)_6Cl_2$

4. How many carbon atoms are there in each of the following?
   a. ethane
   b. heptane
   c. butane
   d. nonane

5. Name these hydrocarbons.
   a. CH_3CH_3
   b. $CH_2{=}CH_2$
   c. $HC{\equiv}CH$

6. Name these hydrocarbons.

   a.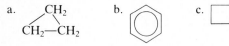
   b. ⬡
   c. □

7. Give the structural formulas of the 4-carbon alkanes (C_4H_{10}). Identify butane and isobutane.

8. Give the structures for these hydrocarbons.
   a. propane
   b. pentane

9. Indicate whether the structures in each set represent the same compound or isomers.
   a. CH_3CH_3 $\underset{\displaystyle CH_3}{CH_3}$

   b. $\overset{\displaystyle CH_3}{CH_3CH_2}$ $CH_3CH_2CH_3$

   c. $CH_3CH_2\overset{\displaystyle CH_3}{CH}CH_2CH_3$ $CH_3\overset{\displaystyle CH_3}{CH}CH_2CH_2CH_3$

   d. $CH_3\overset{\displaystyle CH_3}{CH}CH_2CH_3$ $CH_3CH_2\underset{\displaystyle CH_3}{\overset{\displaystyle CH_3}{CH}}$

e. $CH_3CH_2\overset{\displaystyle CH_3}{CH}{-}\overset{\displaystyle CH_3}{CH_2}$ $CH_2CH_2\overset{\displaystyle CH_3}{CH}CH_3$

10. Indicate whether each compound is saturated or unsaturated.

   a. $CH_3\underset{\displaystyle CH_3}{C}{=}CH_2$
   b. $CH_3{-}\underset{\displaystyle CH_3}{\overset{\displaystyle CH_3}{C}}{-}CH_3$

   c. $CH_3C{\equiv}CCH_3$
   d. ⬡

11. Give the structure for cyclopentane.

12. Propylene, like ethylene, is an alkene. Its molecular formula is C_3H_6. Draw a structure for propylene. Be sure to follow the valence rules.

13. There are three isomeric pentanes (C_5H_{12}). Give a structure for each.

14. Classify the compounds in Problem 10 as alkanes, alkenes, or alkynes.

15. Classify the following pairs as homologs, identical, isomers, or none of these.

   a. $CH_3CH_2CH_3$ and $CH_3CH_2CH_2CH_3$

   b. $\overset{\displaystyle CH_2}{\underset{\displaystyle CH_2{-}CH_2}{}}$ and $CH_3CH_2CH_3$

   c. $CH_3\overset{\displaystyle CH_3}{CH}CH_3$ and $CH_3\underset{\displaystyle CH_3}{CH}CH_3$

   d. $CH_2{=}CHCH_3$ and $\overset{\displaystyle CH_2}{\underset{\displaystyle CH_2{-}CH_2}{}}$

16. Give a structural formula and a condensed structural formula for octane.

17. Complete these structures by adding hydrogen atoms.
   a. $C{-}C{-}C{-}C{-}C{-}C$
   b. $C{-}\underset{\displaystyle C}{C}{-}C{-}C{-}C$

18. The compound $CH_3CH_2CH_2CH_2CH_2CH_3$ is a sol-

vent used to extract oil from soybeans. What is the name of the compound?

19. Indicate whether the compound is aromatic or not.

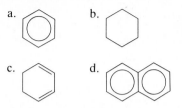

20. What is the meaning of the circle in the hexagon in the modern structure of benzene?
21. What is the physiological effect of methane?
22. What is the physiological effect of gaseous hydrocarbons (other than methane) and of the vapors of liquid hydrocarbons?
23. What is the danger in swallowing liquid alkanes?
24. Distinguish between lighter and heavier liquid alkanes in terms of their effect on the skin.
25. Write a balanced equation for the addition of hydrogen to ethylene to form ethane.

26. Write a balanced equation for the addition of hydrogen to acetylene to form ethane.
27. Write a balanced equation for the complete combustion of methane (reaction with oxygen) to form carbon dioxide and water.
28. Write a balanced equation for the complete combustion of propane to form carbon dioxide and water.
29. The complete combustion of benzene forms carbon dioxide and water.

$$C_6H_6 + O_2 \rightarrow CO_2 + H_2O$$

Balance the equation. What mass of carbon dioxide is formed by the complete combustion of 39 g of benzene?

30. In June of each year, *Chemical and Engineering News* publishes a list of the top 50 industrial chemicals. Look in the most recent June issue and find the annual production of each of the following chemicals.
a. acetylene   b. ethylene
c. benzene   d. toluene
e. propylene   f. cyclohexane

References and Readings

1. Benfey, O. T. "August Kekulé and the Birth of the Structural Theory of Organic Chemistry in 1858." *Journal of Chemical Education*, January 1958, pp. 21–25.
2. Hart, Harold. *Organic Chemistry: A Short Course*, 7th ed. Boston: Houghton Mifflin, 1987.
3. Hill, John W., and Dorothy M. Feigl. *Chemistry and Life*, 3d ed. New York: Macmillan, 1987. Chapter 14 provides a more detailed treatment of hydrocarbons.
4. Julian, Maureen M. "What Compound Was Discovered as a Result of an Insurance Claim?" *Journal of Chemical Education*, October 1981, p. 793.
5. Morrison, Robert T., and Robert N. Boyd. *Organic Chemistry*, 5th ed. Boston: Allyn and Bacon, 1987.

11

Organic Chemistry
Some Hors d'Oeuvres

In the last chapter, we discussed the chemistry of hydrocarbons. Here, and in some of the chapters that follow, we consider the chemistry of other organic compounds—all leading up to the chemistry of life.

As you might guess, the chemistry of life is extremely complex. Rather than jump right into the middle of that complexity, we introduce several simple families of organic compounds. In each, we start with simple compounds and then move on to some that are more complicated and, perhaps, more interesting. Don't assume that you will be bored by the small molecules, though. Some of them have fascinating properties of their own.

The Chlorinated Hydrocarbons: Many Uses, Some Hazards

Now that you know what hydrocarbons are, let's look at some chlorinated hydrocarbons. When chlorine gas (Cl_2) is mixed with methane (CH_4) in the presence of ultraviolet light, a reaction takes place at a very rapid (even explosive!) rate. The result is a mixture of products (Figure 11.1), some of which are probably quite familiar to you.

Methyl chloride is used as a refrigerant and a chemical intermediate. Methylene chloride is an important solvent. It has been used, for example, to extract caffeine from coffee to make the decaffeinated brands. Chloroform was used as an anesthetic in earlier times, but such use is now considered dangerous. The dosage required for effective anesthesia is too close to a lethal dose. Carbon tetrachloride has been used as a dry-cleaning solvent and in fire extinguishers. It is no longer recommended for either use. Exposure to carbon tetrachloride (or most of the other chlorinated hydrocarbons, for that matter) can cause severe damage to the liver. Even the vapor, breathed in small amounts, can cause serious illness when the exposure is prolonged. Use of a carbon tetrachloride fire extinguisher in conjunction with water to put out a fire can be deadly. Carbon tetrachloride reacts with water to form phosgene ($COCl_2$), an extremely poisonous gas. In fact, phosgene was used in poison-gas warfare during World War I. Evidence also indicates that both carbon tetrachloride and chloroform can cause cancer if ingested.

A variety of more complicated chlorinated hydrocarbons are of considerable interest. DDT (dichlorodiphenyltrichloroethane) and other chlorinated hydrocarbons used as insecticides are discussed in Chapter 17. In Chapter 12, we study the PCBs (polychlorinated biphenyls). For now, let's just say that all chlorinated hydrocarbons have similar properties. Most are only slightly polar, and they do not dissolve in water, which is highly polar. Instead, they dissolve (and *dissolve in*, since solubilities are reciprocal) fats, oils, greases, and other substances of low polarity. That is why certain chlorinated hydrocarbons make good dry-cleaning solvents; they remove grease and oily stains from fabrics. That is also why DDT and PCBs cause problems for fish and birds and perhaps for people; the toxic substances are concentrated in fatty animal tissues.

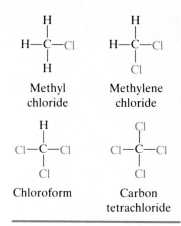

Figure 11.1 The four chlorine derivatives of methane.

The Chlorofluorocarbons: Spray Cans and Skin Cancer

Carbon compounds containing fluorine, as well as chlorine, have been used as the dispersing gases in aerosol cans and as refrigerants. Properly called **chlorofluorocarbons**, they are perhaps best known as **Freons**, from their Du Pont trade name. Structures of three of the Freons are illustrated in Figure 11.2. At room temperature, the chlorofluorocarbons are gases or liquids with low boiling temperatures. They are essentially insoluble in water and inert toward most other substances. These properties make them ideal propellants for use in

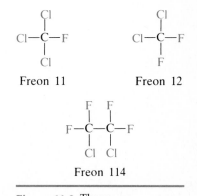

Figure 11.2 Three chlorofluorocarbons.

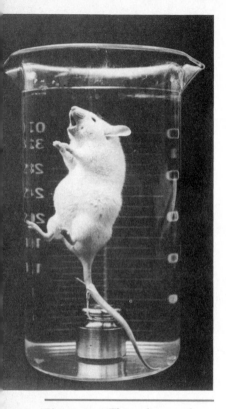

Figure 11.3 This submerged mouse survives by breathing a liquid perfluoro compound saturated with oxygen. [Courtesy of L. C. Clark, Jr.]

aerosol cans of deodorants, hair sprays, and food products. Unfortunately, the inertness of these compounds allows them to persist in the environment. They diffuse into the stratosphere, where they undergo chemical reactions that may lead to the destruction of the ozone layer, which protects the Earth from harmful ultraviolet radiation. A decrease in the amount of ozone in the stratosphere might lead to an increase in the incidence of skin cancer. We look at this problem in more detail in Chapter 15.

Fluorinated compounds have found some interesting uses. Some have been used as blood extenders. Oxygen is quite soluble in **perfluorocarbons**. (The prefix *per-* means that all hydrogen atoms have been replaced by fluorine atoms.) These compounds can therefore serve as temporary substitutes for hemoglobin, the oxygen-carrying protein in blood (Figure 11.3). Fluosol-SA, a mixture of perfluorodecalin and perfluorotripropylamine, has been tested in hundreds of patients in Japan. In the United States, it has been used mainly for those who reject transfusions of normal blood for religious reasons. (Teflon, a perfluorinated polymer, is discussed in Chapter 12.)

The Alcohols: Methyl Alcohol

We have seen that one or more hydrogen atoms of a hydrocarbon can be replaced by chlorine. Since both hydrogen and chlorine are univalent, this is a reasonable substitution. Other atoms, or even groups of atoms, also may substitute for hydrogen. Let's look at one such group of atoms, the **hydroxyl group** (—OH).

Wood alcohol (CH_4O) is made by heating wood (in the absence of air) until it breaks down and by then collecting the vapors formed. This process is called **destructive distillation** (Figure 11.4). A number of other chemical substances are formed as well as wood alcohol. What is the structure of wood alcohol? Following the valence rules, there is only one way in which one carbon, one oxygen, and four hydrogen atoms can be put together.

In condensed form, the structure is simply CH_3OH. The chemical name for wood alcohol is methyl alcohol, or simply methanol. Notice that it is related to methane in that one hydrogen of methane has been replaced by the —OH group.

Methyl alcohol is quite toxic.* As little as a 1-oz (30-mL) shot can cause permanent blindness or death. Many accidents each year are attributed to mistaking this alcohol for its less harmful relative, ethyl alcohol, the intoxicating ingredient of beverages such as beer, wine, whiskey, vodka, and brandy.

Methanol is made commercially mainly by reacting carbon monoxide with hydrogen.

$$CO \ + \ 2\,H_2 \ \longrightarrow \ CH_3OH$$

The reaction is carried out at a high temperature and pressure and in the presence of a catalyst. Despite its toxicity, methanol is a valuable industrial solvent. It is also a raw material for the production of other chemicals.

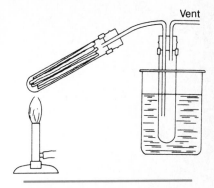

Figure 11.4 An apparatus for the destructive distillation of wood. The wood is heated in an enclosed tube, and alcohol is condensed in the second tube by the cold water in the beaker. Gases formed in the process can be burned as they emerge through the vent tube.

---

* The common term *poison* is applied to a variety of substances, often without any indication of degree of toxicity. To avoid this uncertainty, biochemists use the term LD_{50} to indicate the dosage that would be lethal to 50% of a population of test animals. This dosage is expressed in terms of the amount of drug per kilogram of body weight. The LD_{50} for humans usually can only be estimated.

Extrapolation from toxicity in rats or rabbits to toxicity in people can only be approximate at best. What is toxic in rats may not be as toxic in humans. Or it might be more toxic. However, general trends in toxicity usually can be judged from animal studies. Toxicities of some alcohols are listed in Table 11.1

Table 11.1 Lethal Oral Doses (in Rats) for Some Alcohols

Alcohol	Structure	Boiling Point (degrees Celsius)	LD_{50} (grams per kilogram of body weight)	Uses
Methanol	CH_3OH	64	—*	Solvent, fuel additive
Ethanol	CH_3CH_2OH	78	10.3	Solvent, beverages
1-Propanol	$CH_3CH_2CH_2OH$	97	1.9	Solvent
2-Propanol	$CH_3CHOHCH_3$	82	5.8	Solvent, body rubs
1-Butanol	$CH_3CH_2CH_2CH_2OH$	118	4.4	Solvent
1-Hexanol	$CH_3(CH_2)_4CH_2OH$	156	4.6	—
Ethylene glycol	$HOCH_2CH_2OH$	198	8.5	Antifreeze
Glycerol	$HOCH_2CHOHCH_2OH$	290 (dec)†	>20	Moisturizer

* No LD_{50} is given for methanol. Its acute (short-term) toxicity is not terribly high. However, it is metabolized to formaldehyde (HCHO) in the body, so that the chronic (long-term) toxicity is quite high. The LD_{50} for formaldehyde administered orally to rats is 0.070 g per kilogram of body weight. The LD_{50} for acetaldehyde, the metabolite of ethanol, is 1.9 g/kg.
† Glycerol decomposes.
SOURCE: Martha Windholz (Ed.), *The Merck Index*, 10th ed. Rahway, NJ: Merck and Co., 1983.

Ethyl Alcohol: Booze and Biochemistry

The next member of the homologous series of alcohols is ethyl alcohol (CH_3CH_2OH). Like methanol, ethyl alcohol (also called ethanol or grain alcohol) is toxic. One pint (about 500 mL) of pure ethanol, rapidly ingested, will kill most people. Even the strongest alcoholic beverages, however, are seldom more than 45% ethyl alcohol, or 90 proof.* Excessive ingestion over a long period of time leads to deterioration of the liver and loss of memory, and can lead to strong physiological addiction.

More than two thirds of the adult population in the United States drinks alcoholic beverages at least occasionally. The majority do so responsibly. Unfortunately, the nature of people and of alcohol is such that misuse—and abuse—is all too common.

It is estimated that there are about 10 million alcoholics in the United States. These are people so severely addicted that they are unable to hold a steady job or maintain stable family relationships. In addition, over half the fatal automobile accidents involve at least one drinking driver. Babies born to alcoholic mothers often are small, deformed, and mentally retarded. Some investigators believe that this **fetal alcohol syndrome** can occur even if the mothers drink only moderately. Alcohol is by far the most abused drug in the United States.

Generally, ethanol acts as a mild **depressant**; it slows down both physical and mental activity. Large amounts can produce unconsciousness or even death. Table 11.2 list the effects of various doses. Keep in mind, though, that the effects of alcohol vary with the drinker's weight, the amount of food in the drinker's stomach, the experience of the drinker, and other factors. Although generally a depressant, small amounts of ethanol, however, seem to act as a stimulant, perhaps by relaxing tensions and relieving inhibitions. Studies have shown that moderate drinking (no more than one or two drinks a day) lessens the chance for hospitalization for coronary disease by 40%. Moderate drinkers live longer, on the average, than nondrinkers. Heavy drinking, however, shortens the life span by contributing to diseases of the liver, cardiovascular system, and virtually every other organ of the body.

The mechanism for detoxification of ethanol has been determined. A liver enzyme system called cytochrome P-450 is involved. This system oxidizes foreign materials. This leads to the detoxification of ethanol,

---

* The **proof** is merely twice the percentage of alcohol by volume. The term has its origin in an old seventeenth-century English method for testing whiskey. Dealers were perhaps too often tempted to increase profits by adding water to the booze. A qualitative method for testing the whiskey was to pour some of it on gunpowder and ignite it. If the gunpowder ignited after the alcohol had burned away, that was considered "proof" that the whiskey did not contain too much water.

Table 11.2 Approximate Relationship Between Drinks Consumed, Blood-Alcohol Level, and Behavior for a 70-kg (154-lb) Moderate Drinker

Number of Drinks*	Blood-Alcohol Level (percent by volume)	Behavior
2	0.05	Mild sedation; tranquillity
4	0.10	Lack of coordination
6	0.15	Obvious intoxication
10	0.30	Unconsciousness
20	0.50	Possible death

* Rapidly consumed 30-mL (1-oz) "shots" of 90-proof whiskey, 360-mL (12-oz) bottles of beer, or 150-mL (5-oz) glasses of wine.

some drugs, pesticides, and other substances. The oxidation doesn't necessarily detoxify. Some carcinogens are activated by cytochrome P-450, and other substances (methanol, for instance; see footnote to Table 11.1) are made more toxic by this process.

Large amounts of cytochrome P-450 build up in the liver of the heavy drinker or drug user, enabling the liver to detoxify large amounts of alcohol (or drugs). This accounts for the tolerance that addicts develop. Just as an overdose of insulin requires the diabetic to take some sugar, an overabundance of cytochrome P-450 could require the alcoholic to consume more alcohol. Thus, we are able to propose a chemical mechanism for alcoholism.

Different people have different cytochrome P-450 systems. These differences are probably inherited. These inherited structural differences may explain why some people become more readily addicted than others, and why alcoholism seems to run in certain families.

The Synthesis of Ethanol

Ethyl alcohol for beverages is made by the fermentation of grain or other starchy or sugary materials (Chapter 16). If the sugar is glucose—a simple sugar—the reaction is

$$C_6H_{12}O_6 \xrightarrow{\text{yeast}} 2\ CH_3CH_2OH\ +\ 2\ CO_2$$

In addition to its use as a beverage, ethyl alcohol is used in great quantities as an industrial solvent and as a starting material for other chemical products. Most industrial ethanol is prepared by the reaction of ethylene with water.

$$CH_2{=}CH_2 \;+\; H_2O \;\xrightarrow{\;H^+\;}\; CH_3CH_2OH$$

The H^+ over the arrow indicates that an acid (usually sulfuric) is used to catalyze the reaction. This is a striking example of how chemists can change materials to meet people's needs. Ethylene is a gas that has little use as it is. It is produced in great quantity as a by-product of the cracking of petroleum (Chapter 13). By causing the ethylene to react with water, the chemist converts it to a useful and valuable solvent.

Synthetic ethanol has exactly the same properties as that from fermentation. It is generally cheaper, and, unlike alcoholic beverages, it carries no excise tax. To prevent the consumption of synthetic alcohol as a beverage, poisonous or noxious substances are added. The resulting **denatured alcohol** is unfit for drinking. Common denaturants include gasoline and methanol.

Taxes on alcoholic beverages raise about $6 billion a year in the United States. An estimated 100 000 to 200 000 deaths are attributed annually to alcohol. The total cost to society—from medical treatment, accidents, lost work time, and other factors—is estimated to be $120 billion a year.

More Alcohols: The Functional Group

Methyl alcohol (CH_3OH) and ethyl alcohol (CH_3CH_2OH) are the first two members of a homologous series of alcohols. Adding another CH_2 group gives propyl alcohol ($CH_3CH_2CH_2OH$). With three carbon atoms, however, there is the possibility of another alcohol. This one has the OH group on the middle, or number two, carbon.

$$\begin{array}{c} CH_3CHCH_3 \\ | \\ OH \end{array}$$

It is an isomer of propyl alcohol and is called isopropyl alcohol. Alternative names for the two alcohols are 1-propanol and 2-propanol, where the numbers indicate the location of the OH group, on the end (first carbon) or on the middle (second carbon). Isopropyl alcohol is familiar as rubbing alcohol (70% isopropyl alcohol and 30% water). It is also used in cosmetics such as aftershave lotions.

There are many other alcohols. A few representative ones are shown in Table 11.1 (page 217). All alcohols are characterized by the presence of a hydroxyl group (OH group). This pair of atoms is called the **functional group** of the alcohol; most of the characteristic reactions of alcohols take place at the OH group. Functional groups serve as unifying concepts for much of organic chemistry. Table 11.3 lists some of the more important functional groups.

Table 11.3 Selected Organic Functional Groups

Name of Class	Functional Group	General Formula of Class
Alkane	None	R—H
Alkene	$-\overset{\textstyle\mid}{C}=\overset{\textstyle\mid}{C}-$	$R-\overset{\textstyle R}{\underset{}{C}}=\overset{\textstyle R}{\underset{}{C}}-R$
Alkyne	$-C\equiv C-$	$R-C\equiv C-R$
Alcohol	$-\overset{\textstyle\mid}{\underset{\textstyle\mid}{C}}-O-H$	$R-O-H$
Ether	$-\overset{\textstyle\mid}{\underset{\textstyle\mid}{C}}-O-\overset{\textstyle\mid}{\underset{\textstyle\mid}{C}}-$	$R-O-R$
Aldehyde	$-\overset{\textstyle O}{\overset{\textstyle\|}{C}}-H$	$R-\overset{\textstyle O}{\overset{\textstyle\|}{C}}-H$
Ketone	$-\overset{\textstyle O}{\overset{\textstyle\|}{C}}-$	$R-\overset{\textstyle O}{\overset{\textstyle\|}{C}}-R$
Amine	$-\overset{\textstyle\mid}{\underset{\textstyle\mid}{C}}-\overset{\textstyle\mid}{\underset{\textstyle\mid}{N}}-$	$R-\overset{\textstyle H}{\underset{}{N}}-H$ $R-\overset{\textstyle H}{\underset{}{N}}-R$ $R-\overset{\textstyle R}{\underset{}{N}}-R$
Carboxylic acid	$-\overset{\textstyle O}{\overset{\textstyle\|}{C}}-O-H$	$R-\overset{\textstyle O}{\overset{\textstyle\|}{C}}-O-H$
Ester	$-\overset{\textstyle O}{\overset{\textstyle\|}{C}}-O-\overset{\textstyle\mid}{\underset{\textstyle\mid}{C}}-$	$R-\overset{\textstyle O}{\overset{\textstyle\|}{C}}-O-R$
Amide	$-\overset{\textstyle O}{\overset{\textstyle\|}{C}}-\overset{\textstyle\mid}{N}-$	$R-\overset{\textstyle O}{\overset{\textstyle\|}{C}}-\underset{\textstyle H}{N}-H$ $R-\overset{\textstyle O}{\overset{\textstyle\|}{C}}-\underset{\textstyle H}{N}-R$ $R-\overset{\textstyle O}{\overset{\textstyle\|}{C}}-\underset{\textstyle R}{N}-R$

Alkyl Groups

In addition to the OH group, each alcohol molecule has a carbon–hydrogen portion that is derived from an alkane. For example, the methyl group (CH_3—) of methyl alcohol is considered to be derived from methane (CH_4) by removal of a hydrogen atom. Similarly, the ethyl group (CH_3CH_2—) of ethyl alcohol is derived from ethane (CH_3CH_3). Two groups are derived from propane ($CH_3CH_2CH_3$). One has its attachment through an end carbon atom, the other through the central atom.

CH₃CH₂CH₂—	CH₃CHCH₃
Propyl group	Isopropyl group

Collectively, these groups derived from alkanes are called alkyl groups. Often the letter *R* is used to stand for alkyl groups in general. Since the OH group characterizes an alcohol, the formula ROH is used to indicate alcohols generally or collectively.

Phenols and Ethers

Compounds with a hydroxyl group attached directly to a benzene ring are called phenols. The parent compound is itself called phenol.

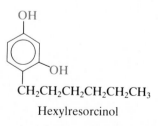

Phenol

Phenol was the first widely used antiseptic. Joseph Lister used it for antiseptic surgery in 1867. Unfortunately, phenol causes severe burns. It has been replaced by other phenolic compounds such as hexyl-resorcinol.

OH

OH

$CH_2CH_2CH_2CH_2CH_2CH_3$

Hexylresorcinol

These newer compounds are more effective as antiseptics and less damaging to the skin.

Compounds that have two hydrocarbon groups attached to an oxygen atom are called ethers. The most important ether is diethyl ether ($CH_3CH_2OCH_2CH_3$). Once widely used as an anesthetic, its main use nowadays is as a solvent. It dissolves many organic substances such as fats. Diethyl ether boils at 36 °C, so it is readily evaporated away, enabling one to recover the dissolved substance. Often called simply "ether," diethyl ether is highly flammable. Great care must be taken to avoid sparks or flames when ether is in use.

Example 11.1   Give the formula for dimethyl ether.
     Dimethyl ether has an oxygen atom joined to two methyl groups. Its formula is CH_3OCH_3.

Methyl *tert*-butyl ether is widely used as an octane booster in gasoline (Chapter 13).

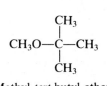

Methyl *tert*-butyl ether

Ethers generally are represented by the formula ROR.

Aldehydes and Ketones: Two Families with a Common Functional Group

Next, we consider two related families of organic compounds, the aldehydes and the ketones. Both families are characterized by the presence of a carbon atom doubly bonded to an oxygen atom. This arrangement is called the **carbonyl functional group**. Aldehydes have a hydrogen atom attached to the carbonyl carbon; ketones have the carbonyl carbon attached to two other carbon atoms. If we let R stand for the hydrocarbon portions of molecules, we can write general formulas for aldehydes and ketones.

The simplest aldehyde is formaldehyde. It is a gas at room temperature but is readily soluble in water. As a 40% solution called *formalin*, it is used as a preservative for biological specimens and in embalming fluid.

Formaldehyde is used to make certain plastics (Chapter 12). It also is used to disinfect homes, ships, and warehouses. Commercially,

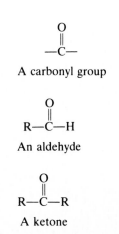

A carbonyl group

An aldehyde

A ketone

formaldehyde is made by the oxidation of methanol. This is the same reaction that occurs in the human body when methanol is ingested.

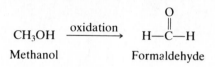

CH₃OH $\xrightarrow{\text{oxidation}}$

Methanol   Formaldehyde

The next member of the homologous series of aldehydes is acetaldehyde, formed by the oxidation of ethanol.

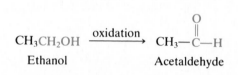

CH₃CH₂OH $\xrightarrow{\text{oxidation}}$

Ethanol   Acetaldehyde

Example 11.2   Give the structure of the next homolog above acetaldehyde in the homologous series of aldehydes.
    The homolog has one more CH₂ unit.

(The compound is called propionaldehyde.)

Benzaldehyde has an aldehyde group attached to a benzene ring.

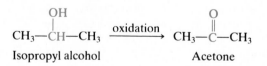

Also called (synthetic) oil of almond, benzaldehyde is used in perfumery and flavoring. There are also many ketones. The simplest is acetone, made by the oxidation of isopropyl alcohol.

$$\underset{\text{Isopropyl alcohol}}{CH_3-\overset{\overset{\displaystyle OH}{|}}{CH}-CH_3} \xrightarrow{\text{oxidation}} \underset{\text{Acetone}}{CH_3-\overset{\overset{\displaystyle O}{\|}}{C}-CH_3}$$

Acetone is a common solvent for such organic materials as fats, rubbers, plastics, and varnishes. It also finds use in paint and varnish removers. It is the major (sometimes the only) ingredient in fingernail polish removers.

The Odorous World of Organic Acids

The familiar strong acids (Chapter 7) such as sulfuric, hydrochloric, and nitric acids are called **mineral acids** because they are derived from inorganic materials. Many of the weak acids are organic ones. They were so named originally because they were derived from plant or animal sources, that is, from organisms. Nowadays, like other carbon compounds, they often are synthesized from coal tar or petroleum.

The functional group of the organic acids is called the **carboxyl group**, and the acids are called **carboxylic acids**. The simplest carboxylic acid is formic acid. It was first obtained by the destructive distillation of ants (Latin *formica*: "ant"). The bite of an ant smarts because the ant injects formic acid as it bites. The stings of wasps and bees also contain formic acid (as well as other poisonous materials).

Acetic acid can be made by the aerobic fermentation of a mixture of cider and honey. This produces a solution (vinegar) that contains about 4 to 10% acetic acid, plus a number of other compounds that give vinegar its flavor. Acetic acid is probably the most familiar *weak* acid used in educational and industrial chemistry laboratories.

The third member of the homologous series of acids, propionic acid, is seldom encountered in everyday life. The fourth member is more familiar, at least by its odor. If you've ever smelled rancid butter, you probably wish you hadn't—but you know what butyric acid smells like. It is one of the most fantastically foul-smelling substances imaginable. Butyric acid can be isolated from butterfat or synthesized in the laboratory. It is one of the ingredients of body odor. Extremely small amounts of this and other chemicals enable bloodhounds to track fugitives.

The acid with a carboxyl group attached directly to a benzene ring is called benzoic acid. In general, carboxylic acids can be represented by the formula RCOOH.

Like the mineral acids, the carboxylic acids form salts. Calcium propionate, sodium benzoate, and other carboxylate salts are widely used as food additives to prevent molds (Chapter 19).

Carboxylic acids are common in nature. Many of the higher homologs are obtained from fats (Chapter 18).

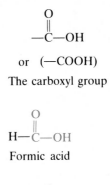

Esters

Esters are derived from carboxylic acids and alcohols. The general reaction involves splitting out a molecule of water.

$$R—\overset{\overset{\displaystyle O}{\|}}{C}—OH \; + \; R'OH \quad \xrightarrow{H^+} \quad R—\overset{\overset{\displaystyle O}{\|}}{C}—OR' \; + \; H_2O$$

An acid   An alcohol   An ester

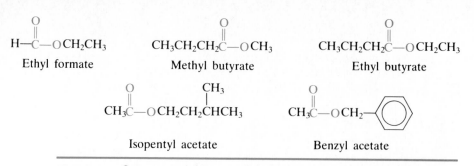

Figure 11.5 Some esters of interest. Ethyl formate is an artificial rum flavor. Methyl butyrate occurs in apples, and ethyl butyrate occurs in pineapples. Isopentyl acetate is banana oil, used as a solvent and in flavoring, while benzyl acetate is oil of jasmine, used in perfumes.

Although the carboxylic acids often stink, the esters usually have pleasant odors. Many esters are used in perfumes. For example, methyl butyrate, made from butyric acid and methanol, is used in artificial fruit flavors. Pineapple oil is a solution of ethyl butyrate in ethanol. Figure 11.5 shows these and other esters of interest.

Amines and Amides: Nitrogen-Containing Organics

Many organic compounds contain nitrogen. We encounter a variety of such compounds in the chapters that follow. In this chapter, we introduce two families that will provide you with a vital background for the material ahead—the amines and the amides.

The **amines** contain the elements carbon, hydrogen, and nitrogen. They are derived from ammonia by replacing one, two, or three of the hydrogen atoms by an alkyl group (Figure 11.6).

The simplest amine is methylamine (CH_3NH_2). The next higher homolog is ethylamine ($CH_3CH_2NH_2$). With two carbon atoms, however, we can also have dimethylamine (CH_3NHCH_3). Note that ethylamine and dimethylamine are isomers; both have the molecular formula

Figure 11.6 Ammonia and three kinds of amines derived from it by replacing one, two, or all three of the hydrogen atoms by alkyl groups.

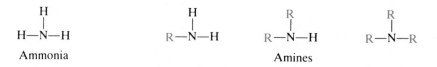

C_2H_7N. With three carbon atoms, there are several possibilities, including trimethylamine (CH_3NCH_3).

$$CH_3$$

Example 11.3 Give structures and names for the other three-carbon amines.

Two are derived from the two propyl groups.

$CH_3CH_2CH_2NH_2$ CH_3CHCH_3
$$NH_2$$

Propylamine   Isopropylamine

The other has one methyl group and one ethyl group.

$CH_3CH_2NHCH_3$

Ethylmethylamine

The amine with an NH_2 group attached directly to a benzene ring has the special name *aniline*.

$$\text{⟨benzene ring⟩} - NH_2$$

The simple amines are similar to ammonia in odor, basicity, and other properties. It is the higher amines that are the most interesting, though. Figure 11.7 includes a variety of these. Notice that each structure contains an —NH_2 unit. This is called the **amino group**.

CH_3
CH_2—C—NH_2
H

Amphetamine

H_2N—$CH_2CH_2CH_2CH_2CH_2CH_2$—NH_2

1,6-Hexanediamine
(an intermediate in the synthesis of nylon)

H_2N—$CH_2CH_2CH_2CH_2CH_2$—NH_2

Cadaverine
(odor of decaying flesh)

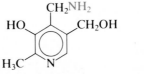

Pyridoxamine
(a B-complex vitamin)

Figure 11.7 Some amines of interest.

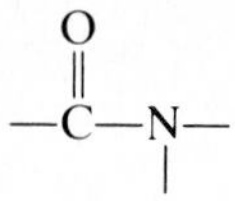

Amide functional group

The other nitrogen-containing compounds we discuss in this chapter—the **amides** (what a difference that one letter makes!)—contain oxygen as well as carbon, nitrogen, and hydrogen. Amides can be made by reacting a carboxylic acid with ammonia or an amine to form a salt, and by then heating the salt to drive off the water.

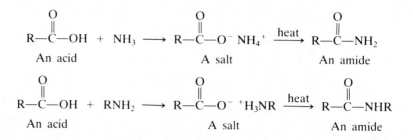

The simple amides are of little interest to us here, but the complex amides are of tremendous importance. Nylon (Chapter 12) is an amide. Even more important are the proteins (Chapter 18), which, quite close to the basis of life itself, are also amides. Remember the amide functional group; you will see it many times in the chapters that follow.

Heterocyclic Compounds: Alkaloids and Others

In Chapter 10, we considered several cyclic hydrocarbons. These compounds feature rings of carbon atoms. Now let's look at some ring compounds that have atoms other than carbon. These **heterocyclic compounds** usually have one or more nitrogen, oxygen, or sulfur atoms.

Example 11.4 Which of the following structures represent heterocyclic compounds?

a. CH_2-CH_2 with O bridging   b. CH_2-CH_2 with S bridging   c. CH_2-CH_2 with CH_2 bridging   d. CH_2-CH_2 with NH bridging

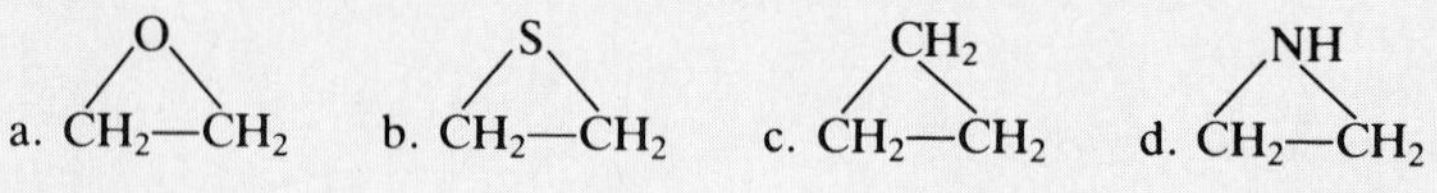

Compounds a, b, and d have oxygen, sulfur, and nitrogen atoms, respectively, in a ring structure; these represent heterocyclic compounds.

Many amines, particularly heterocyclic ones, occur naturally in plants. Like other amines, these compounds are basic. They are called **alkaloids**, which means "like alkalis." Among the familiar alkaloids are morphine, caffeine, nicotine, and cocaine. The action of these compounds as drugs is considered in Chapter 23. Of more immediate interest are pyrimidine, which has two nitrogen atoms in a six-membered ring, and purine, which has four nitrogen atoms in two rings that share a common side.

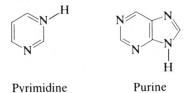

Pyrimidine   Purine

Compounds related to these two play important roles in the chemistry of all living cells.

Nucleic Acids: The Chemistry of Heredity

Complex compounds called nucleic acids are found in every living cell. They serve as the information and control centers of the cell. There are actually two kinds of nucleic acids. **Deoxyribonucleic acid (DNA)** is found primarily in the cell nucleus. **Ribonucleic acid (RNA)** is found in all parts of the cell. Both nucleic acids are long chains of repeating units called **nucleotides**.

Nucleotides themselves consist of three parts: a sugar, a heterocyclic amine base, and a phosphate unit. The sugar is either ribose or deoxyribose.

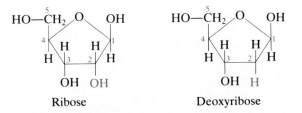

Ribose   Deoxyribose

Note that deoxyribose differs from ribose in that it lacks an oxygen atom on the second carbon atom. Ribose is found in ribonucleic acid (RNA), and deoxyribose is found in deoxyribonucleic acid.

When incorporated into nucleic acids, the hydroxyl group on the first carbon atom is replaced by one of five heterocyclic amine bases.

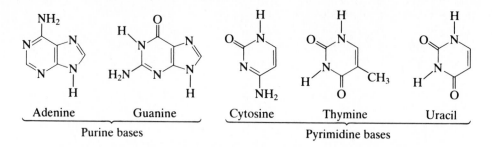

Adenine	Guanine	Cytosine	Thymine	Uracil

Purine bases   Pyrimidine bases

The bases with two fused rings, adenine and guanine, are classified as purines. Cytosine, thymine, and uracil are pyrimidines.

The hydroxyl group on the fifth carbon of the sugar unit is converted into a phosphate ester group. Adenosine monophosphate is a representative nucleotide. In AMP, the base is adenine and the sugar is ribose.

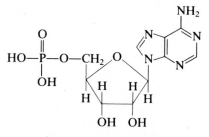

Adenosine monophosphate

Example 11.5 Identify the sugar and the base in the nucleotide thymidine monophosphate.

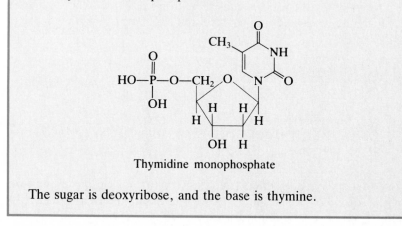

Thymidine monophosphate

The sugar is deoxyribose, and the base is thymine.

Nucleotides are joined to one another through the phosphate group to form nucleic acid chains. The phosphate unit on one nucleotide forms

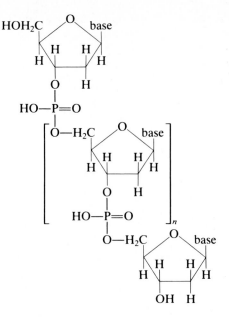

Figure 11.8 The backbone of a deoxyribonucleic acid molecule. The *n* indicates that the unit is repeated many times.

an ester linkage to the hydroxyl group on the third carbon atom of the sugar unit in a second nucleotide. This unit is in turn joined to another nucleotide, and the process repeated to build up the long nucleic acid chain (Figure 11.8). The backbone of the chain consists of alternating phosphate and sugar units. The heterocyclic bases are branched off this backbone.

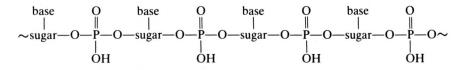

As we have seen, the sugar in DNA is deoxyribose, and that in RNA is ribose. The bases in DNA are adenine, guanine, cytosine, and thymine. Those in RNA are adenine, guanine, cytosine, and uracil (Table 11.4).

A most important feature of the nucleic acid molecule is the

Table 11.4 Components of DNA and RNA

	DNA	RNA
Purine bases	Adenine	Adenine
	Guanine	Guanine
Pyrimidine bases	Cytosine	Cytosine
	Thymine	Uracil
Pentose sugar	Deoxyribose	Ribose
Inorganic acid	Phosphoric acid	Phosphoric acid

sequence of the four bases along the strand. The molecules are huge, with molecular weights ranging into the billions for mammalian DNA. Along these great chains, the four bases may be arranged in essentially infinite variations. That is a crucial feature of these molecules, because it is the base sequence that is used to store the multitude of information needed to build living organisms. Before we examine that aspect of nucleic acid chemistry, let us consider one more important feature of nucleic acid structure.

The Double Helix

In experiments designed to probe the structure of DNA, it was determined that the molar amount of adenine (A) in DNA corresponded to the molar amount of thymine (T). Similarly, the molar amount of guanine (G) is essentially the same as that of cytosine (C). To maintain this balance, the bases in DNA must be paired, A to T and G to C. But how? At the midpoint of this century, it was quite clear that the answer to this question would bring with it a Nobel Prize. Although many illustrious scientists worked on the problem, two scientists who were relatively unknown in the world of science announced in 1953 that they had worked out the structure of DNA. Using data that involved quite sophisticated chemistry, physics, and mathematics, and working

Figure 11.9 James D. Watson and Francis Crick, discoverers of the double helix model of DNA. [Courtesy of Harvard University Biological Laboratories.]

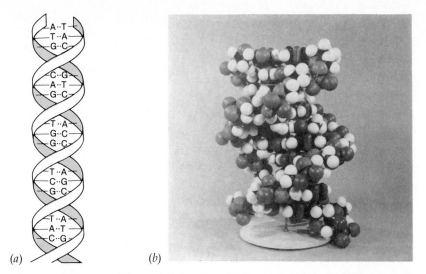

(a)

(b)

Figure 11.10 (a) A schematic representation of the DNA double helix. (b) A model of the DNA molecule. [Photo courtesy of Science Related Materials, Inc., Janesville, WI.]

with models not unlike a child's construction set, James D. Watson and Francis Crick (Figure 11.9) determined that DNA must be composed of two helixes wound about one another. The phosphate and sugar backbone of the polymer chains form the outside of the structure, which is rather like a spiral staircase. The heterocyclic amines are paired on the inside—with guanine always opposite cytosine and adenine always opposite thymine. In our staircase analogy, these base pairs are the steps (Figure 11.10).

Why do the bases pair in this precise pattern, always A to T and T to A, always G to C and C to G? The answer is hydrogen-bonding and a truly elegant molecular design. Figure 11.11 shows the two sets of base pairs. You should notice two things. First, a pyrimidine is paired with a purine in each case, and the long dimensions of both pairs are identical (1.085 nm).

The second thing you should notice in Figure 11.11 is the hydrogen-bonding between the bases in each pair. When guanine is paired with

Figure 11.11 The pairing of bases in the DNA double helix.

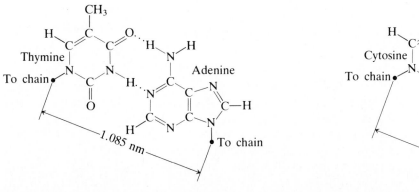

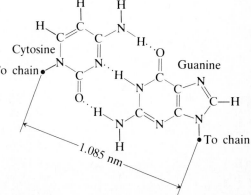

Figure 11.12 RNA occurs as single strands that can form double-helical portions by internal base pairing.

cytosine, three hydrogen bonds can be drawn between the bases. No other pyrimidine–purine pairing will permit such extensive interaction. Indeed, in the combination shown in the figure, both pairs of bases fit like lock and key.

The molecules of RNA consist of single strands of the nucleic acid. Some internal (intramolecular) base-pairing may occur in sections where the molecule folds back on itself. Portions of the molecule may exist in double-helical form (Figure 11.12).

Watson and Crick received the Nobel Prize in 1962 for discovering, as Crick put it, "the secret of life." The structure these two scientists proposed was accepted almost immediately by other scientists around the world because it answers so many crucial questions. It can explain how cells are able to divide and go on functioning, how genetic data are passed on to new generations, and even how proteins are built to required specifications. It all depends on the base-pairing.

DNA: Self-replication

Cats have kittens that grow up to be cats. Bears have cubs that grow up to be bears. How is it that each species reproduces after its own kind? How does a fertilized egg "know" that it should develop into a kangaroo and not a koala?

The physical basis of heredity has been known for a long time. Most higher organisms reproduce sexually. A sperm cell from the male unites with an egg cell from the female. The fertilized egg so formed must carry all the information needed to make the various cells, tissues, and organs necessary for the functioning of a new individual. For human beings, that single cell must carry the information for the making of legs, liver, lungs, heart, head, hair, and hands—in short, all the instruction ever needed for growth and maintenance of the individual. In addition, if the species is to survive, information must be set aside in germ cells—both sperms and eggs—for the production of new individuals.

The hereditary material is found in the nuclei of all cells, concentrated in elongated, threadlike bodies called chromosomes. The number of chromosomes varies with the species. Human body cells have 46 chromosomes. Germ cells carry only half that number. Thus, in sexual reproduction, the entire complement of chromosomes is achieved only when the egg and sperm combine; a new individual receives half its hereditary material from each parent.

Chromosomes are made of nucleoproteins. The nucleic acid in chromosomes is DNA, and it is the DNA that is the primary hereditary material (Figure 11.13). Arranged along the chromosomes are the basic units of heredity, the genes. Structurally, genes are sections of the DNA molecule (some viral genes contain only RNA). When cell division occurs, each chromosome produces an exact duplicate of itself. Transmission of genetic information therefore requires the **replication** (copying or duplication) of DNA molecules. The Watson–Crick double helix provides a ready model for this process. If the two chains of the double helix are pulled apart, then each chain can direct the synthesis of

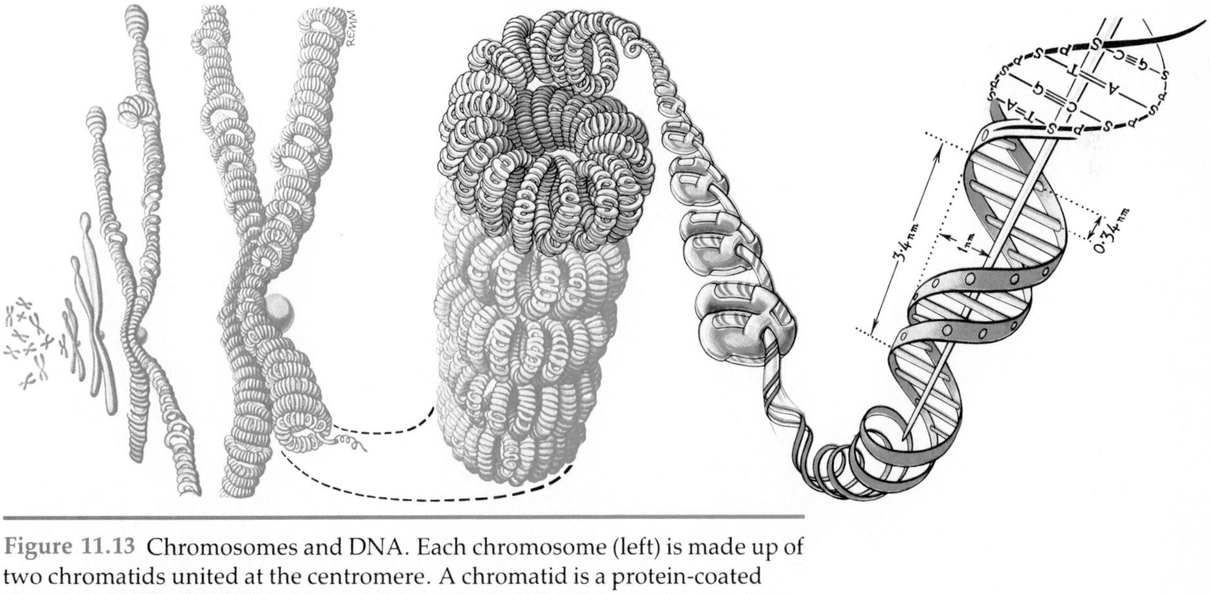

Figure 11.13 Chromosomes and DNA. Each chromosome (left) is made up of two chromatids united at the centromere. A chromatid is a protein-coated strand of multicoiled DNA. [Adapted from Roger Warwick and Peter L. Williams, *Gray's Anatomy*, 36th British ed. (London: Longmans Co. Ltd., 1980), p. 15, with permission of Churchill Livingstone.]

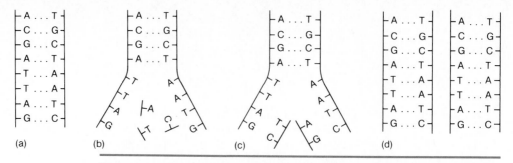

Figure 11.14 DNA replication. (a) The original double helix, flattened out here for clarity. (b) The helix beginning to split. Some free nucleotides from the cell are shown. (c) Nucleotides from the cell beginning to pair with bases on each original strand. (d) The two new double helixes, each identical to the original.

a new DNA chain. In the cellular fluid surrounding the DNA are all the necessary nucleotides. Synthesis begins with the base on a nucleotide pairing with its complementary base on the DNA strand (Figure 11.14). Keep in mind that adenine can pair only with thymine, and guanine only with cytosine. Each base unit in the separated strand can pick up only a unit identical to the one it had paired with before. Each of the separating chains serves as a template, or pattern, for the formation of a new complementary chain.

As the nucleotides align, they react with one another to form the sugar–phosphate backbone of the new chain. In this way, each strand of the original DNA molecule forms a duplicate of its former partner. Whatever information was encoded in the original DNA double helix is now contained in each of the replicates. When the cell divides, each daughter cell gets one of the DNA molecules and all the information that was available to the parent cell.

DNA can be compared to a book containing directions for putting together a model airplane or knitting a sweater. Knitting directions store information as words on paper. Letters of the alphabet are arranged in a certain way (e.g., "knit one, purl two"), and these words direct the knitter to carry out a particular operation with needles and yarn. If all the directions are correctly followed, the ball of yarn becomes a sweater.

How is information stored in DNA? The sequence of bases along the DNA chain encodes the directions for building an organism. Just as *saw* means one thing in English and *was* means another, the sequence of bases C-G-T means one thing, and G-C-T means something else. Although there are only four "letters"—the four bases—in the genetic codes of DNA, their sequence along the long strands can vary so widely that essentially unlimited information storage is available. Each cell carries in its DNA all the information it needs to determine all the hereditary characteristics of even the most complex organism.

RNA: Protein Synthesis and the Genetic Code

DNA carries a message that must somehow be relayed and acted upon in the cell. Because DNA is seldom identified outside of the cell nucleus, its information, or "blueprint" must be transported by something else. It is! In the first step, called **transcription**, DNA transfers its information to a special RNA molecule called **messenger RNA** (mRNA). The base sequence of DNA specifies the base sequence of mRNA. Thymine in DNA calls for adenine in mRNA, cytosine specifies guanine, guanine calls for cytosine, and adenine requires uracil. Remember that in RNA molecules, uracil is used in place of DNA's thymine. Notice the similarity in structure of these two bases (page 230).

DNA Base	Complementary RNA Base
Adenine	Uracil
Thymine	Adenine
Cytosine	Guanine
Guanine	Cytosine

The next step in protein synthesis involves interpretation of the code which has been "xeroxed" by mRNA and **translation** of that code into a protein structure. mRNA travels from the nucleus to structures called ribosomes in the cytoplasm of the cell. Ribosomes also are constructed of nucleoproteins. mRNA becomes attached to the ribosome, and it is here that the genetic code is deciphered.

There is another type of RNA molecule, called **transfer RNA** (tRNA), located in the cytoplasm of cells. The tRNA is responsible for translating the specific base sequence of mRNA into a specific amino acid sequence in a protein. (The function of proteins is dependent on the sequence of their amino acid building blocks. Proteins are discussed in Chapter 18.) The tRNA molecule has a looped structure (Figure 11.15) which contains a critical set of three bases, called a **triplet**, which determine which amino acid will be attached to an end of tRNA. Each molecule of tRNA can carry only one amino acid. The triplet configuration specifies which amino acid tRNA will carry. For example, tRNA molecules with the triplet configuration of CCU always carry the amino acid, glycine, and the base triplet GCG specifies arginine.

Figure 11.15 A given transfer RNA can carry only one kind of amino acid, which is specified by the base triplet in the anticodon.

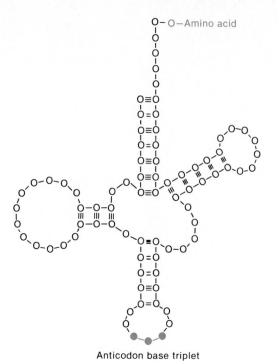

Anticodon base triplet

We are now ready to build proteins. Strung out along some ribosomes in the cytoplasm of a cell is an mRNA molecule. The sequence of bases along this molecule was determined by the DNA of a gene in the cell nucleus. Floating around in the cytoplasm surrounding the mRNA are tRNA molecules, each carrying its own amino acid. Let us suppose that a portion of the mRNA base sequence reads

$$\sim C—G—C—G—G—A—G—G—C\sim$$

The first three bases (CGC) could pair, through hydrogen-bonding, with a nucleic acid that had a GCG sequence. There, floating by, is just such a nucleic acid, a tRNA molecule with the amino acid arginine attached. The triplet of bases of tRNA pairs up with the first three bases of the mRNA.

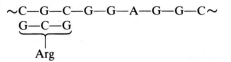

The next three bases of the mRNA, GGA, pair up with a CCU tRNA molecule, which carries the amino acid glycine.

~C—G—C—G—G—A—G—G—C~
 G—C—G   C—C—U
      /
    Arg   Gly

When the two tRNA molecules are appropriately lined up along the mRNA, a bond forms between the two amino acids.

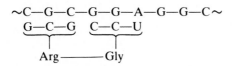

The next three bases on the mRNA (GGC) pair with a CCG tRNA molecule.

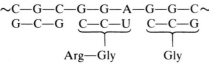

The amino acid carried by the tRNA is then joined to the previously linked amino acids. The first tRNA molecule, having delivered its amino acid, returns to the cytoplasm.

~C—G—C—G—G—A—G—G—C~
       C—C—U   C—C—G
      Arg—Gly—Gly

In this way the protein chain is gradually built up. The chain is released from the tRNA and mRNA as it is formed.

Each base triplet on the mRNA molecule is called a **codon**. The base triplets of the tRNA are called **anticodons**. A codon on mRNA always pairs with its complementary tRNA anticodon. A complete dictionary of the genetic code has been compiled. There are 64 possible codons and only 20 amino acids; there is considerable redundancy in the code. Two of the amino acids each are specified by six different codons. Two others have only one codon each. Three codons are stop signals, calling for termination of the protein chain.

Genetic Engineering

Over 3000 human diseases have a genetic component. Over the last decade or so, researchers have linked specific genes to specific diseases.

Now the ability to use this information to diagnose and cure genetic diseases appears to be within our grasp. By determining the location of a gene on the DNA molecule, scientists have been able to identify and isolate genes with specific functions.

A gene is a rather elusive substance. There are approximately 10 000 genes on each human chromosome, and isolating the one gene that is defective (in the case of genetic disease) is understandably difficult. One detection approach is to treat the DNA with enzymes called **restriction endonucleases**. This method yields segments of genetic material called **restriction fragment length polymorphisms** or RFLPs (pronounced "rif lips"). These segments are much easier to work with because they contain fewer genes. The pattern of RFLPs resulting from enzyme treatment is an inherited characteristic. RFLP patterns characteristic of certain families can be isolated. If the pattern of a relative matches that of a person with a genetic disease, that relative will probably develop the disease. Thus it is possible to identify and even predict the occurrence of a genetic disease.

A future hope of genetic engineering is that we will be able to introduce a functioning gene into a person's cells, thus correcting the action of a defective gene.

Recombinant DNA

All living organisms (except some viruses) have DNA as their hereditary material. It should be possible, then, to place a gene from one organism into the genetic material of another. **Recombinant DNA** (rDNA) technology does just that. First, the gene is identified. Then it is isolated, by the RFLP process, and placed in a separate piece of DNA. The recombined DNA is then transferred into bacteria or other suitable organism. The final step, called **cloning**, produces many copies of the modified bacteria, which can then produce relatively large amounts of the protein coded for by the gene.

By working backward from the amino acid sequence (Chapter 18) of the protein, scientists can work out the base sequence of the gene that codes for the protein. When isolated, the gene is spliced into a special kind of bacterial DNA called a **plasmid**. The recombined plasmid is then inserted into the host organism (Figure 11.16). Once inside the host, the plasmid replicates, making multiple exact copies (or clones) of itself. As the rDNA-engineered bacteria multiply, they become virtual factories for producing the desired protein.

Many valuable materials, difficult to obtain in any other way, are now made using rDNA technology. People with diabetes formerly had to use insulin from pigs or cattle. Now human insulin, a protein coded for by human DNA, is being produced by the cell machinery of bacteria

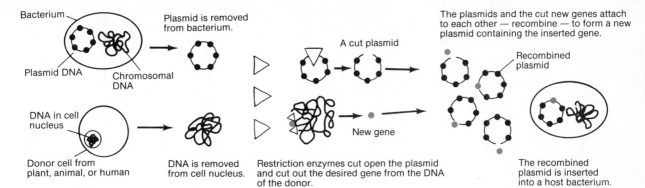

Bacterium

Plasmid DNA

Chromosomal DNA

Plasmid is removed from bacterium.

A cut plasmid

DNA in cell nucleus

Donor cell from plant, animal, or human

DNA is removed from cell nucleus.

New gene

Restriction enzymes cut open the plasmid and cut out the desired gene from the DNA of the donor.

The plasmids and the cut new genes attach to each other — recombine — to form a new plasmid containing the inserted gene.

Recombined plasmid

The recombined plasmid is inserted into a host bacterium.

Figure 11.16 A schematic outline of rDNA technology. [Adapted with permission from *Biotechnology*, ACS Department of Government Relations and Science Policy, December 1985. Copyright 1985, American Chemical Society.]

(Figure 11.17). All newly diagnosed insulin-dependent diabetics in the United States are now treated with rDNA-produced human insulin.

Human growth hormone, used to treat children who fail to grow properly, was formerly available only in tiny amounts obtained from cadavers. Now it is readily available through rDNA technology. This technology also yields interferon, a promising anticancer agent. Scientists have even designed bacteria that will "eat" the oil released in an oil spill.

Concern over the potential for disaster in this type of research has lessened somewhat in recent years. Initially, scientists worried about the possibility of producing a deadly "artificial" organism. What if a gene that causes cancer was spliced into the DNA of a bacterium that normally inhabits our intestine? We would have no natural immunity against such an organism. To protect against such a development, strict guidelines for recombinant-DNA research have been instituted.

The new molecular genetics has already resulted in some impressive achievements. Its possibilities are mind boggling—elimination of genetic defects, a cure for cancer, a race of geniuses, and who knows what else? Knowledge gives power: it does not necessarily give wisdom. Who will decide what sort of creature the human species should be? The greatest problem we shall likely face in our use of bioengineering is that of choosing who is to play God with the new "secret of life."

Figure 11.17 A globule of the first human insulin produced through recombinant DNA technology. [Courtesy of Eli Lilly and Company.]

Problems

1. Give structural formulas for the following compounds.

   a. methyl chloride
   b. chloroform
   c. ethyl chloride
   d. methylene chloride
   e. carbon tetrachloride

2. List several properties of chlorinated hydrocarbons.

3. What are chlorofluorocarbons? How are they used?

4. List several properties of chlorofluorocarbons.

5. What is a blood extender? What kind of compounds are used as blood extenders?

6. Give structural formulas for the following compounds.
   a. methyl alcohol   b. isopropyl alcohol
   c. ethyl alcohol

7. List two ways that methanol can be made.

8. Compare the acute toxicities of methanol and ethanol.

9. List two ways that ethanol can be made.

10. What is fetal alcohol syndrome?

11. What are some of the long-term effects of excessive ethanol consumption?

12. What is denatured alcohol?

13. What is a functional group?

14. Give the chemical names for the substances known by these familiar names.
   a. grain alcohol   b. rubbing alcohol
   c. wood alcohol

15. What is the percent alcohol by volume in 80 proof vodka?

16. What is the proof of a beverage that is 43% alcohol by volume?

17. Why is ethanol, rather than methanol, used as a solvent for internal medicines?

18. Give structures for the following alkyl groups.
   a. ethyl   b. isopropyl

19. Name the following alkyl groups.
   a. $CH_3CH_2CH_2-$ b. CH_3-

20. Give the structure for phenol.

21. Give an important use for phenols.

22. Give the structure for diethyl ether.

23. Give an important historical use for diethyl ether. What is its main use today?

24. Thymol is used as an antimold agent in preserving books.

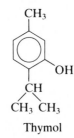

Thymol

To what family does thymol belong? Name the alkyl groups on the benzene ring.

25. Give the structure of the carbonyl functional group.

26. Give the structure of each of the following compounds.
   a. acetaldehyde   b. formaldehyde

27. What is formalin?

28. Name these compounds.

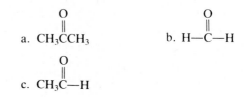

29. What structural feature distinguishes aldehydes from ketones?

30. Give the structural formula for acetone.

31. Give the structure of the alcohols that are oxidized to each of the following.

a. $CH_3\overset{O}{\overset{\|}{C}}CH_3$ b. $H-\overset{O}{\overset{\|}{C}}-H$

c. $CH_3\overset{O}{\overset{\|}{C}}-H$

32. What is a mineral acid?

33. Name three mineral acids.

34. Give the structure of the carboxyl group.

35. Give structural formulas for each of the following.
   a. acetic acid   b. butyric acid
   c. formic acid

36. Name these compounds.

a. $CH_3CH_2CH_2OH$ b. $CH_3CH_2\overset{O}{\overset{\|}{C}}-OH$

37. Give structural formulas for each of the following.
   a. ethyl butyrate   b. methyl butyrate

38. Compare the odors of typical carboxylic acids and esters.

39. For what family is $R\overset{O}{\overset{\|}{C}}OR'$ the general formula?

40. Give structural formulas for the following.
   a. ethylamine   b. methylamine
   c. dimethylamine

41. What is the name of the $-NH_2$ group?

42. Name the following compounds.

a. $CH_3CH_2CH_2NH_2$
b. $CH_3CH_2NHCH_2CH_3$

43. Give the structure of the amide functional group.

44. Name the following compounds.

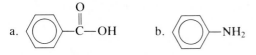

a. b.

45. Which of the following represent heterocyclic compounds?

a. $\begin{array}{c} CH_2-NH \\ | \quad\quad | \\ CH_2-CH_2 \end{array}$ b. $\begin{array}{c} CH_2-CH_2 \\ | \quad\quad | \\ CH_2-CH_2 \end{array}$

c. $\begin{array}{c} CH_2-O \\ | \quad\quad | \\ CH_2-CH_2 \end{array}$ d. $\begin{array}{c} CH_2-CH-NH_2 \\ | \quad\quad | \\ CH_2-CH_2 \end{array}$

46. What is an alkaloid?

47. Name three alkaloids that are familiar drugs.

48. Name the two kinds of nucleic acids. Which is found in the nucleus of the cell?

49. What is the sugar unit in RNA? What sugar is found in DNA?

50. List the heterocyclic bases found in DNA. List those found in RNA.

51. How do DNA and RNA differ in structure?

52. Which of the following are nucleotides?

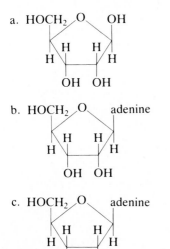

a. $HOCH_2$ ... OH

b. $HOCH_2$ ... adenine

c. $HOCH_2$ ... adenine

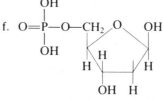

d. ... uracil

e. ... cytosine

f. ... OH

53. For each of the structures in Problem 52, identify the sugar unit as ribose or deoxyribose.

54. Identify the sugar and the base in each of the following nucleotides.

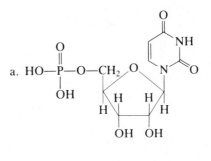

a.

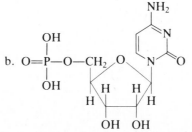

b.

55. Which base is purine and which is pyrimidine?

a. 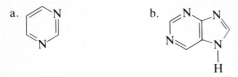 b.

56. Answer the questions for the molecule shown.

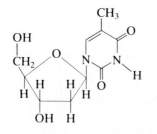

a. Is the base a purine or a pyrimidine?
b. Would the compound be incorporated in DNA or in RNA?

57. Answer the questions posed in Problem 56 for the following compound.

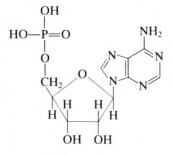

58. What kind of intermolecular force is involved in base pairing?

59. In DNA, which base would be paired with the base listed?
a. cytosine   b. adenine
c. guanine   d. thymine

60. In an RNA molecule, which base would pair with the base listed?
a. adenine   b. guanine
c. uracil   d. cytosine

61. Describe the process of replication.

62. In replication, a parent DNA molecule produces two daughter molecules. What is the fate of each strand of the parent DNA double helix?

63. We say DNA controls protein synthesis, yet most DNA resides within the cell nucleus while protein synthesis occurs outside of the nucleus. How does DNA exercise its control?

64. Explain the role of messenger RNA in protein synthesis.

65. Explain the role of transfer RNA in protein synthesis.

66. Which nucleic acid or acids is (are) involved in the process referred to as transcription?

67. Which nucleic acid or acids is (are) involved in the process referred to as translation?

68. Which nucleic acid contains the codon?

69. Which nucleic acid contains the anticodon?

70. The base sequence along one strand of DNA is ~ATTCG~. What would be the sequence of the complementary strand of DNA?

71. What sequence of bases would appear in the messenger RNA molecule copied from the original DNA strand shown in Problem 70?

72. If the sequence of bases along a messenger RNA strand is ~UCCGAU~, what was the sequence along the DNA template?

73. What are the complementary triplets on tRNA for the following triplets on mRNA?
a. UUU   b. CAU
c. AGC   d. CCG

74. What are the complementary triplets on mRNA for the following triplets on tRNA molecules?
a. UUG   b. GAA
c. UCC   d. CAC

75. What is the relationship between the cell parts called chromosomes, the units of heredity called genes, and the nucleic acid DNA?

76. What are RFLPs? How do RFLP patterns indicate that a person probably will develop a genetic disease?

77. List the four steps in rDNA technology.

78. What is a plasmid?

79. Discuss some applications of genetic engineering.

References and Readings

1. Capindale, J. B. "Organic Chlorine Compounds: A Boon or a Curse?" *Chem 13 News*, January 1984, pp. 6–7.

2. Capindale, J. B. "Organic Chlorine Compounds: Insecticides and Bacteriocides." *Chem 13 News*, February 1984, pp. 4–5.

3. Moses, Susan, and Jean A. Parr. *Biotechnology*. Washington, DC: American Chemical Society, 1985.

4. Gilbert, Walter, and Alison Taunton-Rigby. "Gene Splicing." *Research & Development*, June 1984, pp. 176–181.

5. Hart, Harold. *Organic Chemistry: A Short Course*, 7th ed. Boston: Houghton Mifflin, 1987.

6. Hill, John W., and Dorothy M. Feigl. *Chemistry and Life*, 3d ed. New York: Macmillan, 1987. Chapters 14 through 23 cover this material in more detail than this textbook.

7. Miller, J. A. "First Live Gene-Splice Release: It's Already History." *Science News*, 12 April 1986, p. 228.

8. Moore, Michael. "Within Our Grasp: The New Hope of Genetic Treatment." *Health Sciences*, Summer 1986, pp. 1–6.

9. Ouellette, Eileen M. "The Fetal Alcohol Syndrome." *Contemporary Nutrition*, March 1984, pp. 1–2.

10. Rosenfeld, Albert. "Tippling Enzymes." *Science 81*, April 1981, pp. 24–25.

11. Watson, J. D. *The Double Helix*. New York: New American Library, 1968. Irreverent, gossipy account of the discovery of the structure of DNA.

12. Windholz, Martha (Ed.). *The Merck Index*. 10th ed. Rahway NJ: Merck and Co., 1983.

12

Polymers

Giants Among Molecules

For the most part, chemistry is a study of molecules. Most of the molecules we have discussed so far have been simple ones composed of only a few atoms each. One exception was DNA, the giant molecule that controls heredity (Chapter 11). A single molecule of DNA may contain several million atoms. Many other familiar substances, such as the starches and the proteins (Chapter 18), also are made up of giant molecules. (You should realize that these molecules are giants only when compared to the usual small molecules; most are still invisibly small.)

Chemists call these molecular giants **macromolecules** (from the Greek *makros*: "large" or "long"). The naturally occurring ones just mentioned are the stuff of which living things are made. Macromolecules also have served humanity for centuries in the form of wood, wool, cotton, and silk, which are used for housing and clothing. In the form of starches and proteins, macromolecules are foodstuffs.

It took scientists many years to discover that at least some of the special properties of these substances, particularly their physical properties such as tensile strength and flexibility, were related to the enormous size of their molecules. And it took them even longer to determine the structure of giant molecules. The major finding of the experiments was that the giant molecules are constructed from "building blocks"—smaller, repetitive units, like the bricks that make up a wall.

The temptation to improve on nature has always been great, and it has rarely been resisted. Chemists studying natural macromolecules soon learned to produce their own, which they designed to fit specific purposes and to enhance certain properties that suited those purposes. Quite often, chemists were able to improve on nature's handiwork.

Polymerization: Making Big Ones Out of Little Ones

Macromolecular materials have come to be known to the general public as **plastics**. Not all substances made of giant molecules are plastic, however, at least not in the scientific sense. In scientific terminology, the word *plastic* describes a substance that can be softened by heat and formed by pressure. To the nonchemist, however, the process of manufacture is not readily apparent and the word *plastic* means any of a variety of structural materials, films, and fibers.

Chemists call a macromolecular substance a **polymer** (from the Greek *poly*: "many," and *meros*: "parts"). Polymers usually are made in the laboratory by the combination of many small molecules into much larger molecules with significantly different properties. The small molecules—the building blocks—are called **monomers** (from the Greek *monos*: "one"). A polymer is as different from a monomer as long spaghetti noodles are from the powdery wheat flour spaghetti is made of. For example, polyethylene, the familiar solid, waxy "plastic" in plastic bags, is a polymer prepared from a monomer (ethylene), which is a gas.

Improving on Nature: Billiard Balls and Celluloid Collars

The oldest attempts to improve on nature simply involved chemical modification of the natural macromolecules. The synthetic material **celluloid**, as its name implies, was derived from natural cellulose. When cellulose is treated with nitric acid, a derivative called cellulose nitrate is formed. In response to a contest to find a substitute for ivory for use in billiard balls, John Wesley Hyatt, an American inventor, found a way to soften cellulose nitrate by treating it with ethyl alcohol and camphor. The softened material could be molded into smooth, hard balls. How many of us could afford to have a pool table in our basements if we had

Figure 12.1 Celluloid, a nitrocellulose product, was the first synthetic plastic. Around the turn of the century, many commercial uses were found for celluloid. [Courtesy of E. I. du Pont de Nemours & Company, Inc., Wilmington, DE.]

to buy ivory billiard balls? Hyatt brought the game of billiards into the economic reach of working people.

Celluloid also was used in movie film and for stiff collars (so they didn't require repeated starching) (Figure 12.1). Because of its dangerous flammability (cellulose nitrate also is used as smokeless gunpowder), celluloid was removed from the market when safer substitutes became available. Today, movie film is made from cellulose acetate, another semisynthetic modification of cellulose. And high, stiff collars on men's shirts are out of fashion. Only Ping-Pong balls remain as a reminder of this once important material.

It didn't take long for the chemical industry to recognize the potential of synthetics. Scientists both in and out of industry began making macromolecules from scratch, starting with small molecules rather than modifying large ones. The first such truly synthetic polymers were the phenol–formaldehyde resins, first made in 1910. These complex polymers are discussed later in this chapter. Let's look at some simpler ones first.

Polyethylene: From the Battle of Britain to Bread Bags

$CH_2{=}CH_2$
Ethylene

Polyethylene is the simplest of the synthetic polymers, and it's the cheapest, too. It is familiar today in the plastic bags used for packaging fruit and vegetables, in garment bags for dry-cleaned clothing, in garbage-can liners, and in many other items. Polyethylene is made from ethylene. This unsaturated hydrocarbon (Chapter 10) is produced in large quantities from the cracking of petroleum.

Under high pressure and temperature and in the presence of a catalyst, ethylene molecules are made to join together in long chains. The polymerization can be represented by the reaction of a few monomer units.

$$\cdots\ +\ CH_2{=}CH_2\ +\ CH_2{=}CH_2\ +\ CH_2{=}CH_2\ +\ CH_2{=}CH_2\ +\ \cdots\ \longrightarrow$$

$$\sim CH_2CH_2{-}CH_2CH_2{-}CH_2CH_2{-}CH_2CH_2 \sim$$

The dotted lines and tildes ($\sim$) are like et ceteras; they indicate that the number of monomers and the polymer structure are extended for many units in each direction. A truer picture of the polyethylene molecule is given in Figure 12.2, which shows a segment of a framework model. Still better is the space-filling representation in Figure 12.3. Both models are deficient in that they are much too short. Real polyethylene molecules have varying numbers of carbon atoms, from a few hundred to several thousand.

Polyethylene was invented shortly before the start of World War II.

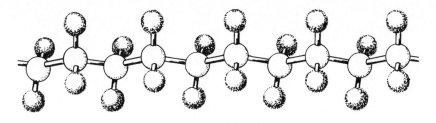

Figure 12.2 Ball-and-stick model of a segment of a polyethylene molecule.

Before long, it was used for insulating cables in a top-secret invention—radar—which helped British pilots spot enemy aircraft before the aircraft became visible to the naked eye. Polyethylene proved to be tough and flexible and an excellent electrical insulator. It could withstand both high and low temperatures. Without polyethylene, the British could not have had effective radar, and without radar, the Battle of Britain might have been lost. The invention of this simple plastic may have changed the course of history.

Today, there are two principal kinds of polyethylene produced by the use of different catalysts and different reaction conditions. **High-density polyethylenes** have largely linear molecules. These linear molecules can assume a fairly ordered, crystalline structure; this gives high-density polyethylenes greater rigidity and higher tensile strength than other ethylene plastics. Linear polyethylenes are used for such things as threaded bottle caps, radio and television cabinets, toys, and large-diameter pipes.

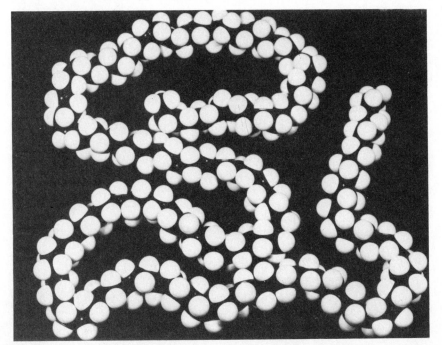

Figure 12.3 Space-filling model of a polyethylene molecule of 150 carbon atoms. [Reprinted with permission from L. Mandelkern, *An Introduction to Macromolecules* (Berlin: Springer-Verlag, 1973). Copyright © 1973 by Springer-Verlag, Inc.]

Figure 12.4 These three bottles made of polyethylene were heated in the same oven for the same length of time. The two that melted have branched polyethylene molecules; the other is composed of unbranched chains. [Photo by Mari Ansari.]

$$CH_2{=}CHCl$$

Vinyl chloride

Figure 12.5 Some "vinyl" objects. [Photo © Zeva Oelbaum, New York.]

Low-density polyethylenes have more branched chains, and the long molecules are somewhat cross-linked. The side chains prevent the molecules from assuming a crystalline structure. Low-density polyethylenes are waxy, semirigid, transparent or translucent plastics that are quite resistant to chemicals. They are used to make plastic bags, refrigerator dishes, insulation for electrical wiring, squeeze bottles, and many other common household articles.

Polyethylene is a **thermoplastic** material—that is, it can be softened by heat and then reformed. (Another class of polymers, called thermosetting polymers, are permanently hardened when formed; we look at some of that type shortly.) Polyethylene is the most important plastic on the market today. Annual production in the United States is about 7 billion kg.

Vinyl Polymers

Would you like a tough synthetic material that looks like leather yet costs only a fraction as much? Would you like a clear, rigid material from which unbreakable bottles could be made? Would you like an attractive, long-lasting floor covering? How about inexpensive phonograph records that give excellent sound reproduction? The solution to each of your desires would be a synthetic material called polyvinyl chloride (PVC). It has all these properties—and more.

The properties of a polymer, like those of other substances, can be changed by varying the molecular architecture. Replacing one of the hydrogen atoms of ethylene with a chlorine atom gives a compound called vinyl chloride. This compound is a gas at room temperature. It is insoluble in water. For polymerization, the gas is suspended under pressure, as tiny droplets in water. The heat generated by the reaction is readily dissipated by the water.

A two-dimensional representation of the structure of a segment of a polyvinyl chloride molecule is illustrated below. The structure of polyvinyl chloride differs from that of ethylene by having a chlorine atom on every other carbon atom.

$$\sim CH_2CH{-}CH_2CH{-}CH_2CH{-}CH_2CH\sim$$
$$\quad\ \ \underset{Cl}{|}\qquad\ \underset{Cl}{|}\qquad\ \underset{Cl}{|}\qquad\ \underset{Cl}{|}$$

Polyvinyl chloride

PVC is readily formed in a variety of shapes. The clear, transparent polymer is used in plastic wrap and in clear plastic bottles. To simulate leather—or simply to make the material more attractive—coloring is added and the surface is textured. United States industries produce about 3 billion kg of vinyl plastics annually. Not only has this synthetic material gone a long way toward replacing scarce and expensive natural materials, but it has also proven to be superior to natural materials for many applications.

Vinyl chloride, the monomer from which vinyl plastics are made, is known to be a carcinogen; it causes cancer in some of the people who work where it is made or used. We discuss cancer-causing chemicals in some detail in Chapter 25.

Addition Polymerization: One + One + One ... Gives One!

There are two general types of polymerization reactions: addition polymerization and condensation polymerization. In **addition polymerization**, the building blocks (or monomers) add to one another in such a way that the polymeric product contains all the atoms of the starting monomers. The polymerization of ethylene to form polyethylene and of vinyl chloride to form polyvinyl chloride are examples. Notice that in the structure of polyethylene (page 248), the two carbons and the four hydrogens of each monomer molecule are incorporated into the polymer structure. In **condensation polymerization**, a portion of the monomer molecule is *not* incorporated in the final polymer but is split out as the polymer is formed. This process is discussed later in this chapter.

There are many familiar addition polymers. Most are made from monomers that can be considered to be derivatives of ethylene in which one or more of the hydrogen atoms are replaced by another atom or group. For example, we saw in the preceding section that vinyl chloride is derived from ethylene by replacing one of the hydrogen atoms with a chlorine atom. Similarly, replacing one of the hydrogen atoms with a benzene ring gives a monomer called *styrene*. Polymerization of styrene produces polystyrene, the plastic from which Styrofoam insulation and other materials are made. A segment of a polystyrene molecule is illustrated below.

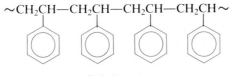

Polystyrene

The polymer structures and polymerization equations we have written are quite cumbersome. There are other ways of writing formulas

Figure 12.6 A giant bubble of tough, transparent plastic film emerges from the die of an extruding machine. The film is used in packaging, consumer products, and food service operations. [Courtesy of Reynolds Metals Company.]

$CH_2{=}CH$

Styrene

Table 12.1 Some Addition Polymers

Monomer	Polymer	Polymer Name	Some Uses
$H_2C=CH_2$	$\left[\begin{matrix} H & H \\ -C-C- \\ H & H \end{matrix}\right]_n$	Polyethylene	Plastic bags, bottles, toys, electrical insulation
$H_2C=CH-CH_3$	$\left[\begin{matrix} H & H \\ -C-C- \\ H & CH_3 \end{matrix}\right]_n$	Polypropylene	Indoor-outdoor carpeting, bottles
$H_2C=CH-\bigcirc$	$\left[\begin{matrix} H & H \\ -C-C- \\ H & \bigcirc \end{matrix}\right]_n$	Polystyrene	Simulated wood furniture, styrofoam insulation and packing materials
$H_2C=CH-Cl$	$\left[\begin{matrix} H & H \\ -C-C- \\ H & Cl \end{matrix}\right]_n$	Polyvinyl chloride (PVC)	Plastic wrap, simulated leather (Naugahyde), phonograph records, garden hoses
$H_2C=CCl_2$	$\left[\begin{matrix} H & Cl \\ -C-C- \\ H & Cl \end{matrix}\right]_n$	Polyvinylidene chloride (Saran)	Food wrap
$F_2C=CF_2$	$\left[\begin{matrix} F & F \\ -C-C- \\ F & F \end{matrix}\right]_n$	Polytetrafluoroethylene (Teflon)	Nonstick coating for cooking utensils, electrical insulation
$H_2C=CH-CN$	$\left[\begin{matrix} H & H \\ -C-C- \\ H & CN \end{matrix}\right]_n$	Polyacrylonitrile (Orlon, Acrilan, Creslan, Dynel)	Yarns, wigs
$H_2C=CH-O-\overset{O}{\overset{\|}{C}}-CH_3$	$\left[\begin{matrix} H & H \\ -C-C- \\ H & O-\overset{\|}{C}-CH_3 \\ & O \end{matrix}\right]_n$	Polyvinyl acetate	Adhesives, textile coatings, chewing gum resin, paints
$H_2C=\overset{CH_3}{\underset{O}{\overset{\|}{C}}}-C-O-CH_3$	$\left[\begin{matrix} H & CH_3 \\ -C-C- \\ H & C-O-CH_3 \\ & O \end{matrix}\right]_n$	Polymethyl methacrylate (Lucite, Plexiglas)	Glass substitute, bowling balls

for polymers and equations for polymerization reactions—ways that do not involve writing out long segments of the polymer chain. We can represent the polymerization of ethylene by the equation

$$n\ CH_2{=}CH_2 \longrightarrow +CH_2CH_2+_n$$

In the formula for the polymer product, the repeating polymer unit (sometimes called the **segmer**, meaning "repeating segment") is placed within brackets with bonds extending to both sides. The subscript n indicates that this molecular fragment is repeated to an unspecified number of times in the full polymer structure. It is certainly easier to write the bracketed segmer than to draw the extended chain—although the latter is probably better at conveying the concept of a polymer as a giant molecule (even though this formula also represents only a small portion of the whole molecule).

The simplicity of the abbreviated formula does facilitate certain comparisons between the monomer and the polymer. Note that the monomer ethylene contains a double bond and polyethylene does not. The double bond of the reactant contains two pairs of electrons. One of these pairs is used to connect one monomer unit to the next in the polymer (indicated by the lines sticking out to the sides in the segmer). That leaves only a single pair of electrons between the two carbons of the polymer segmer—in other words, a single bond. Note that each segmer unit in the polymer has the same composition (C_2H_4) as the monomer.

We could not attempt to discuss all the important addition polymers here. Table 12.1 describes a few of the more interesting ones and some of their uses.

Out of War: The Plastics Industry

The plastics industry really got going during World War II—after the war cut off the Allies' supply of natural rubber. The industry developed, to a large extent, out of the search for synthetic substitutes for rubber.

Natural rubber can be broken down into a simple hydrocarbon called isoprene. The macromolecules from which rubber is made are now known to have the structure

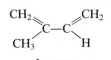

Isoprene

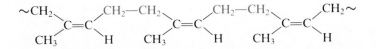

Isoprene is a volatile liquid. Rubber is a semisolid, elastic material.

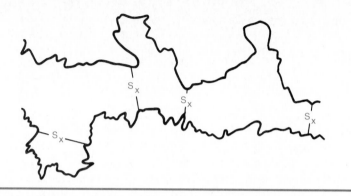

Figure 12.7 Vulcanized rubber. The subscript *x* indicates an indefinite number.

The long-chain molecules that make up rubber can be coiled and twisted and intertwined with one another. The stretching of rubber corresponds to the straightening of the coiled molecules. Natural rubber is soft and tacky when hot. It can be made harder by reaction with sulfur. This process, called **vulcanization**, cross-links the hydrocarbon chains with sulfur atoms (Figure 12.7). Vulcanization was discovered by Charles Goodyear, who was issued U.S. Patent No. 3633 in 1844.

Its three-dimensional cross-linked structure makes vulcanized rubber a harder, stronger substance that is suitable for automobile tires. Surprisingly, cross-linking also improves the elasticity of rubber. With just the right degree of cross-linking, the individual chains are still relatively free to uncoil and stretch. And, when a stretched piece of this material is released, the cross-links pull the chains back to their original arrangement (Figure 12.8). Rubber bands owe their snap to this sort of molecular structure.

(a) (b) (c)

Figure 12.8 Vulcanization of rubber cross-links the molecular chains. (a) In unvulcanized rubber, the chains slip past one another when the rubber is stretched. (b) Vulcanization involves the addition of sulfur cross-linkages between the chains. (c) When the vulcanized rubber is stretched, the sulfur cross-linkages prevent the chains from slipping past one another. Vulcanized rubber is stronger and more elastic than unvulcanized rubber.

Synthetic Elastomers: More Bounce to the Ounce

Synthetic polymers with rubberlike properties are called **elastomers**. Natural rubber is a polymer of isoprene. Some synthetic elastomers are closely related to it. For example, polybutadiene is made from the monomer butadiene. Butadiene differs from isoprene only in that it lacks a methyl group on the second carbon atom. Polybutadiene is made rather easily from the monomer.

$$n \; CH_2{=}CH{-}CH{=}CH_2 \longrightarrow \; {+}CH_2{-}CH{=}CH{-}CH_2{+}_n$$

However, it has only fair tensile strength and poor resistance to gasoline and oils. These properties make it of limited value for automobile tires, the main use of elastomers.

Another synthetic elastomer, polychloroprene (Neoprene), is made from a monomer that is quite similar to isoprene but that has a chlorine in place of isoprene's methyl group.

$$n \; CH_2{=}\underset{\underset{Cl}{|}}{C}{-}CH{=}CH_2 \longrightarrow \; {+}CH_2{-}\underset{\underset{Cl}{|}}{C}{=}\underset{\underset{H}{|}}{C}{-}CH_2{+}_n$$

Neoprene shows better oil and gasoline resistance than other elastomers. It is used to make gasoline hoses and similar items used around automobile service stations.

Another of the synthetic rubbers illustrates the principles of **copolymerization**. In this process, a mixture of two monomers forms a product in which the chain contains units of both building blocks. Such a material is called a **copolymer**. SBR rubber is a copolymer of styrene (25%) and butadiene (75%). A segment of an SBR molecule might look something like this.

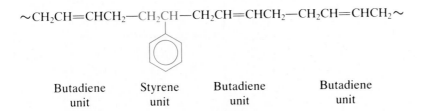

| Butadiene unit | Styrene unit | Butadiene unit | Butadiene unit |

This synthetic is more resistant to oxidation and abrasion than natural rubber but has less satisfactory mechanical properties.

Like natural rubber, the SBR molecules contain double bonds and can be cross-linked by vulcanization. SBR rubber accounts for over 40% of the total production of elastomers. Production in the United States is about 800 million kg of SBR out of a total elastomer production of

$$CH_2{=}CH{-}CH{=}CH_2$$
Butadiene

2.0 billion kg per year. Chemists even have learned to make polysio-prene, a substance identical in every way to natural rubber except that it comes from petroleum refineries rather than plantations of rubber trees.

In addition to elastomers (mainly SBR), tires for automobiles contain many substances. Carbon black is added before vulcanization. It acts in some as yet unknown way to strengthen the elastomer. The tire is further strengthened mechanically by the use of nylon cord, fiberglass, and steel belts.

Painting with Polymers

The development of the plastics industry following World War II revolutionized many aspects of modern society. Not the least affected was the paint industry. Even the process of painting has been dramatically changed by the development of water-based paints.

A paint usually is composed of three components: (1) a binder (or vehicle) that hardens to form a continuous film, (2) a pigment that supplies the desired color, and (3) a volatile solvent that evaporates. In oil-based paints, linseed oil often is used as a binder. The solvent is usually petroleum distillates, a mixture of hydrocarbons. The pigment can be titanium dioxide (TIO_2; white), carbon black (C), chrome yellow ($PbCrO_4$), oxides of iron (brown or red), or organic dyes of various colors. White lead [with an approximate composition of $Pb(OH)_2 \cdot 2PbCO_3$] was once used extensively, but it has been banned for interior use because of its high toxicity. Illnesses in infants and toddlers, particularly in ghetto areas, have been attributed to the ingestion of chips of old, peeling lead paint.

In water-based paints, a synthetic polymer with rubberlike properties is used as the binder. These latex paints are emulsified in water to give a variety of spreading characteristics. Dyes and pigments can be added to produce the desired colors. Among the plastics used in latex paints are polyvinyl acetate, polymethyl methacrylate (or related acrylic latexes), polystyrene, styrene-butadiene copolymers, and polytetra-fluoroethylene (Teflon). Paints can be formulated to have a fantastic variety of colors, textures, and resistancies.

Condensation Polymers: Splitting Out Little Ones to Make Big Ones

The polymers considered so far are all addition polymers. All the atoms of the monomer molecules are incorporated into the polymer molecules. In condensation polymerization, a portion of the monomer molecule is not incorporated in the final polymer.

As an example, let's consider the formation of one type of nylon. (There are several different nylons, each prepared from a different monomer or set of monomers, but all share certain common structural features. We discuss another kind of nylon later in this chapter). The monomer in this type of nylon is a carboxylic acid with an amino group on the sixth carbon atom, 6-aminohexanoic acid. The polymerization involves the reaction of a carboxyl group of one monomer molecule with the amine group of another. This reaction produces an amide bond that holds the building blocks together in the final polymer.

$$\overset{\text{O}}{\overset{\|}{\text{H}_2\text{NCH}_2\text{CH}_2\text{CH}_2\text{CH}_2\text{CH}_2\text{COH}}}$$

6-Aminohexanoic acid

$$\ldots + \text{HO}-\overset{\text{O}}{\overset{\|}{\text{C}}}\text{CH}_2\text{CH}_2\text{CH}_2\text{CH}_2\text{CH}_2\overset{\text{H}}{\underset{|}{\text{N}}}-\text{H} + \text{HO}-\overset{\text{O}}{\overset{\|}{\text{C}}}\text{CH}_2\text{CH}_2\text{CH}_2\text{CH}_2\text{CH}_2\overset{\text{H}}{\underset{|}{\text{N}}}-\text{H} + \ldots \longrightarrow$$

$$\sim\overset{\text{O}}{\overset{\|}{\text{C}}}\text{CH}_2\text{CH}_2\text{CH}_2\text{CH}_2\text{CH}_2\overset{\text{H}}{\underset{|}{\text{N}}}-\overset{\text{O}}{\overset{\|}{\text{C}}}\text{CH}_2\text{CH}_2\text{CH}_2\text{CH}_2\text{CH}_2\overset{\text{H}}{\underset{|}{\text{N}}}\sim + n\,\text{H}_2\text{O}$$

Water molecules are formed as a by-product. It is this formation of a nonpolymeric by-product that distinguishes condensation polymerization from addition polymerization. Note that the formula of the segmer unit is *not* the same as that of the monomer.

As our next example, let's consider Bakelite. These phenol-formaldehyde resins were first synthesized by Leo Baekeland, who received U.S. Patent No. 942 699 for the process in 1909. They are made from formaldehyde and phenol (Chapter 11).

The phenol–formaldehyde resins are formed by splitting out water molecules, the hydrogen atoms coming from the benzene ring, the oxygen atoms from the aldehyde. The reaction proceeds stepwise, with formaldehyde adding first to the 2- and 4-positions of the phenol molecule.

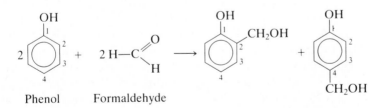

Phenol   Formaldehyde

The substituted molecules then interact by splitting out water (remember that there are hydrogen atoms at all the unsubstituted corners of a benzene ring).

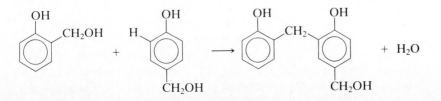

The hookup of molecules continues until an extensive network is achieved.

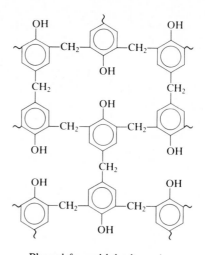

Phenol-formaldehyde resins

Water is driven off by heat as the polymer sets. The structure of the polymer is extremely complex. It exists as a huge, three-dimensional network somewhat like the framework of a giant building. A schematic representation of part of such a molecule is illustrated above. Note that the phenolic rings are joined together by CH_2 units from the formaldehyde.

Recall that polyethylene is thermoplastic; it can be softened by heat and then reformed. Phenol–formaldehyde resins are representative of a group called **thermosetting** plastics. These are fusible at some stage of their production but become permanently hard under the influence of heat and pressure. They cannot be softened and remolded.

Thermoplastic polymers can be recycled rather easily. They can be melted down and remolded. Recently, methods have been developed for the recycling of some thermosetting plastics. These methods work by breaking the large molecules down to smaller ones that can be converted into the original monomers. Both types of recycling consume energy, but the biggest problem in recycling plastics is collection and separation. Not only must be the plastics be separated from other types of solid wastes, but the various kinds of polymers must be separated from each other—PVC from polyethylene, and so on.

Fibers and Fabrics

Let's consider another area in which synthetics have mimicked nature's work successfully. Cotton, wool, and silk have long been spun and woven into fabrics for clothing and other uses. The fibrous nature and, particularly, the great tensile strength of these materials were perfectly suited to such purposes. In the last few decades, synthetic polymers with similar physical properties have revolutionized the clothing industry and have outdone the natural fibers in their resistance to stretching and shrinking (and to moths).

In 1940, only 10% of the fibers used in the United States were synthetic. Today we use 75% synthetics, 80% of those derived from petroleum. Cotton consumption in the United States each year is about 7 kg per capita, or 24% of all fibers used (the other 1% is mainly wool). Could we ever go back to using all natural fibers? It would mean committing a lot of land to cotton and sheep. Besides, would we really want to give up the many synthetics with superior properties?

Both addition and condensation polymerization yield useful fibers. Polyacrylonitriles (Orlon, Acrilan, Creslan, and the like) are addition polymers.

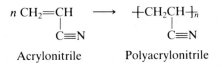

Acrylonitrile   Polyacrylonitrile

These acrylic fibers are used in clothing, carpets, and other products. Polyesters (Dacron) and polyamides (nylon) are condensation polymers. Dacron polyester is made by the condensation of ethylene glycol with terephthalic acid.

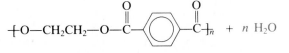

Ethylene glycol   Terephthalic acid

Dacron

Polyester fibers are used in wash-and-wear clothing. The same polymer, when formed as a film rather than fiber, is called Mylar. When magnetically coated, Mylar tape is used in audio and video recording.

The most common polyamide, Nylon 66,[*] also is made by the condensation of two different monomers.

---

[*] Nylons are named according to the number of carbon atoms in each monomer unit. There are six each in 1,6-hexanediamine and adipic acid; hence, the polymer is Nylon 66. The nylon on page 257 is made from a single monomer that contains six carbon atoms, so it is called Nylon 6.

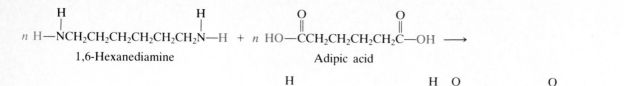

1,6-Hexanediamine   Adipic acid

Silk and wool, which are natural protein fibers, are also polyamides. All these fibers, like high-density polyethylene, owe their strength to the ordered, relatively rigid arrangement of their long molecules. However, the polyamides and polyesters possess features not shared by polyethylene. The former compounds contain polar functional groups, and the interaction of these groups gives these polymers their unique tensile strength.

Bulk Properties of Polymers

There are many other types of polymers, both synthetic and natural. The molecules are large, and variations in their structures—and consequent variations in their properties—are almost unlimited. The development of synthetic polymers had led to a whole new world of plastics, rubber substitutes, and synthetic fibers. There is no sharp line separating these types of polymers. The same basic material can be used in two (or even three) of the categories. Nylon, for example, can be molded to make plastic toys or television cabinets, or it can be spun into a fiber and woven into cloth. The major differences between polymer types are mainly in the organization of the big molecules into bulk materials. The molecules of elastic materials are randomly oriented (**amorphous**) and tangled with one another. On the other hand, molecules in fibers of high tensile strength are highly organized, even crystalline, in arrangement (Figure 12.9).

An important parameter of most polymers is the **glass transition temperature** (T_g). Above this temperature, the polymer is rubbery and tough; below it, the polymer is like glass— hard, stiff, and brittle. Each polymer has a characteristic T_g. We want automobile tires to be tough and elastic, so we use materials with a low T_g. On the other hand, we want plastic substitutes for glass to be glassy. Thus, they have T_gs well above room temperature. We make use of the T_g concept in everyday life. To remove chewing gum from clothing, we apply a piece of ice to lower the temperature of the gum below the T_g of the polyvinyl acetate resin that makes up the bulk of the gum. The cold, brittle resin then crumbles readily and can be removed.

Giant molecules can impart great structural strength. The great strength of wood, for example, is due in large part to the arrangement of

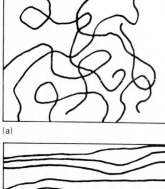

(a)

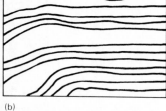

(b)

Figure 12.9 Organization of polymer molecules.
(a) Crystalline arrangement.
(b) Amorphous arrangement.

Figure 12.10 A high glass transition temperature (T_g) gives Plexiglas the glasslike properties well suited for this covered walkway at Indiana University–Purdue University at Indianapolis. Students have irreverently dubbed the walkway "The Gerbil Tube." [Courtesy of Rohm & Haas Company.]

giant cellulose molecules (Chapter 18) in a matrix of lignin. The many synthetic substitutes for wood are attempts at duplicating the structural strength of the natural polymer.

The techniques for designing materials have become so refined that a seemingly impossible combination of properties can be incorporated into a single substance. For example, spandex fibers (used for stretch fabrics [Lycra] in ski pants, girdles, and bathing suits) combine the elasticity of rubber (necessitating coiled, flaccid molecular chains) and the tensile strength of a fiber (necessitating a crystalline arrangement of fairly rigid, highly ordered chains). How can scientists make materials like spandex? By grafting two molecular structures onto one polymer chain. Blocks of components with fibrous character are alternated with blocks of elastomer in the same giant molecule, and the fabric made from the polymer exhibits both sets of properties—flexibility *and* rigidity!

Only a few of the many uses of plastics have been mentioned here. Clothing, containers, kitchenware, synthetic foam rubber for furniture padding, skis, ski boots, bottles and bottle caps, artificial arteries and bone joints, radio and stereo cabinets, windows, windshields, outdoor signs, and protective coatings are only the start of any list of the plastic items that have added so much to our way of life. Plastics play a vital part in our recreation, communication, and personal well-being, and in the decoration of our homes, offices, and commercial buildings. Plastics have become very much a part of our lives—and a part of our problems.

Plasticizers and Pollution

It is difficult to process some plastics, particularly the vinyl types. As formed, they can be rather hard and brittle. But they can be made more flexible and easier to handle by adding chemicals called plasticizers.

Ideally, **plasticizers** are liquids of low volatility. They act by lowering the T_g of the plastic to make it more flexible and less brittle. Undiluted polyvinyl chloride (PVC) cracks and breaks easily, but with proper plasticizers added, it is soft and pliable. Plastic raincoats, garden hoses, and seat covers for automobiles can be made from the modified PVC. Plasticizers are generally lost by diffusion and evaporation as a plastic article ages. The plastic becomes brittle, then cracks and breaks.

One type of plasticizer once used widely but now largely banned is the **polychlorinated biphenyls** (PCBs). These compounds are derived from a hydrocarbon—biphenyl ($C_{12}H_{10}$)—by the replacement of anywhere from 1 to 10 of the hydrogen atoms with chlorine atoms. The structures of the parent compound and some of the PCBs are shown in Figure 12.11. Note the structural similarity of the PCBs to DDT. PCB residues have been found in fish, birds, water, and sediments. The physiological effect of PCBs is similar to that of DDT (Chapter 17), with long-term effects of greater concern than acute toxicity. Poultry and eggs have been found to have concentrations greater than the 5 ppm

Figure 12.11 Biphenyl and some of the PCBs derived from it. These are but a few of the hundreds of possible PCBs. DDT is shown for comparison.

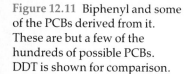

Biphenyl

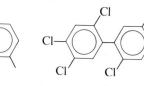

PCB₁   PCB₂

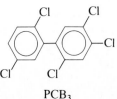

PCB₃

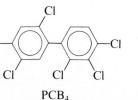

PCB₄

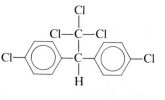

DDT

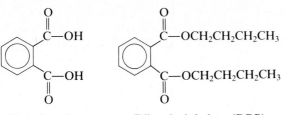

Phthalic acid   Dibutyl phthalate (DBP)

Figure 12.12 Phthalic acid and some derivatives. Dioctyl phthalate is also called di-2-ethylhexyl phthalate (DEHP).

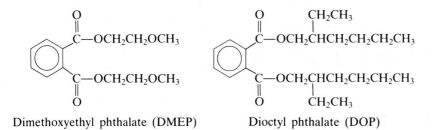

Dimethoxyethyl phthalate (DMEP)   Dioctyl phthalate (DOP)

allowed by the Food and Drug Administration (FDA). Under the law, the contaminated poultry products had to be destroyed. Fish taken from major rivers from coast to coast in the United States also have been found to be contaminated with PCBs, at levels often above the 5 ppm maximum allowed by the FDA.

Not all the PCBs manufactured were used as plasticizers. Some were used in electrical equipment. PCBs have a high electrical resistance, so they are useful as insulating materials in transformers, condensers, and other electrical apparatus. They also have high boiling points, making them useful as heat-transfer media. PCBs have a density greater than that of water and hence they were often added to lubricants intended for underwater use.

The same properties that made the PCBs so desirable as industrial chemicals—particularly their stability and solubility—cause them to be an environmental hazard. They degrade slowly in nature, and their solubility in nonpolar media—animal fat as well as vinyl plastics—leads to their concentration in the food chain. In humans, accidental ingestion has caused skin, liver, and gastrointestinal problems. Experiments with monkeys have shown that levels of less than 5 ppm in food interfere with reproduction. Monsanto Corporation, the only company in the United States that produced PCBs, discontinued production in 1977, but PCBs will remain in the environment for years.

Another class of plasticizers is the **phthalates**. These compounds are esters derived from phthalic acid. The parent acid and some of the common plasticizers derived from it are shown in Figure 12.12. Unlike the PCBs, which are used as a complex mixture, the phthalates are used generally as separate compounds, with individual substances chosen to impart particular levels of flexibility and elasticity to the plastic.

Phthalate plasticizers appear to have low acute toxicity. Their long-term effects are generally unknown. Phthalates have been leached from the PVC bags in which whole blood has been stored. It is thought that these plasticizers may contribute to shock lung, a sometimes fatal condition observed in some patients after a blood transfusion. Phthalates have been found in the heart muscles of cattle, dogs, rabbits, and rats. Very high levels of exposure can induce mutations and birth defects in laboratory animals. What implications, if any, these findings have for human health is not yet known.

Disposal of Plastics

Synthetic polymers have become an important part of our lives, largely because they do so many jobs so well. It may be that in one respect these materials are too good: they last almost forever! Once they are dumped into the environment, some tend to remain there. On land they contribute to the unsightly litter scattered along roadways and elsewhere, and they provide breeding grounds for disease-carrying insects and rodents. Not only are plastic discards also found in all the oceans of the world, but they may represent an immediate danger to some fish. Small fish have been found dead with their digestive tracts clogged by fragments of plastic foam that they had ingested with their food.

Even if everyone pitches in so that, instead of being scattered all over, plastics are collected in refuse containers, there are still problems. Solid wastes are frequently disposed of in sanitary land-fills. We are rapidly running out of dumps to accommodate the tremendous volume of solid waste we produce; and synthetic polymers, though buried, do not rot and go away.

Another approach to the problem is to encourage the reuse or recycling of synthetics (and everything else). Thermoplastic polymers can be melted down and remolded. Recently, methods of recycling thermosetting polymers have been developed. Both procedures require energy, however, something we must also conserve. In addition, the discarded synthetics must first be collected and then separated from other types of solid waste and from each other.

There is another way to dispose of discarded plastics—and that is to burn them. This, however, can lead to an increase in air pollution. Automobile tires burn with a sooty, stinking smoke, and polyvinyl-chloride produces toxic hydrogen chloride gas when it burns. Incinerators are corroded by acidic gases and clogged by those plastics that melt from the heat but don't burn.

Plastics as Fire Hazards

The accidental ignition of fabrics, synthetic or otherwise, has caused untold human misery. The U.S. Department of Health, and Human Services estimates that as many as 150 000 to 200 000 people are injured and several thousand are burned to death each year in accidents associated with flammable fabrics. While some synthetics are particularly flammable, others may hold the answer to the problem.

A good deal of research has been done to develop flame-retardant fabrics. One common approach involves the incorporation of chlorine and bromine atoms within the giant molecules of the polymeric fiber. Federal regulations require that children's sleepwear, in particular, be made of such flame-retardant materials. A major setback to the program occurred in 1977 when the Consumer Product Safety Commission banned a flame-retardant compound called Tris [tris(2,3-dibromo-propyl) phosphate], which was used on sleep wear made from polyester and acetate fabrics. In laboratory tests, the compound had been shown to be both carcinogenic and **mutagenic** (mutation inducing). The search for safe fire-retardants continues.

Another problem associated with plastics and fire is that of toxic gases. In 1972, lethal amounts of cyanide were found in the bodies of plane-crash victims in Chicago. The Illinois Department of Law Enforcement traced the cyanide to burned plastics in the plane's cabin. Firefighters often refuse to enter burning buildings without gas masks for fear of being overcome by fumes from burning plastics. Laboratory tests show that hydrogen cyanide is formed in large quantities by the burning of polyacrylonitrile (Orlon, Creslan, Acrilan, etc.) and other nitrogen-containing plastics. The increasing use of plastics in home furnishings and constructions calls for more research in the area of identifying—and quantifying—the products of the combustion of plastics.

Plastics and the Future

Plastics have provided us with marvelous materials, and they also have presented us with problems. We could do away with plastics—but before you opt for this easy solution, look around. What are you ready to give up? The tires, the upholstery, the vinyl hardtop on your car? The keys on your piano? The artificial turf on your athletic field? Your Styrofoam cooler? Your telephone? The tile on your floor? Most of your clothes? You would have to give up all these items and many more if plastics were banned.

Since the coal and petroleum from which plastics are made are nonrenewable resources, we should use plastics wisely, not wastefully. The thermoplastic materials can be recycled, melted down, and remolded. Thermosetting plastics cannot be recycled easily. Perhaps these should be used only in items designed for long-term use, not for disposable gadgets. But plastics, with their problems as well as their promise, are likely to be a part of our lives for a long time to come.

Problems

1. Define these terms.
   a. macromolecule   b. monomer
   c. polymer   d. segmer
   e. elastomer   f. copolymer
   g. plasticizer

2. What does the word *plastic* mean in the scientific sense?

3. What is celluloid? How is it made?

4. Why is celluloid no longer used to make movie film?

5. Describe the structure of high-density polyethylene. How does this structure explain the properties of this polymer?

6. Describe the structure of low-density polyethylene. How does this structure explain the properties of this polymer?

7. What is a thermoplastic polymer?

8. Give the structure of a segment of polyvinyl chloride that is at least ten carbon atoms long.

9. Give the structure of the monomer from which polyvinyl chloride is made.

10. How does the structure of polyvinyl chloride differ from that of polyethylene?

11. List several uses of polyvinyl chloride.

12. What is addition polymerization?

13. What structural feature usually characterizes molecules used as monomers in addition polymerization?

14. Give the structure of the monomer from which polystyrene is made.

15. What is Styrofoam?

16. In the following equation, identify the parts labeled *a*, *b*, and *c* as monomer, polymer, and segmer.

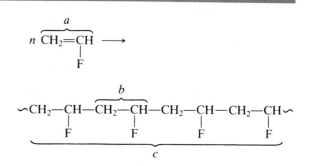

17. What type of polymerization (addition or condensation) is represented in Problem 16?

18. Give the structure of the polymer made from acrylonitrile ($CH_2{=}CH{-}C{\equiv}N$). Show at least four segmer units.

19. Give the structure of the polymer made from vinyl acetate ($CH_2{=}CH{-}O\overset{\overset{\displaystyle O}{\|}}{C}CH_3$). Show at least four segmer units.

20. Give the structure of the polymer made from 1-hexene ($CH_2{=}CHCH_2CH_2CH_2CH_3$). Show at least four segmer units.

21. Give the structure of the polymer made from styrene ($CH_2{=}CH{-}\langle\bigcirc\rangle$). Show at least four segmer units.

22. Give the structure of the polymer made from methyl methacrylate ($CH_2{=}\overset{\overset{\displaystyle CH_3}{|}}{C}{-}\overset{\overset{\displaystyle O}{\|}}{C}OCH_3$). Show at least four segmer units.

23. Give the structure of the monomer from which polytetrafluoroethylene (Teflon) is made.

24. Saran has the structure

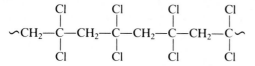

Give the structure of the monomer from which Saran is made.

25. Give the structure of isoprene, the "monomer" of natural rubber.

26. Explain the elasticity of rubber.

27. Describe the process of vulcanization.

28. How does vulcanization change the properties of rubber?

29. Name three synthetic elastomers.

30. Why has white lead been banned as a pigment for paints intended for interior use?

31. Give the structure of butadiene.

32. How does polybutadiene differ from natural rubber in properties? In structure?

33. How does polychloroprene (Neoprene) differ from natural rubber in properties? In structure?

34. How is styrene–butadiene rubber (SBR) made?

35. What substances other than elastomers are used in automobile tires?

36. What are the three components of a paint?

37. List some pigments used in paints.

38. List some polymers used in paints.

39. What is condensation polymerization?

40. What is Bakelite? From what monomers is it made?

41. What is a thermosetting polymer? What structural feature characterizes a thermosetting polymer?

42. What proportion of the fibers used in the United States are synthetic?

43. Nylon 46 is made from these monomers:

$$H_2NCH_2CH_2CH_2CH_2NH_2 \quad \text{and}$$

$$HO-\overset{\overset{\displaystyle O}{\|}}{C}CH_2CH_2CH_2CH_2\overset{\overset{\displaystyle O}{\|}}{C}-OH$$

Give the structure of nylon 46 showing at least two units from each monomer.

44. Is nylon 46 an addition polymer or a condensation polymer? Explain.

45. Give the structure of a polymer made from glycolic acid ($HO-CH_2O-OH$). Show at least four segmer units. *Hint*: Compare with nylon 6 on page 257.

46. What is the glass transition temperature (T_g) of a polymer?

47. For what uses do we want polymers with a low T_g? With a high T_g?

48. How do plasticizers act to make polymers less brittle?

49. What are some problems associated with the use of PCBs as plasticizers?

50. What are pathalates? What problems are associated with their use as plasticizers?

51. What problems arise when plastics are discarded into the environment?

52. What problems arise when plastics are disposed of in landfills?

53. What problems arise when plastics are disposed of by incineration?

54. Discuss plastics as fire hazards.

55. What is Tris? Why was it banned?

56. Dacron has a high T_g. How can the T_g be lowered so that a manufacturer can permanently crease a pair of Dacron slacks?

57. Kodel is a polyester fiber. The monomers are terephthalic acid and 1,4-cyclohexanedimethanol. Write the structure of a segment of Kodel containing at least one of each monomer unit.

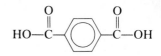

Terephthalic acid

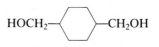

1,4-Cyclohexanedimethanol

58. From what monomers might the following copolymer be made?

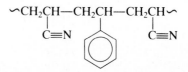

59. Kevlar, a polyamide used to make bulletproof vests, is made from terephthalic acid and phenylenediamine. Write the structure for a segment of the Kevlar molecule.

$$H_2N - \bigcirc - NH_2$$

Phenylenediamine

60. Isobutylene ($CH_2\!=\!CCH_3$) polymerizes to form
$$|$$
$$CH_3$$

polyisobutylene, a sticky polymer used as an adhe-

sive. Give the structure of polyisobutylene. Show at least four segmer units.

61. Copolymerized with butadiene, isobutylene (Problem 60) forms butyl rubber. Give the structure of butyl rubber. Show at least three isobutylene segmers and one butadiene segmer.

62. Where will plastics come from when the Earth's supplies of coal and petroleum are exhausted?

63. Make a list of plastic objects (or parts of objects) that you encounter in your daily life. Try to identify a few of the kinds of polymers used is making the items. Compare your list with those of some of your classmates.

References and Readings

1. Artandi, Charles. "Fibers in Medicine." *Chem Tech*, *August 1981, pp. 476–479.*
2. Flavin, Christopher. *The Future of Synthetic Materials*. Washington, DC: The Worldwatch Institute, 1980.
3. Griffith, James R. "The Chemistry of Coatings." *Journal of Chemical Education*, November 1981, pp. 956–958.
4. Harris, Frank W. 'Introduction to Polymer Chemistry." *Journal of Chemical Education*, November 1981, pp. 837–843.
5. Hecht, Annabel. "Sealing Teeth to Prevent Cavities." *FDA Consumer*, April 1984, pp. 5–7.
6. Marvel, C. S. "The Development of Polymer Chemistry in America—The Early Days." *Journal of Chemical Education*, July 1981, pp. 535–539.
7. Orna, Mary Virginia. "Chemistry and Artists' Colors." *Journal of Chemical Education*, April 1980, pp. 267–269.
8. Smay, V. Elaine. "The Glue Revolution." *Popular Science*, August 1980, pp. 62–65, 106–107.
9. Smith, John K., and David A. Hounshell. "Wallace H. Carothers and Fundamental Research at Du Pont." *Science*, 2 August 1985, pp. 436–442.
10. Zimmt, Werner S. 'Coatings From Acrylic Polymers." *ChemTech*, November 1981, pp. 681–683.
11. Zollinger, Heinrich. "On Shirts and Things." *Chem-Tech*, August 1980, pp. 461–462.

Energy

A Fuels Paradise

Our modern industrial civilization is fueled by fossils. Over 90% of the energy used to sustain our way of life comes from the fossil fuels—coal, petroleum, and natural gas. In this chapter, we consider the origin and chemical nature of these fuels and how they are burned to release energy that plants of ages past captured from rays of sunlight.

Fuels are substances that burn readily with the release of significant amounts of energy. Fuels are *reduced* forms of matter, and the burning process is oxidation (Chapter 8). If an atom already has its maximum number of bonds to oxygen (or to other electronegative atoms such as chlorine or bromine), the atom cannot serve as a fuel. Indeed, such substances can be used to put *out* fires. Figure 13.1 shows some representative fuels and nonfuels. (In Chapter 14, we discuss possible future energy sources.)

To start, let's look at some of the general considerations in energy conversion. Transformations of energy, like changes in matter, are governed by the laws of nature. If you want to understand our current energy problems, you must first learn something about these laws. Then you should consider any scheme for energy conversion in light of these laws.

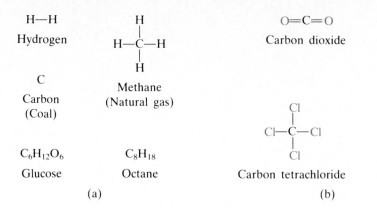

Figure 13.1 (a) Some fuels (reduced forms of matter). (b) Some nonfuels (oxidized forms of matter).

Heavenly Sunlight Flooding the Earth with Energy

No doubt you have heard a lot about nuclear energy—some good things, some not so good. But did you know that life on Earth depends upon nuclear energy? You don't have to worry about problems with the reactor, though. It's about 150 million km away, and it will last, most likely, for another 10 to 15 billion years. That rather ordinary star that we call the sun is the source of nearly all the energy available to us on Earth.

Recall (Chapter 1) that the SI unit of energy is the joule (J). A watt (W), the SI unit for power, is 1 joule per second (J/s). The sun is a nuclear fusion reactor that steadily converts hydrogen to helium (Chapter 4) with a power output of 4×10^{26} watts (W). Although the Earth receives only about 1 part in 50 billion of that energy, it is abundantly bathed in energy, receiving about 1.73×10^{17} W [173 000 terawatts (TW)*] from the sun—an amount equivalent to the output of 115 million nuclear power plants. In three days, the Earth receives energy from the sun equivalent to all fossil fuel reserves.

Only small amounts of energy are available to us from sources other than the sun. Heat coming up from the Earth's interior amounts to only about 32 TW, and the tides produce only about 3 TW. These are almost insignificant quantities; the sun supplies nearly all the energy available in the biosphere (Table 13.1). (The biosphere is the thin film of air, water, and soil in which all life exists. It is only about 15 km thick.)

---

* 1 TW = 10^{12} W.

Table 13.1 The Earth's Energy Ledger (Rough Estimates)

Item	Energy (terawatts)	Approximate Percent
In		
Solar radiation	173 000	99+
Internal heat	32	0.02
Tides	3	0.002
Out		
Direct reflection	52 000	30
Direct heating*	81 000	47
Water cycle*	40 000	23
Winds*	370	0.2
Photosynthesis*	40	0.02

* This energy is eventually returned to space via long-wave radiation (heat).
SOURCE: M. King Hubbert, "The Energy Resources of the Earth." *Scientific American*, September 1971.

Example 13.1 How many joules of energy are consumed by a 100-W bulb burning for 1 hour (3600 seconds)?

$$100 \text{ W} = 100 \text{ J/s}$$

$$\frac{100 \text{ J}}{1 \text{ s}} \times 3600 \text{ s} = 360\,000 \text{ J}$$

Energy and the Life-Support System

Only a small fraction of the energy that the biosphere receives is used to support life. In fact, about 30% of the incident radiation is immediately reflected back into space as short-wave radiation (ultraviolet and visible light). Nearly half is converted to heat, making the third planet a warm and habitable place. About 23% of the solar radiation is used to power the water cycle (Chapter 16), evaporating water from land and seas. The radiant energy of the sun is converted into the potential energy of water vapor, water droplets, and ice crystals in the atmosphere. This potential energy is converted to the kinetic energy of falling rain and snow and of flowing rivers.

A tiny—but most important—fraction of the solar energy is absorbed by green plants, which use it to power photosynthesis. In the presence of green pigments, called **chlorophylls**, this energy is used to convert carbon dioxide and water into glucose, a simple sugar rich in energy.

$$6 \ CO_2 \ + \ 6 \ H_2O \ \xrightarrow[\text{chlorophyll}]{\text{sunlight}} \ \underset{\text{Glucose}}{C_6H_{12}O_6} \ + \ 6 \ O_2$$

This reaction also serves to replenish the oxygen in the atmosphere (see Chapter 15). The glucose can be stored or it can be converted into more complex foods and structural materials. All animals depend on the stored energy of green plants for survival.

Energy and Chemical Reactions

How fast—or how slowly—chemical reactions take place depends on a number of factors. One such factor is *temperature*. Reactions generally take place faster at a higher temperature. For example, coal (carbon) reacts so slowly with oxygen (from the air) at room temperature that the change is imperceptible. However, if coal is heated to several hundred degrees, it reacts at a much more rapid rate. The heat evolved in the reaction keeps the coal burning smoothly.

We make use of our knowledge of the temperature effect on chemical reactions in our daily lives. For example, we freeze foods to retard those chemical reactions that lead to spoilage. When we want to cook our food more rapidly, we turn up the heat.

The effect of temperature on the rates of chemical reactions is explained in terms of the kinetic-molecular theory. At high temperatures, molecules move more rapidly. Thus, they collide more frequently, creating an increased chance for reaction. The increase in temperature also supplies more energy for the breaking of chemical bonds—a condition necessary for most reactions.

For example, for hydrogen and chlorine to react, the bonds between hydrogen atoms and between chlorine atoms must be broken.*

---

* This statement says nothing about the *order* in which the bonds are broken. The accepted **mechanism** for the reaction between hydrogen and chlorine is quite complicated. The first step is presumed to be the breaking down of the chlorine molecule into atoms.

$$Cl_2 \ \rightarrow \ 2 \ Cl$$

This may be accomplished by absorption of heat or light energy. In the second step, it is postulated that chlorine atoms collide with hydrogen molecules and form hydrogen chloride molecules and hydrogen atoms.

$$Cl \ + \ H_2 \ \rightarrow \ HCl \ + \ H$$

In the third step, it is thought that these hydrogen atoms attack a chlorine molecule, form more HCl and regenerate chlorine atoms.

$$H \ + \ Cl_2 \ \rightarrow \ HCl \ + \ Cl$$

The second and third steps then are repeated until virtually all the H_2 and Cl_2 are converted to HCl.

$$H\overset{\xi}{\cdot}H \;+\; Cl\overset{\xi}{\cdot}Cl \;\longrightarrow\; 2\;H\!-\!Cl$$

Once the reaction has started, the energy released by the formation of hydrogen to chlorine bonds more than compensates for that required to break the H—H and Cl—Cl bonds; thus, there is a net conversion of chemical energy to heat energy.

Another factor affecting the rate of a chemical reaction is the *concentration* of reactants. The more molecules there are in a given volume of space, the more likely a collision. The more collisions there are, the more reactions that occur. For example, you could light a wood splint and then blow out the flame. The splint would continue to glow as the wood reacted slowly with the oxygen of the air. If the glowing splint were placed in pure oxygen, the splint would burst into flame, indicating a much more rapid reaction. This factor can be interpreted in terms of the concentration of oxygen: air is about 21% oxygen, so the concentration of O_2 molecules in pure oxygen is about five times as great as in air.

Catalysts (Chapter 8) also affect the rate of chemical reactions. Catalysts are of great importance in the chemical industry. Reactions that otherwise would be so slow as to be impractical can be made to proceed at a feasible rate using appropriate catalysts. Catalysts are even more important in living organisms. Biological catalysts, called **enzymes**, mediate nearly all the chemical reactions that take place in living systems. Enzymes are discussed further in subsequent chapters.

The energy changes associated with chemical reactions are quantitatively related to the *amounts* of chemicals that are changed. For example, burning 1 mol (16 g) of methane to form carbon dioxide and water releases 192 kcal of energy as heat.

$$CH_4 \;+\; 2\;O_2 \;\longrightarrow\; CO_2 \;+\; 2\;H_2O \;+\; 192\;kcal$$

Burning 2 mol (32 g) of methane produces twice as much heat, 384 kcal.

Chemical reactions that result in the release of heat are said to be **exothermic**. The burning of methane, gasoline, and coal are all exothermic reactions. In each case, chemical energy is converted into heat energy. There are other reactions, such as the decomposition of water, in which energy must be supplied.

$$2\;H_2O \;+\; 137\;kcal \;\longrightarrow\; 2\;H_2 \;+\; O_2$$

If energy is supplied as heat, such reactions are said to be **endothermic**. It takes 137 kcal of energy to decompose 36 g of water into hydrogen and oxygen. It should be noted that the same amount of energy is released when enough hydrogen is burned to form 36 g of water.

$$2\;H_2 \;+\; O_2 \;\longrightarrow\; 2\;H_2O \;+\; 137\;kcal$$

Keep that in mind when you hear people suggest hydrogen as a possible replacement for natural gas or petroleum as a fuel.

Example 13.2 Burning 1 mol of propane releases 526 kcal of energy.

$$C_3H_8 + 5\ O_2 \longrightarrow 3\ CO_2 + 4\ H_2O + 526\ kcal$$

How much energy is released when 15.0 mol of propane are burned?

$$\frac{526\ kcal}{1\ mol} \times 15.0\ mol = 7890\ kcal$$

Example 13.3 How much energy is released when 220 g of propane (Example 13.2) is burned?

First calculate the formula weight of propane.

$$(3 \times C) + (8 \times H) = (3 \times 12.0) + (8 \times 1.0) = 36.0 + 8.0 = 44.0$$

One mole of propane weighs 44.0 g. Therefore,

$$220\ g \times \frac{1\ mol}{44.0\ g} \times \frac{526\ kcal}{1\ mol} = 2630\ kcal$$

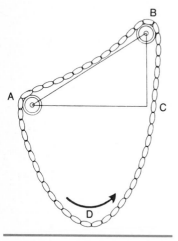

Figure 13.2 A proposed model for perpetual motion. When a uniform chain hangs over two pulleys, the weight of length AD equals the weight of CD. Since the weight of AB is greater than the weight of BC, the chain "should" run around in the direction of the arrow, gaining speed and producing energy—but it doesn't. [From *Perpetual Motion: Electrons in Atoms and Crystals* by Alec T. Stewart. Copyright © 1965 by Alec Stewart. Reprinted by permission of Doubleday, a division of Bantam, Doubleday, Dell Publishing Group, Inc.]

The Energy Enigma: Energy Is Conserved, Yet We're Running Out

We take a look at our present energy sources and our future energy prospects shortly. To do so scientifically, however, let's first look at some of the laws of nature. You should recall that natural laws merely summarize the results of many experiments. We won't trouble you with a recounting of all those experiments here; we will merely state the laws and some of their consequences.

The **first law of thermodynamics** grew out of a variety of experiments during the early 1800s. By 1840, it was clear that, although it can be changed from one form to another, energy is neither created nor destroyed. This law (also called the **law of conservation of energy**) has been restated in a number of ways, including "you can't get something for nothing" and "there is no such thing as a free lunch." Energy can't be made from nothing. Neither does it just disappear, although it may go someplace else.

From the first law of thermodynamics we can conclude that we can't win—in the sense that we can't make a machine that produces energy from nothing. From that law alone, however, we might conclude that we can't possibly run out of energy because energy is conserved. That is true enough, but it doesn't mean that we don't have problems. There is another law with a long arm from which we cannot escape.

Figure 13.3 Energy always flows spontaneously from hot to cold, never the reverse.

Energy and the Second Law: Things Are Going to Get Worse

Despite innumerable attempts, no one has ever built a successful "perpetual-motion" machine. You can't make a machine that will produce more energy than it consumes. Even if an engine isn't doing any work, it loses energy (as heat) due to the friction of its moving parts. In fact, in any real engine, you can't get as much useful energy as you put in. You can't even break even!

If energy is neither created nor destroyed, why do we always need more? Will what we have now not last forever? The answer lies in the fact that energy can be changed from one form to another and not all forms are equal; and high-grade forms of energy are constantly being degraded into low-grade forms. Energy flows downhill. Mechanical energy eventually is changed into heat energy. Hot objects cool off by transferring their heat to cooler objects. There is a tendency toward an even distribution of energy. Energy always flows from a hot object to a cooler one. The reverse never occurs spontaneously.

Observations of heat flow led to the formulation of the **second law of thermodynamics**. In one form (of many), the law states that energy does not flow spontaneously from a cold object to a hot one. Now it is true that we can *make* energy flow from a cold region to a hot one—that's what refrigerators are all about—but we cannot do so without producing changes elsewhere. This sort of reversal of a natural process can only be done at a price. The price, in the case of refrigerators, is the consumption of electricity.

Another way to look at the second law is in terms of disorder. Scientists use the term **entropy** to indicate the randomness of a system. The more mixed up a system is, the higher its entropy. This concept is illustrated in Figure 13.4. Natural processes tend toward a greater entropy—toward disorder.

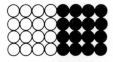

High order
low entropy

ONE WAY

Less order
more entropy

Figure 13.4 Any spontaneous change is toward a less ordered state—toward greater entropy.

The nursery rhyme "Humpty-Dumpty" illustrates the second law of thermodynamics. Once Humpty had fallen and been broken into a more disordered state, "all the king's horses and all the king's men" couldn't put him together again. Once we have taken all the high-grade iron ore (highly ordered iron atoms) and distributed it throughout the world as particles of rust (randomly distributed iron atoms), all the world's geologists and all the world's chemists won't be able to put them together again.

Actually, we sometimes *can* reverse the tendency toward randomness—but only through the expenditure of energy. For example, a new deck of playing cards could be arranged by suit and by rank, a situation of high order (low entropy). If the cards were dropped and scattered randomly about the floor (higher entropy), they could be returned to their ordered state only by the expenditure of energy—bending, stooping, and stacking. Your body would have to convert food energy to muscular energy to do the task. It is quite unlikely that the cards would pick themselves up off the floor and resume their ordered state.

Thus, the second law places restrictions on what can be done. It also has something to say about what we should do about many of our environmental problems, and it points out—rather firmly—a number of things we cannot do.

People Power: Early Uses of Energy

Primitive peoples obtained their energy (food and fuel) by collecting wild plants and hunting wild animals. They expended the energy so obtained in hunting and gathering. Domestication of the horse and the ox increased the availability of energy only slightly. The raw materials used by these work animals were natural—and replaceable—plant materials.

Probably the first device used to convert energy into useful work was the waterwheel. The Egyptians first used waterpower about 2000 years ago, primarily for grinding grain. Later on, waterpower was used for sawmills, textile mills, and other small factories. Windmills were introduced into western Europe during the Middle Ages, primarily for pumping water and grinding grain. More recently, windpower has been used to generate electricity.

Windmills and waterwheels are fairly simple devices for converting the kinetic energy of blowing wind and flowing water into mechanical energy. They were sufficient to power the early part of the industrial revolution, but it was the development of the steam engine that freed the factories from locations along waterways. Since 1850, the water turbine, the internal combustion engine, the steam turbine, the gas turbine, and a variety of other energy-conversion devices have been added to the arsenal. In fact, it has been estimated that the power output of all these conversion devices has increased the energy available for use by a factor of 10 000.

Let's turn our attention now to the fossils that fueled the industrial revolution and still serve as the basis for modern civilization.

Coal: The Carbon Rock of Ages

Coal is a complex material. It is composed principally of carbon, but it also contains smaller percentages of other elements. The quality of coal as an energy source is based on its carbon content; the rankings run from low-grade peat and lignite to high-grade anthracite (Table 13.2). Soft (bituminous) coal is much more plentiful than hard coal (anthracite). Lignite and peat have become increasingly important as the supply of higher grades of coal is depleted.

The supply of coal is limited. Deposits that exist today are less than 600 million years old. For a part of that time, the Earth was much warmer than it is now, and plant life flourished (Figures 13.5 and 13.6). Most plants lived, died, and decayed—playing their normal role in the carbon cycle (Figure 13.7). But a small portion of the plant material became buried under mud and water. There, in the absence of oxygen, it could decay only partially. The structural material of plants is largely cellulose (Chapter 18), a compound of carbon, hydrogen, and oxygen. Under the action of increasing pressure, as the material was buried more deeply, the cellulose molecules broke down. Small molecules rich in hydrogen and oxygen escaped, leaving behind a material increasingly rich in carbon. Thus, peat is a young coal, only partly converted. Anthracite, on the other hand, has been nearly completely carbonized.

Presumably, fossil fuels are still being formed in nature. This process is perhaps most evident in peat bogs. The rate of formation is

Table 13.2 Approximate Composition (Percent by Mass) of Various Typical Grades of Coal (Dry Basis)

Grades of Coal	Carbon	Hydrogen	Oxygen	Nitrogen
Wood (for comparison)	50	6	43	1
Peat	59	6	33	2
Lignite (brown coal)	69	5	25	1
Bituminous (soft coal)	88–89	5	5–15	1
Anthracite (hard coal)	95	2–3	2–3	Trace

SOURCE: M. G. Zabetakis and L. D. Phillips *Coal Mining*. Washington, DC: U.S. Government Printing Office, 1975.

Figure 13.5 Giant ferns, reeds, and grasses that grew during the Pennsylvanian period 300 million years ago helped form the coal we burn today. [Courtesy Field Museum of Natural History, Chicago.]

Figure 13.6 This fossil is part of a fern frond (or leaf) of the genus *Pecopteris*, a common plant in the coal swamps of 300 million years ago. It was found in the shale just above a coal seam in Wilmington Township, Will County, Illinois. [Courtesy of the Smithsonian Institution, Washington, DC.]

extremely slow, however. In fact, it has been estimated that we are using fuels 50 000 times as fast as they are being formed.

People probably have used small amounts of coal since prehistoric times, but, as late as 1760, wood was—for practical purposes—the only fuel being used. In only a few hundred more years, we will have removed from the Earth and burned virtually all the remaining recoverable fossil fuels. Of all that ever existed, we will use about 90% in a period of 300 years. On a geological time scale, that is but a moment. Even on a historical basis, the fossil-fuel feast is but a brief interlude.

Coal as a Fuel: Dirty, Inconvenient, and Invaluable

Coal is an inconvenient fuel. As a solid, it can't be pumped through pipes to get it out of the ground. It must be removed by dangerous deep mining or devastating strip mining. It may be possible to transport coal by grinding it to a powder, mixing it with water, and pumping the slurry through a pipeline. At the present time, though, coal has to be hauled by trucks, trains, and barges. A good deal of energy is consumed in getting coal from the mine to the power plant or the factory where it is to be used.

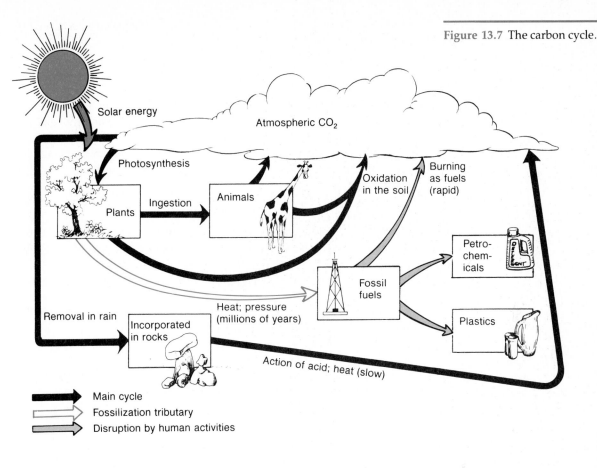

Figure 13.7 The carbon cycle.

Coal is our most plentiful fossil fuel. The United States is estimated to have over 40% of the world's reserves of coal. There is probably enough easily recoverable coal to last us several hundred years. Electric utilities in the United States burn nearly 700 million t of coal each year, generating 1300 TW of electricity (55% of the total). Unfortunately, this coal use is associated with some of our worst environmental problems. It can be obtained most inexpensively by strip mining. In the process, however, vast areas are ripped bare of all vegetation and the exposed soil washes away, filling streams with mud and silt (Figure 13.8). Stripped areas sometimes can be restored, but restoration is expensive. Laws in the United States now require the restoration of most stripped areas and ban stripping on steep slopes where restoration is not feasible. Unfortunately, a lot of damage was done before the laws were passed, and many conservationists still consider the present laws inadequate.

The fuel value of coal depends largely on its carbon content. Complete combustion of the carbon produces carbon dioxide.

$$C + O_2 \longrightarrow CO_2$$

In limited amounts of air, however, large quantities of carbon monoxide and soot are formed.

$$2 \, C \; + \; O_2 \; \longrightarrow \; 2 \, CO$$

The soot is mostly unburned carbon.

Coal, a solid, contains many impurities. Unlike natural gas and petroleum, it often contains a lot of minerals. When the coal is burned, the minerals are left as ash, presenting a major problem in solid-waste disposal. Some minerals enter the air as particulate mattter, constituting a major pollution problem.

Perhaps even worse, most of our remaining coal is high in sulfur. When coal burns, choking sulfur dioxide is poured into the atmosphere. Reacting with oxygen and moisture in the air, the sulfur dioxide is converted into sulfuric acid. This acid slowly eats away steel and aluminum structures, marble buildings and statues, and human lungs. The contribution of coal burning to air pollution is discussed further in Chapter 15.

There *are* ways to clean coal before it is burned. A **flotation method** makes use of the different densities of coal and its major impurities. Coal has a density of about 1.3 g/cm³. Shale, a rock formed from hardened clay, has a density of about 2.5 g/cm³. Pyrite (FeS$_2$), the major source of sulfur in the coal, has a density of about 5.0 g/cm³. By using a solution of the proper density and adding detergents to cause the coal to float, the coal can be floated off, leaving the heavier minerals behind. Unfortunately, the process adds to the cost of the coal. Coal

Figure 13.8 The site of a strip mine in Morgan County, Tennessee. (a) The abandoned mine in 1963. (b) The same mine after it was reclaimed in a demonstration project, 1971. [Courtesy of the Tennessee Valley Authority, Knoxville.]

(a)

(b)

also can be converted to more convenient gaseous and liquid fuels. These processes and other future fuel sources are discussed in the next chapter. Perhaps one or more of these processes will supply most of our energy in the twenty-first century. For many years to come, though, coal—dirty and inconvenient as it is—will remain out best hope for abundant energy.

Coal isn't just a fuel, however. When coal is heated in the absence of air, volatile material is driven off, leaving behind purified carbon called coke (Chapter 9). Part of the volatile material given off in the coking process is condensed into **coal tar**. This gooey mixture is a source of organic chemicals, particularly aromatic compounds (Chapter 10).

Natural Gas: Mostly Methane But Much More

Natural gas is a fossil fuel. It is composed principally of methane, but its composition varies greatly. In North America, natural gas typically contains 60 to 80% methane, 5 to 9% ethane, 3 to 18% propane, and 2 to 14% butane and pentane. The gas, as it comes from the ground, often contains sulfur compounds and nitrogen as impurities.

Natural gas was most likely formed—ages ago—by the action of heat, pressure, and perhaps bacteria on buried organic matter. The gas is trapped in geological formations capped by impermeable rock. It is removed through wells drilled into the gas-bearing formations.

Most natural gas is used as fuel, but it is also an important raw material. In North America, some of the higher alkanes are separated out of the natural gas. Ethane and propane are **cracked**—decomposed by heat in the absence of air—to form ethylene and propylene. These alkenes are intermediates in the synthesis of plastics (Chapter 12) and many other useful commodities.

Cracking breaks chemical bonds and allows new ones to form. Cracking of ethane produces hydrogen and some methane in addition to ethylene. (The equation is not balanced; the proportion of products varies with conditions.)

$$\underset{\begin{array}{c}\text{H}\ \ \text{H}\\|\ \ \ |\\\text{H}-\text{C}-\text{C}-\text{H}\\|\ \ \ |\\\text{H}\ \ \text{H}\end{array}}{}\ \xrightarrow{\text{heat}}\ \underset{\begin{array}{c}\text{H}\ \ \text{H}\\|\ \ \ |\\\text{H}-\text{C}=\text{C}-\text{H}\end{array}}{}\ +\ \underset{\begin{array}{c}\text{H}\\|\\\text{H}-\text{C}-\text{H}\\|\\\text{H}\end{array}}{}\ +\ \text{H}-\text{H}$$

A small amount of higher alkanes also is formed by the combination of fragments formed in cracking. Other products from natural gas include acetylene, hydrogen cyanide ($H-C\equiv N$), chlorinated hydrocarbons, methyl alcohol, and ammonia.

Figure 13.9 Natural gas burns with a relatively clean flame. [Courtesy of the American Gas Association.]

Natural Gas as a Fuel: Clean But Scarce

Natural gas is the cleanest of the fossil fuels. It can contain a variety of impurities as it comes from the ground, but most of its obnoxious impurities can be removed by processing. As a gas, it flows through pipes to wherever it is needed—with only an occasional boost from a pumping station. Very little of its energy is needed to get natural gas from the ground to the consumer.

Composed mainly of methane, natural gas burns with a relatively clean flame, and its products are mainly carbon dioxide and water.

$$CH_4 + 2 O_2 \longrightarrow CO_2 + 2 H_2O$$

As in any combustion in air, some nitrogen oxides are formed. The role of these compounds in air pollution is discussed in Chapter 15. When natural gas is burned in limited amounts of air, carbon monoxide or even elemental carbon (soot) can be major products.

$$2 CH_4 + 3 O_2 \longrightarrow 2 CO + 4 H_2O$$
$$CH_4 + O_2 \longrightarrow C + 2 H_2O$$

Even so, natural gas is the cleanest of the fossil fuels. Unfortunately, it is also the fuel in shortest supply. Our reserves are expected to decline over the next 30 years or so.

Petroleum: Liquid Hydrocarbons

Petroleum is an exceedingly complicated liquid mixture of organic compounds. Petroleum deposits nearly always have associated natural gas, part or all of which is sent through a separation process. The main components of petroleum are hydrocarbons. These are chiefly alkanes, but some are cyclic compounds. Petroleum also has varying proportions of sulfur-, nitrogen-, and oxygen-containing compounds.

We have seen that coal is primarily of plant origin. Recent evidence indicates that petroleum is of animal origin. Most likely it is formed primarily from the fats (Chapter 18) of ocean-dwelling, microscopic animals, for it is nearly always found in rocks of oceanic origin. Fats are made up mainly of compounds of carbon and hydrogen, with a little oxygen (for example, $C_{57}H_{110}O_6$). The removal of the oxygen and the slight rearrangement of the carbon and hydrogen atoms in these fats produces typical hydrocarbon molecules as they occur in petroleum.

Petroleum, as it comes from the ground, is of limited use. To make it better suit our needs, we separate it into fractions by boiling it in a distillation column (Figure 13.10). The lighter molecules come off at the

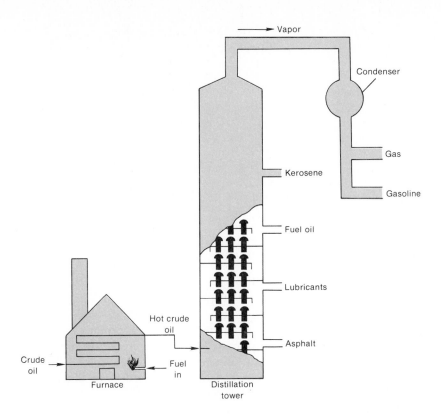

Figure 13.10 The fractional distillation of petroleum.

top of the column, the heavier at the bottom. (The various fractions are described in Table 13.3.)

Gasoline usually is the fraction of petroleum most in demand, and fractions that boil at higher temperatures are often in excess supply. The latter are often converted to gasoline by heating in the absence of air.

Table 13.3   Typical Petroleum Fractions

Fraction	Typical Range of Hydrocarbons	Approximate Range of Boiling Points (degrees Celsius)	Typical Uses
Gas	CH_4 to C_4H_{10}	Less than 40	Fuel, starting materials for plastics
Gasoline	C_5H_{12} to $C_{12}H_{26}$	40–200	Fuel, solvents
Kerosene	$C_{12}H_{26}$ to $C_{16}H_{34}$	175–275	Diesel fuel, jet fuel, home heating; cracking to gasoline
Heating oil	$C_{15}H_{32}$ to $C_{18}H_{38}$	250–400	Industrial heating, cracking to gasoline
Lubricating oil	$C_{17}H_{36}$ and up	Above 300	Lubricants
Residue	$C_{20}H_{42}$ and up	Above 350 (some decomposition)	Paraffin, asphalt

Figure 13.11 Formulas of some of the products formed when $C_{14}H_{30}$ (a typical molecule in kerosene) is cracked. Note the great variety of hydrocarbons with fewer carbon atoms that are formed.

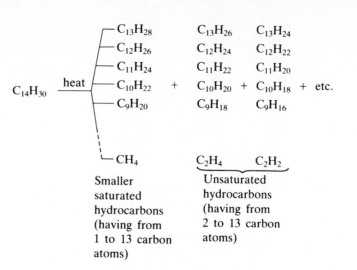

This cracking process breaks down the bigger molecules into the smaller ones. The effect of the process, with $C_{14}H_{30}$ as an example, is illustrated in Figure 13.11. Not only does cracking convert some molecules into those in the gasoline range (C_5H_{12} though $C_{12}H_{26}$), it produces a variety of useful by-products. The unsaturated hydrocarbons are starting materials for the manufacture of a whole host of petrochemicals.

The cracking process described here is a crude, yet illustrative, example of how chemists modify nature's materials to meet human needs and desires. Starting with coal tar or petroleum, a chemist can create a dazzling array of substances with a wide variety of properties. A few of these substances are plastics, pesticides, herbicides, perfumes, preservatives, pain killers, antibiotics, stimulants, depressants, and detergents. Many of these substances are discussed in other chapters. A chemist's ability to create new substances from petroleum products enables us to select materials that more closely suit our specifications. Chemists provide the options; we make the choices.

Gasoline: From the Refinery to the Chemist

Gasoline, like the petroleum from which it is derived, is a mixture of hydrocarbons. The typical alkanes in gasoline have formulas ranging from C_5H_{12} to $C_{12}H_{26}$. Since there are many isomeric forms, particularly for the higher members of the group, we see that gasoline is an exceedingly complex mixture of alkanes. There are also small amounts of other kinds of hydrocarbons present, and even some sulfur- and nitrogen-containing compounds. The gasoline fraction of petroleum as it comes from the distillation column is called **straight-run gasoline**. It doesn't burn very well in modern, high-compression automobile engines. Chemists have learned how to modify it in a variety of ways to make it burn better.

Early in the development of the automobile engine, scientists learned that some types of hydrocarbons burned more evenly and were less likely to ignite prematurely than others. Ignition before the piston was in proper position led to a knocking in the engine. Scientists soon were able to correlate good performance with a branched-chain structure in hydrocarbon molecules. An arbitrary performance standard, called the **octane rating**, was established in 1927. The best performer in a laboratory test engine was found to be a compound that became known as isooctane (it was one of 18 isomeric octanes tested). Isooctane was assigned a value of 100 octane. An unbranched-chain compound, heptane, was found to cause a very bad knock. It was given an octane rating of 0. A gasoline rated 90 octane was one that performed the same as a mixture that was 90% isooctane and 10% heptane.

$$CH_3-\underset{\underset{CH_3}{|}}{\overset{\overset{CH_3}{|}}{C}}-CH_2-\underset{}{\overset{\overset{CH_3}{|}}{CH}}-CH_3 \qquad CH_3CH_2CH_2CH_2CH_2CH_2CH_3$$

Isooctane   Heptane

During the 1930s, chemists discovered that the octane rating of gasoline could be improved by heating gasoline in the presence of catalysts such as sulfuric acid (H_2SO_4) and aluminum chloride ($AlCl_3$). This increase in octane rating was attributed to a conversion (**isomerization**) of a part of the unbranched structures to highly branched molecules. For example heptane molecules could be isomerized to a branched structure.

$$CH_3CH_2CH_2CH_2CH_2CH_2CH_3 \xrightarrow[\text{heat}]{H_2SO_4} CH_3CH_2-\underset{\underset{CH_3}{|}}{CH}-\underset{}{\overset{\overset{CH_3}{|}}{CH}}-CH_3$$

Chemists also are able to combine small hydrocarbon molecules (below the gasoline range) into larger ones more suitable for use as fuel.

Certain chemical substances also were discovered that, when added in small amounts, substantially improved the antiknock quality of gasoline. Chief among these additives was tetraethyllead. This compound, when added in amounts as small as 1 mL per liter of gasoline (about 1 part per thousand), would increase the octane rating from approximately 55 to 90 or more.

Lead is toxic, however. Large amounts of it have entered the environment through the combustion of leaded gasoline in automobiles.

$$C_2H_5-\underset{\underset{C_2H_5}{|}}{\overset{\overset{C_2H_5}{|}}{Pb}}-C_2H_5$$

Tetraethyllead

Further, lead fouls the catalytic converters used in modern automobiles. Thus, to get high octane ratings in unleaded fuels, petroleum refineries use **catalytic reforming** to convert low-octane alkanes to high-octane aromatic compounds. For example, hexane (with an octane number of 25) is converted to benzene (with an octane number of 106).

$$CH_3CH_2CH_2CH_2CH_2CH_3 \xrightarrow[\text{heat}]{\text{catalyst}} \bigcirc + 4\ H_2$$

$$(C_6H_{14}) \qquad\qquad\qquad (C_6H_6)$$

(Air pollution problems associated with automobiles are discussed in Chapter 15.)

To recapitulate briefly, American scientists were able, by World War II, to convert nearly all petroleum fractions into gasoline. Furthermore, they were able to improve octane ratings substantially by rearranging the molecules and adding tetraethyllead. It was this high-performance fuel, as much as or more so than superior machines, that made possible the Allied victory in the air war with the Axis powers (Germany, Italy, and Japan) during World War II.

Petroleum as a Fuel: What There Is, Isn't All Ours

Petroleum is largely a mixture of hydrocarbons, and hydrocarbons burn readily. Complete combustion yields mainly carbon dioxide and water. A representative reaction is that of an octane.

$$2\ C_8H_{18} + 25\ O_2 \longrightarrow 16\ CO_2 + 18\ H_2O$$

Combustion in air also leads to the formation of nitrogen oxides. Incomplete burning produces carbon monoxide and soot. Petroleum usually contains small amounts of sulfur compounds that produce sulfur dioxide when burned.

Efficiently burned, petroleum products are rather clean fuels. Fuel oil, used for heating homes or to produce electricity, can be burned efficiently, while contributing only moderately to air pollution. Gasoline, however, the major fraction of petroleum, is used to power automobiles; and the internal combustion engines in most automobiles are rather inefficient and contribute heavily to air pollution (Chapter 15) (Figure 13.12).

Air pollution isn't the only problem: burning up our petroleum reserves will leave us without a ready source of many familiar materials. Nearly all industrial organic chemicals come from petroleum (smaller

Figure 13.12 Petroleum can be burned rather cleanly to heat homes or to produce electricity. Most of it, however, is burned in internal combustion engines to propel automobiles. These inefficient machines contribute mightily to air pollution.

amounts are derived from coal and natural gas); and plastics, synthetic fibers, solvents, and many other consumer products are made from these chemicals.

Petroleum is more abundant than natural gas. It also contains more impurities. Some of the impurities can be removed by processing, but petroleum-derived fuels are generally dirtier than natural gas. Crude oil is liquid, a convenient form for its transportation. Petroleum is pumped easily through pipelines or hauled across the oceans in giant tankers. Little energy is required to pump petroleum from the ground.

As domestic supplies diminish, however, we will have to expend an increasing fraction of the energy content of petroleum to transport it to where it is needed. Also, as petroleum is pumped out of a well, that which remains behind is increasingly difficult to get out. This is a natural consequence of the second law of thermodynamics: that which is left behind is more scattered, and more energy is required to collect it.

United States petroleum reserves are expected to decline rapidly, Even the massive field on the North Slope of Alaska would last only 2 or 3 years if it was our only source. Offshore deposits along the Atlantic coast of the United States would produce only a few years' supply, and offshore drilling requires a great deal more energy (and money) than drilling on land.

Projections show that in about the year 2005 United States petroleum production will reach a point at which it will take as much energy to obtain a barrel of oil as we get back from burning it as fuel. Discoveries of new reserves may change that date a bit, but probably not by much: a massive field found in 1982 off the coast of California—

with estimated reserves of a billion barrels—extends that date by only 5 months!

The developed countries of North America, western Europe, and Japan depend heavily on oil imports from developing nations, some of which are politically unstable. Such dependence makes the industrial nations vulnerable economically and politically.

The Fossil Fuel Feast: A Brief Interlude

After the steam engine came into widespread use (by about 1850), the industrial revolution was powered largely by coal. By 1900, about 95% of the world's energy production came from burning coal. With the development of the internal combustion engine, petroleum became increasingly important and by 1950 replaced coal as the principal fuel.

The supply of fossil fuels on the Earth is limited. Estimated United States and world reserves are given in Table 13.4. We burn these materials at a rapid rate. Annual United States and world production is

Table 13.4   Estimated U.S. and World Reserves of Economically Recoverable Fuels*

Fuel	United States	World
Coal	183 580 000 000	423 233 000 000
Petroleum	11 420 000 000	111 092 000 000
Natural gas	7 555 000 000	88 731 000 000

* All figures are expressed in metric tons (t) of coal equivalents for ready comparison of energy content. One metric ton of crude petroleum is equivalent to 1.47 t of coal. For natural gas, 1000 m^3 is equivalent to 1.332 t of coal. Coal reserves do not include lignite (brown coal).
SOURCE: Department of Economic and Social Affairs, *Statistical Yearbook 1979–80*. New York: United Nations, 1981.

Table 13.5   Annual United States and World Production of Fossil Fuels*

Fuel	United States	World
Coal	601 142 000	2 299 842 000
Petroleum	618 810 000	3 848 087 000
Natural gas	601 708 000	1 853 520 000

* All figures are expressed in metric tons of coal equivalents.
SOURCE: Department of Economic and Social Affairs, *Statistical Yearbook 1982*. New York: United Nations, 1984.

given in Table 13.5. Estimates of reserves vary greatly, depending on the assumptions made. Even the most optimistic estimates, however, lead to the conclusion that our nonrenewable energy resources are being depleted rapidly. Indeed, in just a century, we will have used up over half the fossil fuels that were formed over the ages. As far as energy sources are concerned, we are in the midst of the "chemical century" (Reference 11).

The ability to modify hydrocarbon molecules enables the petroleum industry to produce increased amounts of whatever fraction they desire. They can, on demand, produce more fuel oil or more gasoline. They can even make gasoline from coal (Chapter 14). They can't, however, increase the amount of fossil fuels aboard Spaceship Earth. Let's hope that scientists soon develop new sources of energy that do not depend on petroleum, so that we can stop what is, in many ways, a profligate waste of resources. Spaceship Earth has abroad it all the supplies it will ever have. We must use them wisely.

Problems

1. What is a fuel?
2. Which of the following are fuels?
   a. C_3H_8 b. C_3F_8 c. C_6H_6
3. What kind of reaction powers the sun?
4. What fraction of the Earth's energy powers photosynthesis?
5. How does temperature affect the rate of a chemical reaction?
6. Why does wood burn more rapidly in pure oxygen than in air?
7. Use the kinetic-molecular theory to explain the effect of temperature on the rate of a chemical reaction.
8. What is an exothermic reaction? Give an example.
9. What is an endothermic reaction? Give an example.
10. State the first law of thermodynamics.
11. State the second law of thermodynamics in terms of energy flow and in terms of order and disorder.
12. What is entropy?
13. What is the principal difference in anthracite, bituminous, and lignite coal?
14. What are the advantages and disadvantages of coal as a fuel?
15. Write the equation for the complete combustion of coal (C).

16. Write the equation for the incomplete combustion of coal (C) to form carbon monoxide.
17. What is soot?
18. Describe the flotation method for cleaning coal before it is burned.
19. What is coal tar?
20. What is the main component of natural gas?
21. What are the main products formed when ethane is cracked?
22. What are the advantages and disadvantages of natural gas as a fuel?
23. Write the equation for the complete combustion of methane.
24. Write the equation for the incomplete combustion of methane to form carbon monoxide.
25. Write the equation for the incomplete combustion of methane to form soot.
26. What is the physical state of each of the three fossil fuels?
27. What is thought to be the origin of petroleum?
28. What are the advantages and disadvantages of petroleum as a source of fuels?
29. Write the equation for the complete combustion of pentane (a hydrocarbon in gasoline).

30. What is straight-run gasoline?

31. What is meant by the octane rating of a gasoline?

32. What are the advantages and disadvantages of tetra-ethyllead as an octane booster?

33. What was the principal fuel used in the United States before 1800?

34. What was the principal fuel used in the United States from about 1850 to 1950?

35. What has been our principal fuel since 1950?

36. What is likely to be our principal fuel in the early decades of the twenty-first century?

37. How have people modified the carbon cycle?

38. Energy is conserved. How can we ever run out of energy?

39. Is entropy increased or decreased when a fossil fuel is burned?

40. Burning 1.00 mol of methane releases 192 kcal of energy.

$$CH_4 + 2 O_2 \rightarrow CO_2 + 2 H_2O + 192 \text{ kcal}$$

How much energy is released by burning 5.00 mol of methane?

41. It takes 137 kcal of energy to decompose 2 mol of water.

$$2 H_2O + 137 \text{ kcal} \rightarrow 2 H_2 + O_2$$

How much energy does it take to decompose 55.5 mol of water?

42. How much energy is released by burning 64.0 g of methane (see Problem 40)?

43. How much energy is released by burning 400 g of hydrogen (see Problem 41)?

44. Energy is stored in a flashlight battery. Eventually the battery runs down. Has the energy been destroyed? Explain your reasoning.

45. A woman uses 1000 kcal in running 10 km in 40 minutes. Calculate her average power output in watts. (1.000 kcal = 4184 J.)

46. Energy requirements of the human brain are about 20% of the total body metabolism. Calculate the power output (in watts) of your brain if you use 2000 kcal of energy per day.

47. In 1981 the U.S. Geological Survey estimated world oil reserves at 723 billion barrels. Present annual use is 20 billion barrels. How long will these reserves last if the present rate of use continues?

48. The Geological Survey study (Problem 47) estimated undiscovered reserves at 550 billion barrels. In what year will we "run out of oil" if both estimates are accurate and if the present rate of use continues?

49. Living organisms seem to violate the second law of thermodynamics by taking materials from a less-ordered to a more-ordered state. How are they able to do that?

50. Can fossil fuels be recycled? If your answer is yes, explain how. If your answer is no, explain why not.

51. What is the ultimate source of nearly all the energy on Earth?

52. Why did the United States shift from coal to petroleum and natural gas when we have much larger reserves of coal than of the two hydrocarbons?

53. How is crude petroleum modified to better meet our needs and wants?

54. What are some possible sources of petrochemicals—for manufacturing drugs, plastics, detergents, and such—when our petroleum reserves are gone?

55. What weight of carbon dioxide is formed by the combustion of 1250 kg of coal when the coal is 40% carbon? The equation is

$$C + O_2 \rightarrow CO_2$$

References and Readings

1. Asimov, Isaac. "In Dancing Flames a Greek Saw the Basis of the Universe." *Smithsonian*, November 1971, pp. 52–57. Discusses heat, from the Greek "element" fire to the first law of thermodynamics.

2. Bailey, Maurice E. "The Chemistry of Coal and Its Constituents." *Journal of Chemical Education*, July 1974, pp. 446–448.

3. Basalla, George. "Energy and Civilization." *EPRI Journal*, July–August 1979, pp. 20–25.

4. Bent, Henry A. "Haste Makes Waste: Pollution and Entropy." *Chemistry*, October 1971, pp. 6–15. This article should be required reading for all science—and nonscience—students.

5. Brown, Theodore L. *Energy and the Environment.*

Columbus, OH: Charles E. Merrill, 1971. Highly recommended.

6. DiLavore, Philip. *Energy: Insights from Physics.* New York: Wiley, 1984.

7. Duncan, William. "Oil: An Interlude in a Century of Coal." *Chemistry and Industry*, 2 May 1981, pp. 311–316.

8. Eckholm, Erik P. "The Firewood Crisis." *Natural History*, October 1975, pp. 7–22. High oil prices lead to a scarcity of wood, the principal fuel of one third of the world's people.

9. Kolb, Doris, and Kenneth E. Kolb. "Chemical Principles Revisited: Petroleum Chemistry," *Journal of Chemical Education*, July 1979, pp. 465–469.

10. Kraushaar, Jack J., and Robert A. Ristinen. *Energy and Problems of a Technical Society.* New York: John Wiley, 1984.

11. Lapp, Ralph E. "The Chemical Century." *Bulletin of the Atomic Scientists*, September 1973, pp. 8–14. Use of fossil fuels will began to decline after the year 2000, so the twentieth century will be unique in the use of chemical energy.

12. Porter, George. "The Laws of Disorder." *Chemistry*. Ten articles appearing between May 1968 and February 1969.

13. Read, John F. "The Entropy Crisis." *Canadian Chemical Education*. January 1975, pp. 6–7. Energy is conserved; entropy is the problem.

14. Stevenson, Kenneth L. "Brief Introduction to the Three Laws of Thermodynamics." *Journal of Chemical Education*, May 1975, pp. 330–331.

14

Energy for the Future
Alternative Sources

Within your lifetime, it is likely that natural gas and petroleum will become so scarce and so expensive that people won't be able to afford them as fuels. At the current rate of use, United States reserves of petroleum will be depleted in 1995, and world reserves will be gone in 2020. Over half the world's petroleum reserves are in the troubled lands of the Middle East. United States reserves of natural gas amount to only about 10 years' supply. Its coal reserves would last 300 years at the present rate of use, but the rate of use is increasing rapidly. Coal use will increase even more dramatically as the other two fossil fuels are depleted.

We could put more reliance on nuclear power. Other nations have (Table 14.1). The public is quite fearful of nuclear power plants, however. That apprehension was exacerbated by the accident at Chernobyl (Chapter 4). The United States has over 100 operating nuclear reactors, with about 30 more on order. But no new plants have been ordered since 1978. It takes about 10 years to build a nuclear power plant. There will have to be a dramatic change in public attitude quite soon if nuclear power is to play a big role in our future.

What shall we do for energy a decade or so from now? In a democracy, the citizens decide. Unfortunately, when energy sources are concerned, the choices have to be made years in advance to avoid traumatic change.

In this chapter, we consider a variety of energy sources—some of them that will supplement fossil fuels for now and others that will contribute to our energy needs in the future. As the "chemical century" draws to a close, one or more of these sources may well replace fossil fuels as our primary source of energy.

Table 14.1 Percentage of Total Electricity Generated by Nuclear Power Plants

Country	Percent
France	65
Sweden	50
Finland	38
West Germany	31
Japan	26
Britain	18
United States	17
Canada	13
U.S.S.R.	11
India	4
Brazil	2

Convenient Energy: Electricity

In Chapter 13, we saw that we can never run out of energy because energy is conserved. We can, however, burn up our fossil fuels. What we really want is *useful* energy. Perhaps even more than that, however, we want *convenient* energy. We also saw in Chapter 13 that the convenience of a fossil fuel depends on its physical state: gases are most convenient, liquids next, solids least convenient. Perhaps the most convenient form of energy of all is electricity (Table 14.2). With it we can have light and hot water and can run motors of all sorts. We can use it to heat or cool our homes and workplaces. When looking at future energy sources, then, we look mainly at ways of generating electricity.

Table 14.2 Extraction, Transportation, Distribution, and Convenience of Fuels in Various Physical States

Physical State	Extraction	Transportation to Cities	Distribution Within a City	In Use Convenience	In Use Cleanliness
Solids (coal, wood)	Shovels, borers, blasting	Trucks, trains, barges (slurry with water in pipe)	Trucks, buckets	Least	Dirtiest
Liquids (gasoline, fuel oil)	Pumps	Pipelines, tankers, barges, trucks	Trucks		
Gases (natural gas)	Pumps	Pipelines	Pipes		
Electricity* (electron flow)	—	Wires	Wires	Most	Cleanest

*Produced by burning any of the primary fuels, and included for comparison.

Figure 14.1 A coal-fired power plant for generating electricity.

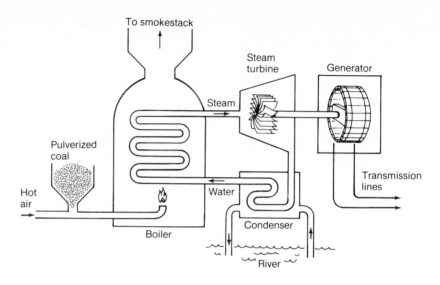

Any fuel can be burned to boil water, and the steam produced (in great enough quantities) can turn a turbine to generate electricity. Figure 14.1 shows a coal-fired steam power plant. At present, 55% of United States electrical energy comes from coal-burning plants. Such facilities are at best only about 40% efficient; 60% of the energy of the fossil fuel is wasted as heat. This energy "loss" is an inevitable

Figure 14.2 Schematic diagram of a nuclear power plant used to generate electricity.

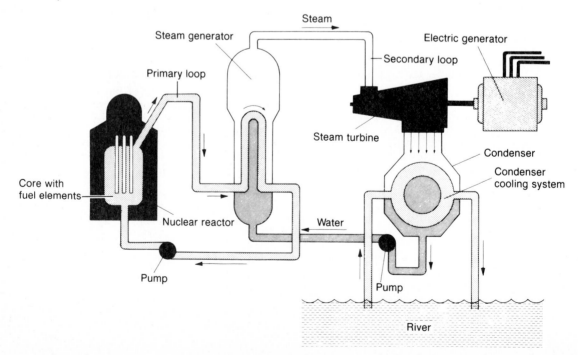

consequence of the second law of thermodynamics. It is part of the price we pay for convenience. (Another part, air pollution, is discussed in Chapter 15.)

We cannot count on fossil fuels forever, so let's look at some alternative systems for producing electricity.

Nuclear Power: Paradise or Perdition?

The fission reactions that power nuclear bombs can be controlled in nuclear reactors. The energy released during fission can be used to generate steam, and the steam can turn a turbine to generate electricity (Figure 14.2).

At present, about 17% of United States electricity comes from nuclear power plants. The eastern seaboard and upper midwestern states, many of which have minimal fossil fuel reserves, are heavily dependent on nuclear power for electricity.

There are actually several types of nuclear power plants. The one in Figure 14.2 is a pressurized water reactor, the principal type now being constructed in the United States. Earlier models were mainly boiling water reactors in which the steam from the reactor was used to power the turbine directly. No attempt is made here to discuss all the possible types of reactors. A summary of the characteristics of several types is provided in Table 14.3, however.

At the dawn of the nuclear age, nuclear power was envisioned by

Table 14.3   Characteristics of Several Types of Nuclear Reactors

Reactor	Abbreviation	Fuel	Moderator	Primary Coolant	Fluid in Secondary Loop
Light water	LWR				
Boiling water	BWR	^{235}U (enriched to 3%)	Water	Water	(No secondary loop)
Pressurized water	PWR	^{235}U (enriched to 3%)	Water	Water	Water
Canadian uranium–deuterium	CANDU	^{235}U (natural abundance)	Heavy water*	Heavy water	Water
High-temperature gas-cooled	HTGR	^{235}U (enriched to 3%)	Graphite	Helium gas	Water
Liquid-metal fast breeder	LMFBR	^{235}U (at start), then ^{238}U converted to ^{239}Pu	—	Molten sodium	Molten sodium†

* Heavy water is water enriched with the heavy isotope of hydrogen called deuterium ($^{2}_{1}H$). Heavy water often is indicated by the formula D_2O. Other references to water are to ordinary, or light, water.
† A third loop in the LMFBR contains water, which is converted to steam to power the turbine.

some to be destined to fulfill the Biblical prophecy of a fiery end to our world. Indeed, as the cold war between the United States and the Union of Soviet Socialist Republics intensified during the 1950s, it was difficult to see how nuclear war could be avoided. If it had come, it could well have been the end of life on Earth.

While some were predicting doom, others saw nuclear power as a source of unlimited energy. During the late 1940s, it was commonly predicted that electricity from nuclear plants would become so cheap that it eventually would not have to be metered. Lewis Strauss, chairman of the Atomic Energy Commission, predicted that lights in public buildings could be left on continuously; on-off switches would not be needed! Nuclear power has not yet brought us either paradise or perdition, but it has become one of the most controversial issues of our time. Our great need for energy indicates that the controversy will continue for years to come.

Nuclear power plants use the same fission reactions as nuclear bombs. A **moderator** (see Table 14.3) is used to slow down the fission neutrons. The reaction is also controlled by the insertion of boron steel or cadmium **control rods**. Boron and cadmium absorb neutrons readily, so the rods prevent the neutrons from participating in the chain reaction. These rods are installed when the reactor is being built. Removing them part way starts the chain reaction; the reaction is stopped when the rods are pushed in all the way.

Nuclear power plants have one great advantage over coal- and oil-burning plants: they do not pollute the air with soot, fly ash, sulfur dioxide, or other noxious chemicals. The plants have disadvantages as well, however. Elaborate and expensive safety precautions must be taken to protect plant workers and the inhabitants of surrounding areas from radiation.

The reactor itself is heavily shielded and housed inside a containment building of metal and reinforced concrete. Because loss of coolant water can result in meltdown of the reactor core, elaborate backup emergency cooling systems have to be constructed. Despite what proponents call the utmost precautions, some opponents of nuclear power still fear a runaway nuclear reaction in which the containment building would be breached and massive amounts of radioactivity would escape into the environment. The chance for such an accident is probably exceedingly small, but if it did occur, thousands could be killed and large areas rendered uninhabitable for centuries. The benefit of nuclear power—abundant electrical energy—is clear, but the small probability of an accident, multiplied by the possible severity of such an occurrence, leads to a highly uncertain risk. It is therefore possible for scientists and others to debate endlessly the desirability of nuclear power.

Another problem with nuclear power is that the fission products are highly radioactive and must be isolated from the environment for centuries. Again, scientists disagree about the feasibility of nuclear waste disposal. Proponents of nuclear power say that such wastes can be safely stored in old salt mines or other geologic formations. Opponents fear that the wastes may arise from their "graves" and eventually contaminate the groundwater. It is impossible to do a million-year experiment in a few years to prove who is right. At any rate, we have nuclear wastes now—from power plants and nuclear weapons production—and we must try to find a solution to this problem.

There also are wastes from the mining of uranium ore. Over 200 million tons of **tailings** (wastes left from uranium ore processing) now plague 10 western states. These tailings are mildly radioactive, giving off radon gas and gamma radiation. Dust from these tailings carry these problems to surrounding areas. Toxic selenium and arsenic compounds are leached into the groundwater.

Still another problem, **thermal pollution**, is unavoidable. As the energy from any material is converted to heat to generate electricity, some of the energy is released into the environment as "waste" heat. Nuclear power plants do generate more thermal pollution than plants that burn fossil fuels, but the difference is not as important, perhaps, as the other problems we have mentioned.

In 1979, a loss-of-coolant accident at the Three Mile Island nuclear power plant near Harrisburg, Pennsylvania, released small amounts of radioactivity into the environment. Although no one was killed or seriously injured, this accident whetted public fear of nuclear power. The accident at Chernobyl was even more frightening. The reactor core meltdown killed several people outright. Others died from radiation sickness in the following weeks and months. Thousands were evacuated. A large area will remain contaminated for decades. Radioactive fallout spread across much of Europe. Thousands, particularly those close to the accident, are at greatly increased risk of cancer from exposure to the radiation. At Three Mile Island, a containment building kept most of the radioactive material inside. The Chernobyl plant had no such protective structure.

An unfounded fear is that nuclear power plants might blow up like the bombs that devastated Hiroshima and Nagasaki (Chapter 4)—but uranium-fueled plants can't explode like nuclear bombs. The uranium is enriched to, at most, 3 or 4% uranium-235. To make a bomb, you need about 90% uranium-235.

There is considerable controversy over most aspects of nuclear power, and there are scientists on both sides. While they may be able to agree on the results of laboratory experiments, scientists obviously do not agree on what is best for society. Those who wish to explore this controversy further are urged to see the references and readings at the end of this chapter.

$$_{92}^{238}U + _{0}^{1}n \longrightarrow _{92}^{239}U$$

$$_{92}^{239}U \longrightarrow _{93}^{239}Np + _{-1}^{0}e$$

$$_{93}^{239}Np \longrightarrow _{94}^{239}Pu + _{-1}^{0}e$$

Plutonium-breeding reactions
(a)

$$_{90}^{232}Th + _{0}^{1}n \longrightarrow _{90}^{233}Th$$

$$_{90}^{233}Th \longrightarrow _{91}^{233}Pa + _{-1}^{0}e$$

$$_{91}^{233}Pa \longrightarrow _{92}^{233}U + _{-1}^{0}e$$

Uranium-233-breeding reactions
(b)

Figure 14.3 Reactions for breeding fissile fuel from nonfissile isotopes.

Breeder Reactors: Making More Fuel Than We Burn

Less than 1% of naturally occurring uranium is the fissionable uranium-235 isotope. Large quantities of uranium-238 are available as a by-product of the production of uranium-235. Uranium-238 can be converted to fissile plutonium-239 by bombardment with neutrons. Uranium-239 is formed initially, but it rapidly decays to neptunium-239. The neptunium quickly decays to plutonium. The reactions are shown in Figure 14.3(a).

A reactor can be built with a core of fissionable plutonium surrounded by uranium-238. As the plutonium fissions, neutrons convert the uranium-238 shield to more plutonium; thus, the reactor breeds more fuel than it consumes. There is enough uranium-238 to last several centuries, so one of the disadvantages of nuclear plants could be overcome by the use of **breeder reactors**.

Breeder reactors have some problems of their own, however. Plutonium is fairly low melting (640 °C), and the plant is limited to fairly cool—and inefficient—operation. Water is not adequate as a coolant; it is necessary to use molten sodium metal in the primary loop. These reactors often are called liquid-metal fast breeder reactors (LMFBR). If an accident occurred in such a breeder, the sodium could react violently with both the water and air.

Since plutonium is low melting, a failure of the cooling system could cause the reactor's core to melt. The radioactive fission products and the plutonium could be released into the environment, with devastating effects. All reactors are required to have an emergency backup core-cooling system. Whether these systems work or not is a principal area of controversy.

Plutonium is highly toxic. It emits alpha particles, a property that makes it especially dangerous when it is ingested. It is estimated that 1 microgram (μg) in the lungs of a human is enough to induce lung cancer. A further hazard is that reactor-grade plutonium, unlike the uranium used in nuclear reactors, could be readily converted into a nuclear bomb. The use of plutonium as a reactor fuel could easily lead to the further spread of nuclear weapons. There is even the possibility that terrorist groups could fashion a crude weapon from stolen plutonium. They then could threaten to destroy an entire city to force authorities to meet their demands.

Plutonium is produced for nuclear weapons. Some plutonium is formed in ordinary (nonbreeder) reactors. Plutonium has a half-life of about 25 000 years. Assuming that it will be essentially gone after 10 half-lives, we can estimate that our descendants would have to contend with plutonium for 250 000 years.

Another possible breeder reaction is that in which thorium-232 is converted to fissile uranium-233. It is thought that it would be rather difficult to make bombs from reactor-grade uranium-233. However, uranium-233 is like plutonium in that it emits damaging alpha particles. And it has an even longer half-life—160 000 years.

The breeder reactor may be an important power source for the future, but not before some substantial technical problems are overcome. To date, our experience with breeder reactors has not been too encouraging. One such plant near Detroit was closed permanently in 1972 after producing little electricity. The next breeder, scheduled to be built on the Clinch River in Tennessee, has been cancelled. European nations, however, are moving ahead rapidly with the development of these reactors.

Nuclear Fusion: The Sun in a Magnetic Bottle

In Chapter 4, we discussed the thermonuclear reactions that power the sun and that have been adapted by scientists and engineers to make hydrogen bombs. If we could find some way to control these fusion reactions and use them to produce electricity, we would have a nearly unlimited source of power. To date, the fusion reactions are useful only for making bombs, although research in the control of nuclear fusion is progressing (Figure 14.4). Controlled fusion would have several advantages over nuclear fission reactors. The principal fuel, deuterium (^2_1H), is plentiful and is obtained easily by the fractional electrolysis (splitting apart by means of electricity) of water. (Only 1 hydrogen atom in 5000 is a deuterium atom, but we have oceans full of water to work with.) The problem of radioactive wastes would be minimized. The end product—helium—is stable and biologically inert. Escape of tritium might be a problem, because this hydrogen isotope would be readily incorporated into organisms. Tritium (^3_1H) undergoes beta decay, with a half-life of 12.3 years. And there is one other problem associated with any production and use of energy: the unavoidable loss of part of the energy as heat. We would still have to be concerned with thermal pollution.

Great technical difficulties would have to be overcome before a controlled fusion reaction could be used to produce energy. Temperatures of 50 000 000 °C would have to be attained, and no material on Earth could withstand more than a few thousand degrees. At a temperature of 50 000 000 °C, no molecule could hold together, nor could the atoms from which the molecules is made. All atoms would be stripped of their electrons, and the nuclei and free electrons would form a mixture called a **plasma**. There is hope that this plasma could be contained by a strong magnetic field, however. Scientists in several

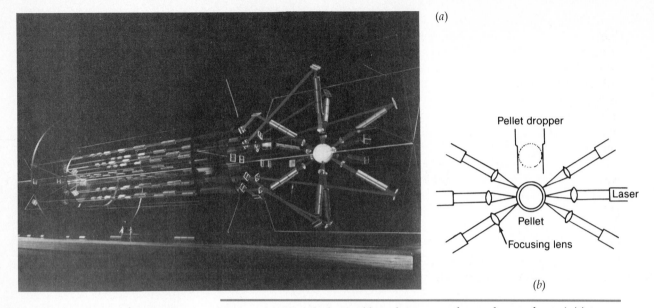

(a)

(b)

Figure 14.4 Sharply focused laser beams may be used some day to initiate nuclear fusion. (a) A model of the 10 000-J high-energy laser with which scientists hope to achieve fusion. (b) How laser energy may be used to compress and rapidly heat a small pellet containing deuterium and tritium; the great heat and pressure applied almost instantaneously cause the nuclei to fuse. [Photo courtesy of the University of California, Lawrence Livermore National Laboratory, and the U.S. Department of Energy.]

Figure 14.5 The sun is the ultimate source of nearly all our energy. [Courtesy of U.S. Department of Energy.]

nations are getting closer to the development of the environment necessary for controlled fusion.

Nuclear fusion may well be our best hope for relatively clean, abundant energy in the future. Much work remains to be done, however. Even when controlled fusion is achieved in the laboratory, it will still be a long time before it becomes a practical source of energy. A prototype power plant will have to be built and tested. If it were successful, commercial plants then could be constructed. Nevertheless, it is unlikely that we will get any significant energy from controlled fusion until well into the next century.

Harnessing the Sun: Solar Energy

At the beginning of Chapter 13, we saw that nearly all the energy on Earth comes from the sun. With all that energy from our celestial power plant, why do we face an energy crisis? The answer lies in the fact that this energy is thinly spread and difficult to concentrate.

Diffuse energy is not very useful. As it arrives on the surface of the Earth, about half of the solar energy is converted to heat. Another 30% is simply reflected back into space. We can increase the efficiency of this conversion rather easily. A black surface absorbs radiation better than a colored one. A simple solar collector can be made by covering a metal surface, painted black, with a glass plate. The glass is transparent to the incoming solar radiation, but it partially prevents the heat from escaping back into space. The hot surface is used to heat water or other fluids, and the hot fluids usually are stored in an insulated tank.

Water heated in this manner can be used directly for bathing, dish washing, and laundry. To heat a building, air can be warmed by being passed around the reservoir and then circulated through the building (Figure 14.6). Even in the cold northern climates, about 50% of home heating requirements could be met by solar collectors. These installations are expensive but could pay for themselves, through fuel savings, in a few years.

Figure 14.6 (a) "Solar One," an experimental house designed by the University of Delaware's Institute of Energy Conversion; the house obtains about 80% of its energy requirement from the sun. (b) A diagram of a solar collector that furnishes hot water and warm-air heat. [Photo courtesy of the University of Delaware.]

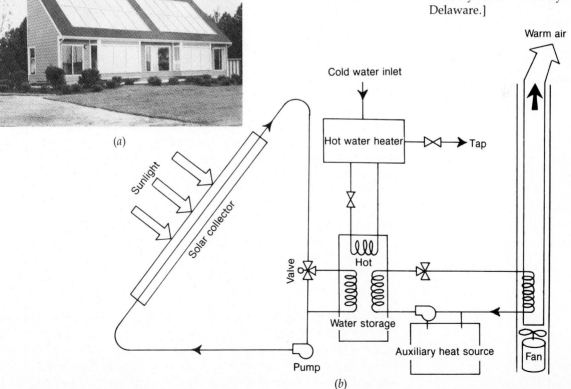

(a)

(b)

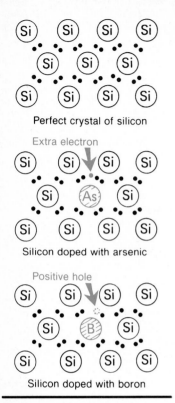

Perfect crystal of silicon

Extra electron

Silicon doped with arsenic

Positive hole

Silicon doped with boron

Figure 14.7 Models of silicon crystals. The crystals doped with impurities are used in solar cells.

Sunlight also can be converted directly to electricity by devices called **photovoltaic cells** or, more simply, **solar cells**. These devices can be made from a variety of substances, but most are made from elemental silicon (Color Plate L). In a crystal of pure silicon, each silicon atom has four valence electrons and is covalently bonded to four other silicon atoms (Figure 14.7). In the manufacture of a solar cell, extremely pure silicon is doped with small amounts of specific impurities and formed into single crystals. One type of crystal has about 1 part per million (ppm) of arsenic added. Arsenic atoms have five valence electrons, four of which are used to form bonds to silicon atoms. The fifth electron is relatively free to move around. This type of crystal, with extra electrons, is called a **donor crystal**. Another crystal is made by adding about 1 ppm of boron. Boron atoms have three valence electrons. These three electrons are used to bond silicon atoms, but there is a shortage of one electron, leaving a **positive hole** in the crystal. This boron-doped crystal is called an **acceptor crystal**.

When the two types of crystals (acceptors and donors) are joined, there is a strong tendency for electrons to flow from the donor to the acceptor. However, the holes near the junction are quickly filled by nearby mobile electrons, and the flow ceases. When sunlight strikes the cell, more electrons are dislodged, creating more mobile electrons and more positive holes. When the two crystals are connected by an external circuit, electrons flow from the donor to the acceptor (Figure 14.8).

An array of solar cells can be combined to form a solar battery. Such solar batteries can produce about 100 W per square meter of surface; it takes a square meter of cells to power one 100-W light bulb. Solar batteries have been used for years to power spaceships. They are now being used to provide electricity for weather instruments in remote areas. Prices are decreasing, and solar batteries are now widely used to power small devices such as electronic calculators (Figure 14.9).

Solar cells are not very efficient. Much of the sunlight striking them is reflected back into space. Their present efficiency of conversion is only about 10%. Generation of enough energy to meet a significant portion of our demands would require covering vast areas of desert land with solar cells. It would require 2000 hectares (ha), or about 5000 acres, of cloud-free desert land to produce as much energy as one nuclear power plant. Research is under way to increase the efficiency of solar cells. It is hoped that an efficiency of about 20% can be attained, thus reducing the land area required to 50% of that now needed.

Using solar energy would require the storage of energy for use at night and on cloudy days. One scheme for this involves storage of energy as heat. While the sun shines, energy would be transferred to tanks of molten salts. Then, heat from these salts would, as needed, be used to generate steam to run a turbine, which would generate electricity. Another possibility is using electricity generated during hours of sunshine to electrolyze water to hydrogen and oxygen.

$$2 \text{ H}_2\text{O} \xrightarrow{\text{electricity}} 2 \text{ H}_2 + \text{O}_2$$

The hydrogen could be transported in tanks or through pipelines (much as natural gas is today) to be used as fuel when and where it is needed. It also could be burned at night to generate electricity.

The technology is available now for the use of solar energy for space heating and for providing hot water. Electricity from solar energy is probably several years away, however, and more research and development are needed.

Biomass: Photosynthesis for Fuel

Why bother with solar collectors and photovoltaic cells to capture energy from the sun when green plants do it every day? Indeed, the idea of growing plants and burning them for energy has been explored. Dry plant material burns quite well. It could be used to fuel a power plant for the generation of electricity. We could start "energy plantations" to grow plants for use as fuel. Burning this **biomass** has some nice advantages: it is a **renewable resource** whose production is powered by the sun (Figure 14.10).

Unfortunately, there are a number of disadvantages to this scheme, too. Most of the available land is needed for the production of food. Even where productive land is available, there are problems. The plants have to be planted, harvested, and hauled to the power plant. Often, the land is far from where the energy is needed, and the overall efficiency is even less than that of solar cells—only about 3% at best.

Plant material does not have to be burned directly, however. Plants high in starches and sugars can be fermented to form ethyl alcohol. Wood can be distilled in the absence of air to produce methyl alcohol (Chapter 11). Both alcohols are liquids and convenient to transport. Both are also excellent fuels that burn relatively cleanly. Bacterial breakdown of plant material produces methane (Chapter 10). Under proper conditions, this process can be controlled to produce a clean-burning fuel similar to natural gas. It should be noted, however, that any of these conversions must result in the loss of a portion of the useful energy. The laws of thermodynamics tell us that we would get the most energy by burning the biomass directly rather than converting it to a more convenient liquid or gaseous fuel.

The stortage of land probably means that we will never obtain a major portion of our energy from biomass. We could, however, supplement our other sources by burning agricultural wastes, fermenting some to ethyl alcohol, producing methyl alcohol from wood where wood is plentiful, and fermenting human and animal wastes to produce methane. The technology for all these processes is readily available.

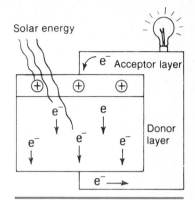

Figure 14.8 Schematic diagram of a solar cell. Electrons flow from the donor crystal to the acceptor crystal through the external circuit.

Figure 14.9 Solar cells are widely used to provide power for small electronic devices such as pocket calculators. [Photo by John Schultz.]

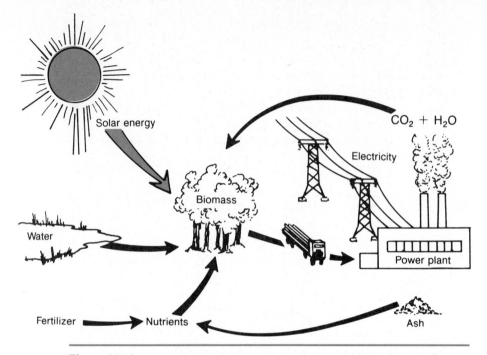

Figure 14.10 Energy from the sun can be converted to eletrical energy by way of green plants. If the ash from the burned plants was not returned to the soil, the soil would become depleted of nutrients and would have to be fertilized. Indeed, most of the land being considered for biomass production is too poor to be productive without massive applications of fertilizer.

Each has been used in the past. Several of the processes now are being used on a limited scale or are being investigated for use in the future.

We could also obtain energy by burning the combustible portion of garbage. This material—mostly paper—could be separated and burned as a supplement to coal for generating electricity. The noncombustible portion of garbage is mostly glass and metal; these can be readily recycled (Chapter 9).

Wind and Water: Once and Future Energy Sources

The sun, by heating the Earth, causes the winds to blow and the water to evaporate and rise in the air, later to fall as rain. The blowing wind and the flowing water can be used as sources of energy. Indeed,

Figure 14.11 The Grand Coulee Dam on the Columbia River in Washington State produces over 5000 MW of electrical energy. [Courtesy of the Bureau of Reclamation, U.S. Department of the Interior; photo by H. S. Holmes.]

they have been used that way for centuries. Water-wheels were used to lift water into the irrigation ditches of ancient Egypt. Windmills were used in tenth-century Persia to grind grain.

Why not use windpower and waterpower to solve the energy crisis? We do use waterpower. About 4% of our current energy production is hydroelectric, most of it in the mountainous western United States. In the modern hydroelectric plant, water is held behind huge dams. Some of this stored water is released against the blades of water turbines. The potential energy of the stored water is converted to the kinetic energy of flowing water. The moving water imparts mechanical energy to the turbine. The turbine drives a generator that converts mechanical energy to electrical energy.

Hydroelectric plants are relatively clean, but most of the best dam sites in the United States have already been used. To obtain more hydroelectric energy, we would have to dam up scenic rivers and flood valuable cropland and recreational areas. Reservoirs silt up over the years, and sometimes dams break, causing catastrophic floods. Even hydroelectric power has its problems.

We could use more windpower. The kinetic energy of moving air is

Figure 14.12 Five giant wind turbine generators near Byron, California, supply enough electricity for 175 homes. [Courtesy of Electric Power Research Institute, Palo Alto, CA.]

Figure 14.13 This wind-driven power unit once furnished 110 V of electricity to a cafe and general store and several surrounding buildings in Duxbury, Iowa. [Courtesy of the *St. Paul Dispatch and Pioneer Press,* St. Paul, MN.]

readily converted to mechanical energy to pump water and grind grain. Windmills have been so used for centuries. Wind also can be used to turn turbines and generate electricity. Giant windmills have been built to test and develop this potential. Smaller units are available for use on farms and for rural dwellings.

Properly developed, windpower could supply about 10% of our energy needs. Wind is clean, free, and abundant. However, one serious drawback is that the wind does not always blow. Some means of energy storage or an alternate source of energy must be available. Land use might become a problem if windpower was used widely. Presumably, land under windmills could be used for farming or grazing, but giant complexes with powerlines would be ugly.

Moon Power: The Tides

The power of the tides is another potential source of energy. One tide-powered generating plant is now in operation in France. Although the tides produce great amounts of energy, there are few appropriate sites available in the United States. Some sites that could be used, such as Passamaquoddy Bay, are prized for their beauty; any attempt to construct a power plant would be strongly opposed by environmentalists (Figure 14.14). Another drawback is that electricity could be generated

Figure 14.14 The tide comes in on a rocky coast. Are we willing to give up the view for a tide-powered electricity-generating plant? [Courtesy of John Schultz/PAR-NYC.]

(a)

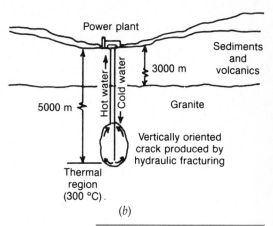

(b)

Figure 14.15 Geothermal energy has the potential of contributing to our energy supply. (a) A commercial geothermal plant at the Geysers in Sonoma County, California. (b) A scheme for extracting energy from dry hot rock 5 km below the surface of the Earth. [Photo courtesy of the U.S. Department of the Interior.]

only when the tide is coming in or going out. There would be breaks in the production of electricity and a system of energy storage would be required.

Earth Power: Geothermal Energy

The interior of the Earth is heated by immense gravitational forces and by natural radioactivity. This heat comes to the surface in some areas through geysers and volcanoes. Geothermal energy has long been used in Iceland, New Zealand, and Italy. It has some potential in the United States—indeed, it is being used now in California (Figure 14.15)—but in the near future this potential could only be realized in areas where steam or hot water is at or near the surface. One drawback of geothermal energy is that the wastewater is quite salty; its disposal could be a major problem.

Oil Shale and Tar Sands: High-Entropy Oil

There are fossil carbon compounds other than coal, petroleum, and natural gas. For example, there are immense reserves of oil shale in Colorado, Utah, and Wyoming and of tar sands in Alberta, Canada. The problem with both resources is that their extraction is difficult and environmental disruption is severe.

Actually, **oil shale** contains no oil. The organic matter is present as a complex material called **kerogen**, which has an approximate composition of $(C_6H_8O)_n$, where n is a large number. When heated in the absence of air, kerogen breaks down, forming hydrocarbon oils similar to petroleum. The problem is that most of the kerogen is thinly distributed through rock. Although it does not seem to be covalently bonded to the rock, it can't be pumped out in the way petroleum can be pumped from the porous rock in which it occurs. It takes a lot of energy to get oil from oil shale, and there is a lot of waste rock left to dispose of. The energy content of the shale oil—by the time it is refined into gasoline and fuel oil—is not a whole lot greater than the energy put into producing it.

Tar sands are quite different from oil shale, but the problems associated with extracting fuel from them are similar. The organic matter of **tar sands** is present as **bitumen**, a hydrocarbon mixture. Separating bitumen from sand is difficult and requires a lot of energy. As in oil shale, the organic matter is thinly distributed in a lot of inorganic waste.

It may be that we will someday get a significant portion of our energy from oil shale and tar sands. It will be expensive energy, though. The thinly distributed, intimately mixed organic matter is high-entropy stuff. To get it into a useful, low-entropy (purified) form will require a lot of energy. The return on the energy invested will probably be rather low.

Coal Gasification and Liquefaction: Convenience and Waste

Quite often, in discussions of future energy supplies, we hear about converting coal to gas or oil. We are running short of gas and petroleum. Why not make them out of coal? The technology has been around for years. Why not get on with it? Gasification and liquefaction do have some advantages: gases and liquids are easy to transport the process of conversion leaves the sulfur behind, thus overcoming one of the serious disadvantages of coal as a fuel.

There are a variety of experimental projects under way in which coal is being converted to a synthetic gaseous fuel. The basic process is one of reduction of carbon by hydrogen.

$$C + 2H_2 \longrightarrow CH_4$$

The hydrogen can be produced by passing steam over hot charcoal.

$$C + H_2O \longrightarrow CO + H_2$$

Coal also can be reacted with hydrogen to form petroleumlike liquids from which gasoline and fuel oil can be made.

Coal can also be converted to methanol, which can be used directly as fuel or converted, in turn, to gasolinelike hydrocarbons. In the first step, the coal reacts with extremely hot steam to form carbon monoxide and hydrogen.

$$C + H_2O \longrightarrow CO + H_2$$

Part of the carbon monoxide is then reacted with water to form more hydrogen.

$$CO + H_2O \longrightarrow CO_2 + H_2$$

Then carbon monoxide is combined with hydrogen to form methanol.

$$CO + 2 H_2 \longrightarrow CH_3OH$$

Finally, the methanol is converted to a hydrocarbon mixture that we represent as C_nH_m.

$$n\ CH_3OH \longrightarrow C_nH_m + H_2O$$

This process was invented by Mobil Oil Corporation. Each step requires an appropriate catalyst.

But both gasification and liquefaction of coal require a lot of energy. The process is inherently wasteful: up to one third of the energy content of the coal is lost in the conversion. Liquid fuels from coal are high in unsaturated hydrocarbons and in sulfur, nitrogen, and arsenic compounds. Their combustion products are high in particulate matter. Coal conversions also require large amounts of water, yet the large coal deposits on which they would be based are in arid regions. Further, the conversions are messy. Without stringent safeguards, the plants would seriously pollute both air and water. To increase our supply of more convenient fossil fuels—oil and gas—we would deplete our third fossil fuel—coal—even more rapidly.

Fuel Cells

When fuels are burned, chemical energy is converted to heat. The heat can be used to produce steam to turn a turbine that powers a generator that produces electricity. A **fuel cell** is a device in which chemical reactions are used to produce electricity directly. These fuel cells differ from the usual electrochemical cell (Chapter 8) in two ways. First, the fuel and oxygen are fed in continously. Second, the electrodes

Figure 14.16 A hydrogen–oxygen fuel cell.

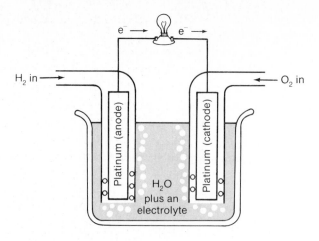

Figure 14.16 A hydrogen–oxygen fuel cell.

are made of an inert material such as platinum that does not react during the process. The electrodes serve to conduct the electrons to and from the solution.

Consider a fuel cell using hydrogen and oxygen (Figure 14.16). Hydrogen is oxidized at the anode; oxygen is reduced at the cathode. The overall reaction is similar to combustion, but not all the chemical energy is converted to heat.

$$2 \ H_2 \ + \ O_2 \ \longrightarrow \ 2 \ H_2O$$

Rather, about 40 to 55% of the chemical energy is converted to electricity.

Fuel cells are used on space craft to produce electricity. The water produced can be used for drinking, so less water need be taken aboard. Fuel cells overall have a weight advantage over storage batteries, an important consideration when launching a spaceship.

On Earth, more research is needed to reduce cost and design long lasting cells. Perhaps someday fuel cells will provide electricity to meet peak needs in large power plant. Unlike huge boilers and nuclear reactors, they can be started and stopped simply by turning the fuel on or off.

Energy: What For?

The United States, with only 6% of the world's population, uses about one third of all the energy presently being generated. What for? Abundant fuel has enabled the United States to build its industrial base and provide its people with one of the highest standards of living in the world.

Industry uses about 25% of all the energy produced in the United States. This energy is used to convert raw materials into the myriad of products our society demands. Utilities use about 34% of the nation's energy production, primarily to generate electricity. Transportation uses another 27%, mainly to power private automobiles. Private homes use about 14%.

Energy lights our homes, heats and cools our living and working spaces, and makes us the most mobile society in the history of the human race. It powers the factories that provide us with abundant material goods. Indeed, energy is the basis of our modern society.

Energy: How Much Is Too Much?

Although there are many problems, science and technology probably will be able to provide us with a plentiful supply of energy for the foreseeable future. This energy will not—indeed, cannot—be pollution free. Any production and use of energy will be accompanied by pollution.

How do we choose the best method of energy production? The task is certainly difficult. The choice should be made by informed citizens who have examined the process from beginning to end. We must know what is involved in the construction of power plants, the production of fuels, and the ultimate use of energy in our homes and factories. We must know that energy is wasted (as heat) at every step in the process. Plentiful power involves substantial thermal pollution in areas where power is generated, for we cannot escape the long arm of the second law of thermodynamics.

Will our profligate consumption of energy affect the Earth's climate? Our activities have already modified the climate in and around major metropolitan areas. The worldwide effects of our expanding energy consumption are harder to estimate.

What can we do as individuals? We can conserve. We can walk more and use cars less. We can reduce our wasteful use of electricity. We can buy more efficient appliances, and we can avoid purchasing energy-intensive products. Indeed, several European nations have higher standards of living than the United States, yet use less energy per capita. Let's learn from them so that we can save energy and better our lives at the same time.

Problems

1. What percentage of United States electricity is generated by coal-burning power plants?

2. Why do we prefer electricity as a form of energy?

3. How does a coal-fired power plant work?

4. What proportion of the chemical energy in coal is converted to electrical energy in a coal-fired plant? What happens to the rest of that chemical energy?

5. What proportion of United States electricity is generated by nuclear power plants?

6. What proportion of the electricity in France is generated by nuclear power plants?

7. What regions of the United States are especially dependent on nuclear power for electricity?

8. How does a boiling water (nuclear) reactor work?

9. How does a pressurized water reactor differ from a boiling water reactor?

10. How does the fuel used by the CANDU reactor differ from that used in pressurized water reactors?

11. Can a nuclear reactor explode like a nuclear bomb? Explain your answer.

12. How does a breeder reactor produce more fuel than it consumes?

13. Write nuclear equations to show how thorium-232 is converted to fissile uranium-233.

14. Write nuclear equations to show how uranium-238 is converted to fissile plutonium-239.

15. What is the function of a moderator in a nuclear reactor?

16. What is the function of control rods in a nuclear reactor?

17. Can nuclear bombs be made from reactor-grade uranium? Explain your answer.

18. List some advantages of nuclear power plants over coal-fired plants.

19. What are some of the problems with the tailings from uranium ore processing?

20. List some disadvantages of using nuclear power.

21. What are some of the disadvantages of breeder reactors?

22. Can a nuclear bomb be made from reactor-grade plutonium?

23. List some possible advantages of a nuclear fusion reactor over a fission reactor.

24. List some possible problems with nuclear fusion reactors.

25. What is plasma?

26. What are some problems associated with the use of solar energy?

27. What is a photovoltaic cell?

28. Define or identify each of the following in relation to solar cells.
    a. acceptor crystal   b. donor crystal
    c. positive hole

29. Describe two ways of storing solar energy for use at night.

30. What is biomass?

31. List some advantages and disadvantages of the use of biomass as a source of energy.

32. Describe how each of the following can be made from biomass.
    a. ethyl alcohol   b. methyl alcohol
    c. methane

34. List two ways that fuel cells differ from electrochemical cells.

35. What are the advantages of fuel cells over storage batteries for space flight?

36. List the advantages, disadvantages, and limitations of each of the following as an energy source.
    a. windpower   b. geothermal power
    c. power from the tides   d. hydroelectric power

37. What is oil shale?

38. List some advantages and disadvantages of obtaining energy from oil shale.

39. What are tar sands?

40. List some advantages and disadvantages of obtaining energy from tar sands.

41. What are some of the advantages and disadvantages of coal gasification?

42. What are some of the advantages and disadvantages of coal liquefaction?

43. Describe the Mobil process for making gasoline from coal.

44. Human and animal wastes can be fermented to produce methane gas. Can you anticipate any problems that are likely to arise from this process?

45. Silicon for electronic devices can be doped by a pro-

cess called neutron transmutation doping. Pure silicon is composed of three isotopes with mass numbers of 28, 29, and 30. Silicon-30, which makes up about 3% of all silicon atoms, can be converted to phosphorus-31 by capturing a neutron followed by radioactive decay. What isotope is formed when ^{30}Si captures a neutron? What particle is given off when that isotope undergoes radioactive decay to form ^{31}P? Write balanced nuclear equations for both processes.

46. Some cities burn rubbish to produce steam for the generation of electricity. Find out if your local government has considered using this source of energy.

47. Your school probably produces a lot of wastepaper. Could this paper be burned to heat the buildings in winter? What problems might be involved?

48. Compare a nuclear power plant with a coal-burning plant. Which would you rather have in your neighborhood? Why?

49. Do you consider thermal pollution a local problem or a problem for the Earth as a whole?

50. Who should decide if and when energy should be rationed?

51. Your children will not have abundant natural gas and petroleum during the next century. What energy source would you choose for them? Why?

52. Which of the following is the best fuel for heating your home?
a. natural gas   b. electricity
c. coal   d. fuel oil
What problems with supply, use, and waste products are involved with each fuel?

53. Where does the electrical energy you use come from? What pollution problems are associated with its generation?

References and Readings

1. Beane, Marjorie. *A Nuclear Waste Primer.* Washington, DC: League of Women Voters, 1980.
2. Beane, Marjorie. *A Nuclear Power Primer.* Washington, DC: League of Women Voters, 1982.
3. Bodansky, David. "Electricity Generation Choices for the Near Term." *Science,* 15 February 1980, pp. 721–727.
4. Bozak, Richard E., and Manuel Garcia, Jr. "Chemistry in the Oil Shales." *Journal of Chemical Education,* March 1976, pp. 154–155.
5. Colombo, Umberto. "Alternative Energy Futures: The Case for Electricity." *Science,* 20 August 1982, pp. 705–709.
6. Commoner, Barry. *The Poverty of Power: Energy and the Economic Crisis,* New York: Knopf, 1976.
7. Electric Power Research Institute. *Electricity.* Los Altos, CA: William Kaufmann, 1982.
8. Gold, Michael. "To Breed or Not to Breed." *Science 82,* May 1982, pp. 34–42.
9. Greenwald, John. "Deadly Meltdown." *Time,* 12 May 1986, pp. 38–52.
10. Kidder, Tracy. "Taming a Star." *Science 82,* March 1982, pp. 55–64.
11. Klass, Donald L. "Renewable Energy." *ChemTech,* October 1984, pp. 610–615.
12. Lihach, Nadine. "Fuel Cells for the 90s." *EPRI Journal,* September 1984, pp. 6–13.
13. "New York City to Test Methanol-Fueled Buses." *Chemical and Engineering News,* 30 June 1986, p. 6.
14. Olson, Steve. "Nuclear Undertakers." *Science 84,* September 1984, pp. 50–59.
15. Owsley, Dennis C., and Jordan J. Bloomfield. "Energy Facts: A Basis for Decision." *ChemTech,* February 1985, pp. 94–98.
16. Rose, Albert. "Solar Energy: A Global View." *ChemTech,* September 1981, pp. 566–571.
17. Skinner, Brian J., and Charles A. Walker. "Radioactive Wastes." *American Scientist,* March–April 1982, pp. 180–181.
18. Stahlkopf, Karl, Robert Williams, and A. B. Carson. "Geologic Disposal of Nuclear Waste." *EPRI Journal,* May 1982, pp. 6–13.
19. Whitaker, Ralph. "New Promise for Photovoltaics." *EPRI Journal,* July–August 1983, pp. 6–15.

15

Air

Breath of Life ... or Death?

We passengers on Spaceship Earth live under a thin blanket of air called the atmosphere. Although other planets in our solar system have atmospheres, the Earth's atmosphere appears to be unique in its ability to support life (Chapter 9). The atmosphere is composed of about 5.2×10^{15} t of air spread over a surface area of 5.0×10^8 km^2.

It is difficult to measure just how deep the atmosphere is. It does not end abruptly, but gradually fades as the distance from the surface of the Earth increases. It is known, though, that 99% of the atmosphere lies within 30 km of the surface of the Earth—a thin layer of air indeed (like the peel of an apple, only relatively thinner).

Without food, we could live about a month. Without water, we could live a few days. But, without air, we would die within minutes. We may run short of food, and we may run short of clean, fresh water, but we aren't likely to run out of air. What we may do is foul the air so that we become sick from it and some of us die prematurely. In some localities, the air eventually may get so bad that a lot of people will die.

It won't happen, you say? Certainly, we all hope it won't. But we better do more than hope. Read on!

The Atmosphere: Divisions and Composition

The atmosphere is divided into layers (Figure 15.1). The layer nearest Earth, the **troposphere**, harbors nearly all living things and nearly all human activity. The next region, the **stratosphere**, is where we find the **ozone layer** that shields living creatures from deadly ultraviolet radiation. Most of this chapter focuses on the troposphere, but we also examine some threats to the ozone layer in the stratosphere. We will not be concerned with higher regions of the atmosphere.

Air is a mixture of gases. Dry air is (by volume) about 78% nitrogen (N_2), 21% oxygen (O_2), and 1% argon (Ar). Water vapor varies from 0% up to about 4%. There are a number of minor constituents, the most important of which is carbon dioxide (CO_2). The concentration of carbon dioxide in the atmosphere is believed to have increased from 296 ppm in 1900 to its present value of 345 ppm. It most likely will continue to rise as we burn more and more fossil fuels (coal, oil, and gas). The composition of the atmosphere is summarized in Table 15.1.

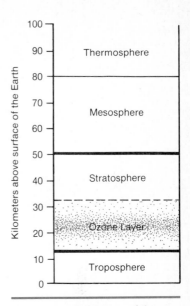

Figure 15.1 Divisions of the atmosphere and approximate location of the ozone layer (color).

The Nitrogen Cycle

Although nitrogen makes up 78% of the atmosphere, the N_2 molecues can't be used directly by higher plants or by animals. They first have to be **fixed**—that is, combined with another element.

Certain types of bacteria convert atmospheric nitrogen (N_2) to nitrates. Other bacteria convert the nitrogen in compounds back to N_2. Thus, a nitrogen cycle (Figure 15.2) is established. Lightning also serves

Table 15.1   Composition of the Earth's Atmosphere (Dry): Molecules of Each Substance per 10 000 Molecules

Substance	Formula	Number of Molecules
Nitrogen	N_2	7800
Oxygen	O_2	2100
Argon	Ar	93
Carbon dioxide	CO_2	3
All others	—	4
Total		10 000

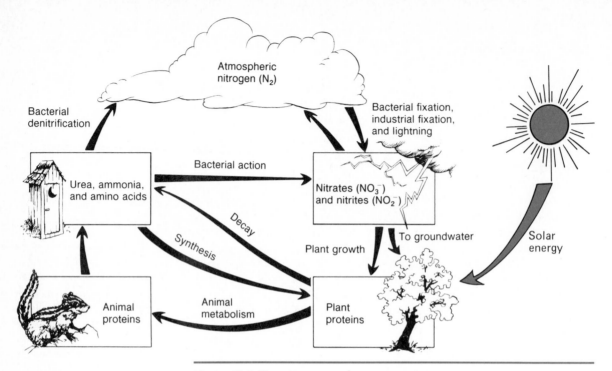

Figure 15.2 The nitrogen cycle.

to fix nitrogen by causing it to combine with oxygen. Nitric oxide (NO) and nitrogen dioxide (NO_2) are formed. The equations are

$$N_2 + O_2 \xrightarrow{\text{lightning}} 2\,NO$$

$$2\,NO + O_2 \longrightarrow 2\,NO_2$$

Nitrogen dioxide reacts with water to form nitric acid (HNO_3).

$$3\,NO_2 + H_2O \longrightarrow 2\,HNO_3 + NO$$

The nitric acid falls in rainwater, adding to the supply of available nitrates in the oceans and the soil.

Humans have undertaken substantial intervention in the nitrogen cycle by fixing nitrogen industrially in the manufacture of nitrogen fertilizers. This intervention has greatly increased our food supply, since the availability of fixed nitrogen is often the limiting factor in the production of food (Chapter 17). Not all the consequences of this intervention have been favorable, however; excessive runoff of nitrogen fertilizer has led to serious water-pollution problems in some areas (Chapter 16).

The Oxygen Cycle

The element oxygen makes up 21% of the Earth's atmosphere. The oxygen supply is constantly being replenished by green plants, including one-celled organisms, called phytoplankton, in the sea. Oxygen is used by both plants and animals in the metabolism of foods. It also is used in the decay and the combustion of plant and animal materials. Some oxygen is consumed in the rusting of metals and in the weathering of rocks. A simplified oxygen cycle is illustrated in Figure 15.3.

In the stratosphere, some oxygen is formed by the action of ultraviolet rays on water. Perhaps more important, some oxygen is converted to ozone.

$$3 \ O_2 \xrightarrow{\text{ultraviolet rays}} 2 \ O_3$$

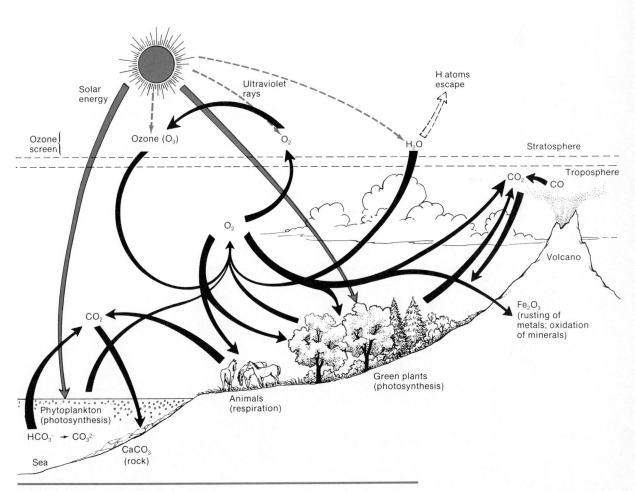

Figure 15.3 The oxygen cycle.

The ozone, in turn, absorbs short-wavelength ultraviolet radiation that would otherwise make any sort of life on Earth close to impossible.

We discuss the atmosphere in more detail in subsequent sections, paying particular attention to changes wrought by human activities.

Doomsday in Donora

The scene is a small industrial town at the bottom of a river valley. The time is late October. A dense smog settles over the valley. The air is acrid with sulfur dioxide. Dust (probably fly ash) settles over the land in a layer so thick that footprints and tire tracks are visible in it. A Halloween parade passes through the streets, the marchers holding handkerchiefs over their faces in an attempt to keep the smog out of their lungs. People become seriously ill. Before the smog lifts, 5 days later, 17 people are dead. Four more, who became ill in October, die by Christmas Eve. Nearly half the population is made ill by smog. One in 10 is seriously ill.

What is this? A scene from a futuristic, nightmarish science-fiction story? No! It is history. The scene actually took place in Donora, Pennsylvania, in 1948. A zinc-manufacturing plant and a huge iron and steel mill contributed to the smog. So did trains, automobiles, and even barges on the Monongahela River. An **atmospheric inversion**—a still, warm upper layer of air over a cool lower layer—kept the smog socked in on the valley (Figure 15.4).

Is Donora a preview of what is in store for us all? Doomsday, U.S.A.? Before we attempt to prophesy, let's look at how air is polluted.

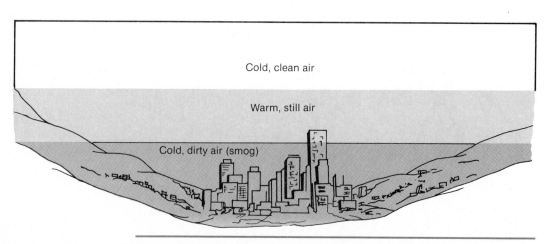

Figure 15.4 An atmospheric inversion traps smog in a valley.

Natural Pollution

Even before there were people, there were cases of air pollution. Volcanoes erupted, spewing ash and poisonous gases into the atmosphere. They still do so. The eruption of Mount St. Helens in 1980 (Figure 15.5) ejected ash that blanketed areas over several states. It spewed out 1000 t of sulfur dioxide per day and is still emitting 160 t per day. El Chichon in Mexico in 1982 ejected 3.3 million t of sulfur dioxide into the stratosphere where it was converted to sulfuric acid mist. Kilauea in Hawaii emits 200 to 300 t of sulfur dioxide per day. Acid rain downwind from this volcano has created a barren region called the Kau desert.

Dust storms, especially in arid regions, add massive amounts of particulate matter to the atmosphere. Dust from the Sahara often reaches the Caribbean and South America. Swamps and marshes emit noxious gases. Nature isn't always benign.

The Air Our Ancestors Breathed

When people appeared on the scene and began their conquest of nature, they also began to pollute the air. They built fires that filled the air with smoke. They cleared land; this made possible even larger dust storms. They built cities, and the soot from their hearths and the stench from their wastes filled the air. The Roman author Seneca wrote in A.D. 61 of the "stink," soot, and "heavy air" of the imperial city. In 1257, the queen of England was forced to move away from the city of Nottingham because the heavy smoke was unendurable. The industrial revolution brought even worse air pollution. Coal was burned to power the factories and to heat the homes. Soot, smoke, and sulfur dioxide filled the air. The good old days? Not in the factory towns. But there were large rural areas relatively unaffected by air pollution.

Pollution Goes Global

With increasing population, the entire world is becoming more urban. It is the huge megalopolises that are most afflicted by air pollution. But rural areas are not unaffected. In the neighborhoods around smoky factories, there is evidence of increased rates of spontaneous abortion and of poor wool quality in sheep, decreased egg production and high mortality in chickens, and increased feed and care required for cattle. Plants are stunted, deformed, and even killed. The giant Ponderosa pines are dying over a hundred miles from the

Figure 15.5 An eruption of Mount St. Helens on 18 May 1980 ejected at least 1 km^3 of material as ash. [Courtesy of the U.S. Geological Survey, Department of the Interior.]

smog-plagued Los Angeles basin. Orbiting astronauts visually traced drifting blobs of Los Angeles smog as far east as western Colorado. Other astronauts, over 100 km up, were able to see the plume of smoke from the Four Corners power plant near Farmington, New Mexico. This was the only evidence they could see from that distance that Earth is inhabited!

Snow in Norway is contaminated by pollutants from England and Germany. Traffic police in Tokyo have had to wear gas masks and take "oxygen breaks"—breathing occasionally from tanks of oxygen. Smog in Athens at times has forced factory closings and traffic restrictions. Acid rain in Canada is spawned by air pollution originating in the United States, contributing to strained relationships between the two countries. Buenos Aires, Sydney, Rome, Tehran, Ankara, Mexico City, and most other cities of the world have had frightening episodes of air pollution.

The main difference between air-pollution problems today and problems in the past is that it is difficult to get away from air pollution now. The problem is worldwide. The air does purify itself, but we are pouring more pollutants into it than it can handle readily.

What is a **pollutant**? It is merely a chemical in the wrong place in the wrong concentration. For example, ozone is a natural and important constituent of the stratosphere, where it shields the Earth from life-destroying ultraviolet radiation. In the troposphere, however, ozone is a dangerous pollutant.

Coal + Fire → London Smog

There are two basic types of smog. One type—consisting mainly of smoke, fog, sulfur dioxide, sulfuric acid, ash, and soot—is called **London smog** (Color Plate M). Indeed, the word **smog** is thought to have originated in England in 1905 as a contraction of the words *smoke* and *fog*.

Probably the most notorious case of smog in history started in London on Thursday, 4 December 1952. A large cold-air mass moved into the valley of the Thames River. A temperature inversion placed a blanket of warm air over the cold air. With nightfall, a dense fog and below-freezing temperatures caused the people of London to heap coal into their small stoves. Millions of these fires burned through the night, pouring sulfur dioxide and smoke into the air. The next day (Friday), the people continued to burn coal when the temperature remained below freezing. The factories added their smoke and chemical fumes to the atmosphere.

Saturday was a day of darkness. For 20 miles around London, no light came through the smog. The air was cold and still. And the coal

Figure 15.6 London smog is often highly visible, with soot and fly ash forming a dark pall of smoke. [Courtesy of Henry H. Valiukas, Minnesota Environmental Control Citizens Association, St. Paul.]

fires continued to burn throughout the weekend. On Monday, 8 December, more than 100 people died of heart attacks while desperately trying to breathe. People tried to sleep sitting up in chairs in order to breathe a little easier. The city's hospitals were overflowing with people with respiratory diseases.

By the time a breeze cleared the air on Tuesday, 9 December, more than 4000 deaths had been attributed to the smog. Other people afflicted at the time died later. The total death count has been estimated at 8000. That is more people than were ever killed in any single tornado, mine disaster, shipwreck, or airplane crash. That is more people than were killed in the Japanese attack on Pearl Harbor in 1941. Air-pollution episodes may not be as dramatic as other disasters, but they can be just as deadly. (The smog at Donora was essentially London smog, with a few added ingredients—chlorides, fluorides, arsenic, lead, and zinc.)

The Chemistry of London Smog

The chemistry of London smog is fairly simple. Coal is mainly carbon, but it contains as much as 3% sulfur. Coal also contains varying amounts of mineral matter. When burned, the carbon in coal is oxidized to carbon dioxide.

$$C + O_2 \longrightarrow CO_2$$

Heat is given off in this process. Not all the carbon is completely oxidized. Some of it winds up as carbon monoxide.

$$2\,C + O_2 \longrightarrow 2\,CO$$

Still other carbon, essentially unburned, ends up as soot.

The sulfur in coal also burns, forming sulfur dioxide, a choking, acrid gas.

$$S + O_2 \longrightarrow SO_2$$

Sulfur dioxide is readily absorbed in the respiratory system. It is a powerful irritant and is known to aggravate the symptoms of people who suffer from asthma, bronchitis, emphysema, and other lung diseases.

As if sulfur dioxide were not bad enough, things get worse. Some of the sulfur dioxide reacts further with oxygen in the air to form sulfur trioxide.

$$2\,SO_2 + O_2 \longrightarrow 2\,SO_3$$

The sulfur trioxide then reacts with water to form sulfuric acid.

$$SO_3 + H_2O \longrightarrow H_2SO_4$$

Sulfuric acid is even more irritating to the respiratory tract than sulfur dioxide.

Usually, London smog also is characterized by high levels of **particulate matter**, solid and liquid particles of greater than molecular size. The large particles often are visible in the air as dust and smoke. Smaller particles of 1 μm or less in diameter—called **aerosols**—are often invisible.

Particulate matter consists in part of soot (unburned carbon). A larger portion is made up of the mineral matter than occurs in coal. These minerals do not burn. Some are left behind as **clinkers**. In the roaring fire of a huge factory or power plant, however, much of this solid material is carried aloft in the tremendous draft. This **fly ash** settles over the surrounding area, covering everything with dust. It is also inhaled, thus contributing to respiratory problems in animals and humans.

Perhaps a more insidious form of particulate matter is the sulfates. Some of the sulfuric acid in smog reacts with ammonia to form a solid material, ammonium sulfate.

$$2\,NH_3 \ + \ H_2SO_4 \ \longrightarrow \ (NH_4)_2SO_4$$

Sulfuric acid, in the form of minute liquid droplets, and the solid ammonium sulfate are easily trapped in the lungs. The harmful effect of these pollutants may be considerably magnified by their interaction. A certain level of sulfur dioxide, without the presence of particulate matter, might be reasonably safe. A certain level of particulate matter, without sulfur dioxide around, might be fairly harmless. But take the same levels of the two together, and the effect might well be deadly. Synergistic effects such as this are quite common whenever chemicals get together (Chapters 9 and 23).

When the pollutants in London smog come into contact with the alveoli of the lungs, the cells are broken down. The alveoli lose their resilience, making it difficult for them to expel carbon dioxide. Such lung damage leads to—or at least contributes to—pulmonary emphysema, a condition characterized by an increasing shortness of breath. Emphysema is the fastest growing cause of death in the United States. The principal factor in the rise of emphysema is cigarette smoking. However, air pollution is known to be a factor, too. For instance, the incidence of the disease among smokers is three times as great in St. Louis, where air pollution is rather heavy, as in Winnipeg, Manitoba, where air pollution is rather mild.

The oxides of sulfur and the aerosol mists of sulfuric acid are damaging to plants. Leaves become bleached and splotchy when exposed to sulfur oxides. The yield and quality of farm crops can be severely affected. These compounds are also major ingredients for the production of acid rain (page 332).

(*a*) (*b*)

Figure 15.7 Northern States Power Company's High Bridge power plant in St. Paul, Minnesota. (a) Before the installation of an electrostatic precipitator. (b) After the installation. [Courtesy of Northern States Power Company.]

What to Do About London Smog

Much research has gone into the prevention and alleviation of London smog. Soot and smog can be removed from smokestack gases in several ways. One method uses electrostatic precipitators (Figure 15.7). These devices induce electrical charges on the particles, which are then attracted to oppositely charged plates and deposited (Figure 15.8).

Another method uses bag filtration, a system that works much like the bag in a vacuum cleaner. The filters are placed in a bag house (Figure 15.9), arranged in a way that allows the filters to be shaken and air blown through periodically in the opposite direction to clean them.

A third device, called a cyclone separator, is arranged so that the stack gases spiral upward with a circular motion. The particles hit the outer walls, settle out, and are collected at the bottom.

Wet scrubbers are also used. These devices remove the particles by passing the stack gases through water. The water is usually sprayed in as a fine mist. The wastewater has to be treated to remove the particulates, and that adds to the cost of this method.

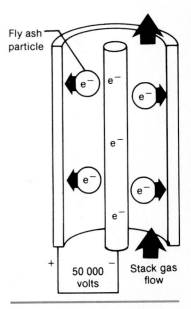

Figure 15.8 Cross section of a cylindrical electrostatic precipitator. Electrons from the negatively charged discharge electrode (in the center), attach themselves to the particles of fly ash, and give them a negative charge. The charged particles then are attracted to the positively charged collector plate, on which they are deposited.

The exact device used depends on the type of coal being burned, the size of the power plant, and other factors. All require energy—electrostatic precipitators use 10% of the plant's output—and the collected ash has to be put somewhere. Ash production in the United States is about 70 million t per year. About 16 million t of this is used. Some replaces a part of the clay in making cement (Chapter 9), and some is melted and blown in air to make mineral wool for insulation. The rest has to be stored; 70% of it goes into ponds and the rest into landfills.

It is harder to remove sulfur dioxide than particulates. Sulfur can be removed from the coal before burning, but both the flotation method (Chapter 13) and gasification or liquefaction processes (Chapter 14) are expensive. Another way to get rid of sulfur is to scrub sulfur dioxide out of stack gases after the coal has been burned. Perhaps the most promising of the scrubbers is the limestone–dolomite process. Limestone ($CaCO_3$) and dolomite (a mixed calcium–magnesium carbonate) are pulverized and heated. Heat drives off carbon dioxide to form calcium oxide (lime).

$$CaCO_3 + heat \longrightarrow CaO + CO_2$$

This basic oxide reacts with sulfur dioxide and oxygen to form solid calcium sulfate.

$$2\,CaO + 2\,SO_2 + O_2 \longrightarrow 2\,CaSO_4$$

This by-product presents a sizable disposal problem. Removal of 1 t of sulfur dioxide produces 2 t of solids.

Sulfur dioxide can also be reacted with hydrogen sulfide and recovered as elemental sulfur.

$$2\,H_2S + SO_2 \longrightarrow 3\,S + 2\,H_2O$$

Sale of the sulfur could partially offset the cost of the process. Great technical problems would have to be overcome, though, before this method could be put into widespread use.

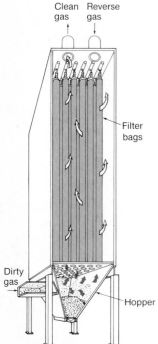
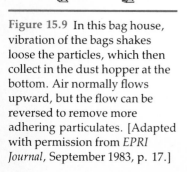

Clean gas   Reverse gas

Filter bags

Dirty gas

Hopper

Figure 15.9 In this bag house, vibration of the bags shakes loose the particles, which then collect in the dust hopper at the bottom. Air normally flows upward, but the flow can be reversed to remove more adhering particulates. [Adapted with permission from *EPRI Journal*, September 1983, p. 17.]

Los Angeles Smog

The other main type of smog is **Los Angeles smog**, or, more properly, **photochemical smog**. Unlike London smog, which accompanies cold, damp air, photochemical smog usually occurs during dry, sunny weather. The principal culprits are unburned hydrocarbons and nitrogen oxides from automobiles. The warm, sunny climate that has

Figure 15.10 Downtown Los Angeles on a smoggy day in 1956. Smog is trapped by a temperature inversion, with a layer of warm air only 100 m above the ground. The upper portion of Los Angeles City Hall can be seen in the clear air above the base of the inversion. [Courtesy of the South Coast Air Quality Management District, Los Angeles, CA.]

drawn so many people to the Los Angeles area is also the perfect setting for photochemical smog.

The chemistry of Los Angeles smog is exceedingly complex. Let's look at the stuff that comes out of an automobile's exhaust pipe and examine the pollutants one at a time.

Carbon Monoxide: The Quiet Killer

When a hydrocarbon burns in sufficient oxygen, the products are carbon dioxide and water. Since both of these substances are normal constituents of air, they are not generally considered to be pollutants. Let's illustrate the combustion process with an octane, one of the hundreds of hydrocarbons that make up the mixture we call gasoline.

$$2\ C_8H_{18}\ +\ 25\ O_2\ \longrightarrow\ 18\ H_2O\ +\ 16\ CO_2$$

When insufficient oxygen is present, another oxide of carbon, carbon monoxide, is formed. Millions of metric tons of this invisible but deadly gas are poured into the atmosphere each year, about 75% of it from automobile exhausts. The United States government has set

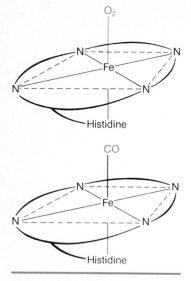

Figure 15.11 Schematic representations of a portion of the hemoglobin molecule. Histidine is an amino acid. Carbon monoxide bonds much more tightly than oxygen, as indicated by the heavier bond line.

danger levels of 9 ppm carbon monoxide (average) over 8 hours and 35 ppm (average) over 1 hour. Even in off-street urban areas, levels often average 7 to 8 ppm. On streets, danger levels are exceeded much of the time. Such levels do not cause immediate death, but, over a long period, exposure can cause physical and mental impairment.

Carbon monoxide is an invisible, odorless, tasteless gas. There is no way for a person to tell that it is around (without using test reagents or instruments). Drowsiness is usually the only symptom, and drowsiness is not always unpleasant. How many auto accidents are caused by drowsiness or sleep induced by carbon monoxide? No one knows for sure. Cigarette smoke also contains a fairly high concentration of carbon monoxide.

Carbon monoxide is not an irritant. It exerts its insidious effect by tying up the hemoglobin in the blood. The normal function of hemoglobin is to transport oxygen (Figure 15.11). Carbon monoxide binds to hemoglobin so strongly that the hemoglobin is prevented from transporting oxygen. Therefore, the symptoms of carbon monoxide poisoning are those of oxygen deprivation. All except the most severe cases of carbon monoxide poisoning are reversible. The best antidote is the administration of pure oxygen. Artificial respiration may help if a tank of oxygen is not available.

Carbon monoxide poisoning impairs the ability of the blood to transport oxygen, and the heart has to work harder to supply oxygen to the tissues. Chronic exposure to even low levels of carbon monoxide may put an added strain on the heart and lead to an increased chance of a heart attack.

Carbon monoxide is a local pollution problem. It is a severe threat in urban areas with heavy traffic, but it does not appear to be a global threat. In laboratory tests, carbon monoxide survives about 3 years in contact with air. But nature is somehow able to prevent its buildup, despite the large amounts being poured into the environment. In fact, on a global basis, it is estimated that up to 80% of the carbon monoxide in the atmosphere comes from natural sources. Except for those highly localized situations (they're bad enough!), nature seems to have carbon monoxide under control.

Nitrogen Oxides: Brown Is the Color of Los Angeles Air

In addition to carbon dioxide, carbon monoxide, and unburned hydrocarbons, automobile exhausts contain oxides of nitrogen. Power plants that burn fossil fuels are another major source of nitrogen oxides. In a reaction similar to the one that occurs in the atmosphere during electrical storms (page 316), nitrogen and oxygen are made to combine in combustion chambers; the main product is nitric oxide (NO).

$$N_2 + O_2 \longrightarrow 2\,NO$$

Oxygen in the atmosphere slowly oxidizes the nitric oxide to nitrogen dioxide (NO_2).

$$2\,NO + O_2 \longrightarrow 2\,NO_2$$

Nitrogen dioxide is an amber-colored gas. Smarting eyes and a brownish haze are excellent indicators of Los Angeles smog (Color Plate N). It is this nitrogen dioxide that plays a vital (villain's) role in photochemical smog. It absorbs a photon of sunlight and breaks down into nitric oxide and reactive oxygen *atoms*.

$$NO_2 + sunlight \longrightarrow NO + O$$

The oxygen atoms react with other components of automobile exhaust and the atmosphere to produce a variety of irritating and toxic chemicals (Figure 15.12).

At present concentrations, nitrogen oxides don't seem particularly dangerous in themselves. Nitric oxide at high concentrations reacts with hemoglobin; as with carbon monoxide poisoning, this leads to oxygen deprivation. Such high levels seldom, if ever, result from ordinary air pollution, but they might be reached in areas close to industrial sources. Nitrogen dioxide is an irritant to the eyes and the respiratory system. Tests with laboratory animals indicate that chronic exposure in the range of 10 to 25 ppm might lead to emphysema and other degenerative diseases of the lungs.

The most serious environmental effect of the nitrogen oxides is

The initiator:

$$NO_2 + sunlight \longrightarrow NO + O$$

Secondary reactions:

$$O + O_2 \longrightarrow O_3$$

$$O + hydrocarbons \longrightarrow aldehydes$$

Tertiary reactions:

$$O_3 + hydrocarbons \longrightarrow aldehydes\ (R-\overset{\overset{\displaystyle O}{\|}}{C}-H)$$

$$Hydrocarbons + oxygen + NO_2 \longrightarrow PAN\ (R-\overset{\overset{\displaystyle O}{\|}}{C}-O-O-NO_2)$$

Figure 15.12 A summary of some of the principal reactions in the formation of photochemical smog. Most reactive intermediates have been omitted from this simplified scheme. (For a complete account, see Reference 1.)

that they produce smog. The gases also contribute to the fading and discoloration of fabrics, however. By forming nitric acid, nitrogen oxides contribute to the acidity of rainwater. They also contribute to crop damage, although their specific effects are difficult to separate from those of sulfur dioxide and other pollutants.

Ozone: Good News, Bad News

Ozone, a natural component of the stratophere, shields the Earth from life-destroying ultraviolet radiation. Ozone is also a familiar constituent of photochemical smog. Inhaled, it is a toxic, dangerous chemical. Ozone is a good example of just what a pollutant is—a chemical substance out of place in the environment. In the stratosphere, it helps make life possible. In the lower troposphere—the part we breathe—it makes life difficult.

In the mesosphere (see Figure 15.1), some oxygen molecules are split into oxygen atoms by short-wavelength, high-energy ultraviolet radiation.

$$O_2 \xrightarrow[\text{radiation}]{\text{ultraviolet}} 2\ O$$

Some of these highly reactive atoms diffuse down to the stratosphere where they react with oxygen molecules to form ozone.

$$\underset{\substack{\text{Oxygen} \\ \text{atom}}}{O} \ + \ \underset{\substack{\text{Oxygen} \\ \text{molecule}}}{O_2} \ \longrightarrow \ \underset{\text{Ozone}}{O_3}$$

The ozone in turn absorbs shorter wavelength, but still lethal, ultraviolet rays, thus shielding us from this harmful radiation. In absorbing the rays, ozone is converted back to oxygen molecules and oxygen atoms, reversing the above reaction.

$$O_3 \xrightarrow[\text{radiation}]{\text{ultraviolet}} O_2 \ + \ O$$

Undisturbed, the concentration of ozone is kept fairly constant by these cyclic processes. We consider in the next section how the stratospheric ozone layer may be disturbed by human activity. First, though, let's look at the effects of ozone down here where we live.

Ozone is a powerful oxidizing agent. At low levels, it causes eye irritation. At high levels, it can cause pulmonary edema, hemorrhage, and even death. The long-term effects of exposure to low levels of ozone are more difficult to evaluate. Inhalation of ozone is particularly dangerous during vigorous physical activity. Members of a New Jersey high school football team had to be hospitalized after collapsing during a severe pollution episode. School children in Los Angeles are not allowed to pay outside when ozone reaches dangerous levels, as it often does. Exposure of animals to 1 ppm of ozone for 8 hours a day for 1 year has produced bronchial inflammation and irritation of fibrous tissues. It is not known if the same thing occurs in humans. It is known that ozone levels have occasionally reached 0.5 ppm in southern California, levels of 0.15 ppm are exceeded frequently.

Los Angeles isn't the only place with a problem. The Northeast, the Texas Gulf Coast, and Chicago are also afflicted. In all, some 70 areas of the United States failed to meet federal health-based ozone standards in 1986.

In addition to adversely affecting health, ozone causes economic damage. It causes rubber to harden and crack, shortening the life of automobile tires and other rubber items. Ozone also causes extensive damage to crops. Tobacco and tomatoes are particularly susceptible.

Figure 15.13 F. Sherwood Rowland.

The Ozone Layer: Chlorofluorocarbons and Nuclear Bombs

We encountered a class of compounds called chlorofluorocarbons in Chapter 11. These compounds are quite useful as propellants in aerosol cans and as refrigerants. A controversy over their use arose in 1974. F. Sherwood Rowland, a chemist at the University of California at Irvine, considered the effect of these compounds on the Earth's protective ozone layer. He speculated that, because of their inertness to ordinary chemical reagents, chlorofluorocarbons might be making their way up through the atmosphere to the stratosphere. There, they may well react with atomic oxygen and ultraviolet light to form chlorine atoms, which catalyze the decomposition of ozone.

$$CF_2Cl_2 + \text{ultraviolet light} \longrightarrow CF_2Cl\cdot + Cl\cdot$$

$$Cl\cdot + O_3 \longrightarrow ClO\cdot + O_2$$

$$\cdot ClO + O \longrightarrow Cl\cdot + O_2$$

Note that the last step results in the formation of another chlorine atom that can break down another molecule of ozone. The second and third steps are repeated many times; thus, the decomposition of one molecule of chlorofluorocarbon can result in the destruction of many molecules of ozone.

The reactions noted here are known to occur in the laboratory under conditions simulating those in the stratosphere. In partial confirmation of Rowland's speculations, chlorofluorcarbons have been detected in the stratosphere. Their effect on the ozone layer has yet to be proven, but the risk involved is rather great. The U.S. National Research Council predicts a 2 to 5% increase in skin cancer for each 1% depletion of the ozone layer.

In 1984, the NRC predicted a 2 to 4% depletion by late in the 21st century. New worries arose in 1986 with the discovery that ozone concentrations over Antarctica plummet each September. This thinning or "hole" in the ozone layer has been getting worse since 1979. No one knows for sure whether this is a natural event or has been caused by human artifacts such as the chlorofluorocarbons. Overall satellite data show that the ozone layer has been thinning at a rate of 0.5% per year since 1978. This, too, may be caused by natural events such as the eruption of El Chichon in 1982. If it is due to chlorofluorocarbons, however, the decrease will continue and perhaps get worse.

Chlorofluorocarbons have been banned in the United States from most aerosol preparations, but they are still used as refrigerants. They still escape into the atmosphere from refrigerators and air conditioners. As propellants in aerosol products, chlorofluorocarbons have been replaced by flammable hydrocarbons such as isobutane (Chapter 10) and by methylene chloride (Chapter 11). Methylene chloride won't burn, but it may be banned as a possible carcinogen.

Perhaps the ultimate threat to the ozone layer is nuclear war. While nuclear explosions near the surface might lead to the ultimate in particulate pollution, bringing on a nuclear winter (Chapter 4), explosions in the stratosphere—set off in an attempt to destroy incoming missiles—would produce massive amounts of nitrogen oxides. The ozone layer might be substantially depleted. Instead of a nuclear winter, we would be burned by an "ultraviolet summer."

Hydrocarbons: Another Culprit

Hydrocarbons are released from a variety of natural sources. Of all hydrocarbons found in the atmosphere, only about 15% are put there by people. In urban areas, though, the processing and use of gasoline is the major source of hydrocarbons in the environment. Gasoline can evaporate anywhere along the line. This simple process contributes substantially to the total amount of hydrocarbons in urban air. The automobile's internal combustion engine also contributes by exhausting unburned and partially burned hydrocarbons.

Certain hydrocarbons, particularly those that contain a double bond, combine with oxygen atoms or ozone molecules to form aldehydes. Such reactions are very complicated, but they can be illustrated by the reaction of ethylene with atomic oxygen to form acetaldehyde.

$$CH_2\!=\!CH_2 \; + \quad O \quad \longrightarrow \quad \overset{\displaystyle O}{\overset{\|}{CH_3C}}\!-\!H$$

$$\text{Ethylene} \qquad \underset{\text{oxygen}}{\text{Atomic}} \qquad \text{Acetaldehyde}$$

Formaldehyde is also formed, as are other more complicated aldehydes. As a class, the aldehydes have foul, irritating odors.

Another complicated series of reactions involving hydrocarbons, oxygen, and nitrogen dioxide leads to the formation of peroxyacetyl-nitrate (PAN). Ozone, the aldehydes, and PAN are responsible for much of the destruction wrought by smog. They make breathing difficult and make the eyes smart and itch. Those who already have respiratory ailments may be severely afflicted. The very young and the very old are particularly vulnerable.

$$\overset{\displaystyle O}{\overset{\|}{CH_3C}}\!-\!O\!-\!ONO_2$$

$$\text{PAN}$$

What to Do About Photochemical Smog

The principal culprits in photochemical smog are nitrogen oxides, hydrocarbons, and sunlight. Reducing the amount of any of these would diminish the amount of smog. It isn't likely that we could—or would want to—reduce the amount of sunlight, however, so let's focus on the other two.

Hydrocarbons have many important uses as solvents. If we could reduce the amount of hydrocarbons entering the atmosphere, the amount of aldehydes and PAN formed would also be reduced. Many industries using hydrocarbons have substantially reduced emissions or switched to other solvents, and new storage and dispensing systems have reduced hydrocarbon emissions from gasoline filling stations. Similarly, modified gas tanks and crankcase ventilation systems have reduced evaporative emissions from automobiles. The main attack, however, has been through the use of **catalytic converters** to reduce hydrocarbon and carbon monoxide emissions in automotive exhausts. The catalyst in these converters is a precious metal such as platinum (Pt). Hydrocarbons and carbon monoxide react rapidly with oxygen on the surface of these metals. With no catalyst, the pollutants are more likely to reach the atmosphere unreacted.

Reducing the amount of nitrogen oxides has proven more difficult. Lowering the operating temperature of an engine helps, but this renders the engine less efficient. Running the engine on a richer (more fuel, less air) mixture lowers nitrogen oxide emissions but tends to raise carbon monoxide and hydrocarbon emissions. You can't get something for nothing, it seems. The one-car–one-person transportation system is inherently inefficient. The best way to reduce emissions from automobiles is to drive less. Walk, ride a bicycle, or use public transportation whenever possible. And if you must drive, use a smaller car.

Acid Rain: Air Pollution → Water Pollution

We have seen (page 321) how sulfur oxides are converted to sulfuric acid. Similarly, nitrogen oxides become nitric acid. These acids fall upon Earth as acid rain or acid snow or are deposited absorbed on particulate matter. **Acid rain** is defined as rain having a pH less than 5.6 (5.6 is the pH of normal rainwater, which is slightly acidic due to dissolved carbon dioxide). Rain with a pH as low as 2.4—that's more acidic than vinegar or lemon juice and over 1000 times as acidic as normal rainwater—has been detected (Figure 15.14). (One pH unit lower [from 5.6 to 4.6, for example] means 10 times as acidic; two pH units lower means 100 times as acidic; three pH units lower means 1000 times as acidic.)

Acid rain is a controversial subject. Scientist and others dispute its causes, its sources, and its degree of harm. The best evidence indicates that it comes from sulfur oxides and nitrogen oxides emitted from power plants, smelters, and automobiles. These acids are carried for hundreds of kilometers before falling as rain or snow.

Acids corrode metals and can even decompose stone buildings and

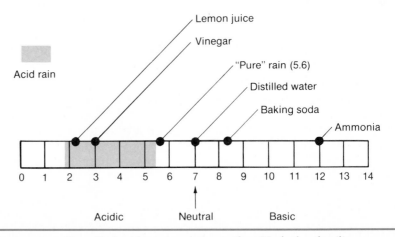

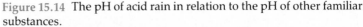

Figure 15.14 The pH of acid rain in relation to the pH of other familiar substances.

(a)

(b)

Figure 15.15 This stone sculpture on the exterior of Herten Castle near Recklinghausen in Westphalia, Germany, shows the effects of acidic smog. The castle was constructed in 1702. (a) The sculpture as it looked in 1908. (b) The sculpture as it looked in 1969. [Reprinted with permission from E. M. Winkler, *Stone: Properties, Durability in Man's Environment* (Berlin: Springer-Verlag, 1973). Copyright © 1973 by E. M. Winkler.]

statues. Sulfuric acid eats away metal to form a soluble salt and hydrogen gas.

$$Fe \quad + \quad H_2SO_4 \longrightarrow FeSO_4 + H_2$$

Iron   Iron(II)
(or steel)   sulfate
                     (soluble)

The reaction shown here is oversimplified. For example, in the presence of oxygen (air), the iron is converted to rust (Fe_2O_3). Marble buildings and statues (Figure 15.15 and Color Plate O) are disintegrated by sulfuric acid in a similar reaction that forms calcium sulfate, a slightly soluble, crumbly compound.

$$CaCO_3 \quad + \quad H_2SO_4 \longrightarrow CaSO_4 + H_2O + CO_2$$

Marble   Calcium
or limestone   sulfate

Color Plate P shows the corrosion of bronze by a natural process.

The effect of acidic waters on plant and animal life is discussed in Chapter 16.

Get the Lead Out

Lead is a heavy-metal poison that affects the functioning of the blood, liver, kidneys, and brain. For years, a lead compound, tetra-ethyllead, has been used to improve the antiknock qualities of gaso-

$$
\begin{array}{c}
CH_2CH_3 \\
| \\
CH_3CH_2\!-\!Pb\!-\!CH_2CH_3 \\
| \\
CH_2CH_3
\end{array}
$$

Tetraethyllead

line. Tetraethyllead is an extremely effective antiknock agent. As little as 0.5 to 1.0 g/L can raise octane rating of gasoline by 10 or more units. It is also quite toxic. Gasoline that contains lead also must have a dye added to warn consumers of its presence.

Modern high-compression engines will not run on straight-run gasoline. As it comes from the distillation tower of a petroleum refinery, gasoline has an octane rating of about 60. Science and technology, through the use of tetraethyllead, have done a fine job of providing high-octane fuel. Unfortunately, the solution to one problem may be the source of another. Large amounts of lead have been spewed into the environment by automobiles. Fairly high concentrations of lead are now found along heavily traveled streets, freeways, and other roads.

Not everyone agrees that this lead presents a health hazard at present levels. Surely, though, its continued used would lead to dangerous concentrations in many areas. At any rate, there is another problem with lead in gasoline. Lead fouls the catalysts in pollution-control devices. Thus, there are two reasons we should "get the lead out." The United States, Western Europe, and Japan are rapidly phasing out the use of leaded gasolines.

To achieve the same octane rating in lead-free gasoline, more branched-chain and aromatic hydrocarbons (Chapter 10) are needed. Even without these extra aromatics, the burning of gasoline produces 3,4-benzpyrene, a chemical that induces cancer in laboratory animals. Some scientists have warned that increasing the aromatic hydrocarbon content of gasoline would probably increase the concentration of this carcinogen in the atmosphere.

Catalytic converters and lead-free gasolines are far-from-perfect answers. Cars with pollution-control devices often are harder to start and run less efficiently than cars without them. They work well only if cars are properly tuned (and most aren't). The catalysts are also expensive, and the precious metals used in them must be imported.

3,4-Benzpyrene

Indoor Air Pollution

You can't necessarily escape air pollution by staying indoors. In high-traffic areas, air indoors has been found to have the same carbon monoxide level as air outside. In some places, office buildings, airport terminals, and apartments have indoor levels that exceed government safety standards. It has been recommended that such buildings be tightly sealed on the lower floors and that they be spaced in such a way that the wind can disperse pollutants around them.

As we try to save energy by installing better insulation and by sealing air leaks around windows and doors, we often make indoor pollution worse. We trap pollutants inside. A kitchen with a gas range often has levels of nitrogen oxides above United States government standards. Those who try to save on the fuel bill by using free-standing kerosene heaters do so at a considerable sacrifice of air quality. These unvented stoves produce levels of nitrogen oxide up to 20 times greater than those permitted by federal regulations for outdoor air. (These regulations do not apply to indoor air.) Another problem, the release of toxic, irritating formaldehyde gas from foamed insulation, led in 1982 to a ban on the use of urea–formaldehyde polymer foams for insulation. (This ban was lifted by the courts in 1983.) Similar polymers are used as adhesives in plywood and particle board; these uses have not been banned.

Perhaps the most prevalent indoor polluter is the cigarette. The smoke-filled room is well known in political mythology as the place where decisions are made. The health effects of smoking on the smoker are also widely known. Smokers face twice the risk of bladder cancer, stroke, and heart attack as those who do not smoke; 4 times the risk of cancer of the esophagus; 7 times the risk of death from chronic bronchitis of emphysema; 8 times the risk of cancer of the larynx; and 10 to 20 times the risk of lung cancer. These risks have been widely publicized in a series of reports released by the Surgeon General of the United States. The first report, published in 1964, led to a ban on television advertisements for cigarettes. A 1972 report included a chapter on the effect of tobacco smoke as an air pollutant; this may well end the era of the smoke-filled room. Studies have shown that the air quality in smoke-filled rooms is poor. The level of carbon monoxide, even in a well-ventilated room, is often equal to, or greater than, the legal limits for ambient air. High levels have been shown to impair time-interval discrimination. In severe cases, performance on psychomotor tests is impaired. People already suffering from heart or lung disease might be severely affected. Even one cigarette can raise the level of particulate matter above government standards.

The nonsmoker is also exposed to significant levels of tar and nicotine. In fact, the average nonsmoker has 5% as much nicotine in the blood as a smoker. Workers at Cornell University Medical College found a glycoprotein (a protein molecule with sugar units attached) in tobacco that causes allergic reactions in many smokers and nonsmokers. The reaction is so severe that it may damage small blood vessels and lead to strokes and heart disease. The risk to nonsmokers from cigarette-related pollutants is not known with certainty. Nevertheless, some states have already banned smoking in meeting rooms, waiting rooms, and other public places.

A most enigmatic indoor air pollutant is radon. A noble gas, radon is colorless, odorless, tasteless, and unreactive chemically. Radon is, however, radioactive. It decays by alpha emission (Chapter 4) with a half-life of 3.8 days.

$$^{222}_{86}Rn \longrightarrow {}^{218}_{84}Po + {}^{4}_{2}He$$

It is released naturally from soils and rocks, particularly granite and shale, and from minerals such as phosphate ores and pitchblend. The ultimate source is uranium atoms found in these materials. Radon is only one of several radioactive materials formed during the multistep decay of uranium. However, it is unique in that it is a gas and escapes. The others are solid and remain in the soil and rock.

Outdoors, the radon dissipates and presents no problems. Trapped inside a house, however, levels build up, sometimes reaching quantities several times the maximum safe level, established by the U.S. Environmental Protection Agency, of 0.15 disintegrations per second per liter of air. At five times that level, the hazard is thought to equal that of smoking two packs of cigarettes a day. Some scientists say that radon threatens 8 million homes in the United States and may cause 20 000 lung cancer deaths a year. The exact threat isn't known, however. The "safe level" was set for people who work in uranium mines. These workers do have high rates of lung cancer, but they also breathe radioactive dust, and many smoke cigarettes. These risk factors are difficult to separate. We need careful studies in order to estimate the risk of radon alone.

Who Pollutes? How Much?

Air pollution causes material damage by dirtying and destroying buildings, clothing, and other material objects. It increases health hazards, especially for the very young, the old, and those already ill. It causes crop damage by stunting or killing green plants. Air pollution reduces visibility, thus increasing auto and air-traffic accidents. With its ugly smoke plumes and unpleasant odors, it is even an aesthetic problem.

Who causes all that pollution? Table 15.2 summarizes the seven major pollutants, their main sources, and their health and environmental effects. Note that motor vehicles are a prominent source. Indeed, they are the source of about one half (by weight) of all air pollutants. Our transportation system accounts for about 80% of carbon monoxide emissions, 40% of hydrocarbon emissions, and 40% of nitrogen oxide emissions. Since most transportation in the United States is by private automobile, we can conclude that the car goes a long way toward doing us in.

On the other hand, most particulate matter comes from power

Table 15.2 The Seven Major Air Pollutants

Pollutant	Symbol	Major Sources	Health Effects	Environmental Effects
Carbon monoxide	CO	Motor vehicles	Interferes with oxygen transport, causing dizziness, death; possibly contributes to heart disease	Slight
Hydrocarbons	C_nH_m	Motor vehicles, industry, solvents	Narcotic at high concentrations; some aromatics are carcinogens	Precursor to aldehydes, PAN
Sulfur oxides	SO_x	Power plants, smelters	Irritate respiratory system; aggravate lung and heart diseases	Reduce crop yields; precursor to acid rain, SO_4^{2-} particulates
Nitrogen oxides	NO_x	Power plants, motor vehicles	Irritate respiratory system	Reduce crop yields; precursor to ozone and acid rain; produce brown haze
Particulate matter	—	Industry, power plants, dust from farms and construction sites	Irritates respiratory system; synergistic with SO_2; contains adsorbed carcinogens, toxic metals	Impairs visibility
Ozone	O_3	Secondary pollutant from NO_2	Irritates respiratory system; aggravates lung and heart diseases	Reduces crop yields; kills trees (synergistic with SO_2); destroys rubber, paint, etc.
Lead	Pb	Motor vehicles, smelters	Toxic to nervous system and blood-forming system	Toxic to all living things

plants (about 40%) and industrial processes (about 45%). Similarly, over 80% of sulfur oxide emissions come from power plants, with an additional 15% coming from other industries. Power plants alone contribute about 55% of nitrogen oxide emissions. Who uses electricity from power plants? You do. A 100-W bulb burning for 1 year uses the electricity generated by burning 275 kg of coal.

What is the *worst* pollutant? Carbon monoxide is produced in huge amounts and is quite toxic. Yet it is deadly only in concentrations approaching 4000 ppm. Its contribution to cardiovascular disease, by increasing stress on the heart, is difficult to measure.

The World Health Organization (WHO) rates sulfur oxides as the worst pollutants. Sulfur oxides are powerful irritants, and, according to WHO, people with respiratory illnesses are more likely to die from exposure to sulfur oxides than to any other kind of pollutant. Sulfur oxides and the sulfuric acid and sulfates formed from them have been linked to as many as 53 000 excess deaths per year in the United States (Reference 8).

Who pollutes? We all do. In the words of Walt Kelley's comic-strip character Pogo, "We have met the enemy, and he is us." Don't despair, however. You can be a part of the solution by conserving fuel and electricity. Many utilities now offer suggestions for saving energy.

Overall, air quality is improving. Sulfur oxide emissions have been reduced somewhat. Particulate emissions have been reduced dramatically, particularly in urban areas. Carbon monoxide levels are down in most cities. Levels of nitrogen oxides have increased, however, and ozone levels have remained essentially unchanged. The air is still polluted at unsafe levels for much of the population at certain times, however, and smog is steadily encroaching on new areas. We still have a long way to go before we all have clean air to breathe, but perhaps we have at last turned the corner.

The Penultimate Pollutants: Carbon Dioxide and Water

No matter how clean an engine or a factory is, as long as it burns coal or petroleum products it produces carbon dioxide and water. The concentration of carbon dioxide in the atmosphere has increased 17% in this century. It continues to increase at an expanding rate because of the increased burning of carbon fuels. We generally don't even consider carbon dioxide and water vapor to be pollutants. Both are natural components of the environment. Certainly their immediate effect upon us is slight. But what about their longterm effects?

Both water vapor and carbon dioxide produce the **greenhouse effect**. These chemicals let the sun's rays (visible light) in to warm the surface of the Earth. When the Earth tries to radiate this heat (infrared energy) back out into space, however, the energy is trapped by water and carbon dioxide molecules (Figure 15.16). Water vapor spewed into the atmosphere soon falls back to Earth as rain. Therefore, it affects the climate mostly at the local level. Carbon dioxide, poured into the air, is not so readily dissipated. It hangs around to affect the climate of the entire world. The concentration of carbon dioxide is now increasing at a rate of 1 ppm per year. Some scientists predict a warming trend that will melt the polar ice caps, and flood coastal cities. The fears have increased in recent years. It has been found that methane and other trace gases also contribute to the greenhouse effect. The concentration of methane in

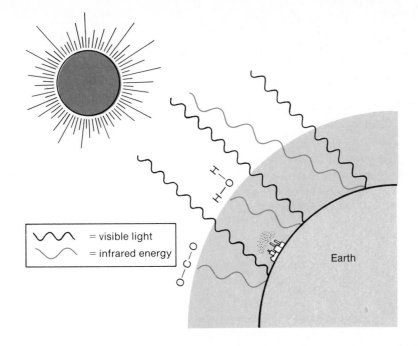

Figure 15.16 The greenhouse effect.

the atmosphere has been increasing since 1977. The oceans are now rising at 3 mm per year. That doesn't seem like much, but even slight rises increase tides and result in higher, more damaging, storm surges.

At present, the atmosphere seems to be cooling, not warming. This may be due mainly to volcanic eruptions spewing ash and other particulates into the atmosphere, but human activities also contribute soot, smoke, and dust. These, perhaps aided by the vapor trails of jet aircraft, screen out the sun's light and tend to lower the Earth's temperature. We may not always be so lucky as to balance one effect against the other, however.

The Ultimate Pollutant: Heat

Electric cars are sometimes advocated as a solution to air pollution. But electric cars require electric power, and electric power requires power plants. Conventional power plants burn coal, gas, and oil. Replacing cars run on fossil fuel with ones run on electricity would serve only to change the *site* of the pollution and perhaps spread it out a bit. Remember that there are losses at every step in the conversion and transmission of energy. More fossil fuels would have to be burned at power plants to provide the same amount of energy for propelling automobiles that we now use.

We can avoid air pollution by using nuclear power plants (Chapter 14). They do not spew soot, smoke, and poisonous chemicals into the atmosphere. But they do introduce some radioactivity into the environment, and there is still no approved way to dispose of radioactive wastes. And all power plants dump vast quantities of residual heat into the environment. This heat may be the ultimate pollutant.

We use fantastic amounts of energy—in cars, factories, homes, schools, and hospitals. We use it everywhere, for everything. The release of energy heats up the environment. Eventually, this heat will change the climate of the Earth, and the ecology of the planet will be affected. As Rene Dubos, world-renowned biologist, put it: "We will destroy our lives by producing more useless, destructive energy to make more and more needless things that do not increase the happiness of people."

Paying the Price

It has been estimated that air pollution costs the United States $16 billion each year. It wrecks our health by causing or aggravating bronchitis, asthma, emphysema, and lung cancer. It destroys our crops

Figure 15.17 Alexander Hamilton's statue suffers soiling from air pollution at the U.S. Treasury Building in Washington, DC. The photo was taken 11 April 1966. [Courtesy of the U.S. Environmental Protection Agency.]

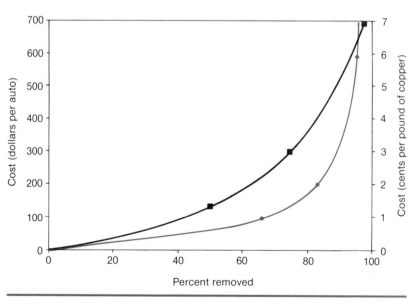

Figure 15.18 The cost of removing pollutants increases exponentially with the percentage of material removed. The color line shows the cost of removing sulfur dioxide from copper smelter fumes. The black line shows the cost of reducing automobile emissions.

and sickens and kills our livestock. It corrodes our machines, blights our buildings, and even destroys our works of art (Figure 15.17).

Elimination of air pollution would not be cheap or easy. It is especially expensive to remove the last fractions of pollutants. Figure 15.18 shows how costs increase as you try to remove a larger percent of pollutants. Note that the curves are exponential: costs soar to infinity as pollutants are reduced to zero. But we could have cleaner air. How clean depends on how much we are willing to pay.

What would we gain by getting rid of air pollution? How much is it worth to see the clear blue sky or to see the stars at night? How much is it worth to breathe clean, fresh air? Whatever your answers, clean air has other, more tangible, benefits. According to the National Resources Defense Council, the Clean Air Act has created 200 000 new jobs and provided $21.4 billion in economic and health benefits annually—compared to a cost of $17 billion.

Problems

1. Which layer of the atmosphere lies nearest the Earth?

2. Which layer of the atmosphere contains the ozone layer?

3. List the three major components of dry air and give the percentage by volume of each.

4. List the two most important variable components of the atmosphere and give the approximate concentration range of each.

5. What is meant by nitrogen fixation? Why is it important?

6. Give the equations by which lightning fixes nitrogen.

7. How has industrial fixation of nitrogen to make fertilizers affected the nitrogen cycle?

8. How is the oxygen supply of Earth's atmosphere replenished?

9. Give the equation for the conversion of oxygen to ozone in the ozone layer.

10. How does the ozone layer protect the inhabitants of Earth's surface?

11. What is an atmospheric inversion?

12. How did the atmospheric inversion contribute to the air-pollution episode in Donora, Pennsylvania, in 1948?

13. Is all air pollution the result of human activity? Explain.

14. What is a pollutant?

15. What is smog?

16. What are the chemical components of London smog?

17. What weather conditions characterize London smog?

18. Write the equation for the oxidation of sulfur to sulfur dioxide.

19. Write the equation for the oxidation of sulfur dioxide to sulfur trioxide.

20. Write the equation for the reaction of sulfur trioxide with water to form sulfuric acid.

21. What is particulate matter?

22. What is an aerosol?

23. What compound is produced when sulfuric acid reacts with ammonia? Write the equation for the reaction.

24. What are clinkers?

25. What is fly ash?

26. Describe the synergistic action of sulfur dioxide and particulate matter.

27. What is an electrostatic precipitator? How does it work?

28. Describe how a bag filter removes particulate matter from stack gases.

29. How does a wet scrubber remove particulate matter? What is a major disadvantage of wet scrubbers?

30. Describe how a limestone scrubber removes sulfur dioxide from stack gases.

31. Give two uses for fly ash.

32. Give the equation for the reaction of sulfur dioxide with hydrogen sulfide to form elemental sulfur.

33. What are the chemical components of photochemical smog?

34. What weather conditions characterize photochemical smog?

35. What is the main source of carbon monoxide in polluted air?

36. What are the health effects of carbon monoxide?

37. How might exposure to carbon monoxide contribute to heart disease?

38. Under what conditions do nitrogen and oxygen combine? Write the equation.

39. Write the equation for the reaction of nitric oxide with oxygen to form nitrogen dioxide.

40. What happens to nitrogen dioxide in sunlight? Write the equation.

41. What are the health effects of ozone in polluted air?

42. List two uses of chlorofluorocarbons.

43. How might chlorofluorocarbons be involved in the depletion of the ozone layer?

44. What health effect might result from depletion of the ozone layer?

45. How might nuclear war bring about an "ultraviolet summer"?

46. What is the main source of hydrocarbons in polluted air?

47. What is PAN? From what is it formed? What are its health effects?

48. What are catalytic converters? How do they work?

49. Give two ways the level of nitrogen oxide emissions from an automobile can be reduced. What pollutants are likely to be increased when these adjustments are made?

50. What is acid rain?

51. How is acid rain formed?

52. What is the effect of acid rain on iron? Give an equation.

53. What is the effect of acid rain on marble? Write the equation.

54. How might acid rain be alleviated?

55. What is tetraethyllead? Why is it used in gasoline?

56. Why is leaded gasoline being phased out?

57. List several indoor air pollutants. What is the source of each?

58. List some risks associated with cigarette smoking.

59. What is the greenhouse effect?

60. Why is zero pollution not possible?

61. Is the electric car a solution to air pollution on a nationwide basis? Within a given city?

62. What are the average and peak concentrations of each of the following in your community?

   a. carbon monoxide   b. ozone
   c. nitrogen oxides   d. sulfur dioxide
   e. particulate matter

63. Does your community have an air-pollution problem? How could it be solved?

64. Do you or your family own a car? If so, what type of pollution-control devices does it have? Are the devices effective?

65. Should students be allowed to smoke in class?

66. Should smoking be allowed in restaurants? In meeting rooms? On buses, trains, and airplanes?

67. The average person breathes about 20 m^3 of air a day. What weight of particulates would a person breathe in a day if the particulate level were 400 $\mu g/m^3$.

68. The atmosphere contains 5.2×10^{15} t of air. How much carbon dioxide (CO_2) is in the atmosphere if the concentration is 345 ppm?

69. The world's termite population is estimated to be 2.4×10^{17}. These termites produce an estimated 4.6×10^{16} g of carbon dioxide, 1.5×10^{14} g of methane, and 2×10^{14} g of hydrogen. How much of each gas is produced by each termite? What percent increase in the total amount of carbon dioxide in the atmosphere (see Problem 68) would the termites cause in 1 year?

References and Readings

1. American Chemical Society. *Cleaning Our Environment: A Chemical Perspective*, 2d ed. Washington, DC: American Chemical Society, 1978. Available from ACS Special Issues Sales, 1155 Sixteenth Street N.W., Washington, D.C. 20036. Section 1 is a comprehensive look at the problem of air pollution, including suggestions for improving the air environment.
2. "America's Acid Rain Problem." *ChemEcology*, November 1986, pp. 2–4.
3. Budiansky, Stephen. "Pandora's Woodbox." *Science 81*, March 1981, pp. 10–11.
4. Ember, Lois R., Patricia L. Layman, Wil Lepkowski, and Pamela S. Zurer. "Tending the Global Commons." *Chemical and Engineering News*, 24 November 1986, pp. 14–64. A special issue on the changing atmosphere, including the threat of global warming and of depletion of the ozone layer.
5. Grove, Noel. "Air: An Atmosphere of Uncertainty." *National Geographic*, April 1987, pp. 502–537.
6. "Kerosene Heaters." *Consumer Reports*, February 1985, pp. 74–75.
7. "The Murky Hazards of Secondhand Smoke." *Consumer Reports*, February 1985, pp. 81–84.
8. Office of Public Awareness. "Air Pollution and Your Health." Washington, DC: U.S. Environmental Protection Agency, June 1979.
9. Postel, Sandra. "Air Pollution, Acid Rain, and the Future of Forests." Washington, DC: Worldwatch Institute, March 1984.
10. Rahn, Kenneth A. "Who's Polluting the Arctic?" *Natural History*, May 1984, pp. 30–38.
11. Revelle, Roger. "Carbon Dioxide and World Climate." *Scientific American*, August 1982, pp. 35–43.
12. "Rising Threat from Methane." *Science Digest*, March 1985, p. 30.
13. Spengler, John D., and Ken Sexton. "Indoor Air Pollution: A Public Health Perspective." *Science*, 1 July 1983, pp. 9–16.
14. "Thomas Seeks Help in Meeting Ozone Standard." *Chemical and Engineering News*, 30 June 1986, p. 5.
15. Weisburd, S. "Stratospheric Ozone: A New Policy Tone." *Science News*, 28 June 1986, p. 404.

16

Water

To Drink and to Dump Our Wastes In

Ours is a watery world, and we, its dominant species, are walking sacks of seawater. Our makeup is not just coincidence: the properties of water make it essential to life, and the properties of life make it dependent on water. The presence of large amounts of liquid water on Earth makes our planet unique in the solar system; most likely, it is the only planet capable of supporting life. If we someday search for life outside our solar system, we probably will look for a watery planet.

Those properties of water that make it uniquely suited for the support of life also make it easy to pollute. Many chemical substances are soluble in water. Thus, they are easily dispersed and eventually are scattered to nearly infinite dilution in the ocean. This scattering is in accord with the natural tendency toward maximum entropy as described by the second law of thermodynamics (Chapter 13). Removing these chemicals from our water supplies, once they are there, often requires enormous expenditures of energy.

In this chapter, we discuss some of the unique properties of water. We also discuss water pollution, wastewater treatment, and water treatment. Don't underestimate the importance of good, clean drinking water. We could live for nearly a month without food. Without water, we would last only a few days.

Water: Some Unusual Properties

Next to air, water is the most familiar substance on Earth. Even so, it is a most unusual compound. It is the only common liquid on the suface of our planet. The solid form of water (ice) is less dense than the liquid. The consequences of this peculiar characteristic are immense for life on this planet. Ice forms on the surface of lakes and insulates the lower layers of water. This enables fish and other aquatic organisms to survive winters in the temperate zones. If ice were more dense than liquid water, it would sink to the bottom as it formed, and even the deeper lakes of the northern United States would freeze solid in winter. Life in the northern lakes and rivers would be very different from what it now is, if indeed there were life in those waters at all.

The same property—ice being less dense than liquid water—has dangerous consequences for living cells. As ice crystals are formed, the expansion ruptures and kills cells. The slower the cooling, the larger the crystals of ice and the more damage there is to the cell. Frozen-food manufacturers take advantages of the properties of water when they freeze foods so rapidly that the ice crystals are kept very small (and thus do minimum damage to the cellular structure of the food); the industry calls such foods "flash frozen."

Water also has a higher density than most other familiar liquids. As a consequence, liquids that are less dense than water and insoluble in water float on the surface of it. A familiar problem in recent years has been the gigantic oil spills that occur when a tanker ruptures or when an offshore well gets out of control. The oil, floating on the surface of the water, often is washed onto beaches, where it does considerable ecological and aesthetic damage (Figure 16.1). If water were less dense than oil, the problem would certainly be of a different nature, though not necessarily less acute.

Figure 16.1 Oil is insoluble in, and less dense than, water. Thus oil spills float on the surface of water affected by the spills. (a) Photo showing crude oil and debris on the San Juan River. (b) A portion of the spill has been enclosed in a boom and is being towed ashore for pickup. [Courtesy of the U.S. Environmental Protection Agency.]

(a)

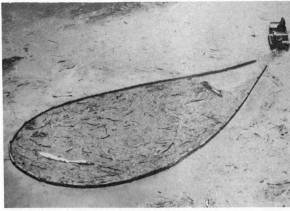

(b)

Table 16.1 Specific Heats of Some Familiar Subtances

Substance	Specific Heat $\left(\dfrac{cal}{g\ °C}\right)$
Aluminum	0.22
Carbon (graphite)	0.17
Copper	0.092
Ethanol	0.59
Diethyl ether	0.54
Iron	0.11
Lead	0.031
Water (ice)	0.50
Water (liquid)	1.00

Another unusual property of water is its high heat capacity. The **heat capacity** of any substance is the quantity of heat that must flow into or out of a system to change its temperature by 1 °C.

$$\text{Heat capacity} = \frac{cal}{°C}$$

Specific heat is the amount of heat required to raise the temperature of 1 g of a substance by 1 °C.

$$\text{Specific heat} = \frac{cal}{g\ °C}$$

Table 16.1 gives the specific heat of a number of familiar substances. Note that it takes almost 10 times as much heat to raise the temperature of 1 g of water by 1 °C as it takes to raise the temperature of 1 g of iron by the same amount. Conversely, much heat is given off by water for even a small drop in temperature. The vast amounts of water on the surface of the Earth thus act as a giant thermostat to moderate daily temperature variations. You need only consider the extreme temperature changes on the surface of the waterless moon to appreciate this important property of water.

Still another way in which water is unique is that it has a high **heat of vaporization**; that is, a large amount of heat is required to evaporate a small amount of water. This is of enormous importance to us for large amounts of body heat can be dissipated by the evaporation of small amounts of water (perspiration) from the skin. This effect also accounts for the climate-modifying property of lakes and oceans. A large portion of heat that would otherwise heat up the land is used to vaporize water from the surface of the lake or the sea. Thus, in summer it is cooler near a large body of water than in interior land areas.

All these fascinating properties of water depend on the unique structure of the water molecule. Recall (Chapter 5) that the water

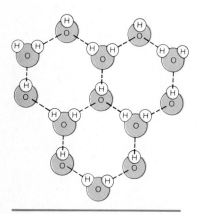

Figure 16.2 The structure of ice has large hexagonal holes. (In this drawing, some of the hydrogen atoms are hidden behind the larger oxygen atoms.)

Figure 16.3 The hexagonal symmetry of the ice crystals in these snowflakes reflects the arrangement of molecules in the crystal structure. [Courtesy of NOAA.]

molecule is highly polar and that in the liquid and solid states, water molecules are strongly associated through hydrogen bonds. In liquid water, the association of molecules is random, and the molecules are close together. When water freezes, its molecules take on a more ordered arrangement with large hexagonal holes (Figure 16.2). This three-dimensional structure extends out for billions and billions of molecules. The large holes account for the fact that ice is less dense than liquid water. The hexagonal arrangement allows water to assume forms of exquisite perfection as snowflakes (Figure 16.3).

The Salt of the Sea

Because of its polar nature, water tends to dissolve ionic substances (Chapter 5). The oxygen ends (negative) of the water molecules are pointed toward the positive ion and the hydrogen ends (positive) are pointed toward the negative ion (see Figure 5.18). This solvent power of water accounts for the saltiness of the sea. Rainwater dissolves minerals (ionic in nature), and these are carried by streams and rivers to the sea. There the heat of the sun evaporates part of the water, leaving the "salts" behind. The oceans grow more salty as the years go by, but because this is a very slow process, the increase in salt content is not noticeable in one lifetime.*

---

* At the present rate of salt accumulation, the seas will become saturated with salt (about 36%) in another 3.5 billion years. If our descendants are still around in that distant age, they will find our oceans much like today's Dead Sea.

Water, Water, Everywhere

Three fourths of the surface of the Earth is covered with water. About 2×10^{22} L of rainwater falls upon the Earth each year. That is indeed a lot of water. Nearly 98% of the water on Earth, though, is seawater—unfit for drinking and not even suitable for most industrial purposes. Much of the rainwater falls into the sea or on areas otherwise inaccessible. Something less than 2% of the water is frozen in the polar ice caps, leaving less than 1% (about 3×10^{17} L) available as fresh water. The average use of water per person (including that used for industrial, agricultural, and other purposes) in the United States is 8 million L per year. At this rate of use (and assuming complete recycling each year), the water supply on Earth could support a population of 40 billion people. But the rate of use is increasing rapidly, and water is not always where the people are. Quite often, too, freshwater supplies are so badly polluted that they are unfit for human use.

Figure 16.4 The water cycle.

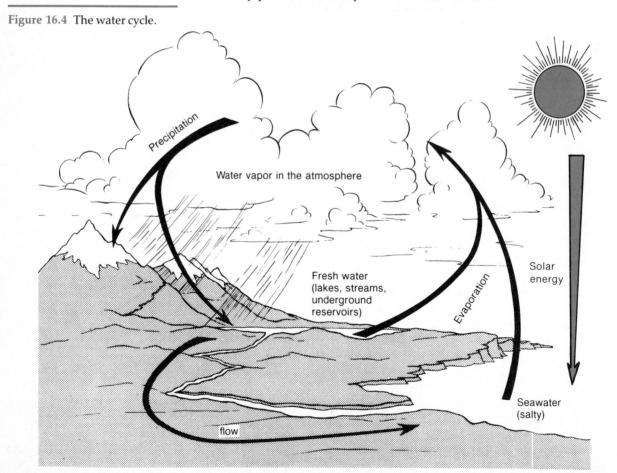

The Water Cycle

The distribution of water between the sea, the ice caps, and fresh-water rivers, lakes, and streams is fairly constant. There is, however, a dynamic cycle of water between these various components (Figure 16.4). Water is constantly evaporating from both water and land surfaces. This water vapor condenses into clouds and returns to Earth in the form of rain, sleet, and snow. This fresh water becomes part of the ice caps, runs off in streams and rivers, and fills lakes and underground reservoirs.

The cycling of water serves to replenish our supply of fresh water. When water evaporates from the sea, salts are left behind. When water moves through the ground, impurities are trapped in the rock, gravel, sand, and clay. This capacity to purify is not infinite, however.

Some Biblical Chemistry

Getting a dependable supply of fresh water has long been a problem (Figures 16.5 and 16.6). When Moses led the Israelites out of Egypt into the wilderness, he encountered a desert area where potable water was scarce. While nearly everyone knows the Biblical account of how Moses

Figure 16.5 An adequate supply of water is essential to life. In this photograph a tube well in a Yemeni village provides water for drinking, crop irrigation, bathing, and laundry. [Courtesy of Food and Agriculture Organization of the United Nations, Rome.]

Figure 16.6 Life becomes difficult without an adequate supply of water. This photograph shows a dry riverbed that became a graveyard for cattle when a drought struck the African Sahel. [World Food Production photo by Florita Botts.]

struck the rock to bring forth water, an incident at Marah is less well known. At Marah, the Israelites couldn't drink the water because it was bitter. According to the account in Exodus, God commanded Moses to throw a tree into the water to purify it.

Recall from Chapter 7 that one property of basic solutions is that they are bitter. The water at Marah was probably basic, or *alkaline*. Such waters are common in desert areas. Attempts have been made to give a chemical explanation of Moses' purification of the brackish water. The tree was probably a dead one, bleached by the desert sun. Such bleaching oxidizes the alcohol groups in cellulose to carboxylic acid groups.

$$\text{CH}_2\text{OH} \quad \text{CH}_2\text{OH} \qquad \longrightarrow \qquad \text{COOH} \quad \text{COOH}$$

Cellulose   Oxidized cellulose

These acidic groups serve to neutralize the alkali in the water. Moses didn't have to understand the science of water purification. He merely applied the appropriate technology.

Biological Contamination: The Need for Clean Water

The pollutants in the waters of Marah were natural ones. Early people did little to pollute the water and the air, if only because their numbers were so few. It was with the coming of the agricultural revolution and the rise of cities that there were enough *Homo sapiens* to pollute the environment seriously. Even then, the pollution was mostly local and largely biological. Human wastes were dumped on the ground or into the nearest stream. Disease organisms were transmitted through food, water, and direct contact.

Contamination of water supplies by microorganisms from human wastes was a severe problem throughout the world until about a hundred years ago. During the 1830s, severe epidemics of cholera swept the Western World. Typhoid fever and dysentery were common. In 1900, for example, there were over 35 000 deaths from typhoid in the United States. Today, as a result of chemical treatment, municipal water supplies in the advanced nations are generally safe. However, water-borne diseases are still quite common in much of Asia, Africa, and Latin America. Indeed, it is estimated that 80% of all the world's sickness is caused by contaminated water. People with water-borne diseases fill half the world's hospital beds and die at a rate of 25 000 people a day. Less than 10% of the people of the world have access to sufficient clean water.

How much water does one really need? Only about 1.5 L a day for drinking: the rest is luxury. In the United States each day, we use about 7 L per person for drinking and cooking; 110 L for cleanliness (bathing, dish washing, laundering, and housecleaning); 80 L for flushing the toilet; and 85 L for swimming pools and lawns. We use much more *indirectly* in agriculture and industry to produce food and other materials; it takes 800 L of water to produce 1 kg of vegetables and 13 000 L of water to produce a steak. We also use water for recreation (for example, swimming, boating, and fishing). For most of these purposes, we need water free of disease-causing bacteria, viruses, and parasitic organisms.

The threat of biological contamination has not been totally eliminated from the developed nations. It is estimated that 30 million people in the United States are at risk because of bacterial contamination of drinking water. Hepatitis, a viral disease occassionally spread through drinking water, at times threatens to reach epidemic proportions, even in the most advanced nations. Biological contamination also lessens the recreational value of water. Swimming is forbidden in many areas (Figure 16.7).

Biological contamination may have reached a peak in the 1960s. Lake Erie was described as "America's Dead Sea" and "the world's largest cesspool." Polluted waters were not unique; unpolluted waters were. The St. Croix River (Figure 16.8), which forms part of the border between Minnesota and Wisconsin, was described as the *only* major unpolluted river near a sizable metropolitan area (Minneapolis–St. Paul).

Figure 16.7 A sign of the times all too common near our lakes and streams. [Courtesy of Ward's Natural Science Establishment, Inc.]

Figure 16.8 The St. Croix River, as yet a relatively unpolluted river. [Courtesy of the Wisconsin Department of Natural Resources, Madison.]

Much improvement has been made in recent years. Fish life thrives again in Lake Erie. Beaches in Lake Michigan near Chicago, closed in 1969 because of pollution, were reopened in 1975. The Mississippi below Minneapolis–St. Paul has been reopened to recreation and swimming after being closed for many years. Shellfish are being harvested again in Maine's Belfast Bay. Shrimp and oysters are returning to Escambia Bay off Pensacola, Florida. Atlantic salmon have returned to the Connecticut River for the first time in 100 years. And, in England, fish once again inhabit the Thames. Despite all the good news, however, pollution from domestic sewage remains the gravest threat to the water supply in the United States. We still have a long way to go.

Chemical Contamination: From Farm, Factory, and Home

The industrial revolution added a new dimension to our water pollution problems. Factories often were built on the banks of streams, and chemicals and other wastes from them were dumped into the streams to be carried away. The rise of modern agriculture also has led to increased chemical contamination as fertilizers and pesticides have found their way into the water system. Transportation of petroleum results in oil spills in oceans, estuaries, and rivers. Acids enter waterways from mines and factories and from acid precipitation. Household chemicals also contribute to water pollution when detergents, solvents, and other chemicals are dumped down drains.

We consider the various aspects of water pollution in later sections of this chapter. First, though, let's look at water as it occurs in nature.

Natural Water Isn't All H_2O

Rainwater carries dust particles from the atmosphere to the ground. Rainwater also dissolves a little oxygen, nitrogen, and carbon dioxide as it falls through the atmosphere. During electrical storms, lightning causes nitrogen, oxygen, and water vapor to combine to form nitric acid. Traces of this, too, are found in rainwater.

As water moves along or beneath the surface of the Earth, it dissolves minerals and matter from decaying plants and animals. Recall that minerals (salts) are ionic, and that ions have either positive or negative charges. The principal positive ions (cations) in natural water are sodium (Na^+), potassium (K^+), calcium (Ca^{2+}), magnesium (Mg^{2+}), and sometimes iron (Fe^{2+} or Fe^{3+}). The negative ions (anions) are usually sulfate (SO_4^{2-}), bicarbonate (HCO_3^-), and chloride (Cl^-).

The presence of calcium, magnesium, and iron salts in the water supply makes the use of soap inconvenient. Water containing these ions is called **hard water**. The positive ions react with the negative ions in soap (Chapter 20) to form a scum that clings to clothes and leaves them dingy looking. The same scum is responsible for the familiar bathtub ring.

Sewage: Some Chemistry and Biology

Pathogenic (disease-causing) **microorganisms** are not the only problem caused by dumping human sewage into our waterways. The breakdown of organic matter by bacteria depletes dissolved oxygen in the water and enriches the water with plant nutrients. A stream can handle a small amount of waste without difficulty, but when massive amounts of raw sewage are dumped into a waterway, undesirable changes occur.

Most organic material can be broken down (degraded) by microorganisms. This biodegradation can be either **aerobic** or **anaerobic**. Aerobic oxidation occurs in the presence of dissolved oxygen. Oxidation of organic matter results in the depletion of oxygen (Figure 16.9). A measure of the amount of oxygen needed for this degradation is the **biochemical oxygen demand** (BOD). The greater the quantity of degradable organic wastes, the higher the BOD. If the BOD is high enough, no life (other than odor-producing anaerobic microorganisms) can survive

Figure 16.9 Common sewage from homes and businesses depletes the dissolved oxygen in water. [Courtesy of Henry H. Valiukas, Minnesota Environmental Control Citizens Association, St. Paul.]

Figure 16.10 An algal bloom on a nutrient-enriched lake. [Reprinted with permission of the Minneapolis Star and Tribune, Minneapolis. MN.]

in the lake or the stream. Flowing streams can regenerate themselves. Rapid ones soon come alive again as oxygen is dissolved by the moving water. Lakes with little or no flow can remain dead for decades.

With adequate dissolved oxygen, aerobic bacteria (those that require oxygen) oxidize the organic matter to carbon dioxide, water, and a variety of inorganic ions [such as nitrates (NO_3^-), nitrites (NO_2^-), phosphates (PO_4^{3-}), sulfates (SO_4^{2-}), and bicarbonates (HCO_3^-)]. The water is relatively clean, but the ions, particularly the nitrates and phosphates, may serve as nutrients for the growth of algae. These algal blooms cause problems, also. When the algae die, they become a part of the organic wastes and increase the BOD (Figure 16.10). This process is called **eutrophication** (Color Plate Q). Algal bloom and die-off are also stimulated by the runoff of agricultural fertilizers (Chapter 17) and by the phosphates in detergents (Chapter 20). This combination leads to dead and dying streams and lakes that an overburdened nature cannot purify nearly as quickly as we can pollute.

When the dissolved oxygen in a body of water is depleted by too much organic matter—whether from sewage, dying algae, or other sources—degradation shifts to an anaerobic process. Anaerobic bacteria thrive in the absence of oxygen. Instead of oxidizing the organic matter, they reduce it. Sulfur is converted to hydrogen sulfide (H_2S) and to foul-smelling organic compounds such as methanethiol (CH_3SH). Nitrogen is reduced to ammonia and to odorous amines (Chapter 11). The foul odors are a good indication that the water is overloaded with organic wastes. No life, other than the anaerobic microorganisms, can survive in such water.

Ecological Cycles

Let's consider water pollution from the point of view of a fish. Take a look at the much simplified ecological cycle in Figure 16.11. Such a cycle—in a more complicated form—could take place in a small lake. Fish in the water produce organic wastes. Bacteria break down these wastes into inorganic materials. These inorganic materials serve as nutrients for the growth of algae. Fish eat the algae, balance is established, and the cycle is complete. This is a biological cycle, but chemistry can lend some insight into it by providing an understanding of the breakdown of materials, the processes of growth, and the role of nutrients.

Now let's look at some of the ways people can disrupt the cycle. We can increase the organic wastes by dumping sewage into the water. In breaking down these wastes, bacteria consume all the dissolved oxygen and the fish die. Chemists can monitor the BOD of the water, thus identifying the problem and gauging its severity. Chemists also can

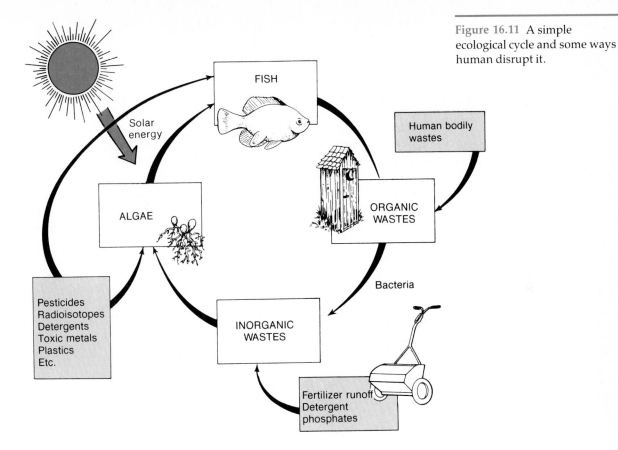

Figure 16.11 A simple ecological cycle and some ways human disrupt it.

contribute to the development of better methods of wastewater treatment, as we see a bit later.

Dumping our sewage into the water isn't the only way we can foul up an ecological cycle. We can and do introduce phosphates from our detergents. Fertilizer runoff and seepage from feedlots add inorganic nutrients to the cycle, and an algal bloom can lead to oxygen depletion and death for the fish. Perhaps the most enigmatic influence of all, though, comes through the introduction of new substances into the ecological water cycle: pesticides, radioisotopes, detergents, toxic metals, plastics.

By creating new materials, chemists have made a number of ecological problems possible. But chemists and chemistry are also necessary to any understanding of our pollution problems. It is only through the development of sophisticated analytical methods and instruments that we are even aware of some of our ecological problems. A stinking lake would be obvious to everyone, but only a person trained in chemistry could determine the level of many dangerous, yet invisible, materials in the water we drink.

Groundwater Contamination

In recent years, another, new water-pollution problem has become apparent: the contamination of groundwater. Half the people in the United States depend on groundwater for drinking, and toxic chemicals have been found in the groundwater supplies for many communities. People living around the Rocky Mountain Arsenal, near Denver, have found their wells contaminated by wastes from the production of nerve gas and pesticides. Wells in St. Louis Park, Minnesota, are contaminated with creosote, a chemical used as a wood preservative. Wells in Wisconsin and on Long Island have been contaminated with aldicarb, a pesticide used on potato crops. Community water supplies in New Jersey have been shut down because of contamination with industrial wastes.

$$\begin{array}{ccc} CH_3 & & O \\ | & & \| \\ CH_3CCH{=}NOCNHCH_3 \\ | \\ CH_3 \end{array}$$

Aldicarb

Chemicals buried in dumps—often years ago, before there was much awareness of environmental problems—have now infiltrated groundwater supplies. Often, as at the Love Canal site in Niagara Falls, New York, schools and houses were built on or near old dump sites. Common contaminants are hydrocarbon solvents such as benzene and toluene (Chapter 10) and chlorinated hydrocarbons such as carbon tetrachloride, chloroform, and methylene chloride (Chapter 11). Especially common is trichloroethylene, widely used as a drycleaning solvent and as a degreasing compound.

$$CCl_2{=}CHCl$$

Trichloroethylene

Except for toluene, all these compounds are suspected carcinogens.

Generally, hydrocarbons and chlorinated hydrocarbons are regarded as insoluble in water. That is a likely conclusion if you try to dissolve one of the compounds in water. However, they do have a slight solubility, often in the parts per million or parts per billion range. It is these slight traces that are found in groundwater. Another property of these compounds is their lack of reactivity. They react only slowly with water or other substances in the ground. They are likely to be there for a long time. The effect of these substances on human health—consumed as they usually are in tiny amounts, but over many years—is as yet unknown.

Another major source of groundwater contamination is **LUST**—an attention-grabbing acronym for leaking underground storage tanks. Gasoline at service stations is usually stored in buried steel tanks. There are perhaps 2.5 million such tanks in the United States. The tanks last an average of about 15 years before they rust through and begin to leak. As many as 200 000 of these tanks may be leaking. Many of them are at stations that went out of business during the fuel shortages of the 1970s.

Gasoline is now being discovered in wells near these tanks in many areas of the country, but the full extent of the problem is not yet known.

Groundwater contamination is particularly alarming because, once contaminated, an underground aquifer may remain unusable for decades or longer. There is no easy way to remove the contaminants. Pumping out the water and purifying it could take years and cost billions of dollars.

The problem of groundwater pollution is serious, but it is often over-dramatized by the news media. The amounts of contaminants are often minute. Chemists now can measure parts per million (ppm) and parts per billion (ppb) on a routine basis. Some contaminants can even be detected at parts per trillion (ppt) or in lesser amounts. Without these sophisticated chemical techniques, we wouldn't even know about some problems. For example, the U.S. Environmental Protection Agency has set a safety limit of 10 ppb for aldicarb—that's 10 mg of aldicarb in 1 000 000 L of water. You would have to drink 32 million L of water to get as much aldicarb as there is aspirin in one tablet! Aldicarb *is* considerably more toxic than aspirin. Contamination of groundwater by it or other toxic substances is serious and should not be taken lightly, but we must keep in mind the fantastically small amounts of toxic materials involved.

Toxic substances in general and toxic wastes in particular are discussed further in Chapter 25.

Acid Waters

Acids pour down upon us from the heavens as acid rain and snow (Chapter 15). These acids corrode metals and limestone and marble edifices, and even ruin the finishes on our automobiles. Acids also are belched up from the bowels of the Earth, draining from abandoned mines (Figure 16.12).

Acidic water is detrimental to life in lakes and streams. Over 300 lakes in New York contain no fish, and 140 lakes in Ontario are devoid of any life. Another 48 000 lakes in Ontario alone are threatened, as are many others in northeastern and north-central United States and eastern Canada. The threat to these areas is presumed to be acid rain originating in the Ohio Valley and Great Lakes regions. The sulfur oxides and nitrogen oxides from power plants, industries, and automobiles in these areas come down hundreds of kilometers downwind as sulfuric and nitric acids. Acid rain has also been linked to decreased soil fertility and declining crop and forest yields.

Because it is difficult to trace pollutants through the atmosphere, the causes of acid rain have been hotly disputed. The effects of acid waters on living organisms are also hard to pin down precisely. Probably

Figure 16.12 Acid forms in water draining from an abandoned mine. [Courtesy of U.S. Environmental Protection Agency.]

the greatest effect of acidity is that it causes the release of toxic ions from rocks and soil. For example, aluminum ions are tightly bound in clays and other minerals. Acids cause the release of the bound ions by a reaction similar to the following.

$$Al_2Si_2O_5(OH)_4 \ + \ 6 \ H_3O^+ \ \longrightarrow \ 2 \ Al^{3+} \ + \ 2 \ SiO_2 \ + \ 11 \ H_2O$$

$$\underset{\text{A clay}}{} \qquad \underset{\text{Acid}}{} \qquad \underset{\text{Aluminum ions}}{} \qquad \underset{\text{Sand}}{}$$

Aluminum ions are not particularly toxic to humans, but they seem to be deadly to young or small fish. Many of the dying lakes have only old fish; none of the young survive. Ironically, lakes destroyed by excess acidity are often quite beautiful. The water is clear and sparkling— quite a contrast to those in which fish are killed by algal blooms.

Acids are no threat to lakes and streams in areas where the rock is limestone (calcium carbonate); the latter neutralizes excess acid.

$$CaCO_3 \ + \ 2 \ H_3O^+ \ \longrightarrow \ Ca^{2+} \ + \ CO_2 \ + \ 3 \ H_2O$$

$$\underset{\text{Limestone}}{} \qquad \underset{\text{Acid}}{}$$

Where rock is principally granite, no such neutralization occurs. There, waters could be treated with pulverized limestone (or other basic substances). A few such attempts have been carried out, but the process is costly and the results seem to be short-lived. If the presumed causes of acid precipitation are correct, an obvious way to remedy the situation is to remove the sulfur from the coal before combustion (Chapter 14) or scrub the sulfur oxides from the smokestack gases (Chapter 15). All such solutions are expensive, however, and their implementation adds to the cost of electricity.

Industrial Water Pollution

United States industries have come a long way toward eliminating water pollution. The chemicals industries, for example, are 98% in compliance with the Water Pollution Control Act, which requires application of the "best practicable technology" to water pollution problems. Nevertheless, in certain areas, industrial water pollution remains a significant problem. We can't examine all the different types of industrial water pollution here, but we can look at a few typical examples.

It takes about 900 kg of steel to make an average-size American automobile. To make that steel, it takes about 100 m^3 of water. Only about 4 m^3 are lost through evaporation. The remainder is polluted with acids (mainly hydrochloric acid), grease and oil, lime, and iron salts.

Chrome plating on bumpers, grills, and ornaments is also a source of pollution. Waste chromium [in the form of chromate ions (CrO_4^{2-})] and cyanide ions (CN^-) are products of this process. In the past, these toxic substances were dumped into waterways. Nowadays, they generally are removed (at least partially) by chemical treatment.

Cyanide is treated with chlorine to form nitrogen gas, bicarbonate ions, and chloride ions.

$$10\ OH^- \ + \ 2\ CN^- \ + \ 5\ Cl_2 \ \longrightarrow \ N_2 \ + \ 2\ HCO_3^- \ + \ 10\ Cl^- \ + \ 4\ H_2O$$

The products of this treatment, bicarbonate and chloride, are still pollutants, but they are much preferable to the highly toxic cyanide. Chromate is removed by reduction with sulfur dioxide. The chromium winds up as the Cr^{3+} ion, the sulfur as sulfate.

$$2\ CrO_4^{2-} \ + \ 3\ SO_2 \ + \ 2\ H_2O \ \longrightarrow \ 2\ Cr^{3+} \ + \ 3\ SO_4^{2-} \ + \ 4\ OH^-$$

Sulfate is still a pollutant, but it is generally not a serious one. Cr^{3+} is relatively insoluble in alkaline solution but is soluble enough in acidic media to constitute a problem.

Now, add in the cost (environmental and economic) of elastomers for tires, fabrics for upholstery, glass for windows, 520 kWh of electricity—the list goes on and on—and it is easy to see that the private automobile is an ecological disaster even before it hits the road. And once on the road, it is a major contributor to air pollution (Chapter 15).

Most other industries also contribute to water pollution. Table 16.2 lists the water required (per metric ton) for the production of a variety of materials. Now let's look at some of the pollutants from various industries.

The textile industry dumps conditioners, dyes, bleaches, and water effluents containing oils, dirt, and other organic debris. Meat-packing plants dump blood, the contents of entrails, and other animal wastes.

Table 16.2 Water Required to Produce Various Materials

Material	Water Required*
Steel	100
Paper	20
Copper	400
Rayon	800
Aluminum	1280
Synthetic rubber	2400

*In cubic meters per metric ton (t). A cubic meter of water weighs 1000 kg, or 1 t.

Other food-processing plants discharge fruit and vegetable skins, seeds, leaves, stems, and other vegetable wastes. Refineries release dyes, oils, acids, brines, sulfur compounds, and other wastes. Chemicals plants produce a variety of waste materials. There are treatment methods available for processing most wastes and most industries are in compliance with the Water Pollution Control Act. Let's turn our attention now to municipal sewage treatment plants, where many are still not in compliance.

Wastewater Treatment Plants

Many communities have **primary sewage treatment plants**. This type of plant has holding tanks or ponds where the sewage is allowed to settle for a while (Figure 16.13). This removes some of the solids as sludge. The primary step involves a very large BOD. Often, all the dissolved oxygen in the pond is used up, and anaerobic decomposition—with its resulting odors—takes over. Effluent from a primary treatment plant contains a lot of dissolved and suspended organic matter, including pathogenic bacteria. A **secondary treatment plant** passes effluent from the primary treatment through sand and gravel filters. There is some aeration in this step, and aerobic bacteria convert most of the organic matter to stable inorganic materials.

A combination of primary and secondary treatment methods, known as the **activated sludge method** (Figure 16.14), is also commonly employed. The sewage is placed in tanks and aerated with large blowers. This causes the formation of large, porous clumps called **flocs**. These serve to filter and absorb contaminants. The aerobic bacteria further convert the organic material to sludge. A part of the sludge is recycled to keep the process going, but huge quantities must be removed for disposal. This sludge is stored on land (where it requires large areas), dumped at sea (where it pollutes the ocean), and burned in incinerators (where it requires energy—such as natural gas—and

Figure 16.13 A diagram of a primary sewage treatment plant.

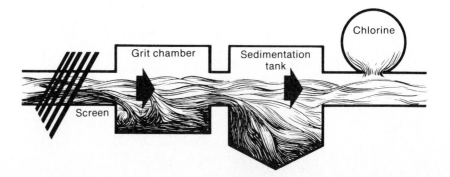

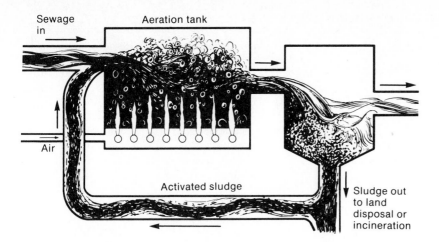

Sewage in

Aeration tank

Air

Activated sludge

Sludge out to land disposal or incineration

Figure 16.14 A diagram of a secondary wastewater treatment plant that uses the activated sludge method.

contributes to air pollution). Sometimes the sludge is used as fertilizer (see next section).

The effluent from sewage plants is treated with chlorine before it is returned to a waterway. Chlorine is added in an attempt to kill any pathogenic microorganisms that might remain. Such treatment has been quite effective in preventing the spread of water-borne infectious diseases such as typhoid fever. Further, some chlorine remains in the water, providing residual protection against pathogenic bacteria. However, chlorination is not effective against viruses, such as those that cause hepatitis. Use of chlorine also was called into question in 1974, when it was shown that chlorination converted dissolved organic compounds into chlorinated hydrocarbons. Many chlorinated hydrocarbons, including such known carcinogens as chloroform and carbon tetrachloride, have been found in the drinking water of several cities that take their water from rivers. The concentration of these materials is in the parts-per-billion range. The threat they pose is probably small, but it's worrisome, nonetheless (not nearly so worrisome, however, as the water-borne diseases that prevail in much of the world where adequate water treatment is not available).

Chlorine is also used to treat drinking water (page 364). Ozone also has been used. Many European cities, including Paris and Moscow, use ozone to treat their drinking water. Ozone is more expensive than chlorine, but less of it is needed. An added advantage is that ozone kills viruses on which chlorine has little, if any, effect. Tests in Russia have shown ozone to be 100 times as effective as chlorine for killing polio viruses.

Ozone (O_3) acts by transferring its "extra" oxygen to the contaminant. The oxidized contaminants are thought to be less toxic than the chlorinated ones. In addition, ozone imparts no chemical taste to the water. Ozone does have the disadvantage that it provides no residual protection against microorganisms. If the problems with chlorine get

Table 16.3 Summary of Wastewater Treatment Methods

Method*	Cost	Material Removed	Percent Removal
Primary			
Sedimentation	Low	Dissolved organics	25–40
		Suspended solids	40–70
Secondary			
Trickling filters	Moderate	Dissolved organics	80–95
		Suspended solids	70–92
Activated Sludge	Moderate	Dissolved organics	85–95
		Suspended solids	85–95
Advanced (tertiary)			
Carbon bed with regeneration	Moderate	Dissolved organics	90–98
Ion exchange	High	Nitrates and phosphates	80–92
Chemical precipitation	Moderate	Phosphates	88–95
Filtration	Low	Suspended solids	50–90
Reverse osmosis	Very high	Dissolved solids	65–95
Electrodialysis	Very high	Dissolved solids	10–40
Distillation	Extremely high	Dissolved solids	90–98

*See Reference 1 for discussion of methods not covered in the text.

worse, though, we may see a shift from chlorine to ozone in the treatment of our drinking water and wastewater. Perhaps ozone could be used to disinfect the water; then just enough chlorine added to provide residual protection.

In many areas, secondary treatment of wastewater is inadequate. **Advanced treatment** (sometimes called **tertiary treatment**) has become increasingly important. A number of advanced processes are in use. One process, charcoal filtration, promises to become increasingly important. Charcoal absorbs organic molecules that are difficult to remove by any other method.

Advanced treatments often are designed to remove phosphates and nitrates. Partial phosphate removal is relatively easy. Aluminum ions from alum [$Al_2(SO_4)_3$] precipitate phosphate ions as insoluble aluminum phosphate.

$$Al^{3+} + PO_4^{3-} \longrightarrow AlPO_4$$

Microorganisms and higher plants also have been used, with some success, to remove phosphates. Nitrate removal is more difficult. Costly processes such as **reverse osmosis** (pressure filtration through a semipermeable membrane) may be needed in some cases. Finding the money to finance adequate sewage treatment will be a major political problem for years to come. (See Table 16.3 for a summary of wastewater treatment methods.)

Back to the Soil: An Alternative Solution

Why do we dump our wastes in the water in the first place? Dumping fouls the water with wastes that contain nutrients that could fertilize the land. Why not return the nutrients to the soil? In many societies, this has always been done for human and animal wastes. It is only in the "overdeveloped" countries that these vital resources are dumped into the waterways on a large scale.

Several pilot projects have been completed on the return of activated sludge to the soil. The sludge can be dried, sterilized, and transported to farmlands. One problem, however, is the high cost of transporting the sludge. This has been partially overcome by piping the suspended sludge directly to the fields. The water in the mixture irrigates the crops; the sludge is a source of nutrients and humus. Another concern is that the pathogens in the sludge will survive and spread disease. The biggest problem, though, seems to be that much of sludge is contaminated with toxic metals. These metals might be taken up by the plants and eventually end up in our food.

Other solutions are possible. A Swedish inventor, Rikard Lindström, has developed a toilet that composts wastes and uses no energy or water. The mild heat of compositing drives off water from the wastes. The system is ventilated to keep the process aerobic. No odors enter the house from a properly installed system. Dried waste is removed about once a year. The initial cost is much higher than that of a flush toilet, however. In the United States, we flush about 50 000 L each of drinking-quality water per person annually. Perhaps we should consider an alternative to flushing our wastes into the water we drink.

A Drop to Drink

The water we drink often comes from reservoirs, lakes, and rivers. A number of cities use water that has been used by other cities upstream. Such water may be badly polluted with chemicals and pathogenic microorganisms. To make it safe and palatable involves several steps of chemical and physical treatment (Figure 16.15).

The water to be purified usually is placed in a settling basin, where it is treated with slaked lime and a flocculant such as aluminum sulfate. These materials react chemically to form a gelatinous mass, aluminum hydroxide.

$$3 \text{ Ca(OH)}_2 + \text{Al}_2(\text{SO}_4)_3 \longrightarrow 2 \text{ Al(OH)}_3 \longrightarrow 3 \text{ CaSO}_4$$

| Slaked lime | Aluminum sulfate | Aluminum hydroxide | Calcium sulfate |

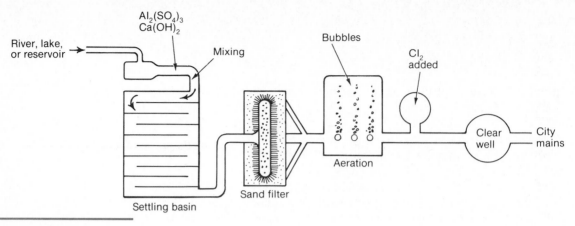

Figure 16.15 A schematic diagram of a municipal water-purification plant.

The aluminum hydroxide carries down dirt particles and bacteria. The water then is filtered through sand and gravel.

Sometimes water is **aerated**—sprayed into the air to remove odors and improve its taste (water without dissolved air tastes flat). Sometimes the water is filtered through charcoal to remove colored and odorous compounds. In the final step, chlorine is added to kill any remaining bacteria. In some communities that use river water, a lot of chlorine is needed to kill all the bacteria, and you can taste the chlorine in the water.

We generally take our drinking water for granted. Perhaps we shouldn't. There are at least 4000, and perhaps as many as 40 000, cases of water-borne illnesses in the United States each year. Twenty million people have no running water at all, and many of these people get water from suspect sources. Another 30 million tap individual wells or springs. Many of these sources are of uncontrolled and unknown quality. Much remains to be done before we can all be assured of safe drinking water.

Poison in the Drinking Water: Fluorides

Fluoride salts are acute poisons in moderate to high concentrations. Indeed, sodium fluoride (NaF) is used as a poison against such pests as roaches and rats. Small amounts of fluoride ion, however, are essential for our well-being. Up to a point, the hardness of our tooth enamel can be correlated with the amount of fluoride present. Tooth enamel is a complex calcium phosphate called hydroxyapatite. Fluoride ions replace some of the hydroxides, forming a harder mineral called fluorapatite.

$$Ca_{10}(PO_4)_6(OH)_2 \ + \ 2\,F^- \ \longrightarrow \ Ca_{10}(PO_4)_6F_2 \ + \ 2\,OH^-$$

Concentrations of 0.7 to 1.0 ppm (by weight) of fluoride (usually as H_2SiF_6 or Na_2SiF_6) have been added to the drinking water of many communities. Evidence indicates that such fluoridation results in a reduction in the incidence of dental caries (cavities) by as much as 50% in some areas. And while only 29% of the nine-year-olds in the United States were cavity-free in 1971–1973, 51% of that age group were cavity-free in 1985. Interpretation of these statistics is complicated, however, because of the varying occurrence of fluorides in the diet and because people retain fluorides at remarkably different rates. One thing you *don't* have to worry about is acute poisoning from fluoridated drinking water. You would have to drink 4000 L of water containing 1.0 ppm of fluoride to approach the lethal dose of 4 g. And that 4000 L would have to be consumed in a short period of time.

There is some concern about cumulative effects of consuming fluorides in drinking water, in the diet, in toothpaste, and from other sources. Excessive fluoride consumption during early childhood can cause mottling of the tooth enamel (Figure 16.16). The enamel becomes brittle in certain areas and gradually discolors. Fluorides in high doses also interfere with calcium metabolism, with kidney action, with thyroid function, and with the action of other gland and organs. Although there is little or no evidence that optimal fluoridation causes problems such as these, fluoridation of public water supplies will most likely remain a subject of controversy.

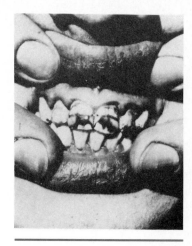

Figure 16.16 Excessive fluoride consumption in early childhood can cause mottling of tooth enamel. A severe case caused by continuous use during childhood of a water supply with an excessive natural concentration of fluoride is shown here. [Courtesy of the National Institute of Dental Research, Washington, DC.]

Poison in the Drinking Water: Nitrates

Some communities (usually smaller ones) and many families obtain water from wells. There's even trouble with well water. Because the water table is dropping in most parts of the country, ever deeper wells are required. And even the well water is contaminated. In many agricultural areas, the concentration of nitrates (NO_3^-) in well water is above the maximum safe level (for infants) of 10 ppm. For adults, the maximum safe level is 50 ppm. Even this level is being approached in some areas.

Excessive nitrates (which are reduced to nitrites in the digestive tract) cause methemoglobinemia, the blue baby syndrome. In parts of Illinois, baby pigs turn blue and die after drinking the water. In California's Imperial Valley, some parents have to buy bottled water for their babies.

Nitrates in the groundwater come from agricultural fertilizers, from decomposition of organic wastes in sewage treatment, and from runoff from animal feedlots. They are highly soluble and thus difficult to remove from water. Only expensive advanced treatment can remove these compounds from water once they are there.

You're the Solution to Water Pollution

The United States Congress has passed laws intended to drastically reduce water pollution. It can't be eliminated entirely, however. To use water is to pollute it. You can do your share by conserving water and by minimizing your use of products that require vast amounts of water for production. And you, the citizen, must be prepared to foot the bill for waste treatment. As our population grows, it will cost plenty just to maintain the present (and often inadequate) water quality. To clean our water up, and then keep it clean will cost even more. Remember that the cost of unclean water is even higher—discomfort, loss of recreation, illness, even death.

Problems

1. Ice is less dense than liquid water. What consequences does this property have for life in northern lakes?

2. Why should foods be flash-frozen rather than frozen slowly?

3. A barge filled with gasoline sinks and breaks open. Would the gasoline dissolve in the water? Would it float or sink?

4. Define heat capacity. Why is the high heat capacity of water important to planet Earth?

5. Why is the high heat of vaporization of water important to our bodies?

6. Why is it cooler near a lake than inland during the summer?

7. Why is ice less dense than liquid water?

8. Why are the seas salty?

9. What proportion of Earth's water is seawater?

10. How is our supply of fresh water replenished by natural processes?

11. List some water-borne diseases. Why are these no longer common in developed countries?

12. What impurities are present in rainwater?

13. List four cations present in groundwater.

14. List three anions present in groundwater.

15. What is hard water? Why is it sometimes undesirable?

16. What is BOD? Why is a high BOD undesirable?

17. What are pathogenic microorganisms?

18. What are the products of the breakdown of organic matter by aerobic bacteria?

19. What is eutrophication?

20. List some of the products of anaerobic decay.

21. List some ways in which groundwater is contaminated.

22. What problems are caused by LUST?

23. List some common industrial contaminants of groundwater.

24. Why do chlorinated hydrocarbons remain in groundwater for such a long time?

25. List two ways by which lakes and streams have become acidic.

26. Why is acidic water especially harmful to fish?

27. List several ways by which the acidity of rain can be reduced.

28. What kind of rocks tend to neutralize acidic waters?

29. How can we restore (at least temporarily) lakes that are too acidic?

30. List two toxic compounds found in wastes from the chrome plating process. How is each removed?

31. Describe a primary sewage treatment plant.

32. What impurities are removed by primary sewage treatment?

33. Describe a secondary sewage treatment plant.

34. What impurities are removed by secondary sewage treatment?

35. Describe the activated sludge method of sewage treatment.

36. What substances remain in wastewater after effective secondary treatment?

37. Why is wastewater chlorinated before it is returned to a waterway?

38. List the advantages and disadvantages of chlorination of wastewater.

39. List the advantages and disadvantages of ozone as a disinfectant of wastewater.

40. What is meant by advanced (or tertiary) treatment of wastewater?

41. What kinds of substances are removed from wastewater by charcoal filtration?

42. How does alum (aluminum sulfate) remove phosphates from water?

43. What are the advantages and disadvantages of spreading sewage sludge on farmlands?

44. Describe an alternative to the flush toilet.

45. Give the equation for the reaction of slaked lime (calcium hydroxide) with alum (aluminum sulfate) to form aluminum hydroxide. What is the purpose of this reaction in water purification?

46. Why are municipal water supplies aerated?

47. How does fluoride strengthen tooth enamel?

48. What are some health effects of too much fluoride in the diet?

49. What is the source of nitrate ions in well water?

50. What is methemoglobinemia? How is it caused?

51. Consult the references or other sources and prepare a brief report on the desalination of seawater by one of the following methods.
    a. distillation   b. freezing
    c. electrodialysis   d. reverse osmosis
    e. ion exchange

52. Radioactive aluminum-26 is used to study mobilization of aluminum by acidic waters. The isotope is transmuted into magnesium-26. What kind of particle is emitted in the process?

References and Readings

1. American Chemical Society. *Cleaning Our Environment: A Chemical Perspective*, 2d ed. Washington, DC: American Chemical Society, 1978.

2. Anderson, Duncan. "Poisoned or Pure?" *American Health*, July–August 1982, pp. 72–75. A guide to water quality around the United States.

3. Broecker, Wallace S. "The Ocean." *Scientific American*, September 1983, pp. 146–160.

4. Cotton, Charles. "Water Treatment Chemicals." *ChemTech*, June 1984, pp. 345–347.

5. Council on Environmental Quality, *Contamination of Groundwater by Toxic Organic Chemicals*. Washington DC: U.S. Government Printing Office, January 1981.

6. Cousteau, Jacques-Ives. *The Cousteau Almanac*. Garden City, NY: Doubleday, 1981.

7. Feder, Bernard. "Is the Flush Toilet Obsolete?" *Science 83*, November 1983, pp. 109–110.

8. Hart, John. *Consider a Spherical Cow: A Course in Environmental Problem Solving*. Los Altos, CA: William Kaufmann, 1985.

9. Noether, Dorit L. "Water Is . . ." *ChemTech*, February 1982, pp. 84–90.

10. "Purifying Water, Cleaning Air: Chemicals Aid Our Environment." *ChemEcology*, March 1982, pp. 2–3.

11. Pye, Veronica I., and Ruth Patrick. 'Groundwater Contamination in the United States." *Science*, 19 August 1983, pp. 713–718.

12. Shell, Ellen Ruppel. "The New Flap Over Fluorides." *American Health*, October 1984, pp. 60–63.

13. Spiro, Thomas G., and William M. Stigliani. *Environmental Issues in Chemical Perspective*. Albany: State University of New York Press, 1980.

14. Steven, James H. "Separations Based on Electrodialysis, Reverse Osmosis, and Ultrafiltration." *Chemistry and Industry*, 2 May 1983, pp. 346–349.

15. Sun, Marjorie, "Ground Water Ills: Many Diagnoses, Few Remedies." *Science*, 20 June 1986, pp. 1490–1493.

16. Wehle, D. H. S., and Felicia C. Coleman. "Plastics at Sea." *Natural History*, February 1983, pp. 20–26.

17

Farm Chemistry

Food for a Hungry World

In 1830, Thomas Robert Malthus, an English clergyman and political economist, made the statement that population increases faster than the food supply. Unless the birthrate was controlled, he said, poverty and war would have to serve as restrictions on the increase.

The population of planet Earth has grown enormously since Malthus's time, reaching 5 billion in 1986. Famine has brought death to millions in Ethiopia and other war-plagued areas. But farmers in developed nations have produced enormous surpluses of food, depressing prices and driving many farmers out of business. Even the developing nations have made great strides. In ten years (1971 to 1981) the population of India, the world's second most populous nation, grew by 136 million people. That increase is equal to the population of Brazil, the sixth most populous country. Yet India increased its food production even more rapidly, becoming a net exporter of grain after decades as an importer.

Despite great surpluses of food in some parts of the world, Spaceship Earth still carries 800 million seriously undernourished passengers.

L. Photovoltaic Cells. Exceptionally pure silicon, doped with traces of other elements such as boron and phosphorus, is used to construct photovoltaic cells. These cells convert the radiant energy of sunlight directly to electrical energy. A small array of these solar cells is shown here. [Courtesy of Exxon Corporation.]

M. London Smog. The smog shown here is derived from the burning of coal in electric power plants and other industrial operations. It is characterized by high levels of sulfur dioxide, particulate matter, and (often) carbon monoxide. This type of smog can be minimized by the use of electrostatic precipitators, scrubbers, and other devices. [Courtesy of Environmental Protection Agency.]

N. Photochemical Smog. This type of air pollution results from the action of sunlight on oxides of nitrogen that are emitted from automobiles and other high-temperature combustion sources. Photochemical smog, often called Los Angeles smog, is characterized by an amber haze like that shown here. [Photograph by Sepp Seitz © 1980; Woodfin Camp & Associates.]

O. The Effects of Smog. The effects of smog, both documented and expected, are numerous. Examples include respiratory ailments, crop and forest damage, and acidified lakes. A particularly striking effect is the action of acidic air pollutants on marble, as this statue shows. [Courtesy of CNRI/French Information Services.]

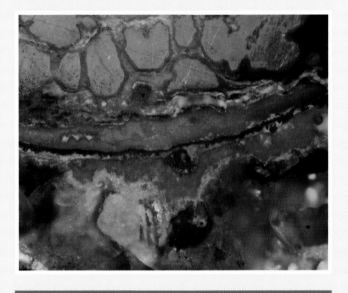

P. Corrosion of Bronze. The corrosion of bronze, a copper alloy, is a natural chemical process. The familiar green patina on copper or bronze objects is a basic copper carbonate.

$$2\,Cu + H_2O + CO_2 + O_2 \longrightarrow Cu_2(OH)_2CO_3$$

In this photomicrograph, magnified approximately $100\times$, of a fragment of ancient bronze found in Thailand, the interface between the metal and its corrosion layer is clearly visible. [Courtesy of William Marin, Jr., Brookhaven National Laboratory.]

Q. Eutrophication of a Lake. The eutrophication of a lake is a
natural process, but the effect can be accelerated greatly by human
effect such as agricultural runoff of fertilizers and the adding of
phosphates to the lake in wastewater. [Courtesy of the Soil Conservation Service, Tom McCabe.]

R. Limestone Caverns. Carbon dioxide from the atmosphere dissolves in rainwater to form an acidic solution. As this acidic rainwater seeps through limestone formations, insoluble calcium carbonate is converted to soluble calcium bicarbonate.

$$CaCO_3 + H_2O + CO_2 \rightleftharpoons Ca(HCO_3)_2$$

Over the centuries, this action can produce a large cave in the limestone formation. The reaction is reversible, however, and as dripping water, saturated with calcium bicarbonate, evaporates, stalactites and stalagmites are formed. [Photograph of Temple of the Sun, Carlsbad Cavern, New Mexico, by Fred E. Mang, Jr.; courtesy of the National Park Service.]

S. Hazardous Materials. This painter's palette holds a number of toxic materials. Popular yellow paints have cadmium sulfide pigments. Red pigments include mercury(II) sulfide and cadmium selenide. Both cadmium and mercury compounds are toxic. For ordinary use, as in house paints, these pigments have been largely replaced by iron oxides, which are relatively safe. [Photograph by Barbara Cushing.]

Figure 17.1 In the developed countries, science has thwarted the Malthusian prediction of hunger. However, starvation is still a fact of life for many people in developing countries, particularly for those in areas wracked by war. In this photo, women and children await emergency relief at Buma, Ethiopia, during the 1983–84 famine. [UNICEF photo of Arild Vollan.]

Half of these are children under five who will carry the physical and mental scars of this deprivation the rest of their lives.

Scientific developments—fertilizers, pesticides, and new genetic varieties of crop plants—have brought abundance to the developed countries and promise the same for developing nations. It was mainly through the increased application of fertilizers that India increased its food production so dramatically.

In this chapter, we examine the agricultural chemicals that help to feed much of the population of planet Earth. Many of these same chemicals are used in public health work to thwart insect-borne diseases. They have helped to keep Malthus's predictions from coming true, at least for now.

Most of the agricultural chemicals are also used around the house, yard, and garden. Although fewer than 3 million people farm in the United States, 83% of our households are engaged in indoor or outdoor gardening. Some 34 million garden for food. About one third of all the

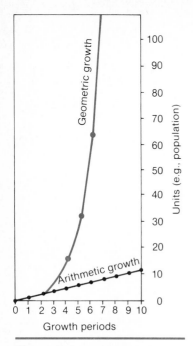

Figure 17.2 Arithmetic and geometric growth through 10 periods (starting with 1 unit).

fertilizers and pesticides used in the United States are applied to lawns, gardens, golf courses, and other nonfarm areas.

Agricultural chemicals have brought us health, wealth, and abundant food. They have also brought us problems. We have already seen (Chapter 1) that Rachel Carson predicted that overuse of pesticides would lead to disaster. We will examine some present and potential problems shortly, but for now let's return to the time of Malthus.

Some Malthusian Mathematics

Malthus's predictions were based on simple mathematics. Population, he said, grows geometrically, while the food supply increases arithmetically. In **arithmetic growth**, a constant amount is added during each growth period. As an example, consider a cookie-jar savings account. The first week, little Lavinia puts in the 25¢ she received for her birthday. Each week thereafter, she adds 25¢. The growth of the savings is arithmetic; it increases by a constant amount (25¢) each week. At the end of the first week, she will have 25¢; at the end of the second, 50¢; the third 75¢; the fourth, $1.00; the fifth, $1.25; the sixth, $1.50; and so on.

Now let's consider an example of **geometric growth**, in which the increment increases in size for each growth period. Again, let's use little Lavinia's bank as an example. The first week she puts in 25¢. The second week she puts in another 25¢ to double the amount (to 50¢). The third week she puts in 50¢ to double this amount again. Each week, she puts in an amount equal to what is already there; she doubles the amount in the bank. At the end of the first week, she has 25¢; the second, 50¢; the third, $1.00; the fourth, $2.00; the fifth, $4.00; the sixth, $8.00; and so on. Before long, little Lavinia will have to start robbing banks to keep up her geometrically growing deposits.

Table 17.1 compares arithmetic and geometric growth through 10 growth periods. These data are shown graphically in Figure 17.2. Note that arithmetic growth is slow and steady; geometric growth starts slowly, then shoots up like a rocket.

Table 17.1 Arithmetic and Geometric Growth Through Ten Periods (Starting with One Unit)

	Growth Period										
	0	1	2	3	4	5	6	7	8	9	10
Arithmetic growth	1	2	3	4	5	6	7	8	9	10	11
Geometric growth	1	2	4	8	16	32	64	128	256	512	1024

For a population growing geometrically, we can calculate the **doubling time** from the **Rule of 70**.* Simply divide the percent of annual growth into 70. For example, the Earth's population is growing 1.7% per year. If it continues to grow at that rate, the population will double in 41 years.

Example 17.1   The population of Mexico City is growing at a rate of 5.5% a year. In 1981, its population was estimated to be 13 million. If growth continues at the same rate, when will the population have doubled to 26 million?
  Divide the annual growth rate into 70.

$$\frac{70}{5.5} = 13$$

The population will reach 26 million in 1994 if the present growth rate continues.

What does all this have to do with chemistry? Well, for one thing, science—with chemistry playing a leading role—made a false prophet of Thomas Malthus, at least for a while. Let's see how it happened. Later on, we look at the prospects for the future.

Plants: Sun-Powered Food-Making Machines

All food comes ultimately from green plants. Although all organisms can transform one type of food into another, only green plants can harness sunlight and use it to convert carbon dioxide and water into the sugars that directly or indirectly fuel all living things.

$$6\ CO_2\ +\ 6\ H_2O\ \xrightarrow{photosynthesis}\ C_6H_{12}O_6\ +\ 6\ O_2$$

This reaction also replenishes the oxygen in the atmosphere (Chapter 15). With other nutrients, particularly compounds of nitrogen and phosphorus, plants can convert the sugars from photosynthesis into proteins, fats, and other chemicals that we use as food.
  The structural elements of plants—carbon, hydrogen, and oxygen—are derived from air and water. Other plant nutrients are taken

---

*The Rule of 70 is derived in the same way as the time for the doubling of savings. Divide 70 by the annual interest rate (compounded daily) to get the number of years in which your money will have doubled.

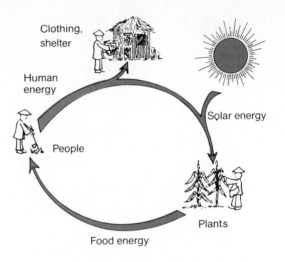

Figure 17.3 Energy flow in a primitive society.

from the soil, and energy is supplied by the sun. In a primitive society, people grow plants for food and obtain energy from the food. Nearly all this energy is reinvested in the production of food, although a portion goes into making clothing and building shelter. Figure 17.3 shows a simplified diagram of the energy flow in a primitive society. Obviously, a real society would be much more complicated than the diagram. Some of the plants, for example, might be fed to animals, and human energy in turn would be obtained from animal flesh or animal products (such as milk and eggs).

In primitive societies, nearly all the energy comes from renewable

Figure 17.4 Flow of nutrients in a primitive system.

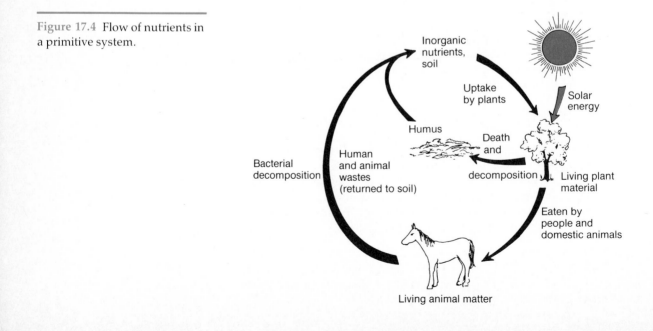

resources. One unit of human work energy, supplemented liberally by energy from the sun, might produce 10 units of food energy. The surplus energy might be used to make clothing or to provide shelter. It also might be used in games or cultural activities.

The flow of nutrients is also rather simple. Unused portions of plants and human and animal wastes are returned to the soil. These are broken down by bacteria to provide the nutrients for the growth of the new plants (Figure 17.4). The humus content of the soil is maintained. Properly practiced, this primitive agriculture could be continued for centuries without seriously depleting the soil. The only problem is that farming at that level doesn't support very many people.

Farming with Chemicals: Fertilizers

To replace nutrients lost from the soil and to increase crop production, modern farmers use a variety of chemical fertilizers. The three primary plant nutrients are nitrogen, phosphorus, and potassium. Let's consider nitrogen first.

Nitrogen, though present in air in the elemental form (N_2), is not generally available to plants. There are some forms of bacteria that are able to fix nitrogen, that is, to convert it to a combined, soluble form. Colonies of bacteria that can perform this vital function grow in nodules on the roots of legumes (plants such as clovers and peas). Thus, farmers are able to restore fertility to the soil by crop rotation. A nitrogen-fixing crop (such as clover) is alternated with a nitrogen-consuming crop (such as corn).

Legumes could still be used to supply nitrogen to the soil, and they still are to some extent. But the modern methods of high-yield production demand chemical fertilizers. Why get a corn crop every other year, when you can have one every year?

Plants usually take up nitrogen in the form of nitrate ions (NO_3^-) or ammonium ions (NH_4^+). These are combined with carbon compounds from photosynthesis to form amino acids, the building blocks of proteins (the polymeric compounds essential to all life processes). For years, farmers were dependent on manure as a source of nitrates. Discovery of deposits of sodium nitrate (called Chile saltpeter) in the deserts of northern Chile led to exploitation of this source as a supplemental source of nitrogen.

A rapid rise in population growth during the late nineteenth and early twentieth centuries led to increasing pressure on the available food supply. This pressure led to an increasing demand for nitrogen fertilizers. The atmosphere offered a seemingly inexhaustible supply—if only it could be converted to a form useful to humans. Every flash of lightning forms some nitric acid in the air. This process probably contributes about 9 kg of nitrogen per hectare of land per year. Henry Cavendish duplicated this natural process in 1776 by passing an electric spark through a mixture of nitrogen and oxygen, but little came of his achievement.

Nitrates and World War I

The first real breakthrough in nitrogen fixation came in Germany on the eve of World War I. The process, developed by Fritz Haber, made possible the combination of nitrogen and hydrogen to make ammonia.

$$3\,H_2 \;+\; N_2 \longrightarrow 2\,NH_3$$

By 1913, one nitrogen-fixation plant was in production and several more were under construction. The Germans were able to make ammonium nitrate (NH_4NO_3), an explosive, by oxidizing part of the ammonia to nitric acid.

$$2\,NH_3 \;+\; 4\,O_2 \longrightarrow 2\,HNO_3 \;+\; 2\,H_2O$$

The nitric acid then was reacted with ammonia to produce ammonium nitrate.

$$HNO_3 \;+\; NH_3 \longrightarrow NH_4NO_3$$

The Germans were interested mainly in ammonium nitrate as an explosive, but it turned out to be a valuable nitrogen fertilizer as well. The same discovery that enabled the Germans to prolong World War I probably helped to postpone for decades Thomas Malthus's predictions of famine.

Figure 17.5 Fritz Haber, the German chemist who invented a process for manufacturing ammonia. [Courtesy of Encyclopaedia Britannica, Inc., Chicago.]

Fritz Haber was awarded the Nobel Prize for chemistry in 1918. There are a number of ironies in this. Alfred Nobel, the Swedish inventor and chemist who died in 1896, endowed the Nobel Prize (including the peace prize) with a fortune derived from his own work with explosives. During his lifetime, he was bitterly disappointed by the fact that the explosives he had developed for excavation and mining were put to such destructive uses in war. And Haber, a man who helped his country during World War I to the extent of going to the front to supervise the release of chlorine in the first poison-gas attack (Chapter 25), was exiled from his native land in 1933. He was a Jew, and Nazi racial laws forced him out of his position as director of the Kaiser Wilhelm Institute of Physical Chemistry. He accepted a post at Cambridge University in England, but his life was ended by a stroke less than a year later.

Figure 17.6 This farmer is applying anhydrous ammonia, a source of the plant nutrient nitrogen, to his fields. [Courtesy of TVA National Fertilizer Development Center, Muscle Shoals, AL (Fred Myers).]

A gas at room temperature, ammonia is easily compressed into a liquid that can be stored and transported in tanks. In this form, called anhydrous (without water) ammonia, it is applied directly to the soil as fertilizer (Figure 17.6).

Some ammonia is reacted with carbon dioxide to form urea, an organic compound that releases the nitrogen slowly to the soil (rather than all at once as the inorganic forms do).

$$2\ NH_3\ +\ CO_2\ \longrightarrow\ NH_2-\overset{\displaystyle O}{\overset{\displaystyle \|}{C}}-NH_2$$
$$\text{Urea}$$

Some ammonia, as we mentioned, is converted to ammonium nitrate, a crystalline solid. Some is converted to solid ammonium sulfate [$(NH_4)_2SO_4$] by reaction with sulfuric acid.

$$2\ NH_3\ +\ H_2SO_4\ \longrightarrow\ (NH_4)_2SO_4$$

These ammonia products (Table 17.2) can be applied separately or combined with other plant nutrients to make a more complete fertilizer.

Table 17.2 Various Forms of Nitrogen Fertilizers Made from Ammonia

Reagent	Product Fertilizer	Formula
(none)	Anhydrous ammonia	NH_3
Carbon dioxide	Urea	NH_2CONH_2
Sulfuric acid	Ammonium sulfate	$(NH_4)_2SO_4$
Nitric acid	Ammonium nitrate	NH_4NO_3
Phosphoric acid	Ammonium phosphate	$(NH_4)_3PO_4$

It is largely through the use of nitrogen fertilizers that we are able to feed the human population that inhabits the Earth. The atmosphere has nearly unlimited nitrogen. Unfortunately, our ways of fixing it require a lot of energy. The hydrogen for synthesis of ammonia comes from natural gas. Petroleum shortages and higher energy prices in the 1970s led to a scarcity of fertilizers and an inflation of prices. Research is now under way that shows promise for transferring nitrogen-fixing genes to nonlegumes. Perhaps we will soon find a way to fix nitrogen gas at ambient temperatures, the way microorganisms do.

Phosphates: A Legacy from Ancient Life

Phosphates have probably been used as fertilizer since ancient times—in the form of bone, guano (bird droppings), or fish meal. However, phosphates as such were not recognized as plant nutrients until 1800. Following that discovery, the great battlefields of Europe were dug up and bones were shipped to plants for processing into fertilizer.

Animal bones are rich in phosphorus, but the phosphorus is tightly bound and not readily available to plants. In 1831, the Austrian chemist Heinrich Wilhelm Köhler treated animal bones with sulfuric acid to convert them to a more soluble form, called superphosphate. In 1843, John Lawes applied the same treatment to phosphate rock. The essential reaction for forming superphosphate is:

$$Ca_3(PO_4)_2 + 2\ H_2SO_4 \longrightarrow Ca(H_2PO_4)_2 + 2\ CaSO_4$$

Phosphate rock or bone (insoluble)   Superphosphate (more soluble)

Phosphate is quite common in the soil but often at concentrations too low for adequate support of plant growth. Fortunately, there are deposits that are more concentrated. The presence of bones and teeth from early fish and other animals in these ores indicates that the deposits are largely the skeletal remains of sea creatures of ages past. Unfortunately, fluorides often are associated with phosphate ores, and they constitute a serious pollution problem from the production of phosphate fertilizers.

Usually phosphorus is absorbed by plants as $H_2PO_4^-$. In plants, phosphates are incorporated into DNA and RNA (Chapter 11). They also are constituents of compounds essential for the conversion of starches to sugars. Phosphates appear to play a vital role in photosynthesis, mainly by involvement in energy-transfer processes. Phosphorus compounds also are essential to the formation of fats and of some proteins and other cell constituents. The availability of phosphorus is often the limiting factor in plant growth.

About 90% of all phosphates are used in agriculture. The United States is the leading producer and user, but the rich phosphate deposits in the United States may be depleted in the 1990s. Recently, however, large offshore deposits have been discovered in the Atlantic Ocean off North Carolina. Worldwide, Morocco has two thirds of the reserves. All the high-grade phosphate reserves may be exhausted in 30 or 40 years. Our use of phosphates scatters them irretrievably throughout the environment.

Potassium: The Simple and the Complex

The third major element necessary for plant growth is potassium. Plants use it in the form of the simple ion K^+. Generally, potassium is abundant and there are no problems with solubility. The precise function of potassium in plant cells is difficult to determine. It seems to be involved in the formation and transport of carbohydrates. Also, it may be necessary for the buildup of proteins from amino acids. Uptake of potassium ions from the soil leaves the soil acidic; each time one potassium ion enters the root tip, a hydronium ion must leave in order for the plant to maintain electrical neutrality (Figure 17.7).

The usual chemical form of potassium in commercial fertilizers is potassium chloride (KCl). Vast deposits of this salt occur in Stassfurt, Germany. For years, this source supplied nearly all the world's potassium fertilizer. With the coming of World War I, the United States sought supplies within its own borders. Deposits at Searles Lake, California, and Carlsbad, New Mexico, now supply much of United States needs. Canada has vast deposits in Saskatchewan and Alberta. Beds of potassium chloride up to 200 m thick lie about 1.5 km below the Canadian prairies (Figure 17.8).

Although reserves are large, potassium salts are a nonrenewable resource. We should use them wisely.

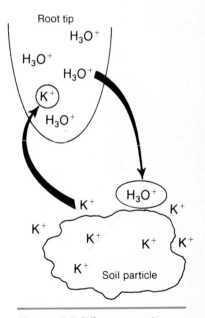

Figure 17.7 When a root tip takes up potassium ions from the soil, hydronium ions are transferred to the soil. Potassium uptake by plants tends to make the soil acidic.

Figure 17.8 Mining potassium chloride about 1.5 km below the prairies of Saskatchewan, Canada. [Courtesy of the Potash Corporation of Saskatchewan.]

Other Essential Elements

In addition to the three major nutrients, a variety of other elements are necessary for proper plant growth. Three secondary plant nutrients—magnesium, calcium, and sulfur—are needed in moderate amounts. Calcium, in the form of lime (calcium oxide), is used to neutralize acidic soils.

$$CaO \ + \ 2\,H_3O^+ \ \longrightarrow \ Ca^{2+} \ + \ 3\,H_2O$$

Calcium ions are also necessary plant nutrients. Magnesium ions (Mg^{2+}) are incorporated into chlorophyll molecules and therefore are necessary for photosynthesis. Sulfur is a constituent of several amino acids, and it is necessary for protein synthesis.

Seven other elements, called **micronutrients**, are needed in very small amounts. These are summarized in Table 17.3. Many soils contain these trace elements in sufficient quantity. However, some soils are deficient in one or more. Production of these soils can be markedly increased by the addition of small amounts of the needed elements.

It is possible that other elements also are required by plants. Some elements that may be needed include sodium, silicon, vanadium, chromium, selenium, cobalt, fluorine, and arsenic. Most of these elements are present in most soils, but their necessity to plant growth has not been established.

Table 17.3 Seven Micronutrients Necessary for Proper Plant Growth

Element	Function	Deficiency Symptoms
Boron	Required for protein synthesis; essential for reproduction and for carbohydrate metabolism	Death of growing points of stems, poor growth of roots, poor flower and seed production
Copper	Constituent of enzymes; essential for reproduction and for chlorophyll production	Twig dieback, yellowing of newer leaves
Iron	Constituent of enzymes; essential for chlorophyll production	Yellowing of leaves, particularly between veins
Manganese	Essential for redox reactions and for the transformation of carbohydrates	Yellowing of leaves, brown streaks of dead tissue
Molybdenum	Essential in nitrogen fixation by legumes and reduction of nitrates for protein synthesis	Stunting, pale green or yellow leaves
Zinc	Essential for early plant growth and maturing	Stunting, seed and grain yields reduced
Chlorine	Increases water content of plant tissue; involved in carbohydrate metabolism	Shriveling

Fertilizers: A Mixed Bag

Most farmers buy **complete fertilizers**, which, despite the name, usually contain only the three main nutrients. There are usually three numbers on fertilizer bags. The first number represents the percent of nitrogen (N); the second, the percent of phosphorus (calculated as P_2O_5); and the third, the percent of potassium (calculated as K_2O). So 20-10-5 means that the fertilizer contains 20% N, 10% P_2O_5, and 5% K_2O. The rest is inert material.

Fertilizers must be water soluble to be used by plants. When it rains, the nutrients from the fertilizers are washed into streams and lakes, where they stimulate blooms of algae. These chemicals, particularly the nitrates, also penetrate the groundwater. In some areas, the nitrates present in wells have reached levels toxic to babies. These and other water pollution problems are examined in detail in Chapter 16.

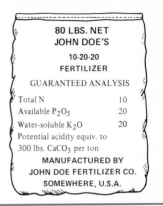

Figure 17.9 The numbers on this bag of fertilizer tell us that the fertilizer is 10% nitrogen, 20% phosphorus, and 20% potassium.

The War Against Pests: In Days of Old

Since the earliest days of recorded human history (and surely even before that), people have been plagued by insect pests. Three of the 10 plagues of Egypt (described in the Book of Exodus) were insect plagues—lice, flies, locusts. The decline of Roman civilization has been attributed in part to malaria, a disease carried by mosquitoes that destroys vigor and vitality when it does not kill. Bubonic plague carried by rats (and by fleas from rats to humans) swept through the Western World repeatedly during the Middle Ages. One such plague (during the 1660s) is estimated to have killed 25 million people—25% of the population in Europe at that time. The first attempt to dig a Panama Canal (by the French during the 1880s) was defeated by an outbreak of malaria.

The use of modern chemical pesticides may be the only thing that stands between us and some of these insect-borne plagues. Pesticides also prevent the consumption of a major portion of our food supplies by insects and other pests. Crop losses to insects in the United States are estimated to be $4 billion a year. When losses to other pests—rodents, fungi, nematodes, and weeds—are included, the losses total $15 billion a year.

In earlier days, people tried to control insect pests by draining swamps, pouring oil on ponds (to kill the mosquito larvae), and using a variety of chemicals. Most of these chemicals were compounds of arsenic. Lead arsenate [$Pb_3(AsO_4)_2$] is a particularly effective poison, since both the lead and the arsenic in it are toxic. A few pesticides, like pyrethrum (used in mosquito control) and nicotine sulfate (Black Leaf 40), are obtained from plant matter.

Table 17.4 Toxicity of Insecticidal Preparations Administered Orally to Rats

Pesticide	LD$_{50}$ (in milligrams of pesticide per kilogram of body weight)
Pyrethrins*	1200
Lead arsenate	800
Carbaryl	540
DDT	113[†]
Lindane	88
Nicotine sulfate[‡]	55
Dieldrin	46
Methyl parathion	17
Endrin	8
Parathion	5

* Active ingredients of pyrethrum.
[†] Estimated LD$_{50}$ for humans is 500 mg/kg.
[‡] Nicotine is much more toxic by injection.

Only a few insects are harmful. Many are beneficial, and others play important roles in ecological systems and are indirectly beneficial. Most poisons are indiscriminate. They kill all insects, not just those we consider pests. Many are also toxic to humans and other animals. Some say we should call such poisons **biocides** (because they kill living things) rather than insecticides. Table 17.4 lists the toxicities of some insecticides.

Example 17.2 How much of the pesticide lindane would it take to kill a 10-kg child if the lethal dose was 88 mg per kilogram of body weight?

$$10 \text{ kg} \times \frac{88 \text{ mg}}{1 \text{ kg}} = 880 \text{ mg or } 0.88 \text{ g}$$

DDT: The Dream Insecticide

Shortly before World War II, a chlorinated hydrocarbon (DDT) was found to be a potent insecticide. The discovery was made in Switzerland by Paul Müller. DDT was tested and found effective against vine-grape pests and against a particularly bad infestation of potato beetles.

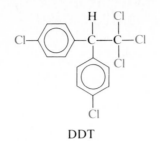

DDT
(Dichlorodiphenyltrichloroethane)

When the war came, the supply of the pesticide pyrethrum was cut off by the Japanese occupation of Southeast Asia and the Dutch East Indies (now Indonesia). Lead, arsenic, and copper compounds, which had been used for insecticides, were needed for armaments and other military purposes. The British and the Americans found themselves in need of an insecticide that would protect soldiers from lice and ticks (bearers of typhus). Those who had to fight in the jungles of the South Pacific needed protection from mosquitoes and other disease-bearing pests.

A few kilograms of DDT were obtained by the Allies and hurriedly tested. Combined with talcum, it made a very effective delousing powder. Clothing was impregnated with DDT and, fortunately, it seemed to have no deleterious effects on humans or other large animals. Allied soldiers were virtually free from lice. The Germans (as they did with the atomic bomb) missed out on DDT, even though it was discovered just across their border. German troops were heavily infested with lice and many were sick with typhus. In wars before World War II, more soldiers probably died from typhus than from bullet wounds.

DDT is easily synthesized from cheap, readily available chemicals. Chlorobenzene and chloral hydrate are warmed in the presence of sulfuric acid.

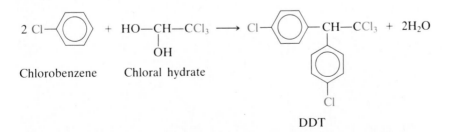

Chlorobenzene   Chloral hydrate   DDT

When the reaction is complete, the mixture is poured into water and the DDT separates out (like other chlorinated hydrocarbons, DDT is essentially insoluble in water).

Pesticides and Public Health

A cheap insecticide effective against a variety of insect pests, DDT came into widespread use after the war. Other chlorinated hydrocarbons were synthesized, tested, and pressed into service in the war against insects. Although invaluable to farmers in the production of food and fiber, the chlorinated hydrocarbons won their most dramatic victories in the field of public health. According to the World Health Organization, approximately 25 million lives have been saved and hundreds of millions of illnesses prevented by the use of DDT and other chlorinated-hydrocarbon pesticides. In India alone, malaria cases were reduced from 75 million to 5 million per year. The average life span in India has been increased by 15 years since the mosquito-eradication program was begun in 1954.

Striking evidence of the value of DDT in public health comes from Sri Lanka (formerly Ceylon), where there were 2.8 million cases of malaria in 1946. By 1963, spraying with DDT had reduced the incidence of malaria to only 17 cases. The spraying program was terminated in 1964, and by 1968, the number of cases had risen to over 1 million per year.

DDT was a dream come true. It seemed that the world would be free at last from insect plagues and insect-borne diseases. Crops would be protected from the ravages of insects and food production would be increased. In recognition of his magnificent discovery, Paul Müller was awarded the Nobel Prize in medicine and physiology in 1948.

Figure 17.10 The first airplane to land on and take off from the shores of the Dead Sea (24 October 1955) was spraying insecticides against mosquitoes near Gora Sofi, Jordan. The spraying was part of a demonstration project sponsored by the U.S. Department of State in the Middle East, Asia, and Africa. The United States is the major supplier of DDT throughout the world. [Courtesy of the U.S. Department of Agriculture, Washington, DC.]

Banning a Dream: The Decline and Fall of DDT

Before Müller even received his prize, however, there were warnings that all was not well. Houseflies resistant to DDT were reported as early as 1946. DDT's toxicity to fish was reported by 1947. By 1948, 11 additional resistant species of insects had been discovered. Such early warnings were largely ignored, and it was assumed that the toxicity would disappear soon after the chemical was discharged into the environment. DDT was used extensively to protect crops and control mosquitoes and to try to prevent Dutch Elm disease. Bird populations began to decline in agricultural areas. By 1962, the year Rachel Carson's book *Silent Spring* appeared, United States production of DDT had reached 76 million kg per year.

Today, in the developed countries, DDT is known mostly for its harmful environmental effects. It interferes with calcium metabolism. Birds are threatened because egg shells are composed mainly of calcium compounds. Eggs of birds that have ingested DDT have thin shells that are poorly formed and easily broken. Even a few parts per billion (ppb) of DDT interfere with the growth of plankton and the reproduction of crustaceans such as shrimp. As bad as DDT sounds, though, it has probably saved the lives of more people than any other chemical substance.

The use of DDT has been largely banned in developed countries, but the chemical is still widely used in less developed areas. Unfortunately, it is no longer effective against many of the pests it once killed readily. Many insects have developed resistance to this and many other pesticides. The resistant insects can now detoxify the chemical compounds that were once deadly to their kind. Resistant flies can tolerate 100 g of DDT per kilogram of body weight, an amount equivalent to a dose of 6 kg for a 60-kg human.

Chlorinated hydrocarbons generally are unreactive. This lack of reactivity was a major advantage of DDT. Sprayed on a crop, DDT stayed there and killed insects for weeks. This **persistence** was also a major disadvantage: the substance did not break down readily in the environment. Although not very toxic to humans and other warm-blooded creatures, DDT is much more toxic to cold-blooded organisms. That includes insects, of course, and fish. DDT's lack of reactivity toward oxygen, water, and components of the soil led to its buildup in the environment, where it threatened fish, birds, and other wildlife.

Chlorinated hydrocarbons are good solvents for fats. The reverse is also true: fats are good solvents for chlorinated hydrocarbons such as DDT. When these compounds are ingested as contaminants in food or water, they are extracted and concentrated in fatty tissues. Their fat-soluble nature causes chlorinated hydrocarbons to be concentrated

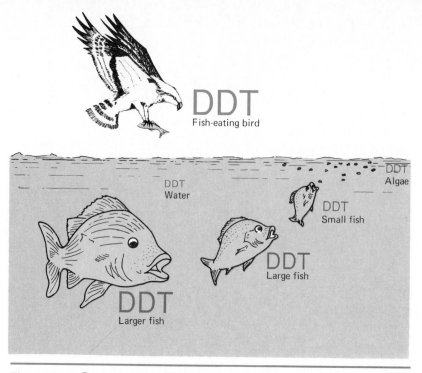

Figure 17.11 Concentration of DDT up the food chain. Animals at the top of the chain have the highest concentration of the insecticide.

up the food chain. This **biological magnification** was graphically demonstrated in California in 1957. Clear Lake, about a hundred miles north of San Francisco, was sprayed with DDT in an effort to control gnats. The water, after spraying, contained only 0.02 ppm of DDT. The microscopic plant and animal life contained 5 ppm—250 times as much. Fish feeding on these microorganisms contained up to 2000 ppm. Grebes, the diving birds that ate the fish, died by the hundreds (Figure 17.11).

DDT and other chlorinated hydrocarbons such as PCBs (Chapter 12) are nerve poisons. Concentrated in the fatlike compounds that make up nerve sheaths, they somehow interfere with the transmission of electrical impulses along these sheaths. DDT also interferes with calcium metabolism, essential to the formation of healthy bones and teeth. Although no harm to humans has ever been conclusively demonstrated, the disruption of calcium metabolism in birds has been disastrous for some species. The egg shells, mainly composed of calcium compounds, are thin and poorly formed. The bald eagle, peregrine falcon, and other birds became endangered species. All have made dramatic recoveries since DDT was banned in the United States in 1972. The ban was not total. DDT may still be used to protect public health.

For example, in case of an outbreak of malaria, DDT could be used to spray mosquito-infested areas. And the *production* of DDT was not banned. DDT is still produced for export and is exported particularly to developing countries, where it still plays a major role in malaria control (Figure 17.12).

The problem of getting a worldwide ban on DDT may solve itself as more and more insect species develop resistance. Its use may be discontinued simply because it is no longer effective. The need for some type of insect control is apparent, however, as was dramatically illustrated in Sri Lanka.

Most chlorinated hydrocarbon insecticides have been banned or severely restricted in developed countries. In following sections, we consider some alternatives—including some nonchemical methods of insect control.

Figure 17.12 Malaria is controlled in developing countries by spraying the walls of houses, where mosquitoes alight after their meal of blood. [Courtesy of the U.S. Agency for International Development, Washington, DC.]

Organic Phosphorus Compounds: More Toxic, Less Persistent

Bans and restrictions on chlorinated hydrocarbon insecticides have led to increased use of organic phosphorus compounds. Typical of these are malathion, diazinon, and parathion.

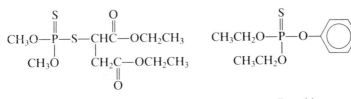

Malathion

Parathion

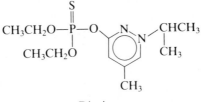

Diazinon

Over two dozen of these insecticides are available commercially (Reference 23). They have been extensively studied and evaluated for effectiveness against insects and for toxicity to people, laboratory animals, and farm animals.

The organic phosphorus compounds are generally much more toxic to mammals than the chlorinated hydrocarbons. Table 17.4 shows that parathion is over 20 times as toxic to rats as DDT (a smaller LD_{50} means greater toxicity). Some 300 000 farmers and fieldworkers suffer from pesticide related illnesses each year in the United States. Organic phosphorus compounds are safer for the environment, though. They break down rapidly, and residues are seldom found in food. Some agricultural leaders have expressed the opinion that we may have gained safety for the peregrine falcon and the bald eagle at the expense of farm workers. The organic phosphorus compounds are nerve poisons. Their mode of action is discussed in Chapter 25.

Carbamates: A Honey of an Insecticide

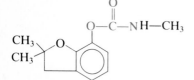

Carbaryl
(Sevin)

Another class of compounds that has gained prominence since the ban on DDT is the carbamates. Typical examples are carbaryl (Sevin), carbofuran (Furadan), and aldicarb (Temik) (Chapter 16). Carbaryl has a low toxicity to mammals, but carbofuran and aldicarb have toxicities similar to that of parathion. Most of the carbamates are directed specifically at one, or a few, insect pests. Thus, they are called **narrow spectrum insecticides**, in contrast to the chlorinated hydrocarbons and organic phosphorus compounds that kill many kinds of insects. The latter are called **broad spectrum insecticides**. Carbamates break down rapidly in the environment. Unfortunately, carbaryl, the most widely used, is particularly toxic to honeybees. Millions of these valuable insects have been wiped out by spraying crops with carbaryl.

The essential feature of the carbamate molecule is the group

Many variations are possible by changing the groups attached to the oxygen and the nitrogen. Several carbamates have achieved commercial importance. The future, however, may belong to alternatives to the synthetic chemical pesticides.

Carbofuran
(Furadan)

The Sex Trap: Pheromones

One promising area of research in insect control is the research into **pheromones**. These interesting chemicals are secreted externally by insects and may serve the function of marking a trail, sending an alarm, or attracting a mate. Perhaps the most interesting pheromones are the insect **sex attractants**, usually secreted by the female to attract males. Chemical research can identify the sex attractants and determine their structures. These compounds then can be synthesized and used to lure males into traps. Alternatively, the attractant can be used in quantities sufficient to confuse and disorient the males, who detect a female in every direction but can't find one to mate with.

Some sex attractants are amazingly simple. Others are complex. The sex attractant for the common housefly is a simple, unsaturated hydrocarbon.

$$CH_3CH_2CH_2CH_2CH_2CH_2CH_2CH_2CH\!=\!CHCH_2CH_2CH_2CH_2CH_2CH_2CH_2CH_2CH_2CH_2CH_2CH_3$$

Although this molecule contains 23 carbon atoms (with a double bond between the ninth and tenth), it is fairly easy to synthesize. Quantities of it are now available for testing and development.

Most research in sex attractants is not easy, however. Some pheromones have complicated structures. All are secreted in extremely tiny amounts. For example, a team of U.S. Department of Agriculture researchers had to use the tips of 87 000 female gypsy moths to isolate a minute amount of a powerful sex attractant. The structure of the compound was determined by using a variety of sophisticated instruments. It was found to be

$$\underset{\underset{CH_3}{|}}{CH_3CHCH_2CH_2CH_2CH_2}\overset{\overset{O}{\diagdown}}{\underset{H}{C}}\!-\!\!\underset{H}{C}CH_2CH_2CH_2CH_2CH_2CH_2CH_2CH_2CH_2CH_3$$

The compound was first isolated in 1967, but it was not until 3 years later that it was synthesized in the laboratory. Field tests showed the synthetic attractant to be effective in concentrations as low as 1 picogram (1×10^{-12} g) in baited traps. (Other pheromones, too, are effective at extremely low concentrations. A male silkworm moth can detect as few as 40 molecules per second. If a female releases a little as 0.01 μg, she can attract every male within 1 km.)

The gypsy moth larvae have defoliated great forests, mainly in the northeastern United States, and have now spread over much of the country. Its pheromone has been used mainly in traps to monitor insect populations.

Pheromones for a variety of insects have been isolated and identified. Most insects use two or three compounds. Some cases are quite unusual. An intriguing example is that of the male cotton bollworm moth. An unsaturated aldehyde called Z-9-tetradecenal causes this moth to try to mate with female tobacco budworm moths. Because their genitals don't fit, the two insects become locked in an amorous embrace and eventually die. Oh, well—all is fair in love and war on insect pests.

Figure 17.13 A male gypsy moth detects odor from a female by using his large antennae. [Courtesy of the U.S. Department of Agriculture, Washington, DC.]

Figure 17.14 Grasshoppers killed by a viral insecticide. [Courtesy of the U.S. Department of Agriculture, Washington DC.]

Much more research has to be done before pheromones will play a major role in insect control. The method is expensive, and research is painstaking and time-consuming. Workers must be careful not to get the attractants on their clothes. Who wants to be attacked on a warm summer night by a million sex-crazed gypsy moths?

Biological Controls: Fighting Bugs with Bugs

Another promising avenue of research is in the use of natural enemies to control pests. Praying mantises and ladybugs are sold commercially and are used to destroy garden pests. Biological controls also include bacteria and viruses, often specific for a given species of pest. *Bacillus thuringiensis*, sold under the trade name Dipel, is widely used by home gardeners against cabbage loopers, hornworms, and other moth larvae. Cost limits the agricultural use of this bacterial pesticide.

Viruses have been tested successfully against a number of insect pests. Final approval was given in 1972 for the use of a virus as a pesticide directed against *Heliothis zea*, the cotton bollworm. The virus is cultured in a "factory" that will be able to supply enough virus to treat a substantial portion of the United States cotton crop. Viral pesticides also show promise against grasshoppers (Figure 17.14) and many other insect pests.

Production of viral pesticides is expensive. Nevertheless, this approach to insect control is intriguing because viral agents are highly specific, generally harmless to people and other animals, and completely biodegradable.

Another biological approach is the breeding of insect- and fungus-resistant plants. Such plants, particularly grain bearers, have contributed considerably to increased crop production in recent years.

Juvenile Hormones

Another approach to insect control is the use of juvenile hormones. Hormones are the chemical messengers that control many life functions in plants and animals. Minute quantities produce profound physiological changes. In the insect world, juvenile hormones control the rate of development of the young. Normally, production of the hormone is shut off at the appropriate time to allow proper maturation to the adult stage.

Chemists have been able to isolate insect juvenile hormones and determine their structure. With knowledge of the structure, they can

Figure 17.15 The use of juvenile hormones in pesticides like methoprene can prevent mosquitoes from maturing. (a) Mosquito larvae. (b) Mosquito larvae trapped in pupae stage. (c) Adult mosquitoes. [Courtesy of Chemical Manufacturers Association, Washington, DC. Reprinted from *ChemEcology*, May 1975.]

synthesize the hormone. Application of this hormone to the ponds where mosquitoes breed keeps mosquitoes in the harmless pre-adult stage. Because only adult insects can reproduce, juvenile hormones appear to be a nearly perfect method of mosquito control.

Chemists have synthesized analogs of juvenile hormones. One such analog, methoprene, was approved by the U.S. Environmental Protection Agency for use against mosquitoes in 1975 and for use against fleas in 1982 (Figure 17.15).

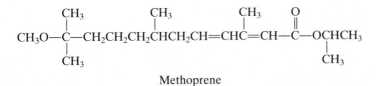

Methoprene

The synthesis of juvenile hormones is difficult (as you could probably guess from their complex structures) and, consequently, quite expensive. The hormones are for use against insects that are pests at the adult stage. Little would be gained by keeping a moth or a butterfly in the caterpillar stage for a longer period of time; caterpillars have voracious appetites and do a great deal of damage to crops. The damage these larvae could do might not be compensated for by their lack of reproduction. The last thing the farmer wants is to postpone their puberty.

Sterilization

A control technique effective on some insects is sterilization. Large numbers of males can be sterilized by radiation, chemicals, or cross-breeding. Then, enough of them can be released so that they far out-number the local fertile males. If a female meets with a sterile male, no offspring are produced.

Radiation sterilization has virtually eliminated the screwworm fly, once a serious pest that affected cattle in the southern United States. Tropical fruitflies also have been eradicated by this method in some areas. (It is likely that the 1981 outbreak of Mediterranean fruitflies in California was caused in part by the accidental release of males that had not been sterilized effectively. Extensive spraying with malathion and the stripping of fruit from trees ended the outbreak.) Sterilization involves raising vast numbers of insects, separating the males from the females, then sterilizing and releasing the males. The great expense and limited applicability of this process probably means that it will not become the major method of insect control. Thus, the search goes on for effective ways to control insects, protect us from disease, and prevent damage to our food supply.

Herbicides and Defoliants

The United States produces about 700 million kg of pesticides annually. Of these, nearly 400 million kg are **herbicides** (chemicals used to kill weeds).

Crops produce more abundant harvests when they have no competition from weeds. Removing the weeds by hand is tedious, back-breaking work. Chemical herbicides have been used for a number of years to kill the unwanted plants. Solutions of copper salts, sulfuric acid, and sodium chlorate ($NaClO_3$) have been used, but it wasn't until the introduction of 2,4-D in 1945 that the use of herbicides became common. Chemically, 2,4-D is 2,4-dichlorophenoxyacetic acid, or one of its derivatives. These chemicals are growth-regulator herbicides and are especially effective against newly emerged, rapidly growing broad-leaved plants.

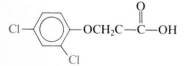

2,4-Dichlorophenoxyacetic acid
(2,4-D)

A relative of 2,4-D, called 2,4,5-T, is especially effective against woody plants; it works by causing the leaves to fall off the plants (**defoliation**). These two chemicals, combined in a formulation called **Agent Orange**, were used extensively in Vietnam to remove enemy cover and to destroy crops that maintained enemy armies. In addition to causing vast ecological damage, 2,4-D and 2,4,5-T were suspected of causing birth defects in Vietnamese children and in babies later born to American soldiers exposed to the herbicides. Laboratory studies show that these compounds, when pure, do not cause abnormalities in fetuses of laboratory animals. Extensive birth defects are caused, though, by contaminants called **dioxins**, once frequently found in the herbicides. These dioxins are also chlorinated compounds. Continuing concern over dioxin contamination led the U.S. Environmental Protection Agency to ban 2,4,5-T in 1985.

Another widely used herbicide is atrazine. Atrazine binds to a protein in chloroplasts in plant cells, shutting off the electron-transfer reactions of photosynthesis. Atrazine is often used on corn crops. Corn plants deactivate atrazine by removing the chlorine atom. Weeds cannot deactivate the compound and are killed.

Another herbicide type is the **preemergents**, such as paraquat. This ionic compound is toxic to most plants, but it is rapidly broken down in the soil. Therefore, paraquat can be used to kill weed plants before the crop seedlings emerge. Paraquat inhibits photosynthesis by accepting the electrons that otherwise would reduce carbon dioxide. It has been used in the United States and Latin America to destroy marijuana crops. Some of the sprayed crop has been harvested anyway, and contaminated marijuana has appeared in United States cities. It is feared that paraquat inhaled with marijuana smoke can cause severe lung damage.

Chemical defoliants facilitate the harvesting of crops. Calcium cyanamide (CaNCN), one of the earliest defoliants used for this purpose, causes cotton plants to lose their leaves when the bolls are mature. This makes it possible to use mechanical cotton pickers. If the leaves were left on, they would be crushed by the machinery and would stain the cotton.

There is no doubt that the use of herbicides has increased the value of agricultural crops by billions of dollars. Roadsides, vacant lots, and industrial areas also have been kept free of weeds. People who suffer from allergies caused by pollen from ragweed and other plant pests have been relieved, and poison ivy and other toxic weeds have been made less common.

With the exception of problems with dioxin contaminants in 2,4,5-T, herbicides generally have a better safety record than insecticides. We still don't know, though, the ultimate effect on the environment of the long-term use of herbicides.

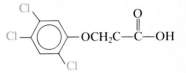

2,4,5-Trichlorophenoxyacetic acid (2,4,5-T)

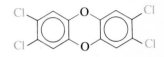

2,3,7,8-Tetrachlorodibenza-para-dioxin (a "dioxin")

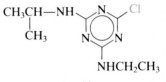

Atrazine

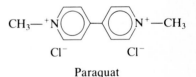

Paraquat

The Antichemical Revolt:
Organic Farming

Problems with pesticides have led to a minor resurgence of organic farming. Many people believe that we can have more nutritious food and avoid environmental problems by banning chemical pesticides and fertilizers and growing crops the natural way.

The movement is admirable in many respects. It seems evident that we should return human and animal wastes to the soil rather than polluting our water and air with them. Organic farming also is less energy intensive. According to a study by the Center for the Biology of Natural Systems at Washington University, comparable conventional farms used 2.3 times as much energy as organic farms. Production on the organic farms was 10% lower, but costs were lower by a comparable percent. Organic farms require 12% more labor than conventional ones, but human labor is a renewable resource, whereas petroleum energy is not.

Most likely, we should practice organic farming to the limit of our ability to do so. But we should not delude ourselves. Despite claims to the contrary, organically grown food has not been shown to be more nutritious than food grown with modern techniques. Chemical analysis can detect no difference between food organically grown and food produced using chemical fertilizers. Anyway, organic fertilizers must be broken down to inorganic chemicals before they can be utilized by plants. Corn doesn't care if nitrate ions are formed by bacteria in manure or by chemists in factories.

Energy and Agriculture

Modern agriculture is energy intensive. Nonrenewable petroleum energy is required for the production of fertilizers, pesticides, and farm machinery. Energy also is required to run the machinery used for tilling, harvesting, and transporting the crops. One third the cost of food in the United States goes to transportation. We spend $6 million and use 3.6 million L of fuel just to transport broccoli from California to New York. It also takes energy to process and package the food (Figure 17.16).

As far as human energy is concerned, United States agriculture is enormously efficient. Each farm worker produces enough food for 78 people. But that productivity is based on fossil fuels. It is estimated that 10 units of petroleum energy are required to produce 1 unit of food energy. Increasing energy prices may well force us to reevaluate our farming methods. If we consider production per hectare, modern farming is marvelously efficient. If we consider the energy used in

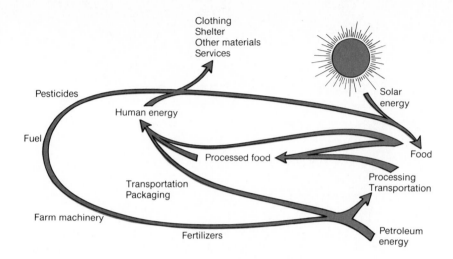

Figure 17.16 Energy flow in modern agriculture.

relation to the energy produced, it is remarkably inefficient. It should be noted, however, that in an energy-efficient, primitive society, nearly all human energy goes into food production. In modern societies, only about 10% of energy is devoted to producing food. The other 90% is used to provide the materials and services that are so much a part of our civilization. We should try to make our food production more energy efficient, but it is unlikely that we will want to return to a primitive way of life.

Can We Feed a Hungry World?

The population of planet Earth reached 5 billion in 1986. At a growth rate of 1.7%, it will reach 10 billion in 2027. Will we able to feed all those people?

Agriculture in developed countries is heavily dependent on petroleum energy and the pesticides and fertilizers derived from petroleum. New plant varieties have increased crop yields, but these crops require increased amounts of pesticide and fertilizer.

We may be able to increase food production substantially through genetic engineering (Chapter 11). We can design plants that produce more food, that are resistant to disease and insect pests, and that grow well in hostile environments. We can design animals that grow larger and produce more meat and milk. Within a decade or so these developments may provide us with food and fiber in unimaginable abundance. But the hungry are with us now, and 81 million more come to dinner each year. The problem isn't just production. Some nations have huge surpluses; others starve. We must find methods to distribute food more efficiently and more equitably.

Figure 17.17 The total supply of arable land is fixed. Continued population growth threatens to outstrip even the most optimistic projections of food production.

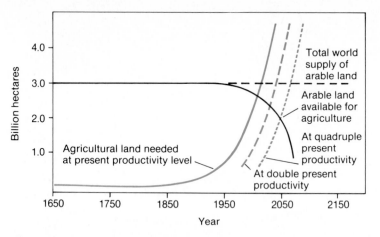

Figure 17.17 shows the relationship between population and arable land available for food production. Even by quadrupling present food production, we could meet our needs only until 2050. Since virtually all the world's arable land is now under cultivation, it should be quite evident to everyone that food production cannot keep up for long with the present rate of population growth.

We seem to be hooked on a high-energy form of agriculture that uses synthetic fertilizers, pesticides, and herbicides and depends on the power of machinery that burns fossil fuels. Ultimately, the only solution to the problem is population control. Obviously, that will come some day. The only questions are when and how. Population control could come through decreasing the birthrate. This could happen voluntarily, but that involves education and cultural change, both usually very slow. It could come through governmental control, but with over 100 sovereign governments in the world such control seems unlikely to be effective. The other possibility for population control is an increase in the death rate—through catastrophic war, famine, pestilence, or the poisoning of the environment by the wastes of an ever-expanding population. The ghost of Thomas Malthus haunts us yet.

Problems

1. What did Thomas Malthus's predict in his famous 1830 statement?

2. Why has Malthus's prediction not come true?

3. What is arithmetic growth? Give an example.

4. What is geometric growth? Give an example.

5. You start a stamp collection, with a plan to buy two stamps a week. How many weeks would it take to acquire 100 stamps? Is the collection growing arithmetically or geometrically?

6. You raise rabbits. Starting with two, you have four at the end of the first month, eight at the end of the second month, and so on; the rabbit population doubles each month. How many rabbits would you

have at the end of 12 months? Is the rabbit population growing arithmetically or geometrically?

7. The hutch for your rabbits (Problem 6) is half full at the end of 12 months. When will it be full?

8. The populations of some Latin American, African, and Asian nations are growing at 3.5% a year. At that rate, how many years of growth will it take for those populations to double?

9. The population of China, now 1.1 billion, is growing at an annual rate of 1.1%. At that rate, how many years of growth will it take for the population to double?

10. The population of India, now 750 million, is growing at an annual rate of 2.0%. At that rate, how many years of growth will it take for the population to double?

11. The number of chemical substances registered by the Chemical Abstracts Service of the American Chemical Society, now 8 million, is increasing at an annual rate of 6.0%. At that rate, how many years will it take for the number of registered compounds to double to 16 million?

12. Where does most of the matter of a growing plant come from?

13. List the three structural elements of plants.

14. In what form is nitrogen used by plants?

15. How is ammonia made? What are the raw materials for ammonia synthesis?

16. List several ways that nitrogen can be fixed.

17. Starting with ammonia, show how ammonium nitrate is made.

18. What is urea? From what is synthetic urea made?

19. What is anhydrous ammonia? What is its use?

20. What is the source of phosphate fertilizers?

21. What is superphosphate? How is it made?

22. In what form is phosphorus absorbed by plants?

23. Why does soil become acidic when potassium ions are absorbed by plants?

24. What is a complete fertilizer? Is it really complete?

25. What is the role of nitrogen in plant nutrition?

26. What is the role of phosphorus in plant nutrition?

27. List the three secondary plant nutrients. What is the role of each in plant nutrition?

28. List three (of the seven) plant micronutrients. What is the role of each in plant nutrition?

29. List the advantages and disadvantages of DDT as an insecticide.

30. What is meant by biomagnification?

31. Why is DDT especially harmful to birds?

32. Describe how chlorinated hydrocarbons become concentrated in a food chain.

33. List the advantages and disadvantages of organic phosphorus compounds as insecticides.

34. What is a narrow spectrum insecticide?

35. List some advantages and disadvantages of carbamate insecticides.

36. What are pheromones?

37. List some advantages and disadvantages to the use of sex attractants as weapons against insect pests.

38. List some advantages and disadvantages of viral and bacterial pesticides.

39. What are juvenile hormones? How are they used against insect pests?

40. Describe the technique of steriliation as a method for controlling insects. Why is it not more widely used?

41. What are herbicides?

42. Name the components of Agent Orange.

43. What is a defoliant? Why are defoliants used on cotton crops?

44. How can atrazine kill weeds without killing corn?

45. What is a preemergent herbicide?

46. How does paraquat kill plants?

47. Compare organic farming with conventional farming. Consider energy requirements, labor, profitability, and crop yields.

48. Discuss the possibility of increasing our food supply through genetic engineering.

49. Malathion residues are found in a sample of bread at a level of 1.0 mg per kilogram of bread (1.0 ppm). How much malathion is there in one 500-g loaf of the bread? There are 20 slices in the loaf. How much malathion is there in one slice of the bread?

50. PCBs are found in a sample of fish at a level of 10 mg per kilogram of fish (10 ppm). How much PCBs are in a 100-g serving of the fish?

51. If you grew a garden, would you use chemical fertilizers? Pesticides? Why or why not?

52. How much DDT would it take to kill a person

weighing 60 kg if the lethal dose was 0.5 g per kilogram of body weight?

53. How much parathion would it take to kill the person in Problem 52 if the lethal dose was 5 mg per kilogram of body weight?

54. Examine the label on a package of yard or garden pesticide. List its trade name and active ingredients.

55. Examine the label of a bag of fertilizer. What is its composition?

56. Use a reference such as *The Merck Index* (Reference 25) to look up each of these pesticides. What is the toxicity of each?

a. methoxychlor   b. lindane   c. aldrin
d. dieldrin   e. heptachlor   f. mirex
g. kepone   h. toxaphene   i. endrin

57. The minimum lethal dose of dioxin is estimated to be 6 μg per kilogram weight. What is the smallest amount that might kill a 70-kg man?

References and Readings

1. "Agent Orange: Identifying the Problem." *The American Legion*, January 1982, pp. 34–37.
2. Barrons, Keith C. *Are Pesticides Really Necessary?* Chicago: Regnery Gateway, 1981.
3. Batie, Sandra S., and Robert G. Healey. "The Future of American Agriculture." *Scientific American*, February 1983, pp. 45–53.
4. Brown, Lester R., and Edward Wolf. "Food Crisis in Africa," *Natural History*, June 1984, pp. 16–20.
5. Ehrlich, Paul R., Anne H. Ehrlich, and John P. Holdren. *Ecoscience: Population, Resources, Environment.* San Francisco: W. H. Freeman, 1977.
6. Foreman, Kim W. "Bringing in the Beneficials." *Organic Gardening*, February 1985, pp. 70–72.
7. Garbett, Mel. "Future Trends in the Lawn and Garden Pesticide Industry." *Chemical Times and Trends*, April 1984, pp. 10–11, 43–46.
8. Gibbons, Boyd. "Do We Treat Our Soil Like Dirt?" *National Geographic*, September 1984, pp. 350–389.
9. Graham, Frank, Jr. *The Dragon Hunter.* New York: Dutton, 1984. Discusses biological controls.
10. Lindsay, David G., and Robert N. Crossett. "Agricultural Chemicals—Inputs and Outputs." *Chemistry and Industry*, 17 September 1984, pp. 645–665.
11. Madson, Chris. "A Smorgasbord of Poisons Found in Plains Wildlife." *Audubon*, January 1983, pp. 121–125.
12. Marshall, Eliot. "The Rise and Decline of Temik." *Science*, 27 September 1985, pp. 1369–1371.
13. Miller, Lois K., A. J. Lingg, and Lee A. Bulla, Jr. "Bacterial, Viral, and Fungal Insecticides." *Science*, 11 February 1983, pp. 715–721.
14. Sanders, Howard J. "Herbicides." *Chemical and Engineering News*, 3 August 1981, pp. 20–35.
15. Sheldon, Richard P. "Phosphate Rock." *Scientific American*, June 1982, pp. 45–51.
16. Shute, Nancy. "What's Bugging You?" *American Health*, July–August 1985, pp. 56–59.
17. Silverstein, Robert M. "Pheromones: Background and Potential for Use in Insect Pest Control." *Science*, 18 September 1981, pp. 1326–1332.
18. Skinner, Karen Joy. "Nitrogen Fixation." *Chemical and Engineering News*, 4 October 1976, pp. 22–35.
19. Storck, William J. "Pesticides Head for Recovery." *Chemical and Engineering News*, 9 April 1984, pp. 35–57.
20. Szulc, Tad. "One Person Too Many?" *Parade Magazine*, 29 April 1984, pp. 16–19.
21. Tangley, L. "A Pair of Star-Crossed Love Bugs." *Science News*, 28 August 1982, p. 135.
22. Torrey, John G. "The Development of a Plant Biotechnology." *American Scientist*, July–August 1985, pp. 354–363.
23. Ware, George W. *The Pesticide Book.* San Francisco: W. H. Freeman, 1978.
24. Ward, John. "Phosphorus: Fertilizer's Middle Name." *Organic Gardening*, June 1985, pp. 40–43.
25. Windholz, Martha (Ed.). *The Merck Index*, 10th ed. Rahway, NJ: Merck and Co., 1983.
26. Worthy, Ward. "Pesticide Chemists Are Shifting Emphasis from Kill to Control." *Chemical and Engineering News*, 23 July 1984, pp. 22–26.

18

Food

Those Incredible Edible Chemicals

Life is sustained by a complex set of chemical processes. Food is the group of chemical reagents that take part in the life-sustaining processes. Food supplies the materials from which living organisms are made. Food supplies the energy that keeps organisms alive. Food furnishes the parts for the growth and repair of tissues. Without food, we would die in a few weeks.

In much of the Western World, the problem often is an over-abundance of food. We eat too much—and often the wrong kind. It is estimated that over 30% of the adults in the United States are overweight. Obesity contributes to poor health, increased risk of cardiovascular disease and diabetes, and perhaps increased susceptibility to other diseases as well. An obese person has a shorter life expectancy than a person of average weight. Our diet (too rich in fat, sugar, salt, alcohol, and cholesterol) has been linked to 6 of the 10 leading causes of death in the United States.

In developing countries, the problem often is too little food or food of inadequate quality. Malnourished children, particularly, are more susceptible to disease. They often die of chicken pox and measles, diseases that present little threat to well-nourished children. Inadequate food leaves children small in stature and stunted mentally. They need more and better food to become strong in body and mind.

Just as life is a lot more than the sum of a set of chemical processes, food is more than just a collection of chemicals. Food often is consumed on social occasions, and it can't be separated from its cultural context.

Chemicals Eat Chemicals

The human body is a conglomeration of chemicals. If we did a chemical analysis of the human body, we would find that it is largely—about two thirds—water. There are also a number of minerals present. The blood and other body fluids contain dissolved salts. The bones and teeth are similar to hydroxyapatite, a mineral found in phosphate rock. A little glucose ($C_6H_{12}O_6$, a simple sugar) is found in the blood. There is fat under the skin and around the internal organs. A form of starch (glycogen) is found in the liver and in muscles. And, in every cell of the body, there are proteins. There are also thousands of other chemicals that play vital roles in the life processes of the human body.

We need food for growth when we are children. At all ages, we need food for the replacement and repair of tissues, and we need food for energy. Normally, these needs are met by the foods we eat. The three main classes of foods used by people and other animals are carbohydrates, fats, and proteins. These, too, are chemicals. For proper nutrition, our diet should consist of balanced proportions of all three types of foods, plus vitamins, minerals, and dietary fibers. In the remainder of this chapter, we see how each of these substances fulfills its function in our bodies.

Carbohydrates: Our Storehouse of Energy

Carbohydrates are compounds of the elements carbon, hydrogen, and oxygen. Usually, the atoms of these elements are present in a ratio expressed by the formula $(C_x(H_2O)_y$. Glucose, which has the formula $C_6H_{12}O_6$, could be written $C_6(H_2O)_6$. Indeed, it is from formulas such as these that the term **carbohydrate** is derived. We use it even though the term is misleading: carbohydrates are not hydrates of carbon.

Sugars, starches, and even celluloses are carbohydrates. Sugars, have been used for ages to make certain foods sweeter and more palatable. Common table sugar is **sucrose** ($C_{12}H_{22}O_{11}$), derived principally from sugarcane and sugar beets. Per capita consumption of sugar in the United States is about 57 kg per year. Much of it is consumed in soft drinks, presweetened cereals, candy, and other highly processed foods with little or no other nutritive value. The empty calories from sweetened foods contribute to tooth decay and obesity.

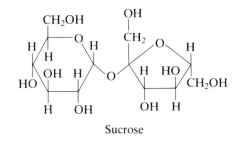

Sucrose

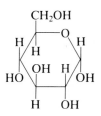

Glucose (dextrose)

Sucrose is broken down in our bodies to two simpler sugars, glucose and fructose. Glucose (also called dextrose) is the most common simple sugar used by the cells of our bodies for energy. **Glucose** is the simple sugar that is circulated in the bloodstream; that is why it is sometimes called **blood sugar**. **Fructose** which occurs in many fruits, is sometimes called **fruit sugar**.

Corn syrup, an increasingly important sweetener, is mainly glucose. It is made by the hydrolysis of starch. **High-fructose corn syrup** is made by using enzymes to convert much of the glucose in the syrup to fructose. Fructose is sweeter than glucose. Foods sweetened to the same degree with high-fructose corn syrup have fewer Calories than those sweetened with glucose.

Lactose is milk sugar. During digestion, it is broken down to two simpler sugars, glucose and galactose. Nearly all human babies have the enzyme necessary to accomplish this breakdown, but many adults do not. People who lack the enzyme get digestive upsets from drinking milk. The indigestible lactose passes unchanged to the colon. There it is acted upon by bacteria. Victims get diarrhea and a bloated feeling from the gases produced. The condition is called **lactose intolerance**. When milk is cooked or fermented, the lactose is at least partially hydrolyzed. People with lactose intolerance may still be able to enjoy cheese, yogurt, or cooked foods containing milk with little or no problem.

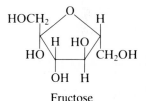

Fructose

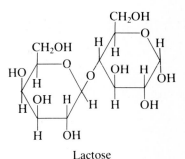

Lactose

Complex Carbohydrates: Starch and Cellulose

As foods, complex carbohydrates have a somewhat better reputation than the simple sugars. Starch is an important part of any balanced diet. Cellulose contributes fiber to our diet. Figure 18.1 shows short segments of starch and cellulose molecules. Notice that both are polymers of glucose.

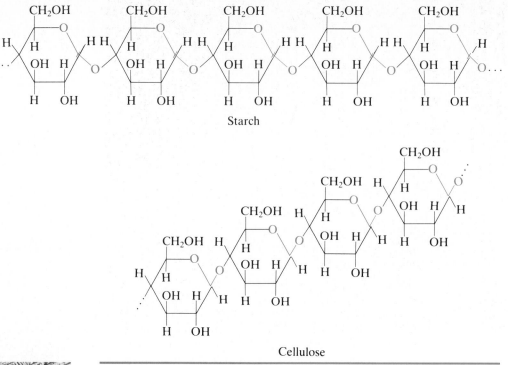

Starch

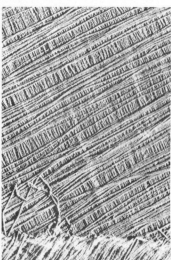

Cellulose

Figure 18.1 Carbohydrates. Both starch and cellulose are polymers of glucose.

The main structural difference between starch and cellulose is in the way the glucose units are hooked together. This may seem to be a minor difference, but all animals can digest and metabolize starch, while humans and most other animals get no food value from cellulose. Bacteria in the digestive tract enable grazing animals and termites to utilize cellulose.

Different linkages also result in different three-dimensional forms for cellulose and starch. For example, cellulose in the cell walls of plants is arranged in **fibrils**, bundles of parallel chains. The fibrils, in turn, lie parallel to each other in each layer of the cell wall (Figure 18.2). In alternate layers, the fibrils are perpendicular; this imparts great strength to the wall.

Actually there are two kinds of starch. One, called **amylose**, has the glucose units joined in a continuous chain, like beads on a string. The other kind, called **amylopectin**, has branched chains of glucose units. These are perhaps best represented schematically, as in Figure 18.3 (each unit is represented by the letters *Glc*).

Animal starch is called **glycogen**. Like amylopectin, it is composed of branched chains of glucose units. Glycogen in muscle and liver tissue is arranged in granules (Figure 18.4). These granules are clusters of

Figure 18.2 Electron micrograph of the cell wall of an alga. The wall consists of successive layers of cellulose fibers in parallel arrangement. [Photo by R. D. Preston, F.R.S., and E. Frei.]

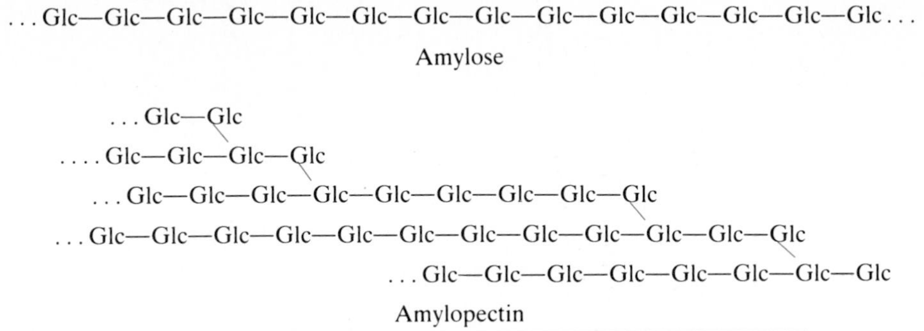

...Glc—Glc—Glc—Glc—Glc—Glc—Glc—Glc—Glc—Glc—Glc—Glc—Glc—Glc...

Amylose

...Glc—Glc

....Glc—Glc—Glc—Glc

...Glc—Glc—Glc—Glc—Glc—Glc—Glc—Glc—Glc

...Glc—Glc—Glc—Glc—Glc—Glc—Glc—Glc—Glc—Glc—Glc—Glc

...Glc—Glc—Glc—Glc—Glc—Glc—Glc—Glc

Amylopectin

Figure 18.3   Schematic representations of amylose and amylopectin; Glc represents a glucose unit. The structure of glycogen is similar to that of amylopectin.

small particles. Plant starch, on the other hand, forms large granules. Plant-starch granules rupture in boiling water to form a paste. On cooling, the paste gels. Potatoes and cereal grains, when cooked in water, form this type of starchy broth.

Starch is broken down to glucose when it is digested. This simple sugar is readily absorbed through the intestinal walls and taken into the bloodstream. The body then metabolizes the glucose, using it as a source of energy. Glucose is broken down through a complex set of more than 50 chemical reactions to produce carbon dioxide and water, with the release of energy.

$$C_6H_{12}O_6 \; + \; 6\,O_2 \longrightarrow 6\,CO_2 \; + \; 6\,H_2O \; + \; energy$$

These reactions are essentially the reverse of photosynthesis. In this way, animal organisms are able to make use of the energy from the sun that was captured by plants in the process of photosynthesis.

The steps in this complicated process are catalyzed by **enzymes**, protein molecules that enable reactions to occur at convenient rates and at much lower temperatures than they would otherwise. Each step involves only a small energy change, whereas the overall reaction releases a large amount of energy. We could obtain the same amount of energy by burning the glucose at elevated temperatures in air without enzymes, but the temperature required and the heat produced in this more direct process would destroy living cells.

Enzymes seem to operate by a sort of lock-and-key effect (Figure 18.5). The enzyme for a particular reaction must fit precisely the molecule it is acting upon. Enzymes that break down starch to glucose don't fit the cellulose molecule. (Note the difference in the way the glucose units are linked together in the two polymers in Figure 18.1; thus, cellulose can't be metabolized by animal organisms.) Some

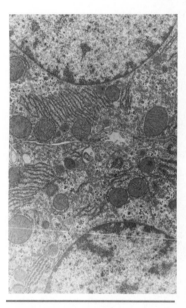

Figure 18.4   Electron micrograph of glycogen granules in a liver cell of a rat. [Photo by Patricia J. Schulz.]

Figure 18.5 The lock-and-key model for enzyme action.

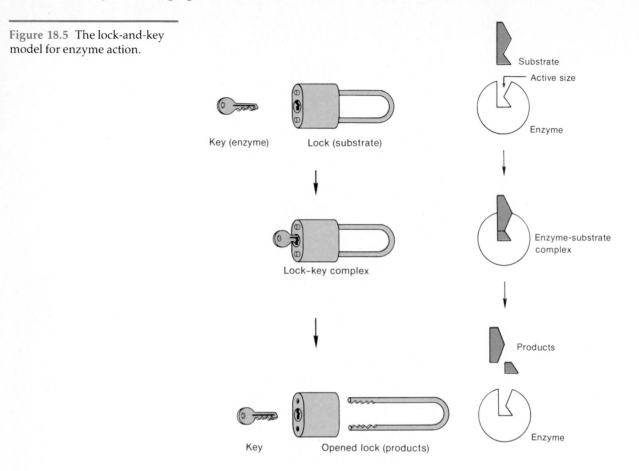

bacteria—such as those inhabiting the digestive tracts of ruminants and termites—do have enzymes for breaking down cellulose to glucose, so these bacteria (and their hosts) are able to obtain food value from cellulose.

Carbohydrates supply about 4 kcal of energy per gram. (The kilocalorie often is simply called a Calorie when used to describe the energy content of foods.) When we eat more than we can use, a small amount of carbohydrate can be stored in our liver and muscle tissues as glycogen. Large excesses, however, are converted to fat for storage.

There is no set dietary recommendation for carbohydrates, but at least 65%, and perhaps as much as 80%, of our Calories should come from them—for these are our bodies' preferred fuels. The majority of our carbohydrate intake should be the complex carbohydrates found in cereal grains, rather than the simple sugars found in many prepared foods.

Fats: Padding, Insulation, and Reserve Energy

Like carbohydrates, fats are compounds of the elements carbon, hydrogen, and oxygen, but they are quite different from carbohydrates. **Fats** are *esters* (Chapter 11). They can be thought of as being derived from the trihydroxy alcohol glycerol and long-chain carboxylic acids (**fatty acids**).

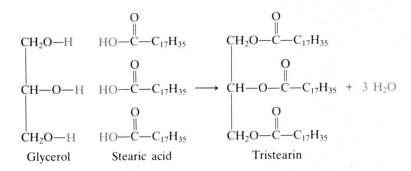

CH_2O-H	$HO-\overset{O}{\underset{\parallel}{C}}-C_{17}H_{35}$	$CH_2O-\overset{O}{\underset{\parallel}{C}}-C_{17}H_{35}$
$CH-O-H$	$HO-\overset{O}{\underset{\parallel}{C}}-C_{17}H_{35}$	$CH-O-\overset{O}{\underset{\parallel}{C}}-C_{17}H_{35} + 3 H_2O$
CH_2O-H	$HO-\overset{O}{\underset{\parallel}{C}}-C_{17}H_{35}$	$CH_2O-\overset{O}{\underset{\parallel}{C}}-C_{17}H_{35}$
Glycerol	Stearic acid	Tristearin

Naturally occurring fatty acids nearly always have an even number of carbon atoms. Representative ones are listed in Table 18.1. Animal fats generally contain both saturated and unsaturated fats, but the former predominate. At room temperature, animal fats are usually solids. Liquid fats, called **oils**, are obtained principally from vegetable sources. Structurally, oils are identical to fats except that they incorporate a higher proportion of unsaturated acid units (see Table 18.1).

Table 18.1 Some Fatty Acids in Natural Fats

Number of Carbon Atoms	Condensed Structure	Name	Source
4	$CH_3CH_2CH_2COOH$	Butyric acid	Butter
6	$CH_3(CH_2)_4COOH$	Caproic acid	Butter
8	$CH_3(CH_2)_6COOH$	Caprylic acid	Coconut oil
10	$CH_3(CH_2)_8COOH$	Capric acid	Coconut oil
12	$CH_3(CH_2)_{10}COOH$	Lauric acid	Palm kernel oil
14	$CH_3(CH_2)_{12}COOH$	Myristic acid	Oil of nutmeg
16	$CH_3(CH_2)_{14}COOH$	Palmitic acid	Palm oil
18	$CH_3(CH_2)_{16}COOH$	Stearic acid	Beef tallow
18	$CH_3(CH_2)_7CH{=}CH(CH_2)_7COOH$	Oleic acid	Olive oil
18	$CH_3(CH_2)_4CH{=}CHCH_2CH{=}CH(CH_2)_7COOH$	Linoleic acid	Soybean oil
18	$CH_3CH_2(CH{=}CHCH_2)_3(CH_2)_6COOH$	Linolenic acid	Fish oils
20	$CH_3(CH_2)_4(CH{=}CHCH_2)_4CH_2CH_2COOH$	Arachidonic acid	Liver

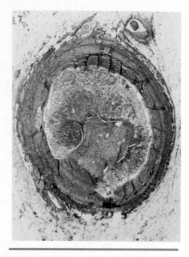

Figure 18.6 Photomicrograph of a cross section of a hardened artery, showing deposits of plaque. The deposits contain cholesterol. [Courtesy of Biomedical Graphics, University of Minnesota Hospitals, Minneapolis.]

Fats are often classified according to the degree of unsaturation of the fatty acids they incorporate. **Saturated fatty acids** contain no double bonds, **monounsaturated fatty acids** contain one double bond per molecule, and **polyunsaturated fatty acids** are those that have two or more double bonds. **Saturated fats** contain a high proportion of saturated fatty acids; the fat molecules have relatively few double bonds. **Polyunsaturated fats** (oils) incorporate mainly unsaturated fatty acids; these fat molecules have many double bonds.

Saturated fats have been implicated, along with cholesterol in one type of **arteriosclerosis** (hardening of the articles). As this condition develops, deposits form on the walls of arteries. Eventually these deposits become calcified (harden), robbing the vessels of their elasticity (Figure 18.6). There is a strong correlation between diets rich in saturated fats and incidence of the disease. It is this correlation that has led to a concern over the relative amounts of saturated and unsaturated fats in our diets. Advertisers who recommend that you buy corn oil margarine (prepared from relatively unsaturated vegetable oil) rather than butter (a relatively saturated animal fat) take advantage of this concern.

There is also statistical evidence that fish oils can prevent heart disease. Researchers at the University of Leiden in the Netherlands have found that Greenlanders who eat a lot of fish have a low risk of heart disease, despite a diet that is high in total fat and cholesterol. The probable effective agents are the polyunsaturated fatty acids such as eicosapentaenoic acid.

$$CH_3CH_2CH{=}CHCH_2CH{=}CHCH_2CH{=}CHCH_2CH{=}CHCH_2CH{=}CHCH_2CH_2CH_2\overset{\displaystyle O}{\overset{\|}{C}}{-}OH$$

Other studies have shown that diets with added fish oil lead to lower cholesterol and triglyceride levels in the blood.

The degree of unsaturation of a fat or an oil usually is measured by the **iodine number**. Chlorine and bromide add readily to carbon–carbon double bonds.

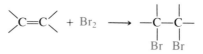

Iodine also adds, but less readily. The iodine number of a fat is the number of grams of iodine that will be consumed by 100 g of fat or oil. The more double bonds a fat contains, the more iodine required for the addition reaction; thus, a high iodine number means a high degree of unsaturation. Representative iodine numbers are listed in Table 18.2. Note that generally lower values for the animal fats (butter, tallow, lard)

compared to those for the vegetable oils. Coconut oil, which is highly saturated, and fish oils, which are relatively unsaturated, are notable exceptions to the general rule.

Fats and oils feel greasy. They are insoluble in water but soluble in organic solvents such as carbon tetrachloride, gasoline, and diethyl ether. Fats and oils are lighter than water; consequently, they float on top of water when they are put into it.

When violently shaken together with water, fats are broken into tiny, submicroscopic particles and dispersed through the water. Such a mixture is called an **emulsion**. Unless a third substance has been added, the emulsion breaks down rapidly. The droplets then recombine and float to the surface of the water. Soap, certain types of gum, or protein can stabilize the emulsion by forming protective coatings on the fat droplets that prevent them from coming together. Milk is an emulsion of butterfat in water. The stabilizing agent in milk is a protein called casein.

Fats are high-energy foods. They yield about 9 kcal of energy per gram. Some fats are "burned" as fuel for our activities. Others are used to build and maintain important constituents of our cells, such as the cells of brain and nerve tissue and the membranes that surround each cell. Fats eaten in excess are stored in the body, where they serve as energy reserves. This stored fat also serves to insulate the body against loss of heat and to protect vital organs from injury by acting as extra padding.

Fats have survival value. Stored fats presumably got our ancestors through the lean times. Our ability to store large amounts of fats probably is in our genes.

Fat in our diet comes from butter, cream, margarine, vegetable oils and shortenings, meat products, and some seeds and nuts. Generally, not more than 30% of our Calories should come from fat. No more than 10% of total Calories should come from saturated fats, with 10% from monounsaturated and 10% from polyunsaturated fats. The average American diet contains too much fat; about 40% of our Calories come from fat, with about 13% of Calories from saturated fats. Most medical authorities recommend that we eat less saturated fats and less total fats than present averages.

Table 18.2 Typical Iodine Numbers for Some Fats and Oils

Fat or Oil	Iodine Number
Coconut oil	8–10
Butter	25–40
Beef tallow	30–45
Palm oil	37–54
Lard	45–70
Olive oil	75–95
Peanut oil	85–100
Cottonseed oil	100–117
Corn oil	115–130
Fish oils	120–180
Soybean oil	125–140
Safflower oil	130–140
Sunflower oil	130–145
Linseed oil	170–205

Proteins: The Stuff of Life

The third class of food, the proteins, is the vital component of all life. No living part of the human body—or of any other kind of animal, for that matter—is completely without protein. There is protein in the blood, the muscles, the brain, and even the tooth enamel. The smallest cellular organisms—the bacteria—contain protein; and viruses, so small that they make bacteria look like giants, are nothing but large molecules of a special type of protein. This special material, called **nucleoprotein**, is found in all cells and is the stuff of life itself. Nucleoproteins are combinations of proteins with nucleic acids. As we saw in Chapter 11, these nucleic acids control the structure and function of cells. This they do through the proteins they cause to be formed.

Each type of cell makes its own kinds of proteins. The proteins serve as the structural material of animals, much as cellulose does for plants. Muscle tissue is largely protein; so are skin and hair. These proteins are made in different forms in different animals. Silk, wool, nails, claws, feathers, horns, and hooves are proteins. All proteins contain the elements carbon, hydrogen, oxygen, and nitrogen. Most also contain sulfur. The structure of a short segment of a typical protein molecule is shown in Figure 18.7.

Proteins are copolymers of about 20 different **amino acids** (Table 18.3). Amino acids have two functional groups. The amino group ($-NH_2$) is on the carbon next to the carbon of the carboxyl group ($-COOH$).

Figure 18.7 (a) Structural formula of a segment of a protein molecule. (b) A space-filling model of a segment of a protein chain. [Courtesy of Science Related Materials, Inc., Janesville, WI.]

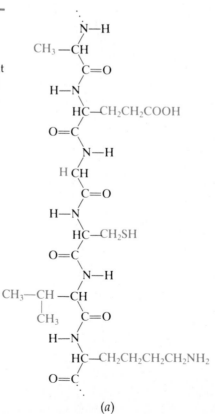

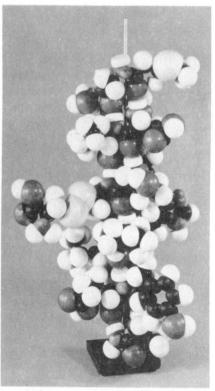

(a)

(b)

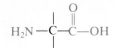

This partial formula indicates the proper placement of these groups, but the structure is not really correct. Acids react with bases to form salts; the carboxyl group is acidic and the amino group is basic. Therefore, these two functional groups interact, the acid transferring a proton to the base. The resulting product is an inner salt, or **zwitterion**, a compound in which the anion and the cation are parts of the same molecule. The amino acids differ in that a variety of other groups (symbolized by the —R) may be attached to the carbon atom that bears the amino group.

$$H_3\overset{+}{N}-\overset{\overset{\displaystyle R}{|}}{\underset{\underset{\displaystyle H}{|}}{C}}-\overset{\overset{\displaystyle O}{\|}}{C}-O^-$$

A zwitterion

Plants are able to synthesize proteins from carbon dioxide, water, and minerals [such as nitrates (NO_3^-) and sulfates (SO_4^{2-})]. Animals, in their digestive tracts, break down proteins from plant matter or animal flesh into the component amino acids. From these, the animals can synthesize proteins for growth and for the repair of body tissues. When a diet contains more protein than is needed for the body's growth and repair, the leftover protein can be used as a source of energy.

The Essential Amino Acids: Eat 'em or Else

The adult human body can synthesize all but eight of the amino acids needed for making proteins. Those eight (indicated in Table 18.3) are called **essential amino acids**. They must be included in our diet. We eat proteins, break them down in our bodies to their constituent amino acids, and then use some of these amino acids to build other protein structures essential for our health. Each of the essential amino acids is a **limiting reagent**. When the body runs out of one of them, it can't make proper proteins.

An **adequate protein** supplies all the essential amino acids in the quantities needed for the growth and repair of body tissues. Most proteins from plant sources are deficient in one or more amino acids. Corn protein is lacking in lysine and tryptophan. People whose diet consists chiefly of corn may suffer from malnutrition even though the Calories supplied by the food are adequate. Protein from rice is short of lysine and threonine. Wheat protein is lacking in lysine. Even soy protein, probably the best nonanimal protein, is lacking in the essential amino acid methionine.

Most proteins from animal sources contain all the essential amino acids in adequate amounts. Lean meat, milk, fish, eggs, and cheese supply adequate proteins. In fact, gelatin is about the only inadequate

Table 18.3   Some Representative Amino Acids

Name	Abbreviation	Essential	Structure
Glycine	Gly	No	$\underset{\underset{\text{}}{+}NH_3}{\overset{\displaystyle CH_2-COO^-}{\mid}}$
Alanine	Ala	No	$CH_3-\underset{\overset{\mid}{+NH_3}}{CH}-COO^-$
Phenylalanine	Phe	Yes	⬡$-CH_2-\underset{\overset{\mid}{+NH_3}}{CH}-COO^-$
Valine	Val	Yes	$CH_3-\underset{\overset{\mid}{CH_3}}{CH}-\underset{\overset{\mid}{+NH_3}}{CH}-COO^-$
Leucine	Leu	Yes	$CH_3CHCH_2-\underset{\overset{\mid}{+NH_3}}{CH}-COO^-$ with $\overset{\mid}{CH_3}$
Isoleucine	Ile	Yes	$CH_3CH_2CH-\underset{\overset{\mid}{+NH_3}}{CH}-COO^-$ with $\overset{\mid}{CH_3}$
Proline	Pro	No	$\begin{array}{c} CH_2-CH_2 \\ CH_2 \quad\quad C \\ +NH_2 \quad H \end{array}$ COO$^-$
Methionine	Met	Yes	$CH_3-S-CH_2CH_2-\underset{\overset{\mid}{+NH_3}}{CH}-COO^-$
Serine	Ser	No	$HO-CH_2-\underset{\overset{\mid}{+NH_3}}{CH}-COO^-$
Threonine	Thr	Yes	$CH_3CH-\underset{\overset{\mid}{+NH_3}}{CH}-COO^-$ with $\overset{\mid}{OH}$
Asparagine	Asn	No	$H_2N-\overset{\overset{\displaystyle O}{\|\|}}{C}-CH_2-\underset{\overset{\mid}{+NH_3}}{CH}-COO^-$

Table 18.3—*continued*

Name	Abbreviation	Essential	Structure
Glutamine	Gln	No	$H_2N-\overset{\overset{\displaystyle O}{\|\|}}{C}-CH_2CH_2-\underset{\underset{\displaystyle {}^+NH_3}{\|}}{CH}-COO^-$
Cysteine	Cys	No	$HS-CH_2-\underset{\underset{\displaystyle {}^+NH_3}{\|}}{CH}-COO^-$
Tyrosine	Tyr	No	$HO-\langle\text{C}_6\text{H}_4\rangle-CH_2-\underset{\underset{\displaystyle {}^+NH_3}{\|}}{CH}-COO^-$
Tryptophan	Trp	Yes	$CH_2-\underset{\underset{\displaystyle {}^+NH_3}{\|}}{CH}-COO^-$ (indole ring)
Lysine	Lys	Yes	$H_3\overset{+}{N}CH_2CH_2CH_2CH_2-\underset{\underset{\displaystyle NH_2}{\|}}{CH}-COO^-$
Arginine	Arg	—*	$H_2N-\underset{\underset{\displaystyle {}^+NH_2}{\|\|}}{C}-NHCH_2CH_2CH_2-\underset{\underset{\displaystyle NH_2}{\|}}{CH}-COO^-$
Histidine	His	—†	$CH_2-\underset{\underset{\displaystyle {}^+NH_3}{\|}}{CH}-COO^-$ (imidazole ring)
Aspartic acid	Asp	No	$HOOC-CH_2-\underset{\underset{\displaystyle {}^+NH_3}{\|}}{CH}-COO^-$
Glutamic acid	Glu	No	$HOOC-CH_2CH_2-\underset{\underset{\displaystyle {}^+NH_3}{\|}}{CH}-COO^-$

* Essential to growing children but not to adult humans.
† Essential to rats but not to humans.

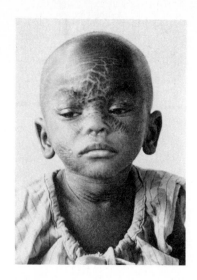

Figure 18.8 A lack of proteins and vitamins causes a deficiency disease known as kwashiorkor. The symptoms include retarded growth, discoloration of skin and hair, bloating, swollen belly, and mental apathy. [Courtesy of Paul Almasy, World Health Organization.]

animal protein. It contains almost no tryptophan and has only small amounts of threonine, methionine, and isoleucine.

Our requirement for protein is about 0.8 g per kilogram of body weight. The requirement for pregnant women is slightly higher. The diet must include all the essential amino acids in sufficient quantity. Diets with inadequate protein are common in much of the world. A protein-deficiency disease called **kwashiorkor** (Figure 18.8) is common at times in parts of Africa where corn is the major food.

Nutrition is especially important in a child's early years. Protein deficiency leads to both physical and mental retardation. The effect on a human's mental capacity is readily apparent from a consideration of Figure 18.9, which shows that the human brain reaches nearly full size

Figure 18.9 Growth of the human brain according to age.

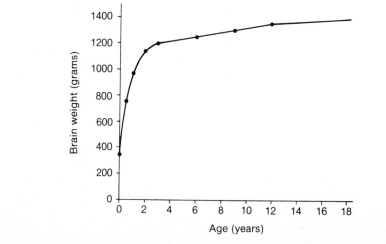

by the age of 2 years. Inadequate nutrition for a child in the mother's womb and at her breast may hopelessly handicap the child for life.

A Myriad of Proteins

The human body contains about 30 000 different proteins. Each of us has his or her own tailor-made set. Proteins are polyamides. The amide linkage (—CONH—) is called a **peptide bond** when it joins two amino acid units. Note that there is still a reactive amino group on the left and a carboxyl group on the right. Each of these can react further to join more amino acid units. This process can continue until thousands of units have joined to form a giant molecule—a polymer called a **protein**.

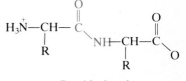

Peptide bond

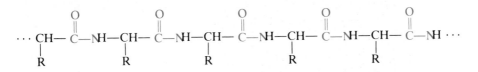

When only two amino acids are joined, the product is a **dipeptide**.

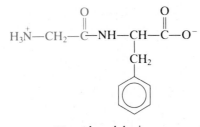

Glycyphenylalanine
(a dipeptide)

When three amino acids are combined, the substance is a **tripeptide**.

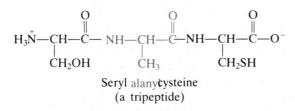

Seryl alanycysteine
(a tripeptide)

Note how these substances are named. Other combinations are named in a similar manner. Those with more than 10 amino acid units are often simply called **polypeptides**. When the molecular weight of a compound exceeds 10 000, it is called a protein. The distinction between polypeptides and proteins is an arbitrary one, and it is not always precisely applied.

The Sequence of Amino Acids

For peptides and proteins to be physiologically active, it is not enough that they incorporate certain amounts of specific amino acids. The order or *sequence* in which the amino acids are connected is also of critical importance. Glycylalanine is different from alanylglycine, for example.

$$\overset{H}{\underset{}{H_3\overset{+}{N}}}-\overset{H}{\underset{|}{CH}}-\overset{O}{\underset{}{\overset{\|}{C}}}-NH-\overset{CH_3}{\underset{|}{CH}}-COO^- \qquad H_3\overset{+}{N}-\overset{CH_3}{\underset{|}{CH}}-\overset{O}{\overset{\|}{C}}-NH-\overset{H}{\underset{|}{CH}}-COO^-$$

Glycylalanine   Alanylglycine

Although the difference seems minor, the two substances behave differently in the body.

When chemists describe peptides (and proteins), they find it much simpler to indicate the amino acids sequence by using the abbreviations for the amino acids (Table 18.3). The sequence for glycylalanine is written Gly-Ala, and that for alanylglycine is Ala-Gly. It is understood from this shorthand that the peptide is arranged with the free amino group to the left and the free carboxyl group to the right.

As the length of a peptide chain increases, the possible sequential variations become almost infinite. And this potential for many different arrangements is exactly what one needs in a material that will make up such diverse things as hair and skin and eyeballs and toenails and a thousand different enzymes. Just as we can make millions of different words with our 26-letter English alphabet, we can make millions of different proteins with the 20 or so amino acids. Just as one can write gibberish with the English alphabet, one can make nonfunctioning proteins by putting together the wrong sequence of amino acids.

Yet, while the correct sequence is ordinarily of utmost importance, it is not always absolutely required. Just as you can sometimes make sense of incorrectly spelled English words, a protein with a small percentage of "incorrect" amino acids may continue to function. It may not function as well, however, as a protein with the correct sequence. And sometimes a seemingly minor change can have a disastrous effect. Some people have hemoglobin with 1 incorrect amino acid unit in about 300. That "minor" error is responsible for sickle cell anemia, an inherited condition that ordinarily proves fatal.

Starvation, Fasting, and Fat Diets

When the human body is totally deprived of food, whether voluntarily or involuntarily, the condition is known as **starvation**. During total fasting, the body's glycogen stores are depleted rapidly and the body calls on its fat reserves. Fat is first taken from around the kidneys and the heart. Then it is removed from other parts of the body. Ultimately, even the bone marrow (which is also a fat-storage depot) is depleted, and it becomes red and jellylike (it is usually white and firm).

Increased dependence on fats as an energy source leads to **ketosis**, a condition characterized by the appearance of *ketones* in the urine. One of the ketones is acetone (Chapter 11). Two other compounds are also prominent in the urine during ketosis: acetoacetic acid and *β*-hydroxybutyric acid. Collectively, these three compounds are called **ketone bodies** (even though one is not a ketone). Note that two of the ketone bodies are acids. Ketosis rapidly develops into **acidosis**; the blood pH drops and oxygen transport is hindered. Oxygen deprivation leads to depression and lethargy. Even the ability to think clearly and the ability to make decisions are impaired.

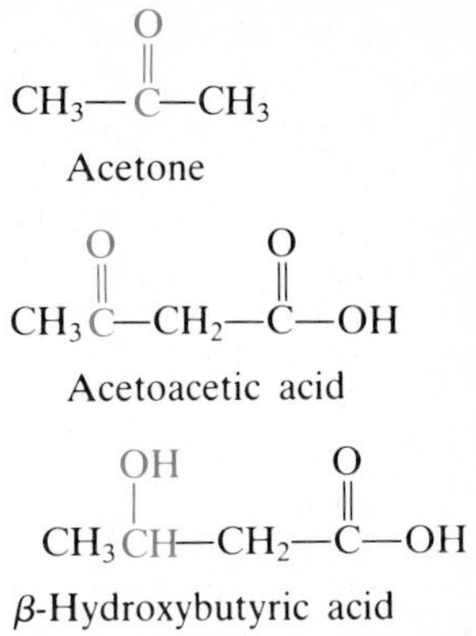

In the more extreme low-carbohydrate diets (for weight loss), ketosis is deliberately induced. Understandably, two of the side effects of these diets are often depression and lethargy. In the early stages of a diet deficient in carbohydrates, the body converts amino acids to glucose. The brain must have glucose; it can't obtain sufficient energy from fats. If there are enough adequate proteins in the diet, tissue proteins are spared. (The liquid protein diets, from which several people died, were based on the inadequate proteins obtained from beef hides and connective tissues.)

In the early stages of a *total* fast, body protein is metabolized at a relatively rapid rate. After several weeks, the rate of protein breakdown slows considerably, as the brain adjusts to using the breakdown products of fatty acid metabolism for its energy source. When fat reserves are substantially depleted, the body must again draw heavily on its structural proteins for its energy requirements. The emaciated appearance of a starving individual is due to the depletion of muscle proteins.

Even low-carbohydrate diets that are high in adequate proteins are hard on the body, which must rid itself of the nitrogen compounds—ammonia and urea—formed by the breakdown of proteins. This puts an increased stress on the liver, where the waste products are formed.

It is interesting to note that contrary to a popular notion, fasting does not cleanse the body. Indeed, quite the reverse occurs. A shift to fat metabolism produces the ketone bodies, and protein breakdown produces ammonia, urea, and other wastes. You can lose weight by fasting, but the process should be carefully monitored by a physician.

Involuntary starvation is a serious problem in much of the world. Even so, starvation is seldom the sole cause of death. Weakened by starvation, the victims of starvation and malnutrition succumb to disease. Even usually minor diseases, such as chicken pox and measles become life-threatening disorders. Barring disease, starvation alone

Figure 18.10 The three ketone bodies.

Table 18.4 Efficiencies of Protein Conversions

Food	Efficiency of Production (%)*
Beef or veal	4.7
Pork	12.1
Chicken or turkey	18.2
Milk	22.7
Eggs	23.3

* Calculated by dividing the weight of edible protein by the weight of the protein feed required to produce it, then multiplying the result by 100. SOURCE: President's Science Advisory Committee, *The World Food Problem*, vol. 2, 1967.

would lead eventually to death from circulatory failure when the heart muscle became too weak to pump blood.

Weight loss through diet and exercise is discussed in Chapter 24.

Thermodynamics, Vegetarians, and Ethnic Dishes

Green plants such as grasses and grains trap a small fraction of the energy of the sun that falls upon them. They use this energy to convert carbon dioxide, water, and mineral nutrients (including nitrates, phosphates, and sulfates) into proteins. Cattle eat the plant protein, digest it, and convert a small portion of it into animal protein. People eat this animal protein, digest it, and reassemble some of the amino acids into human protein. Some of the energy originally trapped by the green plants is lost as heat at every step of the food chain (see Table 18.4). If people ate the plant protein directly, one highly inefficient step could be skipped. Thus, vegetarianism is in harmony with the conservation of energy.

It is interesting to note that a variety of ethnic dishes supply relatively good protein by combining a cereal grain with a legume (beans, peas, peanuts, etc.). The grain is deficient in tryptophan and lysine, but it has sufficient methionine. The legume is deficient in methionine, but it has enough tryptophan and lysine. A few such combinations are listed in Table 18.5.

Although complete proteins can be obtained by a careful mixture of vegetable foods, extreme vegetarianism is dangerous. Even when the diet includes a wide variety of plant materials, an all-vegetable diet is likely to lack vitamin B_{12}, because this nutrient is not found in plants. Other nutrients scarce in all-plant diets include calcium, iron, riboflavin, and (for children not exposed to sunlight) vitamin D. A modified vegetarian diet that includes milk, eggs, cheese, and fish can provide excellent nutrition, with red meat totally excluded.

Table 18.5 Ethnic Foods That Combine a Cereal Grain with a Legume

Ethnic Group	Food
Mexican peasants	Corn tortillas and refried beans
Japanese	Rice and soybean curds (Tofu)
English working classes	Baked beans on toast
American Indians	Corn and beans (succotash)
Children (especially in the United States)	Peanut butter sandwiches
Western Africans	Rice and peanuts (ground nuts)
Cajuns (Louisiana)	Red beans and rice

Minerals: Important Inorganic Chemicals in Our Lives

As might be expected, many of the chemicals in living organisms are organic. But there are also inorganic chemicals that are vital to sustaining life. These inorganic nutrients are called **minerals**. It is estimated that minerals represent about 4% of the weight of a human body. Some of these, such as the chlorides (Cl^-), phosphates (PO_4^{3-}), bicarbonates (HCO_3^-), and sulfates (SO_4^{2-}), occur in the blood and the other body fluids. Others, such as iron (as Fe^{2+}) in the hemoglobin and phosphorus in the nucleic acids (DNA and RNA), are constituents of complex organic compounds.

Minerals essential to one or more living organisms include the following elements.

Figure 18.11 A man affected by goiter. The swollen thyroid gland in the neck results from a lack of the trace element iodine in the diet. [Photo by Joseph J. Mentrikoski, Department of Medical Photography, Geisinger Medical Center.]

Arsenic (As)	Iodine (I)	Selenium (Se)
Calcium (Ca)	Iron (Fe)	Silicon (Si)
Chlorine (Cl)	Magnesium (Mg)	Sodium (Na)
Chromium (Cr)	Manganese (Mn)	Sulfur (S)
Cobalt (Co)	Molybdenum (Mo)	Tin (Sn)
Copper (Cu)	Phosphorus (P)	Vanadium (V)
Fluorine (F)	Potassium (K)	Zinc (Zn)

Together with the structural elements carbon, hydrogen, nitrogen, and oxygen, these make up the 25 chemical elements essential to life. Other elements are sometimes found in body fluids and tissues but are not known to be essential. These include aluminum, lithium, nickel, and boron. Eventually, it may be discovered that one or more of these is essential also.

Minerals serve a variety of functions. The function of iodine is perhaps the most dramatic. A small amount of iodine is necessary for the thyroid gland to function properly. A deficiency of iodine has dire effects, of which goiter is perhaps the best known (Figure 18.11). Iodine is available naturally in seafood, but, to guard against iodine deficiency, a small amount of sodium iodide (NaI) is added to table salt. The use of iodized salt has greatly reduced the incidence of goiter.

Iron(II) ions (Fe^{2+}) are necessary for the proper function of the oxygen-transporting compound hemoglobin. Without sufficient iron, not enough oxygen is supplied to the body tissues. Anemia, a general weakening of the body, results. Foods especially rich in iron compounds include red meat and liver.

Calcium and phosphorus are necessary for the proper development of bones and teeth. Growing children need about 1.5 g of each per day. These elements are available from milk. The needs of adults for calcium and phosphorus are less widely known but are very real just the same.

For example, calcium ions are necessary for the coagulation of blood (to stop bleeding) and for maintenance of the rhythm of the heartbeat. Phosphorus is necessary in order for the body to obtain energy from carbohydrates. Without phosphorus compounds, we couldn't get *any* energy from those quick-energy foods. Compounds containing phosphorus also play many other essential roles in the functioning of the body.

Sodium chloride—in moderate amounts—is essential to life. It is important in the exchange of fluids between cells and plasma, for example. The presence of salt increases water retention, though, and a high volume of retained fluids can cause swelling and high blood pressure (hypertension). There are an estimated 36 million people in the United States who suffer from hypertension. Most physicians agree that our diets generally contain too much salt.

Iron, copper, zinc, cobalt, manganese, molybdenum, calcium, and magnesium are essential to the proper functioning of metalloenzymes, which are life sustaining. The functions of some of the other minerals are quite complex. Something of how they operate is known, but a great deal remains to be learned about the role of inorganic chemicals in our bodies. Bioinorganic chemistry is a flourishing area of research.

The Vitamins: Vital, but Not All Are Amines

Why are British sailors called "limeys"? And what does that have to do with food? Sailors have been plagued since early times by **scurvy**. In 1747, British navy captain James Lind showed that scurvy could be prevented by the inclusion of fresh fruit and vegetables in the diet. A convenient fresh fruit to carry on long voyages (they didn't have refrigeration) was the lime. British ships put to sea with barrels of limes aboard, and the sailors ate a lime or two every day. That is how they came to be known as "lime eaters," or simply "limeys."

In 1897, the Dutch scientist Christiaan Eijkman showed that polished rice lacked something found in the hull. Lack of that something caused the disease **beriberi**, which was quite a problem in the Dutch East Indies at that time. A British scientist, F. G. Hopkins, fed a synthetic diet of carbohydrates, fats, proteins, and minerals to a group of rats. The rats were unable to sustain healthy growth. Again, something was missing.

In 1912, Casimir Funk, a Polish biochemist, coined the word *vitamine* (from the Latin *vita*: "life") for these missing factors. Funk thought all factors contained the amino group. In the United States, the final *e* was dropped after it was found that not all the factors were amines. The generic term became *vitamin*. Eijkman and Hopkins shared the 1929 Nobel Prize in medicine and physiology for their important discoveries relating to vitamins.

Vitamins are specific organic compounds that are required in the diet (in addition to the usual proteins, fats, carbohydrates, and minerals) to prevent specific diseases. Some of the vitamins, their structures, sources, and deficiency symptoms are shown in Table 18.6. The role of vitamins in the prevention of deficiency diseases has been well established. In recent years, massive doses of vitamins have been recommended as preventives or cures for diseases as varied as the common cold and schizophrenia. We discuss this sort of megavitamin therapy in Chapter 22.

As you can see from Table 18.6, the vitamins do not share a common chemical structure. They can, however, be divided into two broad categories: the **fat-soluble vitamins**—including A, D, E, and K—and the **water-soluble vitamins**—made up of the B complex and vitamin C. All the fat-soluble vitamins incorporate a high proportion of hydrocarbon structural elements. There are one or two oxygen atoms present, but the compounds as a whole are nonpolar. In contrast, a water-soluble vitamin contains a high proportion of the electronegative atoms oxygen and nitrogen, which can form hydrogen bonds to water; therefore, the molecule as a whole is soluble in water.

Fat-soluble vitamins dissolve in the fatty tissue of the body, and reserves of these vitamins can be stored there for future use. For example, while on an adequate diet, an adult can store several years' supply of vitamin A. If the diet becomes deficient in vitamin A, these reserves are mobilized, and the adult remains free of the deficiency disease for quite a while. On the other hand, a small child who has not had the opportunity to build up a store of the vitamin soon exhibits symptoms of the deficiency. Many children in developing countries are permanently blinded by vitamin A deficiency. Health workers in these countries often carry injectible solutions of the vitamin for emergency treatment.

Because the fat-soluble vitamins are efficiently stored in the body, overdoses of these vitamins can result in adverse effects. Large excesses of vitamin A cause irritability, dry skin, and a feeling of pressure inside the head. Massive doses of the vitamin administered to pregnant rats result in malformed offspring. Vitamin D, like vitamin A, is fat soluble. The effects of large overdoses are even more severe than with vitamin A. Too much vitamin D can cause pain in the bones, nausea, and diarrhea, and weight loss. Bonelike material may be deposited in kidney tubules, in blood vesssels, and in heart, stomach, and lung tissue. The amounts of both vitamin D and A in nonprescription capsules are regulated by the U.S. Food and Drug Administration.

Even though vitamins E and K also are fat soluble, they are metabolized and excreted. They are not stored to the extent vitamins A and D are.

Table 18.6 Some of the Vitamins

Vitamin	Structure and Name	Sources	Deficiency Symptoms
Fat-Soluble Vitamins			
A	H_3C CH_3 H CH_3 H H CH_2OH H_3C CH_3 H H H CH_3 Retinol	Fish, liver, eggs, butter, cheese; also a vitamin precursor in carrots and other vegetables	Night blindness
D	CH_3 CH_3 CH_3 H_3C H_3C H_2C HO Vitamin D_2 (calciferol)	Cod liver oil, irradiated ergosterol (milk supplement)	Rickets
E	CH_3 CH_3 CH_3 $(CH_2)_3CH(CH_2)_3CH(CH_2)_3CHCH_3$ CH_3 H_3C O CH_3 HO CH_3 α-Tocopherol	Wheat germ oil, green vegetables, egg yolks, meat	Sterility, muscular dystrophy
K	CH_3 CH_3 $CH_2CH=C(CH_2)_3CH(CH_2)_3CH(CH_2)_3CH$ CH_3 CH_3 CH_3 O O Vitamin K_1 (phylloquinone)	Spinach, other green leafy vegetables	Hemorrhage
Water-Soluble Vitamins			
The B Complex	CH_3 Cl^- O $(CH_2)_2OH$ N^+ S C NH_2 CH_2 N N H_3C B_1 (thiamine)	Germ of cereal grains, legumes, nuts, milk, and brewers yeast	Beriberi—polyneuritis resulting in muscle paralysis, enlargement of heart, and ultimately heart failure

418

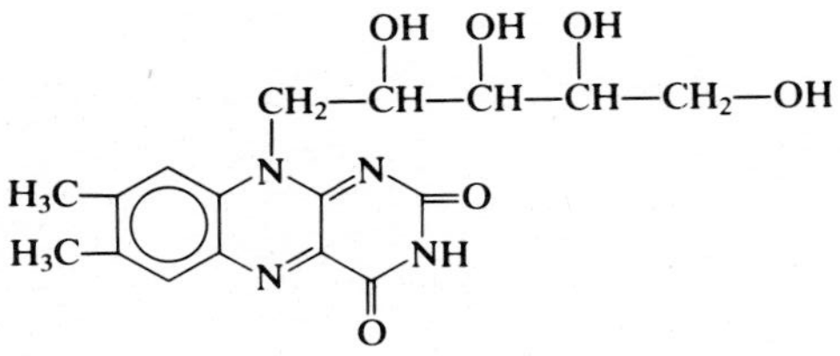

OH OH OH

CH₂—CH—CH—CH—CH₂—OH

H₃C

H₃C

B₂ (riboflavin)

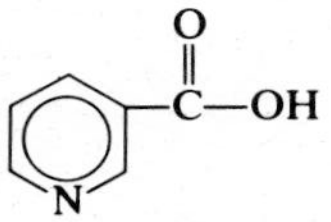

Nicotinic acid   Nicotinamide

Niacin

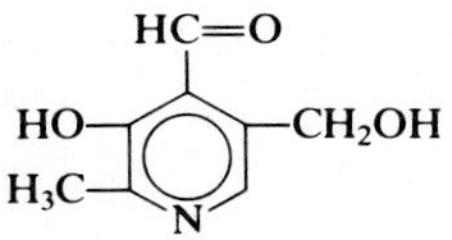

Pyridoxal   Pyridoxol   Pyridoxamine

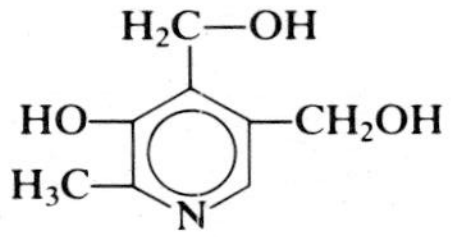

B₆

Pantothenic acid

Biotin

Milk, red meat, liver, egg white, green vegetables, whole wheat flour (or fortified white flour), and fish

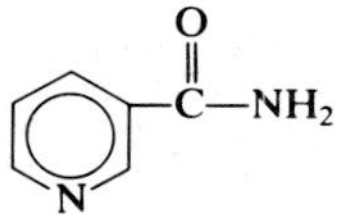

Red meat, liver, collards, turnip greens, yeast, and tomato juice

Eggs, liver, yeast, peas, beans, and milk

Liver, eggs, yeast, and milk

Beef liver, yeast, peanuts, chocolate, and eggs (although this vitamin cannot be synthesized by humans, it is a product of their intestinal bacteria)

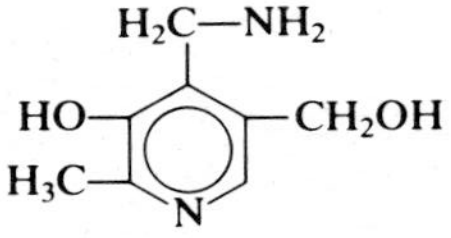

Dermatitis, glossitis (tongue inflammation)

Pellagra—skin lesions, swollen and discolored tongue, loss of appetite, diarrhea, various mental disorders (Figure 18.3)

Dermatitis, apathy, irritability, and increased susceptibility to infections; convulsions in infants

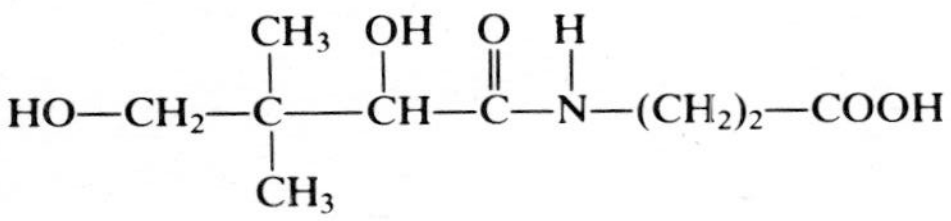

(Possibly) emotional problems and gastrointestinal disturbances

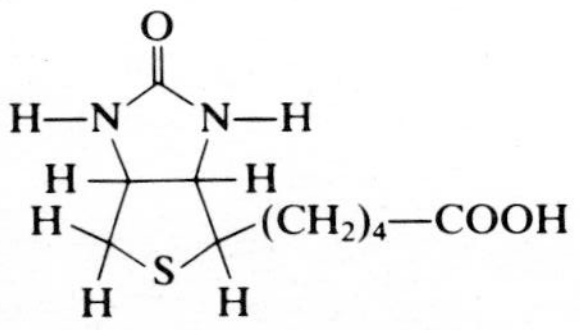

Dermatitis

(continues)

Table 18.6—continued

Vitamin	Structure and Name	Sources	Deficiency Symptoms
	Fat-Soluble Vitamins		
	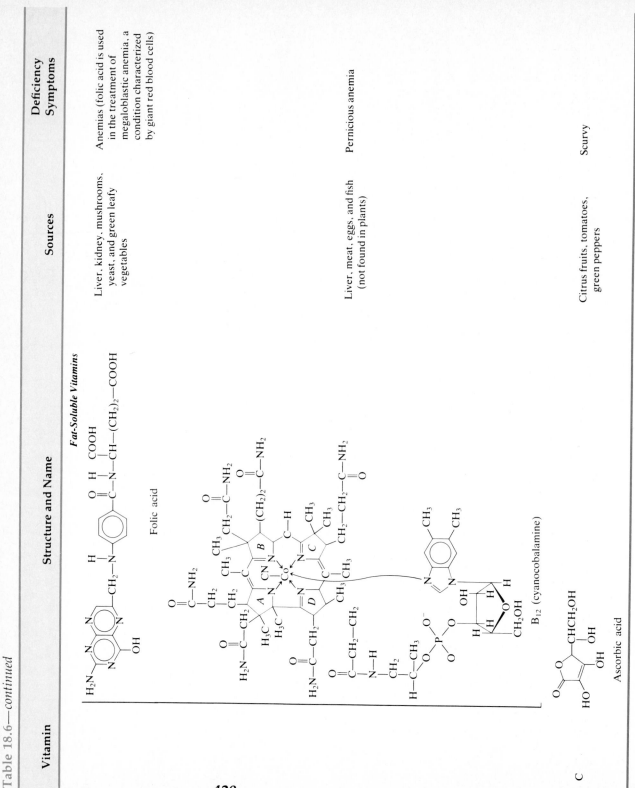 Folic acid	Liver, kidney, mushrooms, yeast, and green leafy vegetables	Anemias (folic acid is used in the treatment of megaloblastic anemia, a condition characterized by giant red blood cells)
	B₁₂ (cyanocobalamine)	Liver, meat, eggs, and fish (not found in plants)	Pernicious anemia
C	Ascorbic acid	Citrus fruits, tomatoes, green peppers	Scurvy

The body has a limited capacity to store water-soluble vitamins. It excretes anything over the amount that can be immediately used. Water-soluble vitamins must be taken at frequent intervals, whereas a single dose of a fat-soluble vitamin can be used by the body over several weeks. A significant portion of the vitamin content of some foods can be lost when the food is cooked in water and then drained. The water-soluble vitamins go down the drain with the water. Because excess water-soluble vitamins are excreted, they generally are not considered toxic. Massive doses can present some hazard, however. For example, people taking huge excesses of vitamin B_6 have suffered nerve damage.

Fiber: The Colon Cancer Connection

The importance of fiber, or roughage, in the diet often was ignored or overlooked until the mid-1970s. Then, some studies were published that got the public excited about fiber. First, it was noted that people in the developed countries are much more likely to get cancer of the colon than are those in the underdeveloped nations. Also, people in the developed countries eat diets rich in highly processed, low-bulk foods, while those in more "primitive" areas eat high-fiber diets. High-fiber diets lead to frequent and robust bowel movements. Cellulose molecules (see Figure 18.1) have lots of —OH groups. These readily form hydrogen bonds to water molecules. Cellulose therefore absorbs a lot of water, leading to softer stools. Low-bulk diets result in less frequent bowel action, with high retention times for feces in the colon.

Bacteria act upon the materials in the colon. (Indeed, bacteria, both living and dead, make up about one third of the dry weight of feces.) With a high-fiber diet, the materials seldom remain in the colon for more than 1 day. With a low-bulk diet, the retention time can be as long as 3 days, allowing for prolonged bacterial activity that produces a high level of mutagenic chemicals. Chemicals that are mutagenic often are also carcinogenic. Thus, a possible chemical link between low-fiber diets and cancer of the colon has been established.

Cellulose makes up the greater part of dietary fiber. It comes only from plant sources. The fibrous portions of plants (stems, peels, and seeds) are rich in fiber. Wheat bran, celery, green beans, apple peels, and fruits eaten with seeds (such as raspberries, strawberries, and figs) are excellent sources of dietary fiber.

Processed Food: Less Nutrition?

Whole wheat is an excellent source of vitamin B_1 and other vitamins. To make white flour, the wheat germ and bran are removed from the grain. The remaining material has few minerals and no vitamins. We eat the starch and use the germ and bran for animal food. At least our cattle and hogs get good nutrition. Similarly, polished rice has had most of its protein and minerals removed, and it has virtually no vitamins. The disease beriberi, mentioned before, became prevalent when polished rice was introduced to Southeast Asia.

When fruits and vegetables are peeled, most of their vitamins, minerals, and fibers are lost. The peels often are dumped (directly or through a garbage disposal) into a nearby stream, where they contribute to water pollution (Chapter 16). The heat used to cook food also destroys some vitamins. If water is used in cooking, part of the water-soluble vitamins (C and B complex) and some of the minerals often are drained off and discarded with the water.

The methods we use for processing and cooking food often are chosen for aesthetic reasons, but sometimes cooking methods help preserve the food. For example, highly milled white flour does keep better than whole wheat flour. Cooked tomatoes keep better than fresh tomatoes, although their vitamin C is partly destroyed by cooking. But tomatoes, potatoes, and apples are peeled merely because they look better that way.

It is estimated that over half of the diet of the average person in the United States consists of processed foods. The "teenager's" diet of hamburgers, potato chips, and colas is lacking in many essential nutrients. Highly processed convenience foods threaten to leave the people of our nation—with its abundance of food—poorly nourished.

Water: With or Without?

In addition to carbohydrates, fats, proteins, minerals, vitamins, and fibers, we need water—about 1.0 to 1.5 L each day. We could satisfy this need by drinking plain water, but often we choose other beverages. In 1985, carbonated soft drinks replaced plain water as the most consumed beverage in the United States. These drinks are flavored and sweetened water with added carbon dioxide. Those sweetened with sugar possibly contribute to obesity and tooth decay. Other than water and sugar they have little or no nutritive value. Some soft drinks contain the stimulant caffeine (Chapter 23). Excessive consumption of these may contribute to hyperactivity in children.

Coffee and tea also contain caffeine. Those who wish to avoid this stimulant can resort to decaffeinated versions or to a variety of herbal teas. Despite some criticism in recent years, centuries of use indicate that coffee and tea are reasonably safe in moderate amounts. So are most, but not all, of the herbal teas. Camomile causes allergic reactions in those sensitive to ragweed. Others, such as jimson weed, pokeroot, and mistletoe, are quite toxic. Some herbal teas have mild medicinal properties, but they are not the magic potions that some of their proponents would have us believe.

Fruit juices generally are nutritious beverages. Many contain vitamins, minerals, and fiber. Fruit *drinks*, however, often contain little or no juice. Whether supplied in cans or as powdered mixes, the drink is mainly flavored sweetened water.

Milk is a nutritious beverage that supplies vitamins, minerals (especially calcium), and high-quality protein. Whole milk also has fat, mostly the saturated variety. Those who wish to reduce their consumption of fats can choose skim or low-fat milk.

We also consume beverages that contain ethyl alcohol (Chapter 11). We focus on those for the remainder of this chapter.

Alcoholic Beverages

The complex chemistry underlying the production of alcoholic beverages (beer, wine, distilled beverages) involves many sequential steps. The primary differences among alcoholic beverages result from the selection of different starting materials. Let's look at beer first.

Beer is made from barley and other grains. Before brewing, the barley is **malted** in a three-step process of steeping, germination, and kilning. The malting process releases enzymes that convert the starch into fermentable sugars. First, the barley kernels are **steeped**—soaked in water. Then, under favorable conditions of humidity and temperature, **germination** takes place. During germination, proteins and starches are broken down and enzymes are released.

Barley is the ideal grain for beer because it provides the enzymes, called **amylases**, required for efficient conversion of barley starches and adjuncts to fermentable sugars. The resulting product is green malt, which is then kiln-dried to reduce the water content. The temperature at which **kilning** takes place determines the ultimate color (light or dark) of the malt. Higher temperatures produce dark malts used for porters, stout, ales, and other dark beers. Lower temperatures yield lighter-colored beers (Table 18.7).

Table 18.7 Types of Beer

Type	Alcohol Content	Color	Flavor Characteristics
Ale	Up to 6%	Medium light	Heavy, high hops
Bock	Variable	Very dark (caramel added?)	Heavy
Cream ale	4–6%	Light	Low hops
Lager	3–4%	Light	Variable
Light	2.5–3.3%	Light	High hops
Malt liquor	4–5%	Medium light	Heavy, low hops
Munich	Up to 5%	Dark brown	Aromatic, sweet, high hops
Porter	4–6%	Dark	Heavy, malty, low hops
Stout	5–7%	Very dark	Heavy, malty, high hops

European beers are made with 100% malt. The German Purity Law specifies that beer must be made from pure water, barley malt, hops, and yeast; nothing else may be added. In the United States, usually only 35 to 60% malt is used. The remainder is unmalted starches such as corn grits or rice. These starches, called **adjuncts**, are added to increase the production of fermentable sugars.

After malting, the roots of the germinated barley are removed, and the roasted grain is mashed. In **mashing**, the malt is crushed and cooked to break down the starches into glucose, maltose, and dextrins. The cooked mash is filtered, yielding a liquid called **wort**. Remember that starch is composed of amylose and amylopectin, linear and branched-chain polymers of glucose. This is where the amylases enter the picture. Amylases can catalyze the cleavage of linear chains to glucose and maltose (two glucose units joined together). The glucose and maltose account for the sweetness of beer. However, amylases cannot cleave the amylopectin molecules at the point of branching. This results in unfermentable carbohydrates, called **limit dextrins**, that contain two to six glucose units. These limit dextrins remain in the finished beer where they contribute Calories and perhaps add to the stability of the foam head on a mug of beer.

Hops—blossoms of the hop plant—are added to the mash. The bitter flavor components of hops counteract the sweetness of the residual sugars. Hops also help to kill wild yeasts that might impart undesirable flavors to the beer.

The process that produces alcohol is called **fermentation**. Enzymes produced by yeast change glucose into ethyl alcohol and carbon dioxide.

$$C_6H_{12}O_6 \longrightarrow 2\ CH_3CH_2OH\ +\ 2\ CO_2$$

After the beer is fermented, it is **lagered**. This is a 2- to 6-month process that allows certain flavors to develop. After lagering, the beer is chillproofed. **Chillproofing** is a process in which enzymes are added to the beer to act upon and make soluble any proteins that might precipitate in cooling. After the beer is chillproofed, carbon dioxide is added and the brew is pasteurized and packaged for sale.

Light beers are made in ways that give a lower alcohol content or fewer carbohydrates in the finished product. One way to accomplish the latter is to add enzymes that break down the limit dextrins so they can be fermented. Another way is to use sugar as an adjunct; sugar yields no limit dextrins.

Wine, another popular alcoholic beverage, is made by fermenting fruits (rather than grain). Wine is almost always fermented grape juice, although the juice of other fruits can be fermented. The process for wine making dates back to ancient times. First, the grapes are crushed (crushed grapes are called **must**). The fermenting process in wine making differs only slightly from the process used for beer. The chemical process in which sugars (glucose) are converted to ethanol and carbon dioxide is the same. The difference comes from natural yeasts found on grape skins. These wild yeasts sometimes cause the wine to spoil. For this reason, gaseous sulfur dioxide (SO_2) is pumped into the must to prevent the growth of harmful bacteria. Sulfur dioxide also combines with acetaldehyde, a substance that can cause an off flavor in the wine. A third function of the compound is that it serves as an antioxidant to prevent the oxidation of ethyl alcohol to acetic acid. Acetic acid is a component of vinegar, and it gives a sour taste to wine.

Wines are classified by alcohol content, sweetness, dryness, and color. Alcohol content is an issue in wine making, because it provides a basis for determining taxation. Table wines have, by definition, 14% or less alcohol content; dessert wines have an alcohol content above 14%. Some dessert wines have a sweeter taste, because the fermentation process is stopped while some sugar remains.

Wines need at least a 9% ethanol content for stability. Fermentation can produce up to about 14% alcohol. Dessert wines are made by adding more alcohol (usually brandy), a process called **fortification**. Wines get their color from grape skins. Red wines are colored by pigments from the skins. White wines are made from green-colored grapes or from grapes without their skins.

Sparkling wines and champagne can be made in two ways. One is simply to cork the bottle while a little sugar remains. The carbon dioxide that is produced as fermentation continues is trapped in the liquid. In the other method, sugar is added after primary fermentation has ceased, and the bottle is corked during a secondary fermentation. Champagne was actually discovered by accident. Long ago in the wine cellars of France, when the wine makers checked their products in the spring, they noted that some of the bottles contained bubbly wine. Worse, the added pressure from these bubbles sometimes caused bottles to burst. Today, champagne is the bubbly wine of celebration and romance, carefully cultivated from just the right blend of 20 to 30 wines.

Light wines are made by fermenting unripe grapes (which have less sugar) or by distilling off part of the alcohol. Nonalcoholic wines have essentially all the alcohol removed.

Strong alcoholic beverages are usually made by fermenting grains. After fermentation, the mash is heated to boiling and distilled. **Distillation** involves boiling off the volatile compounds (such as the alcohol and part of the water), leaving the solids and high-boiling compounds as waste. By varying procedures and combinations of grains, a vast array of alcoholic beverages can be produced. Distilled beverages such as whiskey, vodka, and gin are known as **hard liquor**. They tend to be strong tasting and are typically (and advisably) consumed in small amounts.

Like many processes, fermentation cannot be 100% complete. By-products of fermentation, and other unconverted species collected en route to the production of ethanol, are called **congeners**. Some of these survive distillation and appear in the final product. Congeners include oils, acids, esters, aldehydes, tannins, and other compounds, some of which may be contaminants from aging or packaging processes. Often such flaws go totally undetected by the consumer. Indeed, in tiny amounts they are normal and desirable chemical components of alcoholic beverages.

As with other foods, a variety of substances may be added to some alcoholic beverages. In the United States, as many as 70 additives—preservatives, dyes, foam enhancers, stabilizers—also may be used. Such additives are less common in European beers and wines.

Problems

1. List the three major types of food.
2. What are carbohydrates?
3. What is the role of carbohydrates in the diet?
4. What is the chemical name of the following sugars?
   a. blood sugar   b. table sugar   c. fruit sugar
5. What is the principal sugar in corn syrup?
6. How is high-fructose corn syrup made?
7. What is the main chemical difference in starch and cellulose?
8. What is the dietary difference in starch and cellulose?
9. What is the difference in amylose and amylopectin?
10. What sugar is formed when starch is digested?
11. What are enzymes?
12. Describe the lock-and-key model of enzyme action.
13. Both starch and cellulose are polymers of glucose. Why are the enzymes that aid in the breakdown of starch ineffective in breaking down cellulose?
14. What are the ultimate products formed when cells metabolize glucose?
15. How much energy is supplied by one gram of carbohydrates?
16. What proportion of our diet should be carbohydrates?
17. What is the chemical nature of fats?
18. What are fatty acids?
19. What functions do fats serve?
20. What are polyunsaturated fats?
21. How do animal fats differ from vegetable oils?

22. What is meant by the iodine number of a fat?
23. Which of the following fatty acids are saturated?
    a. linolenic acid   b. linoleic acid
    c. oleic acid   d. palmitic acid
    e. stearic acid
24. Which of the fatty acids in problem 23 are unsaturated?
25. How many carbon atoms are there in a molecule of each of the compounds in problem 23?
26. List some of the properties of fats and oils.
27. What is an emulsion?
28. How much energy is supplied by one gram of fat?
29. What is the maximum percentage of Calories in our diet that should come from fats?
30. In the determination of the iodine number of a fat, what part of the fat molecule reacts with the reagent?
31. Which would you expect to have the higher iodine number, corn oil or beef tallow? Explain your reasoning.
32. Which would you expect to have the higher iodine number, hard or liquid margarine? Explain your reasoning.
33. In what parts of the body are proteins found?
34. What are nucleoproteins?
35. What tissues are largely protein?
36. How does the elemental composition of proteins differ from that of carbohydrates and fats?
37. What functional groups are found on amino acid molecules?

38. How many different amino acids are incorporated into proteins?

39. What is a zwitterion?

40. What are essential amino acids?

41. What is a limiting reagent?

42. What is an adequate protein?

43. List some foods that contain adequate proteins.

44. What is kwashiorkor?

45. At what age does the human brain reach nearly full size?

46. Our requirement for protein is about 0.8 g per kilogram by body weight. What is the protein requirement of a 50-kg (110-lb) person?

47. What is a peptide bond?

48. What is a polypeptide?

49. Of what importance is the sequence of amino acids in a protein molecule?

50. Is the dipeptide represented by Ser-Ala the same as the one represented by Ala-Ser? Explain.

51. What is starvation?

52. In fasting, which stores are depleted first, fats or glycogen?

53. What is ketosis?

54. What is acidosis?

55. From what substances does the brain obtain glucose when carbohydrates are lacking in the diet?

56. What are some of the problems associated with low-carbohydrate diets?

57. In general, what are the problems associated with a strict vegetarian diet?

58. Explain why a diet high in meat products makes less efficient use of the energy originally captured by plants through photosynthesis.

59. Vitamins and minerals are discussed in this chapter. Which is organic and which is inorganic?

60. Compare and contrast vitamins and minerals with respect to effective amounts.

61. Indicate a biological function for each of the following.
    a. iodine   b. iron
    c. calcium   d. phosphorus

62. Which of the following minerals would you expect to find in relatively large amounts in the human body? Explain your reasoning.
    a. Ca   b. Cl   c. Co   d. Mo
    e. Na   f. P   g. Zn

63. Match the compound with its designation as a vitamin.

Compound	Designation
Ascorbic acid	Vitamin A
Calciferol	Vitamin B_{12}
Cyanocobalamine	Vitamin C
Retinol	Vitamin D
Tocopheról	Vitamin E

64. Which of the following are B vitamins?
    a. folic acid   b. riboflavin
    c. β-hydroxybutyric acid   d. thiamine
    e. niacin

65. Identify the vitamin deficiency associated with these diseases.
    a. scurvy   b. rickets
    c. night blindness

66. In each case, identify the deficiency disease associated with a diet lacking in the indicated vitamin.
    a. vitamin B_1 (thiamine)
    b. niacin
    c. vitamin B_{12} (cyanocobalamine)

67. What is the structural difference between water-soluble and fat-soluble vitamins?

68. Identify the following vitamins as water soluble or fat soluble.
    a. vitamin A   b. vitamin B_6
    c. vitamin B_{12} d. vitamin C
    e. vitamin K

69. Identify the following vitamins as water soluble or fat soluble.
    a. calciferol   b. niacin
    c. riboflavin   d. tocopherol

70. Structural formulas for two vitamins are shown below. Identify each of them as water soluble or fat soluble.

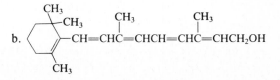

71. Is an excess of a water-soluble vitamin or a fat-soluble vitamin more likely to be dangerous? Why?

72. Could a "one-a-month" vitamin pill satisfy all human requirements? Explain your answer.

73. What is the role of fiber in the human diet?

74. Which essential amino acids are likely to be lacking in corn? In beans?

75. A new bread is made by adding pea flour to wheat flour. Would the bread provide adequate protein? Why or why not?

76. Graham crackers are 8% protein. Assume that your Recommended Daily Allowance (RDA) is 60 g of protein. What percent of your RDA of protein would you receive if you ate 150 g of graham crackers?

77. Assume that 1 cup of skim milk contains 225 g. Skim milk is 3.6% protein. How many grams of protein are there in 4 cups of skim milk?

78. Define or give an example of each of the following terms in relationship to alcoholic beverages.
    a. adjuncts   b. chillproofing
    c. congeners   d. fortification
    e. hops   f. kilning
    g. lagering   h. limit dextrins
    i. malting   j. must
    k. steeping

79. How is light beer made?

80. How is light wine made?

81. Give three reasons sulfur dioxide is added to wines.

82. What is sparkling wine? How is it made?

83. One gram of pure ethyl alcohol provides 7 kcal of energy. How many Calories (kcal) are there in one 30-mL shot of 100-proof (50% alcohol) vodka. The density of ethyl alcohol is 0.79 g/mL. Assume that all the Calories in the vodka are from alcohol.

84. The label on a bottle of light wine indicates that 100 mL of the wine furnishes 70 [food] Cal, 0.2 g of protein, 5.77 g of carbohydrates, and 0.0 g of fat. Assuming that carbohydrates and proteins furnish 4 Cal/g each, alcohol furnishes 7 Cal/g, and that no other caloric nutrients are present, carry out the following calculations.
    a. How many Calories are provided by the alcohol in a 100-mL serving of the wine? What percentage of the total Calories is provided by alcohol?
    b. How many grams of alcohol are there in each 100-mL serving?
    c. The density of alcohol is 0.79 g/mL. How many mL of alcohol are there in each 100-mL serving? What is the percentage alcohol by volume?

85. A 355-mL can of light beer provides 96 kcal and has 2.8 g of carbohydrates, 0.9 g of protein, and no fat. Calculate:
    a. the kcal of alcohol
    b. the grams of alcohol (1 g of alcohol provides 7 kcal of energy)
    c. the volume of alcohol (the density of alcohol is 0.79 g/mL)
    d. the percent by volume of alcohol

86. When Russian refugee Alexander Jourjine escaped from the U.S.S.R. by hiking for 23 days across Finland to Sweden, he carried 9 lb of lard and cheese, several loaves of dried black bread, tea, sugar, and 12 chocolate bars. His diet was supplemented with berries picked along the way. Comment on his choice of food items, especially the lard.

References and Readings

1. "Applying de Minimis at the FDA." *ChemEcology*, April 1987, p. 7. Methylene chloride is safe as a solvent for decaffeinating coffee, but its use in hairspray is risky.

2. Brody, Jane. *Jane Brody's Nutrition Book.* New York: Bantam Books, 1981. An excellent general source.

3. Committee on Chemistry and Public Affairs. *Chemistry and the Food System.* Washington, DC: American Chemical Society, 1980.

4. Eaton, S. Boyd, and Marjorie Shostak. "Fat Tooth Blues." *Natural History*, July 1986, pp. 6–15. What fats do to us.

5. Ensminger, Audrey H., et al. *Foods and Nutrition Encyclopedia.* Clovis, CA: Pegus Press, 1983.

6. Feigl, Dorothy M., and John W. Hill. *General,*

Organic, and Biological Chemistry: Foundations of Life, 2nd ed. New York: Macmillan, 1986. Chapters 14–18 treat these topics in more detail.

7. Grady, Denise, and Sana Siwolop. "An Anti-Cancer Diet?" *Discover*, June 1984, pp. 23–26.

8. Grosser, Arthur E. *The Cook Book Decoder or Culinary Alchemy Explained*. New York: Beaufort Books, 1981.

9. Larkin, Tim. "Herbs Are Often More Toxic Than Magical." *FDA Consumer*, October 1983, pp. 5–11.

10. Martin, Charles, "Hi-Tech Wine Making," *Discover*, March 1983, pp. 86–88.

11. McGee, Harold. *On Food and Cooking: The Science and Lore of the Kitchen*. New York: Charles Scribner's Sons, 1984.

12. Mertz, Walter. "The Essential Trace Elements." *Science*, 18 September 1981, pp. 1332–1338.

13. Prebluda, Harry. "McCollum and the Rat." *Chem-Tech*, May 1984, pp. 266–268. A history of the use of rats for nutritional studies.

14. Wallis, Claudia, "Hold the Eggs and Butter." *Time*, 26 March 1984, pp. 56–63. Cholesterol is the culprit.

19

Food Additives
Bane or Blessing?

The label reads "egg whites, vegetable oils, nonfat dry milk, lecithin, mono- and diglycerides, propylene glycol monostearate, xanthan gums, sodium citrate, aluminum sulfate, artificial flavor, iron phosphate, niacin, riboflavin, and irradiated ergosterol." You recognize only a few of the ingredients. You still want to know what's in the package. Most of the substances listed are **food additives**: substances other than basic foodstuffs that are present in food as a result of some aspect of production, processing, packaging, or storage.

There are two major categories of food additives: **intentional additives** are put in a product on purpose to perform a specific function; **incidental additives** get in accidentally during production, processing, packaging, or storage. Pesticide residues, insect parts, and antibiotics added to animal feeds are examples of the incidental additives that sometimes get into our food. There are about 2800 intentional additives. Sugar, salt, and corn syrup are used in the greatest amounts. These three, plus citric acid, baking soda, vegetable colors, mustard, and pepper, make up over 98% (by weight) of all intentional additives. There are perhaps 10 000 incidental additives.

In this chapter, we consider mainly the intentional additives and how they affect our enjoyment of food, our health, and our well-being.

430

Why Food Additives?

We saw in Chapter 18 that processing results in the removal or destruction of certain essential food substances. Therefore, it is sometimes necessary to add additives to prepared food to increase its nutritional value. Other chemicals are added to enhance color and flavor, to retard spoilage, to provide texture, to sanitize, to bleach, to ripen (or to prevent ripening), to control moisture or dryness, and to prevent (or enhance) foaming.

In the United States, food additives are regulated by the Food and Drug Administration (FDA). The original Food, Drug, and Cosmetic Act was passed by Congress in 1938. Under that act, the FDA had to prove that an additive was unsafe before its use could be prevented. The Food Additives Amendment of 1958 shifted the burden of proof to industry. Now, a company that wishes to use a food additive must first furnish proof to the FDA that the additive is safe for the intended use. The FDA can also regulate the amount of chemicals that can be added.

Food additives are not a recent development. Salt has been used to preserve meat and fish since the time when people lived in caves. Spices have been used since earliest recorded history to flavor and preserve foods. Other additives have been used throughout the centuries. The movement of the population from farms to cities in recent years has increased the necessity of using additives to preserve foods. An increased desire for convenience foods also has led to greater use of additives. Competition for sales has caused industries to use more additives that make their food products more tasty and more attractive. These and other factors have led to an increased consumption of food additives in the United States.

Our increased consumption and our worry over problems caused by some food additives have led to a lot of anguish. This worry often is expressed as concern about "chemicals in our food." As we said in Chapter 16, though, food itself is chemical. Table 19.1 shows the chemical composition of a typical breakfast. Many of the chemicals in the breakfast would be harmful in large amounts but are harmless in the trace amounts that occur naturally in foods. Indeed, some make important contributions to the flavors and aromas that make food so delightful.

Our bodies are also collections of chemicals. Broken down to its elements (Table 19.2), your body would be worth only a few dollars as chemicals. It is the unique combination of the elements in every human body, though, that makes you different from everyone else and makes each individual's value beyond measure. Since food is chemical and we are chemical, we shouldn't have to worry, in general, about chemicals in our food. Perhaps we should be concerned, however, about *some* of the

Table 19.1   Your Breakfast—As Seen by a Chemist*

Chilled Melon

Starches	Anisyl propionate
Sugars	Amyl acetate
Cellulose	Ascorbic acid
Pectin	Vitamin A
Malic acid	Riboflavin
Citric acid	Thiamine
Succinic acid	

Scrambled Eggs

Ovalbumin	Lecithin
Conalbumin	Lipids (fats)
Ovomucoid	Fatty acids
Mucin	Butyric acid
Globulins	Acetic acid
Amino acids	Sodium chloride
Lipovitellin	Lutein
Livetin	Zeazanthine
Cholesterol	Vitamin A

Sugar-Cured Ham

Actomyosin	Adenosine triphosphate
Myogen	(ATP)
Nucleoproteins	Glucose
Peptids	Collagen
Amino acids	Elastin
Myoglobin	Creatine
Lipids (fats)	Pyroligneous acid
Linoleic acid	Sodium chloride
Oleic acid	Sodium nitrate
Lecithin	Sodium nitrite
Cholesterol	Sodium phosphate
Sucrose	

Coffee

Caffeine	Acetone
Essential oils	Methyl acetate
Methanol	Furan
Acetaldehyde	Diacetyl
Methyl formate	Butanol
Ethanol	Methylfuran
Dimethyl sulfide	Isoprene
Propionaldehyde	Methylbutanol

Cinnamon Apple Chips

Pectin	Propanol
Cellulose	Butanol
Starches	Pentanol
Sucrose	Hexanol
Glucose	Acetaldehyde
Fructose	Propionaldehyde
Malic acid	Acetone
Lactic acid	Methyl formate
Citric acid	Ethyl formate
Succinic acid	Ethyl acetate
Ascorbic acid	Butyl acetate
Cinnamyl alcohol	Butyl propionate
Cinnamic aldehyde	Amyl acetate
Ethanol	

Toast and Coffee Cake

Gluten	Methyl ethyl
Amylose	ketone
Amino acids	Niacin
Starches	Pantothenic acid
Dextrins	Vitamin D
Sucrose	Acetic acid
Pentosans	Propionic acid
Hexosans	Butyric acid
Triglycerides	Valeric acid
Sodium chloride	Caproic acid
Phosphates	Acetone
Calcium	Diacetyl
Iron	Maltol
Thiamine	Ethyl acetate
Riboflavin	Ethyl lactate
Mono- and Diglycerides	

Tea

Caffeine	Phenyl ethyl
Tannin	alcohol
Essential oils	Benzyl alcohol
Butyl alcohol	Geraniol
Isoamyl alcohol	Hexyl alcohol

*The chemicals listed are those found normally in the foods. No food additives are itemized, and the chemical listings are not necessarily complete.
SOURCE: Manufacturing Chemists Association, Washington, DC.

Table 19.2 Approximate Elemental Analysis of the Human Body

Element	Percent by Weight in Human Body
Oxygen	65
Carbon	18
Hydrogen	10
Nitrogen	3
Calcium	1.5
Phosphorus	1
Potassium	0.35
Sulfur	0.25
Chlorine	0.15
Sodium	0.15
Magnesium	0.05
Iron	0.004
Total	100

specific chemicals in our food. Most of all, we should try to evaluate the potential hazards of food additives that are used now and may be used in the future—and the hazards of some chemicals that occur naturally in some foods.

Additives That Improve Nutrition

Only a small amount of iodine is necessary for the proper functioning of the thyroid gland. Iodine is plentiful in seafood, but people in inland areas often suffer from iodine deficiency. A particularly striking result is goiter, a swelling of the thyroid often symptomized by an unsightly swollen neck (see Figure 18.11). The first nutrient supplement approved by the Bureau of Chemistry of the U.S. Department of Agriculture, potassium iodide (KI), was added to table salt in 1924 to reduce the incidence of goiter. (The Bureau of Chemistry later became the FDA.)

A number of other chemicals have been added to foods specifically to prevent deficiency diseases. Nutrient supplements are still important today. The addition of vitamin B_1 (thiamine) to polished rice is essential in the Far East, where beriberi still is a problem. The replacement of the B-complex vitamins thiamine, riboflavin, and niacin (which are removed in processing) and the addition of iron (usually ferrous carbonate [$FeCO_3$]) to flour is called **enriching**. Enriched bread made from this flour still isn't as good a food as bread made from whole wheat. It lacks vitamin B_6, pantothenic acid, folic acid, and the minerals zinc and magnesium, nutrients usually provided by whole grain flour.

An investigator in Texas fed enriched white bread to rats. Most of them died of malnutrition in fewer than 60 days. The staff of life, in its modern form, is unable to sustain life. The enrichment of bread, corn meal, and cereals has served to virtually eliminate pellagra, however, a disease that once plagued the southern United States (Figure 19.1).

Vitamin C (ascorbic acid) is added frequently to fruit juices, flavored drinks, and beverages. Although our diets generally contain enough ascorbic acid to prevent scurvy, a number of scientists recommend a much larger intake than basic minimum requirements. We discuss the use of massive dosages of C and other vitamins in Chapter 22.

Vitamin D is added to milk in developed countries. This use of fortified milk has lead to the virtual elimination of rickets. Similarly, vitamin A is added to margarine. (This vitamin occurs naturally in butter; it is added to margarine so that the substitute more nearly matches the nutritional quality of butter.)

If we ate a balanced diet of foods fresh from the farm, we probably wouldn't need nutritional supplements. With our usual diets rich in highly processed foods, however, we need the nutrients provided by vitamin and mineral food additives.

Figure 19.1 A case of acute pellagra. Niacin is called the antipellagra vitamin because its deficiency in humans appears to be the chief cause of the disease. [Courtesy of the World Health Organization.]

Molecular Flavors: Chemicals That Taste Good

If you like spice cake, soda pop, gingerbread, and sausage, you like food additives. These and many other foods depend almost totally on spices and other flavorings for their flavor. Cloves, ginger, cinnamon, and oregano are examples of natural spices. Natural flavors also can be extracted from fruits and other plant materials. Vanilla extract is a familiar example.

Chemists sometimes analyze a natural flavor and determine its components. Then they synthesize the components and make a mixture that may closely resemble the natural product. The major components of natural and artificial flavors are often identical. For example, both vanilla extract and imitation vanilla flavoring owe their flavor mainly to vanillin (Figure 19.2). The natural flavor is usually more complex than the imitation because the natural product contains a wider variety of chemicals than the imitation. Indeed, some scientists say that the imitation, if anything, is safer than the natural flavoring because it contains fewer chemicals—and the chemicals in the imitation flavoring are identical to those in the natural product. Flavors, whether natural or synthetic, probably present little hazard when used in moderation, and they contribute considerably to our enjoyment of food.

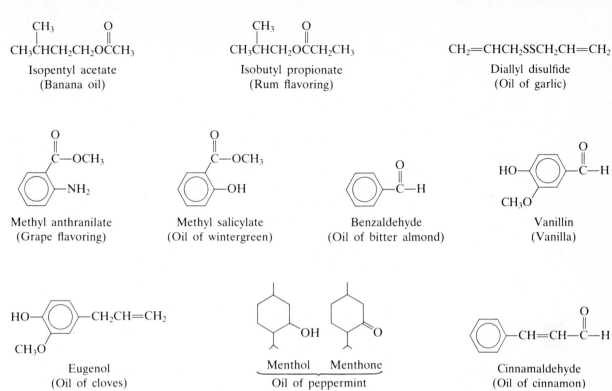

Figure 19.2 Some molecular flavorings.

Additives That Enhance Flavors

Some chemical substances, though not particularly flavorful themselves, are used to enhance other flavors. Common table salt (sodium chloride) is a familiar example. Salt seems to increase sweetness but helps to mask bitterness and sourness.

Salt is a necessary nutrient in moderate amounts. Many physicians and scientists contend that the average person's diet in the United States contains too much salt. Snack foods (crackers, potato chips, pretzels) are particularly heavy in salt. There is evidence that too much salt contributes to high blood pressure (hypertension); three out of four people between 65 and 80 years old suffer from this condition.

Another popular flavor enhancer is monosodium glutamate (MSG). Glutamic acid is one of about 20 amino acids that occur naturally in proteins. MSG is the sodium salt of glutamic acid. It is used in many convenience foods and is used heavily in many Chinese foods. Although glutamates are found naturally in protein, there is evidence that excessive amounts are harmful. A peculiar disease that baffled physicians for a long time finally was traced to the regular consumption of Chinese foods. Individuals who seem particularly susceptible to MSG develop headaches and feel weak after eating in Chinese restaurants. The condition became known as the **Chinese-restaurant syndrome**.

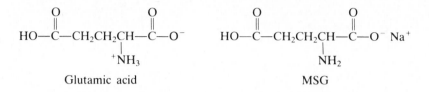

Glutamic acid   MSG

Individuals susceptible to this syndrome may find it difficult to avoid MSG. MSG often is an ingredient of foods for which the FDA has established **standards of identity**. Such foods must contain certain ingredients in certain proportions. They may, however, contain certain *other* substances as well. Mayonnaise and salad dressing are examples of standard foods. Standard foods may contain MSG even if it is not included in the list of ingredients on the labels.

For many years, MSG was added to baby foods. This was done, presumably, to please the parents, because babies do not have a highly developed sense of taste and wouldn't have known the difference. Public pressure—which came after MSG was found to numb portions of the brains of laboratory animals and to exert a possible teratogenic effect (i.e., to cause severe fetal defects)—led to the processors' voluntary elimination of MSG from baby foods.

$$CH_3CH_2C\overset{O}{\overset{\|}{-}}OH$$

Propionic acid

$$CH_3CH_2C\overset{O}{\overset{\|}{-}}O^- \ Na^+$$

Sodium propionate

$$CH_3CH=CHCH=CHC\overset{O}{\overset{\|}{-}}OH$$

Sorbic acid

$$CH_3CH=CHCH=CHC\overset{O}{\overset{\|}{-}}O^- \ K^+$$

Potassium sorbate

Benzoic acid

Sodium benzoate

Figure 19.3 Some spoilage inhibitors.

Additives That Retard Spoilage

Food spoilage can result from the growth of molds (fungi) or bacteria. Propionic acid and its sodium and calcium salts are added to bread and cheese to act as mold and yeast inhibitors. Sorbic acid, benzoic acid, and their salts are also used. The structures of some of these inhibitors are shown in Figure 19.3

Some inorganic compounds are also used as spoilage inhibitors. Sodium nitrite ($NaNO_2$) is used in meat curing and to maintain the pink color of smoked hams, frankfurters, and bologna. Nitrites also contribute to the tangy flavor of processed meat products.

Nitrites are particularly effective as inhibitors of *Clostridium botulinum*, the bacterium that produces botulism poisoning. However, only about 10% of the amount used to keep meat pink is needed to prevent botulism. Sodium nitrate ($NaNO_3$) also has been used in curing meat. The FDA has banned the use of nitrates since they have no advantage over nitrites. At any rate, bacteria in our stomachs readily reduce nitrates to nitrites.

Nitrites have been investigated as possible causes of cancer of the stomach. In the presence of the hydrochloric acid (HCl) in the stomach, nitrites are converted to nitrous acid.

$$NaNO_2 \ + \ HCl \ \longrightarrow \ HNO_2 \ + \ NaCl$$

The nitrous acid then may react with compounds called secondary amines to form nitroso compounds.

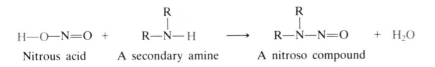

The R— groups may be alkyl groups such as methyl (CH_3—) or ethyl (CH_3CH_2—), or they may be more complex. In any case, these nitroso compounds are among the most potent carcinogens known. By eating foods containing nitrates and nitrites, we may be giving our stomachs the raw materials for cancer. It is a fact that the rate of stomach cancer is higher in countries that use prepared meats than in the backward nations where people eat little or no cured meat. The incidence of stomach cancer is *decreasing* in the United States, however, and we aren't quite sure why. We do know that the reaction between nitrous acid and amines to form nitrosamines is inhibited by ascorbic acid (vitamin C). Perhaps a similar inhibition occurs in our stomachs. Problems with nitrites have led the FDA to approve sodium hypophosphite (NaH_2PO_2) as a substitute.

Sulfur dioxide is another inorganic food additive. A gas at room temperature, sulfur dioxide serves as a disinfectant and preservative, particularly for dried fruits such as peaches, apricots, and raisins. It is also used as a bleach to prevent the browning of wines, corn syrup, jelly, dehydrated potatoes, and other foods. Sulfur dioxide seems safe when ingested with food. However, it is a powerful respiratory irritant when inhaled, and is a damaging ingredient of polluted air in some areas (Chapter 15). Sulfite salts once were used in restaurants to keep salad vegetables appearing crisp and fresh. They were shown to cause severe allergic reactions in some people. The FDA banned this use of sulfites in 1986. Their use in other foods must be indicated on the label.

Figure 19.4 The addition of sorbates to cottage cheese can extend the product's shelf life by 30 or more days. The upper sample, containing 0.1% potassium sorbate, was stored for 3 days at 27°C. The other, stored for the same period at the same temperature, contained no sorbate. [Courtesy of Monsanto Co., St. Louis, MO.]

The Butylated World of Antioxidants

Antioxidants are added to foods to prevent fats and oils from turning rancid, thus making the food unpalatable. Packaged foods that contain vegetable oils or animal fats (bread, potato chips, sausage, dry breakfast cereals) most likely have antioxidants added.

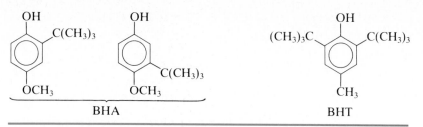

Figure 19.5 Two common antioxidants. BHA is a mixture of two isomers.

Two compounds often used as antioxidants are butylated hydroxytoluene (BHT) and butylated hydroxyanisole (BHA) (Figure 19.5). Both compounds are phenols (Chapter 11).

Fats turn rancid, in part, through oxidation. This process occurs through the formation of molecular fragments called **free radicals**, which have an unpaired electron as a distinguishing future. (Recall from Chapter 5 that covalent bonds are shared pairs of electrons.) We need not concern ourselves with the details of the structures of radicals, but we can summarize the process. First, a fat molecule reacts with oxygen to form a free radical (which we will call Rad·).

$$\text{Fat} + O_2 \longrightarrow \text{Rad·}$$

Then, the radical reacts with another fat molecule to form a new free radical that can repeat the process. A reaction such as this, in which intermediates are formed that keep the reaction going, is called a **chain reaction**. One molecule of oxygen can lead to the decomposition of many fat molecules.

To preserve foods containing fats, processors package the products to exclude air. It isn't possible to exclude air completely, however, so chemical antioxidants are used to stop the chain reaction. They do so by reacting with the free radicals.

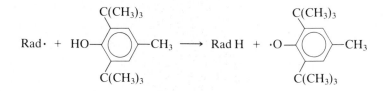

The new radical formed from BHT is rather stable. The unpaired electron doesn't have to stay on the oxygen atom but can move around in the electron cloud of the benzene ring. The BHT radical doesn't react with fat molecules, and the chain is broken.

Why the butyl groups? Without them, the phenols would simply couple when exposed to an oxidizing agent.

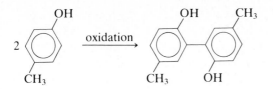

With the bulky butyl groups aboard, the rings can't get close enough together for coupling. They are free, then, to trap free radicals from the oxidation of fats.

Many food additives have been criticized as harmful to people. BHA and BHT are no exceptions. These additives have been reported to cause allergic reactions in some people. Pregnant mice fed a diet containing 0.5% of either BHA or BHT give birth to offspring with chemical abnormalities in their brains.

Not all reports about antioxidants have been unfavorable, however. BHT may have a serendipitous effect. When relatively large amounts of BHT were fed to rats daily, their life spans were increased by a human equivalent of 20 years. One theory about aging is that it, like the rancidity of fats, is caused in part by the formation of free radicals. BHT retards this chemical breakdown in the cells in the same way it retards spoilage in foods. The discovery of the secret of aging might lead to longer life spans for humans. Perhaps even more promising, however, is the possibility of eliminating some of the more deleterious effects of aging. In the laboratory, BHT also has shown antitumor activity. Some scientists have speculated that the increased use of antioxidants as food additives has contributed to a decline in the incidence of stomach cancer in the United States.

BHA and BHT are synthetic chemicals, but there are natural antioxidants. Perhaps the most notable of these is vitamin E. The structure of this compound is given in Table 18.6. Note that vitamin E is also a phenol, with lots of substituents on the benzene ring. Presumably, its action as an antioxidant is quite similar to that of BHT.

Rats deprived of vitamin E become sterile. Because of this, vitamin E is called the antisterility vitamin. Some food faddists promote the ingestion of large amounts of vitamin E to increase sexual prowess, combat wrinkled skin, and prevent heart attacks. There is no unambiguous evidence for any of these claims. Just because vitamin E, in small amounts, prevents sterility in rats does not mean that we can assume that massive dosages will enhance sexual activity in humans.

As we saw in Chapter 18, vitamin E is available in wheat germ oil, green vegetables, vegetable oils, egg yolks, and meat. Most nutritionists contend that it would be nearly impossible to eat a diet deficient in vitamin E. Vitamin E is fat soluble and is stored in the body. Large doses may waste money, but they do not seem to have harmful effects. A belief in the efficacy of vitamin E, however, might lead some people to postpone needed medical treatment.

A vitamin E deficiency can lead to a vitamin A deficiency. Vitamin A can be oxidized to a nonactive form when vitamin E is no longer present to act as an antioxidant. Indeed, the loss of vitamin E's antioxidant effect is believed to be responsible for all of the symptoms of vitamin E deficiency. Polyunsaturated fatty acids, especially, are oxidized at increased rates. Muscular dystrophy, sterility, and other symptoms manifested by animals deficient in vitamin E are believed to result from this "simple" change in body chemistry.

Food Colors: From Carrots to Cancer

Some foods are naturally colored. For example, the yellow compound β-carotene (read *beta*-carotene) occurs in carrots. β-Carotene is used as a color additive in other foods, such as butter and margarine. (Our bodies convert β-carotene to vitamin A; thus, it is a vitamin additive as well as a color additive.) Other natural food colors include beet juice, grape-hull extract, and saffron (from autumn-flowering crocus flowers).

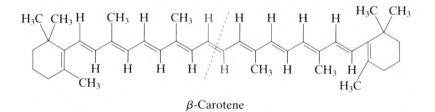

β-Carotene

We have come to expect many foods to have characteristic colors. The food industry, to increase the attractiveness and acceptability of its products, has used synthetic food colors for decades. Since the Food and Drug Act of 1906, the FDA has regulated the use of these coloring chemicals and set limits on their concentrations. But the FDA is not infallible. Colors once on the approved list later were shown to be harmful and were removed from the list. In 1950, a candy company tried to duplicate in Halloween candy the bright orange color of pumpkins. They used a large amount of Food, Drug, and Cosmetic (FD&C) Orange No. 1. Although this color had been safe in amounts previously used, it caused severe gastrointestinal upset in a number of trick-or-treaters. Thus, it was banned by the FDA.

A few years later, two more dyes were banned. FD&C Yellow Nos. 3 and 4 were found to contain small amounts of β-naphthylamine, a carcinogen. Furthermore, the dyes reacted with acids in the stomach to produce more β-naphthylamine. This compound induces cancer of the bladder in laboratory animals. Any chemical shown to induce cancer in

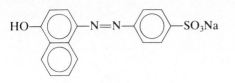

FD&C Orange No. 1

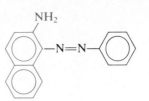

FD&C Yellow No. 3

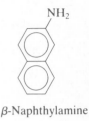

β-Naphthylamine

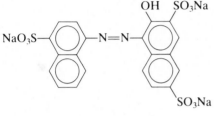

FD&C Red No. 2

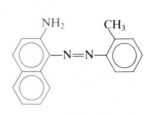

FD&C Yellow No. 4

Figure 19.6 Four synthetic food colors. Note that the two yellow dyes are related to β-naphthylamine, a carcinogen.

laboratory animals is automatically banned under the 1958 **Delaney Amendment** to the Food and Drug Act. Several other dyes have since been banned, including FD&C Red No. 2, which was "delisted" in 1976 after it was shown to cause cancer in laboratory animals. Structures of these synthetic food colors are shown in Figure 19.6.

It should be noted that food colorings, even those that have been banned, present an extremely low hazard. They have been used for years with apparent safety. Even if the hazard is low, however, there is essentially no benefit from food colors, other than an aesthetic one. Perhaps it would be best to ban these chemicals rather than to assume the risk, however small.

There are only a few artificial food colors still approved by the FDA. Several of these are being studied for possible harmful effects. Any foods that contain artificial colors are supposed to say so on the label, and you can avoid them if you want to.

Sweet Chemicals: Sugar Substitutes

Recall from Chapter 16 that sugars are polyhydroxy compounds. Indeed, many compounds with hydroxyl groups on adjacent carbon atoms are sweet. Even ethylene glycol ($HOCH_2CH_2OH$) is sweet, though it is quite toxic. Glycerol, obtained from the hydrolysis of fats (Chapter 18), is also sweet. It is used as a food additive principally for its properties as a **humectant** (moistening agent), however, not as a sweetener.

CH$_2$OH
|
CHOH CH$_2$OH
| |
CHOH CHOH
| |
CHOH CHOH
| |
CHOH CHOH
| |
CH$_2$OH CH$_2$OH

Sorbitol Xylitol

Other polyhydroxy alcohols are used as sweeteners, too. The most common of these is *sorbitol*, made by the reduction of glucose. Another is *xylitol*, which has five carbon atoms with a hydroxyl group on each. These compounds, though not strictly carbohydrates, have about the same Calorie content per gram. They have an advantage over sugars in that they are not broken down in the mouth and thus do not contribute to tooth decay. This makes these alcohols useful in sugar-free chewing gums. In larger amounts, as in candies, sorbitol and xylitol often cause diarrhea.

Sweet Chemicals: The Artificial Sweeteners

Obesity is a major problem in most of the developed countries. Presumably, we could reduce the intake of Calories by replacing sugars with noncaloric sweeteners. It seems, though, that most people who drink diet pop instead of that sweetened with sugar replace those Calories with others and still fail to lose weight. Although there is no evidence that artificial sweeteners are of any value in controlling obesity, they have still become part of our culture.

For many years, the major artificial sweeteners were saccharin and cyclamates. In 1970, after studies showed that cyclamates caused cancer in laboratory animals, those sweeteners were banned. Subsequent studies have failed to confirm the original finding that cyclamates cause cancer. Nevertheless, the FDA has not lifted the ban.

In 1977, saccharin was shown to cause bladder cancer in laboratory animals. The FDA's move to ban saccharin was blocked by Congress because saccharin was the last approved artificial sweetener. Its ban would have meant the end of diet soft drinks and low-Calorie products.

In 1981, the FDA approved a second artificial sweetener, aspartame, the methyl ester of the dipeptide aspartylphenylalanine. Aspartame is about 160 times sweeter than sucrose. Aspartame has largely replaced saccharin as the artificial sweetener of choice. There are anecdotal reports of problems with aspartame, but repeated studies have shown it to be safe, at least in moderate amounts. An exception is that it is not safe for people with phenylketonuria, an inherited condition in which phenylalanine cannot be metabolized properly.

Structures of the three artificial sweeteners are given in Figure 19.7.

Table 19.3 compares the sweetness of a variety of substances. What makes a compound sweet? There is little structural similarity among the compounds. Saccharin, the cyclamates, and compound P-4000 bear little resemblance to the sugars. Even more baffling are the tastes of two compounds that closely resemble P-4000: compound I, which has the

Table 19.3 Sweetness of Some Compounds Relative to Sucrose

Compound	Relative Sweetness*
Glucose	74
Fructose	173
Lactose (milk sugar)	16
Sucrose	100
Maltose (malt sugar)	33
P-4000	400 000
Saccharin	50 000
Aspartame	16 000

* Sweetness is relative to sucrose at a value of 100.

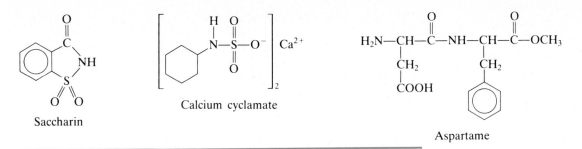

Saccharin   Calcium cyclamate

Aspartame

Figure 19.7 Three artificial sweeteners.

NO$_2$ and NH$_2$ groups reversed, is tasteless; compound II, which has two NO$_2$ groups, is extremely bitter (Figure 19.8). P-4000 is one of the sweetest substances known. In 1974, its use in the United States was banned because of possible toxic effects.

Removing Additives from the GRAS List

Some food additives have been used for many years without apparent harmful effects. In 1958, the United States Congress established a list of additives "generally recognized as safe." This compilation became known as the **GRAS list**. Recent developments have brought to light some deficiencies in the original testing procedures. New research findings—and greater consumer awareness—have led the FDA to begin to reevaluate all the chemicals on the list.

Saccharin, the cyclamates, and several of the food colors now banned were once on the GRAS list, but improved instruments and better experimental designs have revealed possible harm where none was thought to exist. Most of the newer experiments have involved feeding massive doses of additives to laboratory animals, and they have been criticized in that regard. We discuss these testing procedures further in Chapter 25.

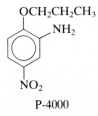

P-4000

Compound I
(tasteless)

Compound II
(bitter)

Figure 19.8 Taste response to these three compounds is dramatically different.

Poisons in Your Food

Oxalic acid

People have been trying to deal with poisons in their food for millennia. Early foragers learned—probably by the painful process of trial and error—that some plants and animals were poisonous. We are all familiar with the fact that some mushrooms are poisonous. Rhubarb leaves contain toxic *oxalic acid*. The Japanese relish a variety of puffer fish that contains a deadly poison in its ovaries and liver. More than a hundred Japanese die each year from improperly prepared puffers.

The most toxic substance known is the toxin produced by the bacterium *Clostridium botulinum*. This organism grows in improperly canned food by a perfectly natural process. If the food isn't properly sterilized before it is sealed in jars or cans, the microorganism flourishes in the anaerobic (without air) conditions. The poison it produces is so toxic that 1 g of it could kill more than a million people. The point to all this is that a food is not inherently good simply because it is natural. Neither is it necessarily bad because a synthetic chemical substance has been added to it.

There's no doubt that we would get better nutrition if we ate nothing but fresh food. For people in large cities, however, that might be impossible; and few people anywhere want to spend all the time it takes to gather fresh food and prepare meals from scratch. Convenience foods are indeed convenient, and food additives make them easier to fix, more attractive, and (in some cases) more nutritious.

Cancer in Your Diet?

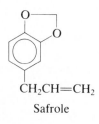

3,4-Benzpyrene

Safrole

There is little chance that we will suffer acute poisoning from approved food additives. But what about cancer? Could all those chemicals in our food increase our risk of cancer? The possibility exists, even though the risk may be low.

We should recognize, though, that carcinogens occur in food naturally, too. A charcoal-broiled steak contains 3,4-benzpyrene, a carcinogen also found in cigarette smoke and automobile exhaust fumes.* Cinnamon and nutmeg contain safrole, a carcinogen that has been banned as a flavoring in root beer.

Among the most potent carcinogens are the **aflatoxins**, compounds

---

* It should be recognized that there is a great differences between ingesting carcinogens (with food) and inhaling them (with cigarette smoke or polluted air). A compound that could induce lung cancer might be harmless in the stomach.

Figure 19.9 Aflatoxins, highly carcinogenic compounds, are produced by molds that grow on peanuts and stored grains. [Courtesy of Waters Associates, Milford, MA.]

produced by molds growing on stored peanuts and grains (Figure 19.9). Aflatoxin B_1 is estimated to be 10 million times as potent a carcinogen as saccharin, and there is no way to keep it completely out of our food. The FDA sets a tolerance of 20 ppb for aflatoxins.

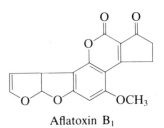

Aflatoxin B_1

It is estimated that we consume 10 000 times as much natural carcinogens as synthetic ones.

Should we ban steaks, spices, peanuts, and grains because they contain naturally occurring carcinogens? Probably not. The risk is slight, and life is filled with more serious risks. Should we ban food additives that have been shown to be carcinogenic? Probably yes. Why take the risk, however slight, when it is easy to avoid?

Incidental Additives

A variety of chemical substances may be present in foods as a result of carelessness or accident in some phase of production, processing, storage, and packing. These incidental additives often receive wide publicity. In 1959, the sale of cranberries was forbidden after some shipments were found to be contaminated by the herbicide aminotriazole, a compound shown to be carcinogenic in tests on laboratory animals. In 1969, coho salmon taken from Lake Michigan were shown to contain DDT above the tolerance level. Sale of these fish was banned, also. In 1970, fish taken from the Detroit River and Lake St. Clair (between Michigan and Ontario) were found to be contaminated with mercury. Mercury compounds also were found in fish taken from the Wisconsin River and other inland waters. Commercial fishing was banned in areas where substantial mercury contamination was found.

Aminotriazole

Polychlorinated biphenyls (PCBs, Chapter 12) have been found in poultry and eggs. These products, too, were seized and destroyed. Related compounds, polybrominated biphenyls (PBBs), meant to be used as fire-retardants, were accidentally mixed with animal feed in western Michigan. Many farm animals were destroyed, and still PBBs got into the food supply. Nearly all the residents of Michigan now have PBBs in their bodies.

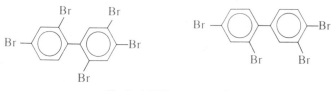

Typical PBB compounds

Chemical substances in animal feed often show up in the meat we eat. Antibiotics are added to animal feed to promote weight gain. In fact, 40% of the 9 million kg of antibiotics produced annually in the United States go into low-level dosages in animal feeds. Residues of these have been found in up to 25% of the animals slaughtered. These residues may result in sensitization of individuals who eat the meat, thus hastening the development of allergies. Antibiotic residues in meat also may hasten the process by which bacteria become drug resistant.

Diethylstilbestrol (DES), a synthetic female hormone (Chapter 22), also has been added to animal feeds to promote weight gain. It was banned in 1973 after evidence showed that it caused vaginal cancer in the daughters of women who had taken DES during pregnancy. The ban was lifted by court action, but it was reinstated in 1979. Recently, DES has also been shown to cause testicular cancer in sons of women who took DES while pregnant.

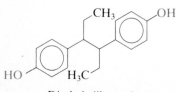

Diethylstilbestrol

Note that the effect of DES did not show up for 15 years. Even then it appeared in the *offspring* of the women who took the drug. This points out some of the problems involved in evaluating a chemical for its possible harmful effects.

A World Without Food Additives?

Could we get along without food additives? Some of us could. But food spoilage might drastically reduce the food supply in an already hungry world. And diseases due to vitamin and mineral deficiencies might flourish again. Foods might cost more and be less nutritious. Food additives seem to be a necessary part of modern society. It is true that there are hazards associated with the use of some food additives, but the major problem with our food supply still is contamination by rodents, insects, and harmful microorganisms (Table 19.4). Indeed, there are 9100 deaths and 6 million illnesses annually in the United States from food poisoning caused by bacterial toxins.

What should we do about food additives? We should be sure that the FDA is staffed with qualified personnel to ensure the adequate testing of proposed food additives. People trained in chemistry are necessary for the control and monitoring of food additives and the detection of contaminants. Research in the analytical techniques necessary for the detection of trace quantities is vital to adequate consumer protection. We also should work for laws adequate to prevent the unnecessary and excessive use of pesticides and other agricultural chemicals that might contaminate our food. Above all, we should be alert and informed about these problems so vital to our health and well-being.

Table 19.4   Destruction, Diversion, or Detention of Food by U.S. FDA in 1976

Problem	Destruction or Diversion*		Import Detentions	
	Million Pounds	Percent	Million Pounds	Percent
Sanitation violations (caused by rats, mice, insects, etc.)	32.3	69	50.3	44
Microbiological contaminants (botulism, salmonellosis, etc.)	8.6	18	36.2	31
Other (chemicals, metals, natural poisons, glass, etc.)	6.1	13	28.5	25
Total	47.0	100	115.0	100

* Diversion to food for nonhuman consumption.
SOURCE: Compiled from data in the *FDA Consumer*.

Problems

1. What is a food additive?

2. What are the two major categories of food additives? Give an example of each.

3. What seven substances make up 98% of all intentional food additives?

4. What United States government agency regulates the use of food additives?

5. What must be done before a food company can use a new food additive?

6. List five functions of food additives.

7. Why are more food additives used today than in 1900? Give three reasons?

8. What is the function of each of the following food additives?
   a. potassium iodide   b. vanilla extract
   c. MSG   d. sodium nitrite

9. What is the function of each of the following food additives?
   a. $FeCO_3$ b. SO_2
   c. potassium sorbate

10. What is the purpose of each of the following food additives?
    a. BHA   b. FD&C Yellow No. 5
    c. saccharin

11. What is the purpose of each of the following food additives?
    a. aspartame   b. vitamin D
    c. sodium hypophosphite

12. What is the difference between vanilla extract and imitation vanilla flavor?

13. What is enriched bread? Is it equal in nutritional value to bread made from whole grain?

14. What is MSG?

15. What is Chinese-restaurant syndrome?

16. What is meant by standards of identity?

17. What kinds of chemical compounds are used as mold inhibitors?

18. Why is sodium nitrite added to cured meats?

19. What is botulism?

20. What are antioxidants?

21. How does BHT work as an antioxidant?

22. What is a chain reaction?

23. What vitamin serves as a fat-soluble antioxidant?

24. What dual role does β-carotene serve as a food additive?

25. What is the Delaney amendment to the Food and Drug Act?

26. What is a humectant?

27. Name three artificial sweeteners. Which are approved for current use?

28. What is the chemical nature of aspartame?

29. What is the GRAS list?

30. What are aflatoxins?

31. List some incidental additives that have been found in foods.

32. What are the major problems associated with the safety of our food supply?

33. Consult a recent issue of the *FDA Consumer* and make a list of incidental food additives that are reported on in that issue. What are the most common causes of contamination?

34. Examine the label on a sample of each of the following.
    a. a can of soft drink   b. a can of beer
    c. a dried soup mix   d. a can of soup
    e. a can of fruit drink   f. a cake mix

    Make a list of the food additives in each. Try to determine the function of each additive.

35. Maraschino cherries are bleached (with sulfur dioxide) and then dyed with an organic food coloring. Do you think consumers would buy the uncolored cherries?

36. One step in the production of raisins involves the treatment of grapes with sodium hydroxide (NaOH). This compound is also known as caustic soda, or lye. Some raisins then are bleached with sulfur dioxide to remove their color. Would you expect to find these "golden" raisins in an organic food store?

37. Fructose is found in many "natural foods" stores. (It even is promoted as a diet aid!) Fructose is made from sucrose by hydrolysis (reaction with water) and separation from the coproduct glucose, or by the catalytic isomerization of glucose (from corn syrup). Is fructose made in one of these ways "natural"? Explain your answer.

38. Which additional food additives (if any) should be banned? Which should be further restricted? Explain your answers fully.

39. Campbell's Chunky Vegetable Beef Soup is available in regular and low-sodium versions. The regular has 935 mg of Na^+ and the low-sodium version has 90 mg of Na^+ per 10.75-oz portion. What percentage of the recommended daily maximum of 3300 mg of Na^+ would you get from one portion of each soup?

40. One large egg supplies 275 mg of cholesterol. The American Heart Association recommends a maximum of 300 mg of cholesterol per day. What percentage of the maximum recommendation is met by one egg?

41. Honey is often promoted as more nutritious than sugar. Honey does contain 0.2 mg of iron, 0.014 mg of riboflavin, and 4 mg of calcium per tablespoon (T). The USRDAs are 18 mg iron, 1.7 mg riboflavin, and 1000 mg of calcium. Calculate the amount of honey you would have to eat to get the USRDA for each nutrient.

References and Readings

1. Adcock, Louis H. "Fredrick Christian Accum: A Chemist for All Seasonings." *Chemistry*, May 1973, pp. 16–18. Accum exposed hazards in the food industry—150 years ago.
2. Corwin, Emil, and Wayne L. Pines. "Why FDA Banned Red No. 2." *FDA Consumer*, April 1976, pp. 18–23.
3. Guild, Walter, Jr. "The Theory of Sweet Taste." *Journal of Chemical Education*, March 1972, pp. 171–173.
4. Holmes, Alan. "Role of Food Additives." *Chemistry and Industry*, 6 February 1984, pp. 104–107.
5. Larkin, Timothy. "Food Additives and Hyperactive Children." *FDA Consumer*, March 1977, pp. 19–21.
   A connection is hard to prove or disprove.
6. Leveille, Gilbert A. "Food Fortification—Opportunities and Pitfall." *Food Technology*, January 1984, pp. 58–63.
7. Megos, Harry N. "Colors—Key Food Ingredients." *Food Technology*, January 1984, pp. 70–74.
8. Rodricks, Joseph V. "Aflatoxins: Hazards from Nature." *FDA Consumer*, May 1978, pp. 16–19.
9. Ross, Walter S. "Artificial Sweeteners—Are They Safe?" *Readers Digest*, December 1985, pp. 140–144.
10. "Salt and Your Health." *Consumer Reports*, January 1984, pp. 17–22.
11. Simpson, Lance L. "Deadly Botulism." *Natural History*, January 1980, pp. 12–24.
12. Smith, R. Jeffrey. "Aspartame Approved Despite Risks." *Science*, 28 August 1981, pp. 986–987.
13. Winter, Ruth. *A Consumer's Dictionary of Food Additives*. New York: Crown Publishers, 1972.
14. Ziporyn, Terra. "The Food and Drug Administration: How 'Those Regulations' Came to Be." *Journal of the American Medical Association*, 18 October 1985, pp. 2037–2046.

20

Household Chemicals

Helps and Hazards

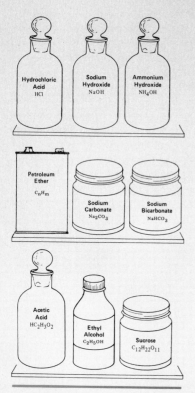

Figure 20.1 A well-stocked chemistry laboratory has a variety of chemicals.

Would you like to work in a place where poisonous chemicals are used nearly every day? Where toxic vapors and harmful dusts are a common hazard? Where corrosive acids and alkalies are frequently used? Where highly flammable liquids and vapors sometimes ignite and cause painful burns and extensive property damage? You wouldn't? Well, friend, welcome to the typical American home!

Many people regard a chemistry laboratory as a dangerous place stocked with poisons, explosives, and corrosive chemicals. This fear is the reason often given for not taking a laboratory course in school. Yet most people are practicing chemists. There are perhaps half a million chemical products available for use in the American home. These range from ammonia for cleaning to zeolites for water softening. The list includes waxes, wax removers, paints, paint removers, toothpaste, rodenticides, detergents, insecticides, spot removers, aspirin, rubbing alcohol, bleaches, baking soda, disinfectants, polishes, abrasives, and many other products.

Many of the chemicals in household products are quite harmless. Others contain highly corrosive acids and alkalis. Some are quite toxic and should be used with extreme caution. Still others are highly flammable and present considerable fire hazards. Some give off toxic vapors or produce harmful dusts. Many cause undesirable changes in the environment when used or discarded.

Extensive use of chemicals in the home has led to an increasing number of accidents. The chemistry laboratory is probably a safer place than many homes. In the laboratory, chemicals usually are used under carefully controlled conditions. By contrast, studies have shown that chemicals often are used in the home without regard to the directions or the precautions given on their labels. Indeed, all too frequently the labels aren't read at all! It is this *misuse* of household chemicals that often ends in tragedy. It would be nice if everyone knew a lot of chemistry. At the very least, everyone should read and follow with great care all directions for using household chemicals.

We have already discussed the chemicals that make up our food (Chapters 18 and 19). Many of the chemicals used in agriculture (Chapter 17) are used in and around the home, too, especially in the yard and garden. We discuss the chemicals we put on our skin and hair (cosmetics) in the next chapter. Chapters 22 and 23 are devoted to the chemicals we use as drugs. There is indeed a great variety of household chemicals. Of all, though, the detergents and related cleaning compounds make up the greatest volume.

Figure 20.2 A modern home is stocked with a variety of chemical products. The products shown here contain the same chemicals shown in the corresponding positions in Figure 20.1.

A Dirty History of Cleaning

In primitive societies, even today, clothes are cleaned by beating them with rocks in the nearest stream. Sometimes plants, such as the soapworts of Europe or the soapberries of tropical America, are used as cleansing agents. The leaves of the soapwort and soapberries contain **saponins**, chemical compounds that produce a soapy lather. These saponins were probably the first detergents.

Ashes of plants contain potassium carbonate (K_2CO_3) and sodium carbonate (Na_2CO_3). The carbonate ion, present in both of these compounds, reacts with water to form an alkaline solution.

$$CO_3^{2-} \; + \; H_2O \longrightarrow \; HCO_3^- \; + \; OH^-$$

Carbonate ion	Bicarbonate ion	Hydroxide ion (basic)

The basic solution has detergent properties. These alkaline plant ashes were used as cleansing agents by the Babylonians at least 4000 years ago. Europeans were using plant ashes to wash their clothes as recently as 100 years ago. Sodium carbonate is still sold today as washing soda.

The Romans, with their great public baths, probably did not use any sort of soap. They covered their bodies with oil, worked up a sweat in a steam bath, and then had the oil wiped off by a slave. They finished by taking a dip in a pool of fresh water. And the slaves? They probably didn't bathe at all.

Cleanliness of body and clothes was not very important during the Middle Ages. Neither master nor serf bathed frequently, if at all. Lords and ladies did sometimes mask their body odors with perfumes, and clean outer garments hid the filth of underclothes. Though soap was known, it was used as a medicine—when it was used at all. The discovery of disease-causing microorganisms and subsequent public-health practices brought about increased interest in cleanliness by the late eighteenth century. Soap was in common use by the middle of the nineteenth century.

Grandma's Lye Soap

The first written record of soap is found in the writings of Pliny the Elder, the Roman who described the Phoenicians's synthesis of soap by using goat tallow and ashes. By the second century A.D., sodium carbonate (produced by the evaporation of alkaline water) was heated with lime (from limestone or seashells) to produce sodium hydroxide (lye).

$$Na_2CO_3 \ + \ Ca(OH)_2 \ \longrightarrow \ 2 \ NaOH \ + \ CaCO_3$$

The sodium hyroxide was heated with animal fats or vegetable oils to produce soap (Figure 20.3). (Note that soap is a salt of a long-chain, carboxylic acid (Chapter 11). The American pioneers made soap in much the same manner. Lye was added to animal fat in a huge iron kettle. The mixture was cooked over a wood fire for several hours. The soap rose to the surface and, upon cooling, solidified. The glycerol remained as a liquid on the bottom of the pot. Both the glycerol and the soap often contained unreacted alkali, which ate away the skin. Grandma's lye soap is not just a myth!

In modern commerical soapmaking, the fats and oils often are hydrolyzed with superheated steam. The fatty acids then are neutralized to make soap. Toilets soaps usually contain a number of additives such as dyes, perfumes, creams, and oils. Scouring soaps contain abrasives, such as silica, pumice, and oatmeal(!). Many soaps claim to have deodorant action, but few, if any, have any active deodorant (Chapter 21) other than the soap itself. The deodorant is just a cover-up perfume that quickly fades from the skin. Some soaps have air blown in before they solidify to lower their density so that they float. Some toilet soaps

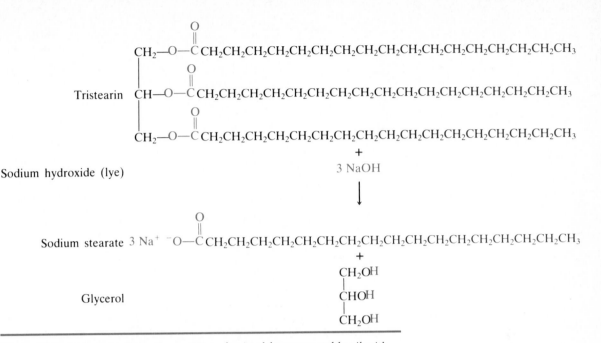

Figure 20.3 Soap is made by the reaction of animal fat or vegetable oil with sodium hydroxide. Animal fats yield hard soaps. Vegetable oils, with unsaturated carbon chains, produce soft soaps. Coconut oils, with shorter carbon chains, yield soaps that are more soluble in water.

contain—or are made completely of—synthetic detergents. Their action is similar to that of soap.

Sometimes soaps are made with cations other than sodium ion. Potassium soaps (see margin) are softer, and they produce a finer lather. They are used alone, or in combination with sodium soaps, in liquid soaps and shaving creams. Soaps also are made by reacting fatty acids with triethanolamine. These substances are used in shampoos and other cosmetics (Chapter 21).

$CH_3(CH_2)_{14}COO^- \ K^+$

Potassium palmitate

$CH_3(CH_2)_{14}COO^- \ (HOCH_2CH_2)_3\overset{+}{N}H$

Triethanolammonium palmitate

How Soap Works

Dirt and grime usually adhere to skin, clothing, and other surfaces, because they are combined with greases and oils—body oils, cooking fats, lubricating greases, and a variety of similar substances—that act a little like sticky glues. Since oils are not miscible with water, washing with water alone does little good.

A soap molecule has a split personality. One end is ionic and

Figure 20.4 Sodium palmitate, a soap. (a) Structural formula. (b) A schematic representation.

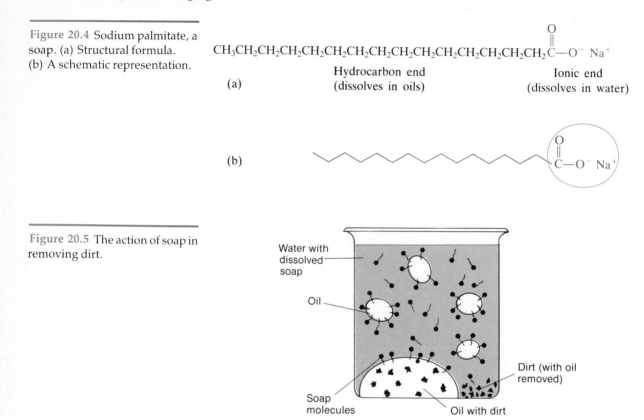

(a)

Hydrocarbon end (dissolves in oils)

Ionic end (dissolves in water)

(b)

Figure 20.5 The action of soap in removing dirt.

dissolves in water. The other end is like a hydrocarbon and dissolves in oils (Figure 20.4). If we represent the ionic end of the molecule as a circle and the hydrocarbon end as a zigzag line, we can illustrate the cleansing action of soap schematically (Figure 20.5). The hydrocarbon "tails" stick into the oil. The ionic "heads" remain in the aqueous phase. In this manner, the oil is broken into tiny droplets and dispersed throughout the solution. The droplets don't coalesce because of the repulsion of the charged groups (the carboxyl anions) on their surfaces. The oil and water form an emulsion, with soap acting as the emulsifying agent. With the oil no longer "gluing" it to the surface, the dirt can be removed easily.

Disadvantages of Soap

For cleaning clothes and for many other purposes, soap has been largely replaced by synthetic detergents, because soaps have two rather serious shortcomings. One of these is that, in acidic solutions, soaps are

(a)

$$CH_3CH_2CH_2CH_2CH_2CH_2CH_2CH_2CH_2CH_2CH_2COO^-Na^+ + H^+ \longrightarrow$$

A soap   An acid

$$CH_3CH_2CH_2CH_2CH_2CH_2CH_2CH_2CH_2CH_2CH_2COOH + Na^+$$

A fatty acid

(b)

$$2\ CH_3CH_2CH_2CH_2CH_2CH_2CH_2CH_2CH_2CH_2CH_2COO^- + Ca^{2+} \longrightarrow$$

Soap anion

$$(CH_3CH_2CH_2CH_2CH_2CH_2CH_2CH_2CH_2CH_2CH_2COO^-)_2\ Ca^{2+}$$

Bathtub ring
(insoluble)

Figure 20.6 Soap suffers two disadvantages. Acids convert soap molecules to fatty acids (a) Hard water ions such as Ca^{2+} precipitate soap as insoluble curds (b) Neither the fatty acids nor the curds have detergent action.

converted to free fatty acids (Figure 20.6a). The fatty acids, unlike soap, don't have an ionic end. Lacking the necessary split personality, they can't emulsify the oil and dirt; that is, they do not exhibit any detergent action. What's more, these fatty acids are insoluble in water and separate as a greasy scum.

The second, and more serious, disadvantage of soap is that it doesn't work well in hard water. Hard water is water that contains certain metal ions, particularly magnesium, calcium, and iron ions. The soap anions react with these metal ions to form greasy, insoluble curds. These deposits make up the familiar bathtub ring. They leave freshly washed hair sticky and are responsible for "tattle-tale gray" in the family wash.

Water Softeners

To aid the action of soaps, a variety of water-softening agents and devices have been developed (Figure 20.7). An effective water softener is washing soda, sodium carbonate ($Na_2CO_3 \cdot 10H_2O$). It makes the water basic (preventing the precipitation of fatty acids) and removes the hard water ions calcium and magnesium. These jobs are performed by the carbonate ion. The ion reacts with water to raise the pH (that is, to make the solution more basic).

$$CO_3^{2-} + H_2O \longrightarrow HCO_3^- + OH^-$$

The carbonate ion also reacts with the ions that cause hard water and removes the ions as insoluble salts.

$$Mg^{2+} + CO_3^{2-} \longrightarrow MgCO_3$$
$$Ca^{2+} + CO_3^{2-} \longrightarrow CaCO_3$$

Trisodium phosphate (Na_3PO_4) is another water-softening agent. Like washing soda, it makes the wash water basic and precipitates calcium and magnesium ions.

$$PO_4^{3-} + H_2O \longrightarrow HPO_4^{2-} + OH^-$$
$$2 PO_4^{3-} + 3 Mg^{2+} \longrightarrow Mg_3(PO_4)_2$$

In addition, phosphates seem to aid in the cleaning process in some other way that is not yet well understood.

Water-softening tanks are also available for use in homes and businesses. These tanks contain an insoluble polymeric material that attracts and holds the calcium, magnesium, and iron ions to its surface,

Figure 20.7 Sudsing quality of hard water versus soft water. From left: detergent in hard water, soap in hard water, soap in soft water, and detergent in soft water. Note that the sudsing of detergent differs little in hard and soft waters. Note also the absence of sudsing and the formation of insoluble material when soap is used in hard water. [Photo by Lawrence Scott.]

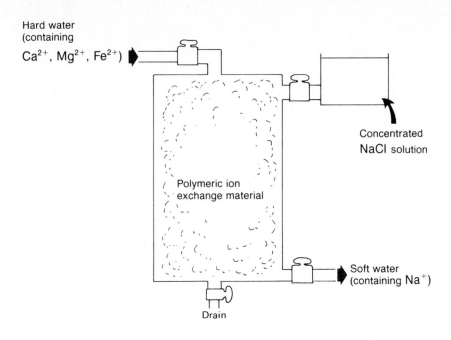

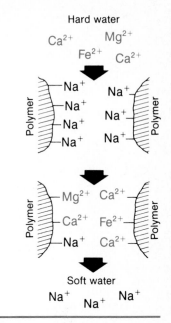

Figure 20.8 One type of water-softening tank contains a polymeric material on which hard-water ions are exchanged for sodium ions. When the material becomes saturated with calcium, magnesium, and iron ions, it is regenerated by flushing with a saturated salt solution.

thus softening the water (Figure 20.8). After a period of use, the polymer becomes saturated and must be discarded or regenerated.

Before leaving the subject of soap, let's mention that soap has some advantages. It is an excellent cleanser in soft water, it is relatively nontoxic, it is derived from renewable resources (animal fats and vegetable oils), and it is biodegradable.

Synthetic Detergents

A second technological approach to the problems with soap was to develop a new synthetic detergent. The molecules of the synthetic detergents were enough like those of soap to have the same cleaning action, but different enough to resist the effects of acids and hard water. Sodium lauryl sulfate is typical of the first (but fairly expensive) synthetic detergents.

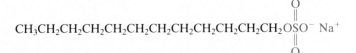

Sodium lauryl sulfate
(sodium dodecyl sulfate)

(a) *Reduction:*

$$\begin{matrix} CH_2OCO(CH_2)_{10}CH_3 \\ | \\ CHOCO(CH_2)_{10}CH_3 \\ | \\ CH_2OCO(CH_2)_{10}CH_3 \end{matrix} + 6\,H_2 \xrightarrow{\text{Ni (catalyst)}} 3\;CH_3(CH_2)_{10}CH_2OH + \begin{matrix} CH_2OH \\ | \\ CHOH \\ | \\ CH_2OH \end{matrix}$$

Trilaurin   Lauryl alcohol   Glycerol
(Dodecyl alcohol)

(b) *Reaction with sulfuric acid:* $CH_3(CH_2)_{10}CH_2OH + H_2SO_4 \longrightarrow CH_3(CH_2)_{10}CH_2OSO_3H + H_2O$

(c) *Neutralization:* $CH_3(CH_2)_{10}CH_2OSO_3H + NaOH \longrightarrow CH_3(CH_2)_{10}CH_2OSO_3{}^- \; Na^+ + H_2O$

Figure 20.9 Early synthetic detergents were made from fats by a three-step process.

These first detergents were derived from fats by reduction with hydrogen, followed by reaction with sulfuric acid, then neutralization (Figure 20.9). Sodium lauryl sulfate is still used in toothpastes, shampoos, and other cosmetics (Chapter 21).

Within a few years, cheap synthetic detergents were produced from petroleum products. **Alkylbenzenesulfonate** (ABS) **detergents** were made from the alkene propylene ($CH_3CH{=}CH_2$), benzene, sulfuric acid, and a base (usually sodium carbonate).

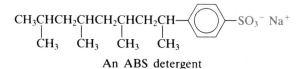

An ABS detergent

Sales of ABS detergents soared. For a decade or more, nearly everyone was happy. Then suds began to appear in sewage treatment plants. Foam piled high in the rivers (Figure 20.10). People in some areas even got a head of foam on their drinking water (Figure 20.11). It was found that the branched-chain structure of ABS molecules was not readily broken down by microorganisms in the sewage treatment plants. The whole supply of groundwater was threatened. Public outcries caused laws to be passed and industries to change their processes. Biodegradable detergents were quickly put on the market, and nondegradable detergents were banned.

The degradable detergents (called linear alkylsulfonates, LAS)

Figure 20.10 Foaming rivers were quite common during the early 1960s. The problem was solved by the development of biodegradable detergents. [Courtesy of the U.S. Department of Agriculture, Washington, DC.]

have linear chains of carbon atoms. Microorganisms can break down LAS molecules by producing enzymes that degrade the molecule two (and only two) carbon atoms at a time (Figures 20.12 and 20.13). The branched chain blocks this enzyme action, preventing the degradation of ABS molecules. Thus, technology has solved the problem of foaming rivers.

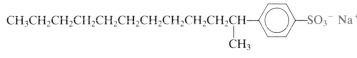

An LAS detergent

It should be pointed out that just because something is degraded, it doesn't simply disappear. Matter is conserved. Everything has to go somewhere. A completely degraded LAS molecule winds up as carbon dioxide, water, and sulfate (SO_4^{2-}). Barry Commoner (Reference 2) contends that LAS degradation leads to phenol, a toxic material, and that degradable detergents are more likely to kill fish than are nondegradable ones. These allegations are denied by the detergent industry.

The cleansing action of synthetic detergents is quite similar to that of soaps. The synthetics work better in acidic solution and in hard water, though. Their calcium and magnesium salts, unlike those of soap, are soluble and do not separate out, even in extremely hard water. Thus, the cleansing action of the synthetic detergent is little affected by hard water.

Figure 20.11 A glass of suds. [Courtesy of Bergwall Productions, New York.]

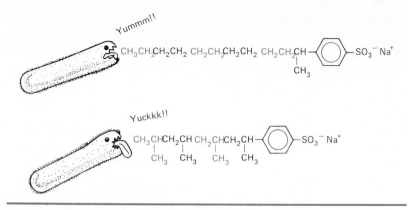

Yummm!!

$CH_3CH_2CH_2CH_2$ $CH_2CH_2CH_2CH_2$ CH_2CH_2CH—⟨ ⟩—SO_3^- Na^+
|
CH_3

Yuckkk!!

CH_3CHCH_2CH CH_2CHCH_2CH—⟨ ⟩—SO_3^- Na^+
| | | |
CH_3 CH_3 CH_3 CH_3

Figure 20.12 Microorganisms can readily digest LAS molecules, but are unable to break down the branched chains of ABS molecules.

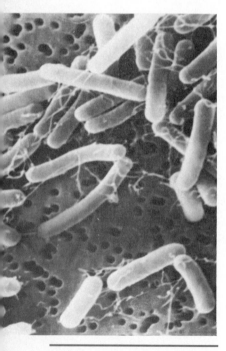

Figure 20.13 Microorganisms, such as the flagellated *Escherichia coli* (shown here magnified 42 500 times by scanning electron microscopy) are able to degrade LAS detergents. [Courtesy of Ethyl Corporation, Baton Rouge, LA.]

Laundry Detergent Formulations: Builders

Even if Barry Commoner is wrong about the toxicity of the degradation products of biodegradable detergents, the laundry detergents we use do affect streams and lakes. The products for use in homes and commercial laundries usually contain much more than LAS molecules. The LAS is a surface-active agent, or surfactant. (Any agent, including soap, that stabilizes the suspension of nonpolar substances— such as oil and grease—in water is called a **surface-active agent**.) In addition to LAS, modern detergent formulations contain a number of other substances to improve detergency, to bleach, to lessen the redeposition of dirt, to brighten, or simply to reduce the cost of the formulation.

Any substance added to surfactant to increase its detergency is called a **builder**. Common builders, once widely used but now banned or restricted in many areas, are the phosphates. An example is sodium tripolyphosphate ($Na_5P_3O_{10}$). It ties up Ca^{2+} and Mg^{2+} in soluble complexes—thus softening the water—and produces a mild alkalinity, providing a favorable environment for detergent action.

We have seen (in Chapter 16) how phosphates speed the eutrophication of lakes. In some areas, phosphates from detergents have been shown to contribute to this process. The degree of their contribution compared to phosphates that occur naturally in sewage remains open to question, however, in other locations.

Several state and local governments have banned the sale of

detergents containing phosphates, and the detergent industry has offered a variety of replacements; the most prominent are sodium carbonate and complex aluminosilicates called *zeolites*. The first acts by precipitating the calcium ions, thus softening the water (page 455).

$$Ca^{2+} + CO_3^{2-} \longrightarrow CaCO_3$$

The $CaCO_3$ precipitate appears to be harmful to automatic washing machines. Further, excess carbonate ions form strongly basic solutions, that is, solutions that contain an excess of OH^- ions.

$$CO_3^{2-} + H_2O \longrightarrow HCO_3^- + OH^-$$

The zeolites are perhaps the most promising of the substitutes. The zeolite anions trap calcium ions by exchanging them for their own sodium ions.

$$Ca^{2+} + Na_2Al_2Si_2O_7 \longrightarrow 2\ Na^+ + CaAl_2Si_2O_7$$

The calcium ions are held in suspension by the zeolites (rather than being precipitated). Further, zeolite solutions are not strongly basic and are therefore less likely than sodium carbonate and sodium silicate solutions to irritate the skin and eyes.

Brighter Than Bright

A variety of other additives are used in detergent formulations. Many contain **optical brighteners**. These compounds, called blancophors (or colorless dyes), absorb the invisible ultraviolet component of sunlight and reemit it as visible light at the blue end of the spectrum. The fabric appears brighter and the blue light camouflages any yellowing. A diagram of this action, along with the structure of one such dye, is shown in Figure 20.14. Clothes treated with an optical brightener on the surface may be dirty underneath, but they look "whiter and brighter than new." These brighteners are also used in cosmetics, paper, soap, plastics, and other products.

Optical brighteners have no known immediate toxic effect on humans. There is some possibility that these compounds might cause cancer or genetic disease. Bjorn Gillberg of the Royal Agricultural College of Sweden has found that brighteners cause some mutations in microorganisms. Brighteners have been shown to cause skin rashes. Their effect on aquatic systems is largely unknown, yet large amounts are entering our waterways. The only possible benefit of these com-

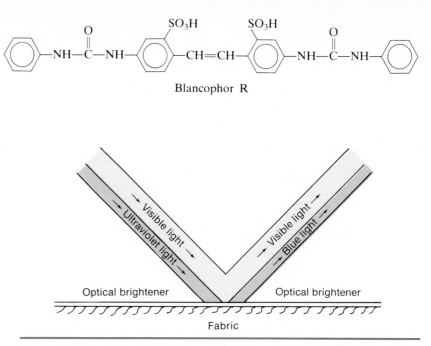

Blancophor R

Figure 20.14 An optical brightener, such as the one illustrated here, converts invisible ultraviolet light to visible blue light, making the fabric look brighter and masking yellowing.

pounds is a cosmetic one: the appearance of our laundry may be more pleasing. Chemists can make optical brighteners, but it is up to the consumer to decide whether the benefit obtained by their use outweighs their monetary, environmental, and health costs.

Liquid Laundry Detergents

Rapidly rising in the home laundry market are liquid detergent formulations. There are two basic types: those built with phosphates or other additives and unbuilt ones that are high in surfactant but contain no phosphates or other builders.

LAS is the cheapest and most widely used surfactant in liquid laundry detergents. In unbuilt formulations, LAS usually is used as the sodium or triethanolamine salt. In built varieties, LAS often is present as the potassium salt. The second most popular liquids are the alcohol ether sulfates. These contain a hydrocarbon portion derived from an alcohol (or alkylphenol), a polar portion derived from ethylene

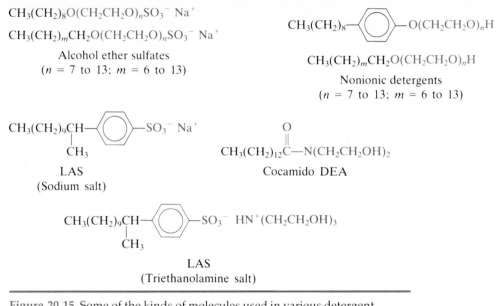

Figure 20.15 Some of the kinds of molecules used in various detergent formulations.

oxide, and a sulfate salt portion. These surfactants are efficient but expensive.

All the surfactants discussed so far, including soap, are **anionic surfactants**; the working part of the molecules is an anion with a nonpolar part and an ionic end. Some liquid detergents contain **nonionic surfactants**. Examples are the alcohol ethoxylates and alkylphenol ethoxylates. The oxygen atoms, by their attraction for water molecules, make their end of the molecule water soluble, just like the ionic end of an anionic surfactant. Structures of several of these detergent molecules are given in Figure 20.15. Nonionic surfactants are great for removing oily soil from fabrics. They are not as good as the anionic surfactants at keeping dirt particles in suspension. The alcohol ethoxylates have the unusual property of being more soluble in cold water than in hot. This makes them particularly suitable for cool-water laundering.

Ethylene oxide

Dishwashing Detergents

Liquid detergents for washing dishes by hand generally contain one or more surfactants as the only active ingredients. Perfumes, colors, and additives supposed to soften and smooth hands are frequently added.

Their main function, though, seems to be the establishment of a base for exaggerated advertising claims. Surfactants used include LAS as the sodium salt or the triethanolamine salt (or both). Some use nonionic surfactants, such as those above. Another nonionic type is the amides made from fatty acids and diethanolamine. Cocamido DEA is an example. Few contain phosphates or other builders. Those that do usually have only small amounts.

Most liquid dishwashing detergents differ significantly only in the concentration or effectiveness of the surfactant. In any formulation, the surfactant loosens the greasy food residues so they can be easily removed. It also traps the oily mess so it doesn't redeposit on dishes. The detergent eventually becomes saturated, though, and breaks down, releasing its greasy load. The weaker the detergent formulation, the sooner it will break down—or the more you will have to use. See Reference 3 for a rating of the efficiency of dishwashing liquids.

Detergents for automatic dishwashers are quite another matter. They often are quite caustic and should never be used for hand dish washing. They contain sodium tripolyphosphate ($Na_5P_3O_{10}$), sodium metasilicate (Na_2SiO_3), sodium sulfate (Na_2SO_4), a chlorine bleach, and only small amounts of surfactant, usually a nonionic type. They depend mainly on their strong alkalis and the vigorous agitation of the machines for the removal of soil.

Bleaches: Whiter Whites

Bleaches are oxidizing agents (Chapter 8). The familiar liquid laundry bleaches are all 5.25% sodium hypochlorite (NaOCl) solutions. They differ only in price. Purex, Clorox, store brands, and generic versions are widely available. Hypochlorite bleaches release chlorine rapidly, and the high concentrations of corrosive chlorine can be quite damaging to fabrics. These bleaches do not work well on polyester fabrics, often causing a yellowing rather than the desired whitening.

Other bleaches are available in solid forms that release chlorine slowly in water. Symclosene is an example of a cyanurate-type bleach. The slow release of chlorine minimizes damage to fabrics because the concentration of chlorine is never very high.

The oxygen-releasing bleaches are usually sodium perborate ($NaBO_2 \cdot H_2O_2$). As indicated by the formula, this compound is a complex of $NaBO_2$ and hydrogen peroxide (H_2O_2). Above 65 °C, the hydrogen peroxide is liberated and acts as a bleach. It decomposes in turn to liberate oxygen.

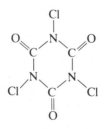

Symclosene

$$2\ H_2O_2 \longrightarrow 2\ H_2O + O_2$$

Borates are somewhat toxic. Perborate bleaches are less active than chlorine bleaches and require higher temperatures, higher alkalinity, and higher concentrations to do an equivalent job. They are used mainly for bleaching white, resin-treated polyester–cotton fabrics. These fabrics last much longer with oxygen bleaching than with chlorine bleaching. Also, properly used, the oxygen bleaches get fabrics whiter than do chlorine bleaches.

All-Purpose Cleaning Products

A variety of all-purpose cleaning products are available for use on walls, floors, countertops, appliances, and other tough, durable surfaces. Those for use in water may contain surfactants, sodium carbonate, ammonia, solvent-type grease cutters, disinfectants, deodorants, and other ingredients. Be sure you check the labels on such products. Some are great for certain jobs but not very good for others. They damage some surfaces and work especially well on others. Most important, they may be harmful to you when they are used improperly.

Household ammonia solutions, straight from the bottle, are good for loosening baked-on grease or burned-on food. Just soak the object overnight. Diluted with water, household ammonia cleans mirrors, windows, and other glass surfaces. Because both ammonia and water are volatile, no residue remains. Mixed with detergent, ammonia rapidly removes wax from linoleum. Ammonia vapors are highly irritating. This cleanser should never be used in a closed room. Ammonia should not be used on asphalt tile, wood surfaces, or aluminum, because it may stain, pit, or erode these materials.

Baking soda (sodium bicarbonate, $NaHCO_3$), straight from the box, is a mild abrasive cleanser. It absorbs food odors readily, making it good for cleaning the inside of a refrigerator. Vinegar (acetic acid CH_3COOH) cuts grease film. It should not be used on marble, because it reacts with marble, pitting the surface.

$$CaCO_3 + 2\ CH_3COOH \longrightarrow Ca^{2+} + 2\ CH_3COO^- + CO_2 + H_2O$$

Most powdered cleansers contain a hard abrasive such as silica (silicon dioxide, SiO_2) for rubbing dirt loose from a surface. They also usually contain a surfactant to dissolve grease. Some brands also feature a chlorine-releasing bleach. These abrasive cleansers may scratch the porcelain on appliances, plastic countertops, and other surfaces. They may even scratch the surface of sinks, toilet bowls, and bathtubs. Dirt gets in the scratches and makes cleaning even more difficult. Use these cleansers (with care) only on surfaces that can withstand the abrasion or on surfaces on which scratches won't matter.

Organic Solvents in the Home

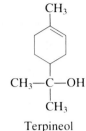

Terpineol

A variety of solvents are used in the home. They may be used to remove paint, varnish, adhesives, waxes, and other materials. Some cleansers also contain organic solvents. Perhaps the best known are those containing pine oil. This oil consists mainly of terpenes, compounds with 10 carbon atoms that occur widely in nature. The pine-oil terpenes usually have a ring structure and one or more alcohol functions. An example is terpineol. Pine oil acts as a mild disinfectant, and it helps to dissolve grease. In moderate concentrations, it has a pleasant odor.

Petroleum distillates are added to some all-purpose cleansers as grease cutters. These are hydrocarbons—much like gasoline—derived from petroleum. They dissolve grease readily, but like gasoline, are highly flammable. They are also deadly when swallowed. The lungs become saturated with hydrocarbon vapors, fill with fluid, and fail to function.

Most of the organic solvents used around the home are volatile and flammable. Many have toxic fumes; nearly all are narcotic at high concentrations. Some people, trying to get their kicks from sniffing glue or other solvents, have died of heart failure. Such solvents should be used only with adequate ventilation. They should never be used around a flame. Be sure to read—and heed—all precautions before you use any solvent. And never use gasoline for cleaning; it is too hazardous in too many ways.

Quaternary Ammonium Salts: Dead Germs and Soft Fabrics

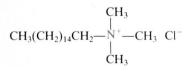

Hexadecyltrimethylammonium
chloride
(a cationic surfactant)

Earlier in this chapter, we discussed anionic detergents and nonionic surfactants. There is a third type, called **cationic surfactants**, in which the working part of the substance is a positive ion. The most common of these are called quaternary ammonium salts, because they have four groups attached to a nitrogen atom that bears a positive charge. An example of such a cationic surfactant is hexadecyltrimethylammonium chloride.

These cationics are not good detergents, but they have a degree of germicidal action. Sometimes they are used along with nonionic surfactants as cleansers and disinfectants in the food and dairy industries. Cationics cannot be used with anionic surfactants: the ions of opposite charge would clump together and precipitate from the solution, and neither surfactant could act as a detergent.

Another kind of quaternary salt with *two* long carbon chains and two smaller groups on nitrogen is used as a fabric softener. An example is dioctadecyldimethylammonium chloride. These compounds are strongly absorbed by the fabric, forming a film on the fabric's surface one molecule thick. The long hydrocarbon chains lubricate the fibers, imparting increased flexibility and softness to the fabric.

We could extend our discussion of household chemicals indefinitely, but all good things must come to an end. Fabric softeners are the "final touch" to this chapter.

$$CH_3(CH_2)_{16}CH_2 - \overset{\overset{\displaystyle CH_3}{|}}{\underset{\underset{\displaystyle CH_3(CH_2)_{16}CH_2}{|}}{N^+}} - CH_3 \quad Cl^-$$

Dioctadecyldimethylammonium
chloride

Problems

1. What are saponins?
2. How do carbonate ions make water basic?
3. What is a soap?
4. What products are formed when tristearin reacts with lye (sodium hyroxide)?
5. List some ingredients found in toilet soaps.
6. What kind of ingredients are used in scouring soaps?
7. How does a potassium soap differ from a sodium soap?
8. Give the structural formula for each of the following.
   a. potassium stearate   b. sodium palmitate
9. How does soap (or detergent) clean a dirty surface?
10. Why does soap fail to work in acidic water?
11. What is hard water?
12. How does hard water affect the action of soap?
13. How does trisodium phosphate soften water? (List two ways.)
14. How does a water-softening tank work?
15. What are some advantages of soap?
16. What are the advantages of synthetic detergents (such as sodium lauryl sulfate)?
17. What is an ABS detergent?
18. Why were ABS detergents banned?
19. What is an LAS detergent?
20. What is a surface-active agent (surfactant)?
21. What is a (detergent) builder?
22. How does sodium tripolyphosphate aid the cleansing action of a soap or detergent?

23. How do zeolites aid the cleansing action of a soap or detergent?
24. List two disadvantages of sodium carbonate as a replacement for phosphates as a builder in detergent formulations.
25. What advantages do zeolites have over carbonates as a builders?
26. What is an optical brightener? How do these compounds work?
27. What is an anionic surfactant?
28. What is an nonionic detergent?
29. What ingredients are used in liquid dishwashing detergents (for hand dishwashing)?
30. How do detergents for automatic dishwashers differ from those used for hand dishwashing?
31. How do cyanurate bleaches work?
32. How does sodium perborate work as a bleach?
33. List some typical ingredients of all-purpose cleansers.
34. List some uses of undiluted household ammonia.
35. List some uses of diluted household ammonia.
36. What safety precautions should be followed when using ammonia?
37. What sort of surfaces should not be cleaned with ammonia?
38. List two properties of baking soda that make it useful as a cleaning agent.
39. List a use of vinegar in cleaning.

40. Why should marble surfaces not be cleaned with vinegar?

41. List some uses of organic solvents in the home.

42. List two useful properties of pine oil cleansers.

43. What are petroleum distillates? What is their function in cleansers.

44. List two disadvantages of petroleum distillates in cleansers.

45. What are the main hazards of using solvents in the home?

46. Should gasoline be used as a cleaning solvent? Why or why not?

47. What are cationic surfactants?

48. Cationic surfactants are not particularly good detergents, yet they are widely used, especially in the food industry. Why?

49. Cationic surfactants often are used along with nonionic surfactants, but seldom are used with anionic surfactants. Why not?

50. Give an example of a compound that acts as a fabric softener.

51. How do fabric softeners work?

52. List the essential ingredients in each of the following household products.
   a. scouring cleaner   b. window cleaner
   c. oven cleaner   d. toilet bowl cleaner
   e. liquid drain cleaner   f. solid drain cleaner
   g. paint remover

Problems 53–57 refer to the following structures.

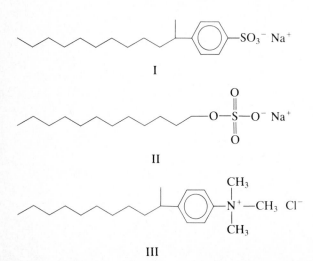

I

II

III

53. Which are biodegradable?

54. Which is an LAS detergent?

55. Which is sodium lauryl sulfate?

56. Which are anionic detergents?

57. Which is a cationic detergent?

58. Give the structures of the products formed in the following reaction.

$$
\begin{array}{l}
\text{CH}_2\text{O}\!-\!\overset{\displaystyle O}{\overset{\|}{\text{C}}}(\text{CH}_2)_{10}\text{CH}_3 \\[4pt]
\text{CH}\!-\!\text{O}\!-\!\overset{\displaystyle O}{\overset{\|}{\text{C}}}(\text{CH}_2)_{10}\text{CH}_3 \;+\; 3\,\text{NaOH} \longrightarrow \\[4pt]
\text{CH}_2\text{O}\!-\!\overset{\displaystyle O}{\overset{\|}{\text{C}}}(\text{CH}_2)_{10}\text{CH}_3
\end{array}
$$

59. Give the structures of the products formed in the following reaction.

$$
\begin{array}{l}
\text{CH}_2\text{O}\!-\!\overset{\displaystyle O}{\overset{\|}{\text{C}}}(\text{CH}_2)_{6}\text{CH}_3 \\[4pt]
\text{CH}\!-\!\text{O}\!-\!\overset{\displaystyle O}{\overset{\|}{\text{C}}}(\text{CH}_2)_{4}\text{CH}_3 \;+\; 3\,\text{NaOH} \longrightarrow \\[4pt]
\text{CH}_2\text{O}\!-\!\overset{\displaystyle O}{\overset{\|}{\text{C}}}(\text{CH}_2)_{8}\text{CH}_3
\end{array}
$$

60. The Clearwater Chemical Company has announced the development of a biodegradable detergent. What tests, if any, should be made before the detergent is marketed?

61. Maxisuds, Inc., has announced that it has found a replacement for phosphate builders in detergents. What tests, if any, should be made before the builder is used in detergent formulations? Should Maxisuds be allowed to market the builder until it is proven harmful? Can a product ever be proven safe?

References and Readings

1. "All-Purpose Cleaners." *Consumer Reports*, October 1982, pp. 519–521.

2. Commoner, Barry. *The Closing Circle*. New York: Bantam Books, 1972. See pp. 151–156.

3. "Dishwashing Liquids." *Consumer Reports*. January 1977, pp. 8–10.

4. Drozd, Joseph C. "An Introduction to Light Duty (Dishwashing) Liquids." *Chemical Times and Trends*, July 1984, pp. 29–32.

5. Fowkes, Frederick M. "Principles of Detergency." *Chemical Times and Trends*, January 1983, pp. 31–33, 53.

6. "Hand and Bath Soaps." *Consumer Reports*, January 1985, pp. 52–55.

7. Hart, J. Roger. "EDTA-Type Chelating Agents in Everyday Consumer Products: Some Medicinal and Personal Care Products." *Journal of Chemical Education*, December 1984, pp. 1060–1061.

8. Layman, Patricia L. "Brisk Detergent Activity Changes Picture for Chemical Suppliers." *Chemical and Engineering News*, 23 January 1984, pp. 17–49.

9. Layman, Patricia L. "Detergents Shift Focus of Zeolites Market." *Chemical and Engineering News*, 27 September 1982, pp. 10–15.

10. Murphy, Lillian S. "Don't Buy From the Shelf When You Can Make It Yourself." *Modern Maturity*, October–November 1982, pp. 17–18.

11. "Soaps and Detergents." New York: Soap and Detergent Association, 1981.

21

Cosmetics

The Chemistry of Charm

Ages ago, primitive people used materials from nature for cleansing, beautifying, and otherwise altering their appearance. Evidence indicates that Egyptians, 7000 years ago, used powdered antimony (Sb) and the green copper ore malachite as eye shadow. Egyptian pharaohs used perfumed hair oils as far back as 3500 B.C. Claudius Galen, a Greek physician of the second century A.D., is said to have invented cold cream. Dandy gentlemen of seventeenth-century Europe used cosmetics lavishly, often to cover the fact that they seldom bathed. Ladies of eighteenth-century Europe whitened their faces with lead carbonate ($PbCO_3$), and many died from lead poisoning.

The use of cosmetics has a long and interesting history, but nothing in the past comes close to the amounts and varieties of cosmetics used by people in the modern industrial world. Each year we spend billions of dollars on everything from hair sprays to toenail polishes, from mouthwashes to foot powders.

What is a cosmetic? The United States Food, Drug, and Cosmetic Act of 1938 defines **cosmetics** as "articles intended to be rubbed, poured, sprinkled or sprayed on, introduced into, or otherwise applied to the human body or any part thereof, for cleansing, beautifying, promoting attractiveness or altering the appearance...." Soap, although obviously used for cleansing, is specifically excluded from coverage by the law. Also excluded are substances that affect the body's structure or functions. Antiperspirants, products that reduce perspiration, are legally classified as drugs. So are antidandruff shampoos. The main difference between drugs and cosmetics is that drugs must be proven "safe and effective" before they are marketed; cosmetics generally do not have to be tested before they're marketed. Most brands of a given type of cosmetic contain the same (or quite similar) active ingredients. Thus, advertising is usually geared to selling sex, smell, and status rather than the actual components.

In this chapter, we take a look at a variety of cosmetics. Emphasis is on those that you or another member of your family are most likely to use.

Figure 21.1 Toothpastes are available under many brand names. The only essential ingredients in any toothpaste are a detergent and an abrasive.

Toothpaste: Soap with Grit and Flavor

After soap (which doesn't count, because the law says it isn't a cosmetic), toothpaste is the most important cosmetic product. The only essential components of toothpaste are a detergent and an abrasive (Figure 21.1). Soap and sodium bicarbonate would do the job quite well but would be rather unpalatable. The ideal abrasive should be hard enough to clean the teeth but not hard enough to damage the tooth enamel. Abrasives frequently used in toothpaste are listed in Table 21.1. Some have been criticized as being too harsh.

Table 21.1 Abrasives Commonly Used in Toothpaste

Name	Chemical Formula
Precipitated calcium carbonate	$CaCO_3$
Insoluble sodium metaphosphate	$(NaPO_3)_n$
Dicalcium phosphate	$CaHPO_4$
Titanium dioxide	TiO_2
Tricalcium phosphate	$Ca_3(PO_4)_2$
Calcium pyrophosphate	$Ca_2P_2O_7$
Hydrated alumina	$Al_2O_3 \cdot nH_2O$
Hydrated silica	$SiO_2 \cdot nH_2O$

Table 21.2 A Typical Recipe for Toothpaste

Ingredient	Function	Amount
Precipitated calcium carbonate	Abrasive	46 g
Castile soap or sodium dodecyl sulfate	Detergent	4 g
Glycerol (glycerin)	Sweetener	20 g
Gum tragacanth or gum cellulose	Thickener	1 g
Oil of peppermint (or peppermint extract)	Flavoring	1 mL
Water	—	28 mL

A typical detergent is sodium dodecyl sulfate (sodium lauryl sulfate).

$$CH_3CH_2CH_2CH_2CH_2CH_2CH_2CH_2CH_2CH_2CH_2CH_2OSO_3^- \ Na^+$$

Any pharmaceutical grade of soap or detergent probably would work satisfactorily. Most toothpastes today are full of minty flavors, colors, aromas, and sweet tastes. Ingredients include sweeteners such as sorbitol, glycerol (gylcerin), and saccharin (Chapter 19); flavors such as peppermint oil and mint; thickeners such as cellulose gum and polyethylene glycols (PEGs); and preservatives such as sodium benzoate (Chapter 19).

A variety of formulas for toothpaste are available (see Reference 2). The formula for a cosmetic is not a chemical formula (Chapter 6) because cosmetics are mixtures rather than pure compounds. Instead, it is more properly called a recipe, that is, a list of materials and directions for preparing a product. Table 21.2 gives a typical recipe for toothpaste.

Tooth decay is caused primarily by bacteria that convert sugars to sticky dextrans or plaque and to acids such as lactic acid ($CH_3CHOHCOOH$). Acids dissolve tooth enamel. Brushing and flossing remove plaque and thus prevent decay. Decay can be minimized by eating sugars only at meals rather than in snacks, and by brushing immediately after eating. Acids such as the phosphoric acid (H_3PO_4) in beer and pop may also dissolve the teeth enamel of people who consume these beverages in large amounts.

$Ca_{10}(PO_4)_6(OH)_2$

Hydroxyapatite

$Ca_{10}(PO_4)_6F_2$

Fluorapatite

Many modern toothpastes contain stannous fluoride (SnF_2), a compound shown to be effective in reducing the incidence of tooth decay. The enamel of teeth is similar to hydroxyapatite in composition. Fluoride from drinking water and toothpaste converts a part of the enamel to fluorapatite. Fluorapatite is a stronger material than hydroxyapatite and is more resistant to decay.

The major cause of tooth loss in adults is gum disease. A toothpaste formulation purported to prevent gum disease appeared in 1982. Its effective ingredients are salt (NaCl), baking soda ($NaHCO_3$), and hydrogen peroxide (H_2O_2).

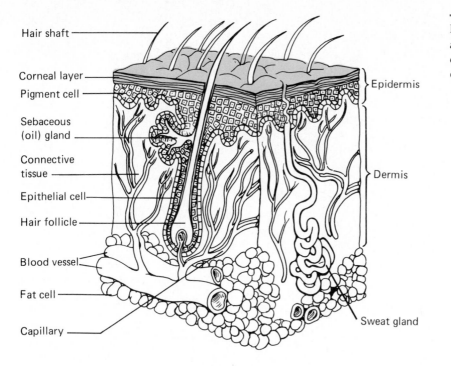

Hair shaft

Corneal layer

Pigment cell

Sebaceous
(oil) gland

Connective
tissue

Epithelial cell

Hair follicle

Blood vessel

Fat cell

Capillary

Epidermis

Dermis

Sweat gland

Figure 21.2 Cross section of an area of skin. Cosmetics affect only the outer corneal layer of dead cells.

Skin Chemistry: Creams and Lotions

Skin is a complex organ that encloses our bodies (Figure 21.2). The outer layer of skin is called the **epidermis**. The epidermis, in turn, is divided into two parts: dead cells on the outside (the corneal layer), and living cells beneath the corneal layer that continually replace corneal cells that are sloughed off. Cosmetics are applied to the dead cells of the corneal layer. About $20 billion a year is spent in the United States on various preparations applied to the skin.

The corneal layer is composed mainly of a tough, fibrous protein called **keratin**. Keratin has a moisture content of about 10%. Below 10% moisture, the skin is dry and flaky. Above 10%, conditions are ideal for the growth of harmful microorganisms. Skin is protected from loss of moisture by **sebum**, an oily secretion of the sebaceous glands. One interesting property of skin is that it is insoluble in water (otherwise we would dissolve in the shower). It is, however, slightly permeable to water and light.

Exposure to sun and wind may leave the skin dry and scaly. Washing too often also removes natural skin oils. A variety of lotions and creams are available to treat the skin. A **lotion** is an emulsion

(Chapter 18) of tiny oil droplets dispersed in water. A **cream** is the opposite; tiny water droplets are dispersed in oil. The essential ingredient of each is a fatty or oily substance that forms a protective film over the skin. Typical ingredients are mineral oil and petroleum jelly; sometimes both are used in the same preparation. These are mixtures of alkanes obtained from petroleum. Petroleum jelly is a higher-boiling fraction than mineral oil. The former is a semisolid; the latter, a viscous liquid. Other ingredients include natural fats and oils, perfumes, waxes, water, and emulsifiers (compounds that keep the oily portions from separating from the water)). Natural materials used on the skin include lanolin, a fat obtained from sheep's wool, and olive oil. Often, beeswax is added to harden the product.

Some creams have been formulated with hormones, queen bee jelly, and other strange ingredients. None of these has been found to confer any particular benefit. Creams and lotions protect the skin by providing a protective coating and by softening it, much as plasticizers soften plastics (Chapter 12). Such skin softeners are called **emollients**. You could just as well use petroleum jelly (one trade name is Vaseline) or a good grade of white mineral oil (sometimes called baby oil) as the fancy creams.

It may seem strange that gasoline, a mixture of alkanes, dries out the skin, and that the higher alkanes in mineral oil and petroleum jelly soften it. Keep in mind, though, that gasoline is a thin, free-flowing liquid. It dissolves the natural skin oils and carries them away. The higher alkanes are viscous. They stay right on the skin and serve as emollients.

Skin **moisturizers**? The term is undefined. Components called moisturizers are usually petroleum jelly, a mineral oil, or a similar substance. They may serve to keep the skin softer, in part, by physically preventing the loss of moisture through the protective film.

Skin is damaged by the ultraviolet radiation from the sun. The shorter wavelengths are more energetic and are especially harmful. **Sunscreen lotions** are used to block this radiation while letting through the less energetic long-wave ultraviolet rays that promote tanning. The active ingredient in many of these preparations is *para*-aminobenzoic acid. Various concentrations of this compound in the lotion provide the **skin protection factor** (SPF) ratings. SPF values vary from 2, which offers little protection, up to 35, which allows people to stay in the sunlight without burning up to 35 times as long as they could with unprotected skin.

The tanning of skin involves the production of the pigment melanin. This dark material protects the deeper layers of skin from burning. Melanin is made from the amino acid tyrosine. Its production is stimulated by the long-wave ultraviolet rays.

Excessive exposure to the short-wave rays can cause premature aging of the skin and can lead to skin cancer. Authorities predict that an epidemic of skin cancer will result in about 20 years from the current popularity of suntanned skin.

Cigarette smoking also leads to premature aging of the skin. Nicotine causes the constriction of the tiny blood vessels that feed the skin. Repeated constrictions several times a day over the years cause the skin to lose its elasticity and become wrinkled. If you want nice skin, don't smoke—and avoid excessive exposure to sun and water.

Lipsticks: Castor Oil and Color

Lipstick is quite similar to skin creams in composition. It is made of an oil and a wax. To keep the lipstick firm, a higher proportion of wax is used than in creams. Dyes and pigments provide color. The oil is frequently castor oil. Waxes often employed are beeswax, carnauba, and candelilla. Perfumes are added to cover up the unpleasant fatty odor of the oil. Antioxidants are also employed to retard rancidity. Bromo acid dyes such as tetrabromofluorescein, a bluish red compound, are responsible for the color of most modern lipsticks. These compounds often are adhered to metal ions to form colored complexes called **lakes**.

Because it has little in the way of protective oils, the skin of the lips is easily dried out, leading to chapped lips. With or without coloring, lipsticks do protect the lips from drying.

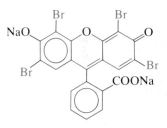

Tetrabromofluorescein

Chemistry: The Eyes Have It

A variety of chemicals are used to decorate the eyes. **Mascara** is used to darken eyelashes. It is composed of a base of soap, oils, fats, and waxes. Mascara is colored brown by iron oxide pigments, black by carbon (lampblack), green by chromium (III) oxide (Cr_2O_3), or blue by ultramarine (a silicate that contains some sulfide ions). A typical composition is 40% wax, 50% soap, 5% lanolin, and 5% coloring matter. Eyebrow pencils, which have about the same base, are colored by lampblack or iron oxides.

Eye shadow has a base of petroleum jelly with the usual fats, oils, and waxes. It is colored by dyes or made white by zinc oxide (ZnO) or titanium dioxide (TiO_2) pigments. A typical composition is 60% petroleum jelly, 10% fats and waxes, 6% lanolin, with the remainder dyes or pigments or both.

Some people have allergic reactions to ingredients in eye makeup. A more serious problem is eye infections from bacterial contamination.

Perfumes, Colognes, and Aftershaves

No one wants to smell bad. Many people like to give off the pleasant aroma of a fruit or a flower, perhaps moderated a bit to avoid overwhelming a neighbor's nose. Perfumes are among the most ancient and the most widely used of the cosmetics. Their chemistry, however, is exceedingly complex.

Originally, perfumes were extracted from natural sources. Nowadays, chemists have identified many of the components and synthesized them in the laboratory. The best perfumes, perhaps, are still made from natural materials, because chemists have so far been unable to identify all the many important, but minor, ingredients.

A good perfume may have a hundred or more constituents. Often the components are divided into three categories, called **notes**, based on differences in volatility. The most volatile fraction (that which vaporizes most readily) is called the **top note**. This fraction, made up of relatively small molecules, is responsible for the odor when a perfume is first applied. The **middle note** is intermediate in volatility. It is responsible for the lingering aroma after most of the top-note compounds have vaporized. The **end-note** fraction has low volatility and is made up of compounds with large molecules.

Several compounds with flowery or fruity odors are synthesized in large quantities for use in perfumes. Some of these, with their approximate odors, are listed in Table 21.3. Odors vary with dilution. A

Table 21.3   Compounds with Flowery and Fruity Odors Used in Perfumes

Name	Structure	Odor
Citral	CH_3C=$CHCH_2CH_2C$=CHC—H (with CH_3, CH_3, O groups)	Lemon
Irone	(cyclohexene ring with CH_3, CH_3, CH_3, CH_3 groups and CH=CH—C=O side chain)	Violet
Jasmone	(cyclopentenone ring with —CH_2CH=$CHCH_2CH_3$ and CH_3 groups, O)	Jasmine
Geraniol	CH_3C=$CHCH_2CH_2C$=$CHCH_2OH$ (with CH_3, CH_3 groups)	Rose

Table 21.4 Unpleasant Compounds Used to Fix Delicate Odors in Perfumes

Compound	Structure	Natural Source
Civetone	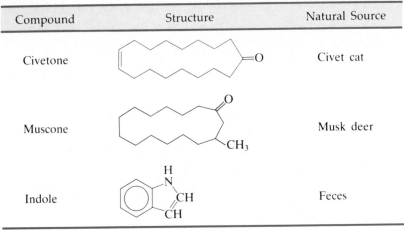	Civet cat
Muscone		Musk deer
Indole		Feces

concentrated solution (lots of compound in a small amount of water or other solvent) may be unpleasant, yet a dilute solution (a small amount in lots of solvent) of the same compound may have a pleasant aroma. Most of the compounds exist in several isomeric forms.

Many flowery or fruity odors are sickeningly sweet, even in dilute solutions. Compounds such as the musks often are added to moderate the odor. Musks and similar compounds have extremely disagreeable odors when concentrated, but often are pleasant at extreme dilution. Several of these compounds are described in Table 21.4.

The civet cat, from which the compound civetone is obtained, is a skunklike animal of eastern Africa. Its secretion, like that of the skunk, is a defensive weapon. The secretion from musk deer is thought to be a pheromone (Chapter 17) used as a sex attractant. Indole and its homolog skatole are principally responsible for the characteristic odor of human feces. These compounds (and others) are now available in synthetic forms.

Perhaps the ultimate in perfumes is Andron by Jovan. Studies hint that α-androstenol may act as a sex attractant betwen male and female humans. α-Androstenol is a steroid (Chapter 22) that occurs naturally in human hair and urine. The evidence is far from conclusive, but Jovan puts a minute amount of the compound in Andron. Jovan Claims that α-androstenol is active at concentrations as low as 6 ppm. Competitors call the whole thing an advertising ploy.

It is of some interest that sex attractants have been identified in primates. Indeed, the same set of pheromones (a mixture of organic acids) that operate as sex attractants in rhesus monkeys has been isolated from the vaginal secretions of human females. It is doubtful, however, that sex attractants have much influence on human behavior.

Skatole

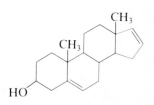

α-Androstenol

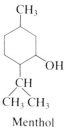

Menthol

$Al_2(OH)_5Cl \cdot 2H_2O$

Aluminum chlorohydrate

We probably have overriding cultural constraints. Anyway, we seem to prefer the sex attractant of the musk deer.

A perfume usually consists of 10 to 25% odorous compounds and fixatives. The remainder is ethyl alcohol, which serves as a solvent. **Colognes** are diluted perfumes. They often contain only 1 or 2% perfume essence. Thus, colognes are about 10% as strong as perfumes. Dilution can be made with ethyl alcohol alone or with alcohol–water mixtures.

Aftershave lotions are similar to colognes. Most are about 50 to 70% ethanol with the remainder water, a perfume, and food coloring. Some have menthol added for a cooling effect on the skin; others add an emollient of some sort to soothe chapped skin.

Perfumes are the source of many of the allergic reactions associated with cosmetics and other consumer products. **Hypoallergenic cosmetics** are those that purport to cause fewer allergic reactions than regular products. The term has no legal meaning, but most "hypoallergenic" cosmetics leave out the perfume.

Deodorants and Antiperspirants

Deodorants are formulations of perfumes designed to mask body odor. **Antiperspirants** stop or retard perspiration. Both serve cosmetic functions, but the antiperspirants, which work by modifying a body function, are technically drugs. The active ingredient in nearly all antiperspirants is aluminum chlorohydrate. This compound is an **astringent**: it acts by constricting the openings of sweat glands, thus reducing the amount of perspiration that escapes. Astringents can be dissolved in alcohol for use in aerosol sprays, and they can be formulated with other ingredients as creams and lotions.

Sweating is a natural, healthy function of the body. Stopping it on a routine basis is probably not very wise. Body odors arise largely from the hydrolysis of sebum, the natural body oil secreted by sebaceous glands in the skin. The breakdown of sebum forms foul-smelling fatty acids such as butyric acid. Other compounds, including amines and organic compounds of sulfur, also are formed on the skin. Healthy people who bathe and change clothes daily have little need for deodorants or antiperspirants.

Some Hairy Chemistry

Like skin, hair is composed of the fibrous protein keratin. Recall (from Chapter 18) that proteins are chains of amino acids. The long protein molecules in keratin are held together by four types of forces—

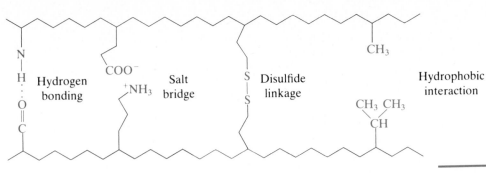

Figure 21.3 The four types of forces holding protein chains together.

hydrogen bonds, ionic forces called salt bridges, covalent disulfide linkages, and dispersion forces. Dispersion forces, the only forces operating between nonpolar side chains, are sometimes called hydrophobic interactions. The various forces are illustrated in Figure 21.3.

The hydrogen bonds of greatest importance are those that involve an interaction between the atoms of two peptide bonds. Thus, the carbonyl oxygen of a peptide link on one chain can form a hydrogen bond to an amide N—H on a neighboring chain.

Some amino acids, such as aspartic acid and glutamic acid, have acidic groups in their side chains (see Table 18.3). Others, such as histidine, lysine, and arginine, have basic groups in their side chains. If the amino acids containing these groups are incorporated into a protein, a proton transfer can occur when the groups come into close contact.

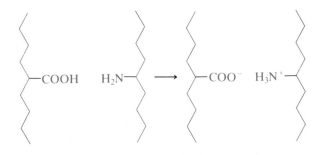

Proton transfer results in opposite charges, which then attract one another. These interactions can occur between chains and are called **salt bridges.**

Disulfide linkages are formed when two cysteine units (whether on the same chain or on two different chains) are oxidized to form a single cystine unit.

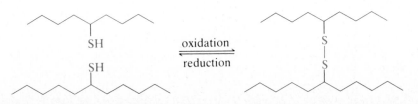

The disulfide bond is a covalent bond that is much stronger than a hydrogen bond. The larger number of hydrogen bonds does compensate somewhat for individual weakness, however. Still weaker are the hydrophobic interactions between nonpolar side chains. We do not consider these weak interactions further here.

Hair Care and Conditioning

The keratin of hair has five or six times as many disulfide linkages as the keratin of skin. When hair is washed, the keratin absorbs water and is softened and made more stretchable. The water disrupts hydrogen bonds and some of the salt bridges. Acids and bases are particularly disruptive to salt bridges, making control of pH important in hair care. The maximum number of salt bridges exist at a pH of 4.1.

The visible portion of hair is dead; only the root is alive. The hair shaft is lubricated by sebum. Washing the hair removes this oil and dirt adhering to it.

Before World War II, the cleansing agent in shampoos was soap. Soap-based shampoos worked well in soft water but left a dulling film on the hair in hard water. Often, people removed the film by using a rinse containing vinegar or lemon juice. Such a rinse usually is not needed with today's products.

Modern shampoos use a synthetic detergent as a cleansing agent. In shampoos for adults, the detergent is often an anionic type, such as sodium dodecyl sulfate. (Yes, that is the same detergent used in many toothpastes.) For shampoos used on babies and children, the detergent is often an amphoteric type that is less irritating to the eyes. Amphoteric detergents react with both acids and bases. The molecule

$$CH_3CH_2CH_2CH_2CH_2CH_2CH_2CH_2CH_2CH_2CH_2CH_2CH_2CH_2CH_2CH_2\overset{\overset{\displaystyle H}{|}}{\underset{\underset{\displaystyle H}{|}}{N}}—CH_2\overset{\overset{\displaystyle O}{\|}}{C}—O^-$$

is an example. In acidic solution, it can accept a proton on the negatively charged oxygen. In basic solution, it can give up one of the protons from nitrogen.

The only essential ingredient in shampoo is a detergent of some sort (Figure 21.4). What, then, is all the advertising about? You can buy shampoos that are fruit and herb flavored, protein enriched, pH balanced, and made for oily and dry hair. Let's have a look at some of the gimmicks.

Hair is protein with acidic and basic groups on the protein chain. It stands to reason that the acidity or basicity of a shampoo would affect hair. Hair and skin are slightly acidic. Highly basic (high-pH) or strongly acidic (low-pH) shampoos would damage the hair. More important, such products would irritate the skin and eyes. Most shampoos, however, have pH values between 5 and 8 (Reference 3), a difference too slight to affect the hair or scalp in any significant way. Shampoos that irritate the eyes usually do so because of other ingredients, not because of a too-high or a too-low pH.

Because hair is protein, protein in shampoos does give the hair more body—or so the advertisements claim. Some white glues are protein, too. Protein in shampoos literally glues split ends together and coats the hair, making it thicker. If that is what you want, then protein shampoo is for you.

Shampoos for oily or dry hair differ in the relative amounts of detergent in given volumes. Those for oily hair presumably are more concentrated. There seems to be little standardization in formulations from one brand to another, however (Reference 11). Shampoos for oily or dry hair seem to be mainly an advertising gimmick.

How about all those flavors and fragrances? Ample evidence indicates that such "natural" ingredients as milk, honey, strawberrries, herbs, cucumbers, and lemons add nothing to the usefulness of shampoos or other cosmetics. Why are they there? Smells sell, and there is an appeal to those taken by the back-to-nature movement. There is one hazard to the use of such fragrances: bees, mosquitoes, and other insects like fruit and flower odors, too. Using such products before going on a picnic or a hike could lead to a bee in your bonnet.

Figure 21.4 Shampoos are available in many forms and colors and under many brand names. The only essential ingredient in any shampoo is a detergent.

More on Hair: Color

The color of hair and skin is determined by the relative amounts of two pigments: **melanin**, a brownish black pigment, and **phaeomelanin**, a red-brown pigment that colors the hair and skin of redheads. Brunettes have lots of melanin. Blondes have little of either pigment. Brunettes who would like to become blondes can do so by oxidizing the colored pigments in their hair to colorless products. Hydrogen peroxide (Chapter 8) is the oxidizing agent most frequently employed to bleach hair.

Hair dyeing is a good deal more complicated than bleaching. The color may be temporary; water-soluble dyes that can be washed out the next time the hair is washed are sometimes used. More permanent dyes penetrate the hair and remain there. These dyes often are used in the form of a water-soluble, often colorless, precursor that soaks into the hair. The chemical then is oxidized by hydrogen peroxide to a colored compound.

Permanent dyes often are derivatives of an aromatic amine called

para-Phenylenediamine

para-phenylenediamine. Variations in color can be obtained by placing a variety of substituents on this molecule, *para*-Phenylenediamine itself produces a black color. The derivative *para*-aminodiphenylaminesulfonic acid is used in blonde formulations. Intermediate colors can be obtained by the use of other derivatives. One such derivative, *para*-

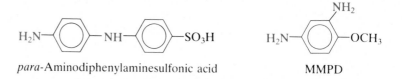

para-Aminodiphenylaminesulfonic acid   MMPD

methoxy-*meta*-phenylenediamine (MMPD), and its sulfate salt have been shown to be carcinogenic when fed to rats and mice. The hazard to those who use these dyes to color their hair is not yet known. It is interesting to note that one substitute for MMPD was its homolog, *para*-ethoxy-*meta*-phenylenediamine (EMPD). Screening revealed that EMPD causes mutations in bacteria. Such mutagens often are also carcinogens. The FDA cannot ban a cosmetic without proof of harm, but testing for carcinogenicity would cost at least $500 000 and take two years. We consider tests of this sort in Chapter 25.

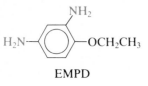

EMPD

The chemistry of colored oxidation products is quite complex. The products probably include quinones and nitro compounds, among others. These are well-known products of the oxidation of aromatic amines. It should be noted that even permanent dyes affect only the dead outer portion of the hair shaft. New hair, as it grows from the scalp has its natural color.

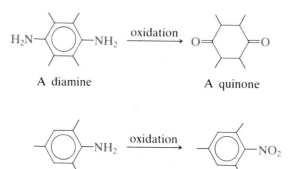

A diamine   A quinone

An amine   A nitro compound

Hair treatments, such as Grecian Formula, that develop color gradually use a rather simple chemistry. A solution containing colorless lead acetate [$Pb(CH_3COO)_2$] is rubbed on the hair. As it penetrates the hair shaft, the Pb^{2+} ions react with sulfur atoms in the hair to form black lead sulfide (PbS). Repeated applications lead to darker colors as more lead sulfide is formed.

Some Chemistry That Will Curl Your Hair

The chemistry of curly hair is interesting. Hair is protein, and adjacent protein chains are held together by disulfide linkages. To put a permanent wave in the hair, you use a lotion containing a reducing agent such as thioglycolic acid. This wave lotion ruptures the disulfide linkages (Figure 21.5), allowing the protein chains to be pulled apart. Then, the hair is set on curlers or rollers and treated with a mild oxidizing agent such as hydrogen peroxide. Disulfide linkages are formed in new positions to give shape to the hair.

The same chemical process can be used to straighten naturally curly hair. The change in hairstyle depends only on how you arrange the hair after the disulfide bonds have been reduced and before the linkages have been restored. As with permanent dyes, permanent curls grow out as new hair is formed.

Hair can be held in place by using **resins**, solid or semisolid organic materials that form a sticky film on the hair. A common resin used on hair is polyvinylpyrrolidone (PVP).

PVP is dissolved in solvent and sprayed on the hair, where the solvent evaporates. Aerosol sprays using chlorofluorocarbon propellants (Chapter 15) were once widely used. Other propellants are now employed.

Holding resins are also available as mousses. A **mousse** is simply a foam or froth. The active ingredients, like those of the hair sprays, are resins such as PVP. Coloring agents and conditioners are also available as mousses.

Cosmetics: Economics and Advertising

Most cosmetics are formulated from inexpensive ingredients. Many highly advertised cosmetics are sold at high prices. Are they worth it? That's a judgment that lies outside of the realm of chemistry. If a product makes you look better or feel better, what value can be placed on it? Only you can decide.

No attempt has been made here to tell you everything there is to know about cosmetics. Many volumes that discuss these interesting chemicals in much more detail are available in libraries. We hope,

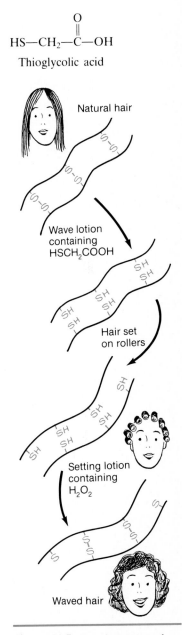

Figure 21.5 The chemistry of the "permanent" waving of hair.

however, that the knowledge you gained here, coupled with that required on cosmetic labels. will make you a better informed consumer. You don't have to pay a lot for extra ingredients that contribute little or nothing to the function of a cosmetic.

Problems

1. What is the legal definition of a cosmetic?
2. Is soap a cosmetic?
3. Why are some items that we use as cosmetics legally classified as drugs?
4. Must cosmetics be proven "safe and effective" before they are marketed?
5. What are the essential ingredients in a toothpaste?
6. What is the principal material of tooth enamel?
7. How do fluorides strengthen tooth enamel?
8. What is the principal cause of tooth decay?
9. What are the best ways to prevent tooth decay?
10. What is the principal material of the corneal layer of skin?
11. What is sebum?
12. What is an emollient?
13. What is the ideal moisture content of skin?
14. What are the principal components of creams and lotions?
15. What is a skin moisturizer?
16. How does lipstick differ from a skin cream?
17. What materials are used to color lipsticks?
18. What is a perfume?
19. What are the three fractions of a perfume? How do these fractions differ at the molecular level?
20. What is a musk? Why are musks added to perfumes?
21. What are colognes?
22. What are the principal ingredients of an aftershave lotion?
23. Why is menthol added to some aftershave lotions?
24. What is a deodorant?
25. What is an antiperspirant?
26. What is an astringent?
27. What chemical substance is the sole active ingredient of most antiperspirants?
28. What is the principal material of hair? How does it differ from the principal material of skin?

29. What are the four types of forces that hold protein molecules to one another in hair?
30. What is the effect of water on hair?
31. What is the only essential ingredient of shampoos?
32. What type of detergent is used in baby shampoos? Why?
33. What chemical substances determine the color of hair and skin?
34. What chemical compound is used most often to bleach hair?
35. What kind of chemical reaction is involved in bleaching hair?
36. What is the difference between a temporary and a permanent hair dye? Why is a permanent dye not really permanent?
37. From what kind of chemical compound are permanent hair dyes often derived?
38. What kind of chemical reagent is used to break disulfide linkages in hair?
39. What kind of chemical reagent is used to restore disulfide linkages in hair?
40. What are resins?
41. How do the hair sprays that hold hair in place work?
42. What is a mousse? List some mousses used in cosmetics.

Problems 43–46 refer to the structures at the top of page 485.
43. Which are synthetic detergents?
44. Which are soaps?
45. Which is an amphoteric detergent?
46. Which is a principal ingredient in baby shampoos?
47. Read the labels of five brands of toothpaste. List the ingredients of each. If you can, classify each ingredient according to its function—that is, as a detergent, an abrasive, and so on. You may find a reference book such as *The Merck Index* (Reference 14) to be helpful.

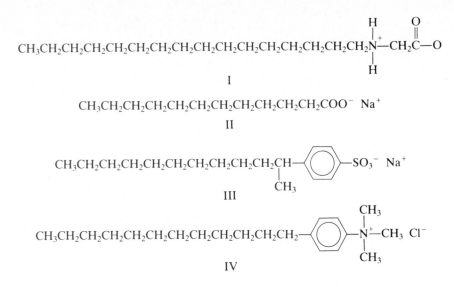

$$CH_3CH_2CH_2CH_2CH_2CH_2CH_2CH_2CH_2CH_2CH_2CH_2CH_2CH_2CH_2CH_2\overset{\overset{\displaystyle H}{|}}{\underset{\underset{\displaystyle H}{|}}{N}}{}^{+}\!\!-CH_2\overset{\overset{\displaystyle O}{\|}}{C}-O$$

I

$$CH_3CH_2CH_2CH_2CH_2CH_2CH_2CH_2CH_2CH_2CH_2COO^-\ Na^+$$

II

$$CH_3CH_2CH_2CH_2CH_2CH_2CH_2CH_2CH_2CH_2CH_2\underset{\underset{\displaystyle CH_3}{|}}{CH}-\!\!\bigcirc\!\!-SO_3^-\ Na^+$$

III

$$CH_3CH_2CH_2CH_2CH_2CH_2CH_2CH_2CH_2CH_2CH_2CH_2-\!\!\bigcirc\!\!-\overset{\overset{\displaystyle CH_3}{|}}{\underset{\underset{\displaystyle CH_3}{|}}{N}}{}^{+}\!\!-CH_3\ Cl^-$$

IV

48. Read the labels on five brands of lotions or creams (those designed for use on the skin, whether face, hands, or body). List the ingredients in each brand. If you can, tell the function of each ingredient.

49. Read the labels on five brands of lipstick. List the ingredients in each brand. If you can, tell the function of each ingredient.

50. Read the labels on five brands of aftershave lotion. What are the two major ingredients in each brand?

What other ingredients are present? What is the function of each?

51. Read the labels on five brands of antiperspirants. List the ingredients in each brand. What is the function of each ingredient?

52. Read the labels on five brands of shampoos. List the ingredients in each brand. Which ingredients are detergents? If you can, tell the function of each nondetergent ingredient.

References and Readings

1. "Chemistry of Cosmetics." *Journal of Chemical Education*, December 1978, pp. 802–803.
2. Goldschmiedt, Henry. *Practical Formulas for Hobby and Profit*. New York: Chemical Publishing Company, 1973. Chapter 3 gives recipes for cosmetics.
3. Griffin, John J.; Robert F. Corcoran, and Kenn K. Akana. "The pH of Hair Shampoos." *Journal of Chemical Education*, September 1977, pp. 553–554.
4. "Hand Lotions and Creams." *Consumer Reports*, August 1977, pp. 448–451.
5. Hopkins, Harold. "What Does Shampoo Do?" *FDA Consumer*, June 1984, pp. 8–11.
6. Lamberg, Lynne. "To Hold Back the Sun." *American Health*, June 1984, pp. 48–53.
7. Layman, Patricia L. "Cosmetics." *Chemical and Engineering News*, 29 April 1985, pp. 19–46.
8. "Mouthwashes." *Consumer Reports*, March 1984, pp. 143–146.
9. Niemark, Jill. "Dew Gooders." *American Health*, October 1984, pp. 22–23. Discusses moisturizers.
10. Rinzler, Carol Ann. "The New and Improved Chemistry of Cosmetics." *Science 82*, April 1982, pp. 54–61.
11. "Shampoos." *Changing Times*, April 1979, pp. 19–20.
12. Traven, Beatrice. *Here's Egg on Your Face*. New York: Pocket Books, 1971. Tells you how to make cosmetics with ingredients from the grocery store and drugstore.
13. "Toothpastes." *Consumer Reports*, March 1984, pp. 138–141.
14. Windholz, Martha (Ed.). *The Merck Index*, 10th ed. Rahway, NJ: Merck, 1983.

22

Chemotherapy
From Colds to Cancer

We do not only practice chemistry in our homes (Chapter 20) and on our skin and hair (Chapter 21), but we also use our bodies as chemical laboratories. There are over 300 000 over-the-counter drugs and more than 25 000 prescription medicines available in the United States. The average person in the United States consumes 5 kg of medicines in a lifetime.

Ours is not the first society to use drugs. The use of chemicals in attempts to relieve pain and cure illnesses dates to prehistoric times. In some societies, the use of drugs was excessive but the variety of drugs was limited. Most societies used ethyl alcohol, a depressant. The use of marijuana (*Cannabis sativa*) goes back at least 3000 B.C. The narcotic effect of the opium poppy was known to the Greeks in the third century B.C. (Figure 22.1). The Indians of the Andes Mountains have long chewed the leaves of the coca plant (*Erythroxylon coca*) for the stimulating effect of the cocaine and the other alkaloids in its leaves.

A scientific rationale for the use of chemical substances to treat diseases was developed early in this century. In 1904, Paul Ehrlich, a German chemist, realized that certain chemicals were more toxic to disease organisms than to human cells. These chemicals could be used to control or cure infectious diseases. Ehrlich coined the term **chemotherapy**, a shorter version of the term "chemical therapy." Ehrlich found that certain dyes used to stain bacteria to make them more visible under a microscope also could be used to kill the bacteria. He used dyes against the organism that causes African sleeping sickness and synthesized an arsenic compound effective against the organism that causes syphilis. Ehrlich was awarded the Nobel Prize in 1908.

The use of drugs is not new. What is new is the vast array of drugs available to people in modern society. Many of these drugs are available without a prescription from the shelves of supermarkets and drugstores. They are "pushed" on television with the latest Madison Avenue techniques. Illegal drugs in great variety are readily available in most places. Even with prescription drugs, prepared and tested under the supervision of the FDA, there are serious problems. Physicians prescribe drugs that had not yet been discovered when they were in medical school. New drugs are being discovered, tested, and placed on the market each year.

The consumer is faced not only with a nearly infinite variety of products but also with confusing and even contradictory advertising. Is there nothing the buyer can do to sort fact from fiction? It isn't as hopeless as it might seem. There may well be 300 000 over-the-counter drugs, but they are formulated from only about 250 significant active ingredients. Armed with a few chemical principles, a willingness to read labels, and a reference work such as *The Merck Index* (Reference 20), consumers can make intelligent decisions about the drugs they use.

In this chapter, we discuss some of the drugs used to relieve mild pain or discomfort and to treat diseases that affect the body. Drugs that act primarily on the mind—as depressants, stimulants, or "mind-benders"—are left for Chapter 23.

Figure 22.1 This ancient Greek coin shows an opium poppy capsule. Some cults in ancient Greece worshiped opium. [Courtesy of the Bureau of Narcotics and Dangerous Drugs, Washington, DC.]

Pain Relievers: Structure and Properties

Freedom from pain has long been a human goal. Alcohol, opium, cocaine, and Indian hemp (marijuana) were used as medicines for pain relief in some early societies. The first successful synthetic pain relievers were derivatives of salicyclic acid (Figure 22.2). Salicylic acid was first isolated from willow bark in 1860, although an English clergyman,

Figure 22.2 Salicylic acid and some of its derivatives.

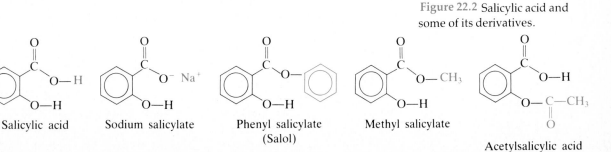

Salicylic acid · Sodium salicylate · Phenyl salicylate (Salol) · Methyl salicylate · Acetylsalicylic acid

Edward Stone, had reported to the Royal Society as early as 1763 that an extract of willow bark was useful in reducing fever. Salicylic acid is itself a good **analgesic** (pain reliever) and **antipyretic** (fever reducer), but it is sour and irritating when taken orally. Chemists sought to modify the structure of the molecule to remove this undesirable property while retaining (or even improving) the desirable properties.

The first modification was simple neutralization of the acid (Chapter 7). The salt sodium salicylate was first used in 1875. It was less unpleasant to swallow, but it proved to be highly irritating to the lining of the stomach. Phenyl salicylate (salol) was introduced in 1886. It passed unchanged through the stomach. In the small intestine, it was hydrolyzed to the desired salicylic acid, but phenol, which is rather toxic, also was formed.

Acetylsalicylic acid, which the German Bayer Company assigned the trade name Aspirin, was first introduced in 1899. It soon became the best-selling drug in the world. Over 55 billion aspirin tablets (under many trade names) are now consumed annually in North America. Another derivative of salicylic acid, methyl salicylate (oil of wintergreen), is used as a flavoring agent. It is also used in rubbing compounds. It causes a mild burning sensation when applied to the skin, thus serving as a counterirritant for sore muscles (Chapter 24).

These derivatives are an excellent example of how variations in structure lead to variations in properties. Some of the new properties are desirable, some are undesirable. We must accept the bad to take advantage of the good. In the meantime, chemists continue to search for more effective pain relievers, with the goal of minimizing undesirable side effects and maximizing beneficial properties.

The Best Aspirin Ever Made

Since aspirin is the most widely used drug in the world, let's take a close look at it. Aspirin is a chemical compound. Like other compounds, its properties are invariant. Each aspirin *tablet* usually contains 325 mg of the compound, held together with an inert binder (usually starch). Aspirin has been tested extensively. The conclusions of impartial studies are invariably the same: the only significant difference between brands is price. In fact, nearly all of the 14 million kg of aspirin produced annually in the United States is made by two companies, Dow Chemical and Monsanto. Only two drug companies, Sterling Drug (which makes Bayer) and Norwich-Eaton, make their own product.

"Buffered" aspirin contains antacids but is not truly buffered. A **buffer** consumes either acid or base and keeps the pH of a solution essentially constant. The antacids consume acid only; they do not buffer. The experts who evaluated Bufferin, the most highly advertised of these products, for the FDA, found that there was no basis for the advertising claim that Bufferin is "twice as fast as aspirin." Neither did they find evidence that Bufferin "helps prevent the stomach upset often caused by aspirin." Some people experience mild stomach irritation when they take aspirin on an empty stomach. Eating a little food first or drinking a full glass of water with aspirin is just as effective as taking "buffered" tablets.

"Extra-strength" formulations simply have 500 mg of aspirin rather than the usual 325 mg. They have no other active ingredients. Simple arithmetic tells us that three plain aspirin tablets are equal to two extra-strength tablets in dosage and are usually much lower in price.

Aspirin relieves minor aches and pains, reduces fever, and suppresses inflammation. It works by reducing the number of pain impulses to the brain. It doesn't cure whatever is causing the pain; it merely kills the messenger. Aspirin inhibits the production of **prostaglandins**, compounds that are involved in increased blood pressure and in the contractions of smooth muscle. Inflammation results from an overproduction of prostaglandin derivatives. The **anti-inflammatory** action of aspirin results from its inhibitory function.

Aspirin also acts as an **anticoagulant**; it inhibits the clotting of blood. Studies indicate that there is some intestinal bleeding every time aspirin is ingested. The blood loss is usually minor (0.5 to 2.0 mL per two-tablet dose), but it may be substantial in some cases. Aspirin should not be used by people facing surgery, childbirth, or other hazards involving the possible loss of blood (nonuse should start a week before the hazard). On the other hand, small doses seem to lower the risk of coronary heart attack and stroke, presumaby by the same anticoagulant action that causes bleeding in the stomach.

Aspirin is not effective for severe pain—for example, pain from a migraine headache. Prolonged use of aspirin, as for arthritic pain, can lead to gastrointestinal disorders. Like all drugs, aspirin is somewhat toxic. It is the drug most often involved in the accidental poisoning of children. The toxicities of aspirin and other drugs are listed in Table 22.1.

Still another hazard associated with aspirin use is allergic reaction. In some people, an allergy to aspirin can cause skin rashes, asthmatic attacks, and even loss of consciousness. Some doctors claim that the allergic reaction may be delayed 3 to 5 hours, so the victim may not associate the reaction with aspirin. Susceptible individuals must be careful to avoid aspirin by itself or in any combination with other drugs.

Table 22.1 Toxicities of Chemicals Presently or Formerly Used in Over-the-Counter Drugs*

Chemical Compound	LD_{50}
Acetaminophen	338†
Acetanilide	800
Aspirin	1500
Caffeine	355
Diphenhydramine	500
Ibuprofen	1050
Methyl salicylate	887
Phenacetin	1.65

* LD_{50} values are for oral administration of the drug in rats in milligrams per kilogram of body weight (unless otherwise noted).
† Orally in mice.
SOURCE: Martha Windholz (Ed.), *The Merck Index*, 10th ed. Rahway, NJ: Merck and Co., 1983.

Aspirin Substitutes

Some people who are allergic to aspirin may safely take a substitute compound called acetaminophen. This compound gives relief of pain and reduction of fever comparable to the action of aspirin. Unlike

CH_3CHCH_2—⟨◯⟩—$CHCOH$ with CH_3 below the first group, CH_3 below CH, and $\overset{O}{\overset{\|}{}}$ above C

Ibuprofen

aspirin, however, it is not effective against inflammation, and it doesn't induce bleeding. Thus, it is of limited use to people with arthritis, but it is often used to relieve the pain that follows minor surgery. The fact that acetaminophen does not promote bleeding probably accounts for its more frequent use than aspirin in hospitals. Acetaminophen usually costs more than aspirin, especially in the form of highly advertised brands such as Tylenol, Panadol, and Anacin-3. Regular acetaminophen tablets are 325 mg; "extra-strength" forms are 500 mg. Overuse of acetaminophen has been linked to liver and kidney damage.

Another analgesic is ibuprofen, approved by the U.S. FDA in 1984 for nonprescription use. Ibuprofen has long been available by prescription under the trade name Motrin in tablets of 300, 400, and 600 mg. The FDA action led to the introduction of ibuprofen under the trade names Advil and Nuprin, each of which consists of 200 mg tablets. Ibuprofen is anti-inflammatory and antipyretic. Unfortunately, people who are allergic to aspirin also are sensitive to ibuprofen, and the drug is considerably more expensive than aspirin.

Combination Pain Relievers

Much of what we spend on analgesics is spent on combinations of aspirin and other drugs. Is our money well spent? What are those other drugs? Do they really add anything to the effectiveness of the medication? Must the consumer be at the mercy of misleading advertising? These are questions that can be answered rather simply. Let's look first at the chemical compounds in some of the more familiar "combination pain relievers."

For many years, the most familiar combination was aspirin, phenacetin, and caffeine (APC). This combination was available under a variety of trade names, or it could be purchased as APC Tablets U.S.P., usually at a lower price than the proprietary medications. Phenacetin has about the same effectiveness as aspirin in reducing fever and relieving minor aches and pains. It has been implicated in damage to the kidneys, in blood abnormalities, and as a possible carcinogen, however. The U.S. FDA banned further use of phenacetin in 1983.

Anacin, which, along with Empirin and Excedrin, was once an APC formulation, now contains only aspirin and caffeine. Caffeine is a mild stimulant found in coffee, tea, and cola syrup. There is no reliable evidence that caffeine enhances the effect of aspirin in any significant way. In fact, recent evidence indicates that for fever reduction, caffeine *counteracts* the action of aspirin. Combinations containing caffeine are therefore *less effective* than plain aspirin for this use. A tablet of Anacin does contain a little more aspirin (400 mg) than a regular (325-mg) aspirin tablet. The usual two-tablet dose of Anacin provides 800 mg of aspirin. You could get the same dose—at a lower price—from two and a half regular aspirin tablets.

Both phenacetin and acetaminophen are related structurally to acetanilide. Acetanilide was itself once used widely to reduce fever and to relieve pain. It is highly toxic, however, and has been largely replaced by safer derivatives and by aspirin.

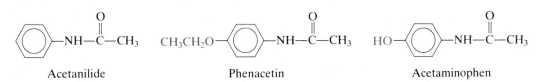

| Acetanilide | Phenacetin | Acetaminophen |

The highly advertised product Excedrin contains both aspirin and acetaminophen. Excedrin also contains caffeine. The advertising claim once was made that Excedrin is more effective than plain aspirin for pain other than headache. It fails to mention that the pain is that suffered by women who have just had babies!

Many other combination products are available. Brands and formulations change frequently. Extensive studies show repeatedly that, for most people, plain aspirin is the cheapest, safest, and most effective product.

Chemistry, Chicken Soup, and the Common Cold

We spend a lot of money for cold remedies, but there is no cure for the common cold. Colds are caused by as many as 100 related viruses. No antibiotic or other drug is effective against any of them. If we can't cure colds, why do we use so much medicine for them?

Most cold remedies treat the symptoms. Some, perhaps, give worthwhile relief. None prevent or cure the cold or shorten its duration. Included in the arsenal of weapons used against the cold are cough suppressants, expectorants, bronchodilators, anticholinergics, nasal decongestants, and antihistamines. An advisory panel to the FDA reviewed these compounds in 1976 and came up with some interesting results. Let's take a brief look at each class of medication.

Cough suppressants (antitussives) are available in great variety, but the FDA's panel found only three ingredients safe and effective. Two are narcotics: codeine (Chapter 23) and dextromethorphan. The other, diphenhydramine, is an antihistamine. The three effective compounds are available as free bases or as salts. Used as directed, they are safe and effective, but abuse of the two narcotics can lead to addiction.

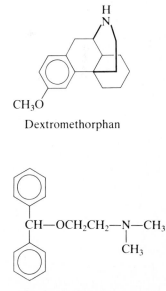

Dextromethorphan

Diphenhydramine

Expectorants are supposed to help bring up mucus out of the bronchial passages. The panel found none of the products on the market to be safe and effective. On the other hand, a variety of **nasal decongestants** *were* found to be safe and effective when they are used as directed. An example is ephedrine (and its salts). Another nasal decongestant is phenylephrine hydrochloride. Both ephedrine and phenylephrine are bronchodilators.

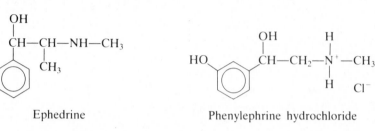

Ephedrine   Phenylephrine hydrochloride

Several **antihistamines** were judged safe and effective when used as directed. Antihistamines relieve the symptoms of allergies: sneezing, itchy eyes, and runny nose. They were judged *not* to be effective against colds. Examples of safe and effective antihistamines are diphenhydramine hydrochloride, chlorpheniramine, and promethazine hydrochloride.

Antihistamines do relieve the symptoms of allergies. When an **allergen** (a substance that triggers an allergic reaction) binds to the surface of certain cells, it triggers the release of histamine. The histamine causes the redness, swelling, and itching associated with allergies. Antihistamines block the release of histamine, but most also enter the brain and act upon the cells controlling sleep. Terfenadine (Seldane), an antihistamine approved for prescription use in 1985, blocks the release of histamine, but cannot enter the brain and thus does not cause drowsiness.

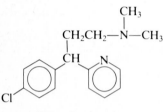

Chlorpheniramine

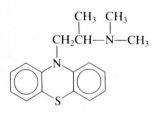

Promethazine

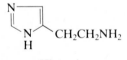

Histamine

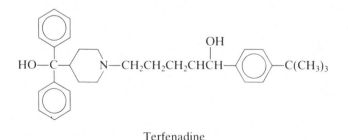

Terfenadine

What should you take for the common cold? Chicken soup is probably as good as anything. Plenty of liquids, plenty of rest, and a week or so of patient suffering are still the best cures. You can gain some relief from the symptoms by using some of the over-the-counter drugs. For an analysis of them by brand name, see Reference 5.

Linus Pauling, Vitamin C, and the Common Cold

In Chapter 18, we discussed the use of vitamin C (ascorbic acid) to prevent scurvy. Only about 40 to 75 mg is needed daily to prevent this deficiency disease.

In 1970, though, Linus Pauling, winner of two Nobel Prizes (for chemistry in 1954 and for peace in 1962), suggested that this vitamin could be used as a weapon against the common cold. He recommended daily doses of 250 to 15 000 mg, depending on the person and the circumstances. Vitamin C, he said, could prevent colds or at least lessen their severity. Pauling's claims were greeted with skepticism—and even ridicule—by many in the scientific and medical communities. Nevertheless, sales of ascorbic acid zoomed as people rushed to try his plan.

Who was right, Pauling or his critics? The issue still has not been settled. Much research has been undertaken since Pauling's original announcement. Some of it seems to confirm his claims; much of it does not. For example, a Canadian study showed that massive doses of vitamin C did not lessen the incidence of colds, but those taking the vitamin did miss fewer days of work than those taking a placebo. (A **placebo** is a substance that is made to look and taste like the real thing but that has no active ingredients.) Studies reported in 1973 showed that no matter how much vitamin C a person swallowed, a maximum of about 200 mg would be retained; the remainder would be excreted. Pauling, however, claims that some people, notably schizophrenics, can absorb incredible amounts of vitamin C.

Pauling and his followers are trying to start a new kind of medicine called **orthomolecular** ("right molecules") **medicine**. He believes that "having the right molecules in the right amounts in the right place at the right time" is essential to good health. Diets and megavitamin dosages are determined on the basis of extensive chemical tests on an individual's blood and urine.

The effectiveness of such therapy is difficult to measure. Each patient is given an individualized regimen to follow. Testing such procedures on large groups to get statistically valid data is impossible. It may be years before we know for sure whether Pauling is right or wrong.

Figure 22.3 Linus Pauling, winner of two Nobel prizes, advocates massive doses of vitamin C to prevent the common cold. [Photo by Margo Moore, San Francisco.]

Antibacterial Drugs

There is no cure for colds, but colds seldom kill. Less than a century ago, however, infectious diseases were the principal cause of death (Figure 22.4); today only 1 infectious disease is among the 10 leading causes. Many of these diseases have been brought under control by the

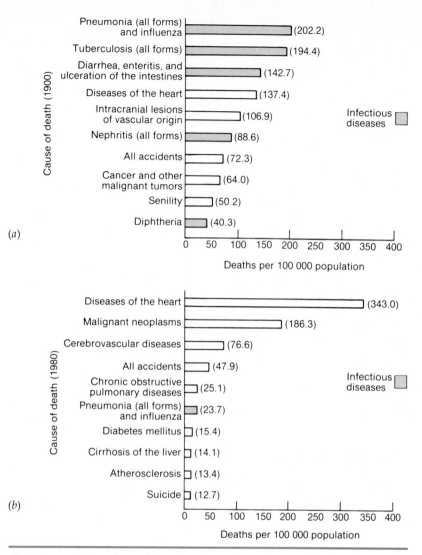

Figure 22.4 (a) In 1900, 5 of the 10 leading causes of death in the United States were infectious diseases. (b) By 1980, only 1 infectious disease was among the top 10.

use of antibacterial drugs. The first of these were the sulfa drugs, the prototype of which was discovered in 1935 by the German chemist Gerhard Domagk. Sulfa drugs were used extensively during World War II to prevent infection in wounds. Many soldiers lived who, with similar wounds, would have died in earlier wars.

The action of sulfanilamide, the simplest of the sulfa drugs, was one of the first to be understood at the molecular level. Its effectiveness is based on a case of mistaken identity. Bacteria need *para*-aminobenzoic

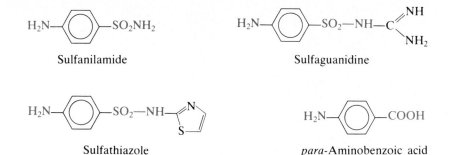

Figure 22.5 Sulfanilamide and two other sulfa drugs. *para*-Aminobenzoic acid has a similar structure.

acid (PABA) to make folic acid. Folic acid is essential for the formation of certain compounds the bacteria require for proper growth. But the bacterial enzymes can't tell the difference between sulfanilamide and PABA, because the substances are similar in structure (Figure 22.5). When sulfanilamide is applied to an infection in large amounts, the bacteria incorporate it into a pseudofolic acid. These false molecules cannot perform the growth-enhancing function of folic acid; hence, the bacteria cease to grow.

Sulfanilamide analogs have been developed and tested by the thousands. Only a few are used today. The structures of two common ones are given in Figure 22.5. Introduction of a heterocyclic ring generally increases the activity of a sulfa drug. Some of the sulfa drugs tend to cause damage to the kidneys by crystallizing there. Some present other toxicity problems. For many purposes, sulfa drugs have been replaced by newer antibacterial drugs.

Penicillins and Cephalosporins

The next important discovery was that of penicillin, an antibiotic. **Antibiotics** are soluble substances (derived from molds or bacteria) that inhibit the growth of other microorganisms (Figure 22.7). Penicillin was first discovered in 1928, but it was not tried on humans until 1941. Alexander Fleming, a Scottish microbiologist then working at the University of London, first observed the antibacterial action of a mold, *Penicillium notatum*. Fleming was studying an infectious bacterium, *Staphylococcus aureus*. One of his cultures became contaminated with blue mold. Contaminated cultures generally are useless. Most investigators probably would have destroyed the culture and started over, but Fleming noted that the bacterial colonies had been destroyed in the vicinity of the mold.

Figure 22.6 Sir Alexander Fleming, discoverer of penicillin. [Courtesy of the Smithsonian Institution, Washington, DC.]

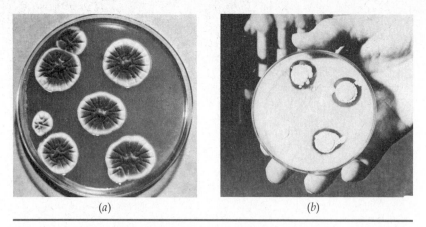

(a)   (b)

Figure 22.7 (a) Penicillin molds. These symmetrical colonies of green mold are *Penicillium chrysogenum*, a mutant form of which now produces almost all of the world's commercial penicillin. (b) Screening antibiotics for inhibiting effects, scientists placed small discs of blotting paper saturated with an experimental antibiotic solution on the surface of petri dishes seeded with germs. The potency of the antibiotic could be detected by measuring the clear (dark) rings around the discs, which showed that the antibiotic inhibited the germs' growth. [Courtesy of Pfizer, Inc.]

Fleming was able to make crude extracts of the active substance. This material, later called penicillin, was further purified and improved by Howard Florey (an Australian) and Ernst Boris Chain (a refugee from Nazi Germany). Florey and Chain were working at Oxford University. Fleming, Florey, and Chain shared the 1945 Nobel Prize in physiology and medicine for their work on penicillin.

It soon became apparent that penicillin was not a single compound but a group of compounds with related structures. Chemists recognized that by deliberately designing molecules with different structures they could vary the properties of the drugs. The penicillins that resulted from these experiments vary in effectiveness. Some must be injected; others can be taken orally. Bacteria resistant to one penicillin may be killed by another. The structures of several common penicillins are shown in Figure 22.8.

A knowledge of the structure of drug molecules enables chemists to be more efficient when they design drugs. A knowledge of the structure of the drug molecule is not sufficient for an understanding of the molecular basis of the drug's action, however. It is also necessary for chemists to understand the structure of the molecules of the human body, of bacteria, and of viruses.

The mode of action of penicillin has been unraveled. In certain bacteria, the cell walls are made up of mucoproteins, polymers in which

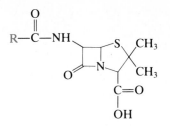
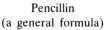

Pencillin
(a general formula)

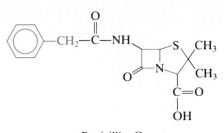

Penicillin G
(by injection; side effects include
diarrhea, possible allergic reactions)

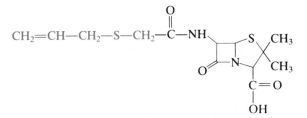

Penicillin O
(less allergenic than penicillin G)

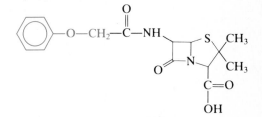

Penicillin V
(may be taken orally;
side effects similar to penicillin G)

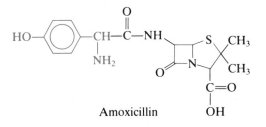

Amoxicillin

Figure 22.8 Some common penicillins.

amino sugars are combined with protein molecules. Penicillin prevents cross-linking between these large molecules and thereby prevents the synthesis of cell walls in bacteria. Cells of higher animals do not have mucoprotein walls. Instead, they have external membranes that differ in composition from the cell walls of bacteria. These membranes are not affected by penicillin. Thus, penicillin can destroy bacteria without harming human cells. Many people—perhaps as many as 5% of the population—are allergic to penicillin, however.

During their early history, antibiotics came to be known as miracle drugs. The number of deaths from blood poisoning, pneumonia, and other infectious diseases was reduced substantially by the use of antibiotics. Only four decades ago, a person with a major infection almost always died. Today, such deaths are rare. Four decades ago, pneumonia was a dread killer of people of all ages. Today, it kills only the very old or those ill from other causes. The antibiotics have indeed worked miracles in our time, but even miracle drugs are not without problems. It wasn't long after the drugs were first used that disease organisms began to develop strains resistant to the drugs. Before erythromycin (an antibiotic obtained from *Streptomyces erythreus*) was used very much, all strains of staphylococci could be handled readily by the drug. After it had been put into extensive use, resistant strains began to appear. Staph infections became a serious problem in hospitals. People who had gone to the hospital to be cured got serious bacterial infections instead. Some even died from staph infections picked up at hospitals.

In a race to stay ahead of resistant strains of bacteria, scientists continue to seek new antibiotics. The penicillins now have partially been displaced by related compounds called cephalosporins. Perhaps the most notable of these is cephalexin (Keflex). Unfortunately, some strains of bacteria already show resistance to cephalexin.

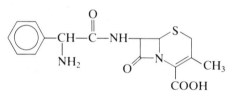

Cephalexin

Tetracyclines: Four Rings and Other Things

Another important development in the field of antibiotics was the discovery of a group of related compounds called **tetracyclines**. The first of these, called aureomycin, was isolated in 1948 by Benjamin Duggor from *Streptomyces aureofaciens*. A group of scientists at Pfizer Laboratories isolated terramycin from *Streptomyces rimosus* in 1950 after testing 116 000 different soil samples. Both drugs later were found to be derivatives of tetracycline, a compound now obtained from *Streptomyces viridifaciens*. These three compounds (Figure 22.9) are called **broad spectrum antibiotics** because they are effective against a wide variety of microorganisms.

Tetracyclines bind to bacterial ribosomes. This inhibits bacterial protein synthesis and thus blocks bacterial growth. The tetracyclines do not bind to mammalian ribosomes and thus do not affect protein synthesis in host cells. Several microorganisms have developed strains resistant to tetracyclines.

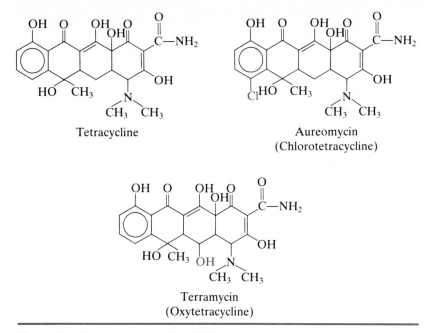

Figure 22.9 Three tetracycline antibiotics.

Antiviral Drugs

Antibiotics have taken much of the terror out of bacterial infections such as tuberculosis and diphtheria. Problems with resistant strains of bacteria are still worrisome, but seldom insurmountable. A variety of viral diseases still plague us, however. Viral diseases are not amenable to treatment with the usual antibiotics, so we suffer from a variety of viral infections ranging from colds to genital herpes and acquired immune deficiency syndrome (AIDS).

Some viral infections—such as poliomyelitis, mumps, measles, and smallpox—can be prevented by vaccination. Influenza vaccines are available, but because there are so many strains of flu viruses (with new ones appearing periodically), the vaccines have limited effectiveness.

In recent years, drugs have been developed that are effective against some viruses. Amantadine has had some success in preventing influenza A infections. Acyclovir (Zovirax) has shown promise against the herpes viruses that cause genital sores, chicken pox, shingles, mononucleosis, and cold sores. Azidothymidine was approved in 1987 to treat AIDS. Before cures for viral diseases become the rule instead of the exception, however, we will have a lot to learn about viruses and how they cause disease.

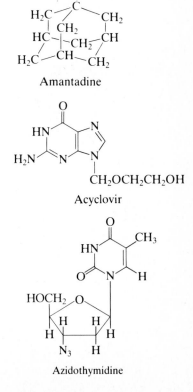

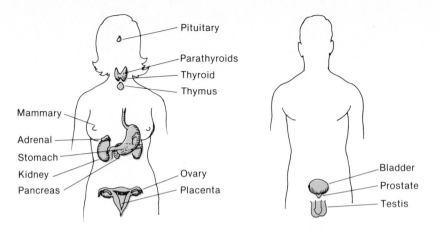

Figure 22.10 The approximate location of the endocrine glands in the human body.

Hormones: The Regulators

Before we can discuss the next group of drugs, we must take a brief look at the human endocrine system and some of the chemical compounds, called **hormones**, that the system produces. Hormones are chemical messengers produced in the endocrine glands (Figure 22.10). Those released in one part of the body signal profound physiological changes in other parts of the body. By causing reactions to speed up or slow down, they control growth, metabolism, reproduction, and many other functions of body and mind.

Some of the more important human hormones and their physiological effects are listed in Table 22.2. Some of the hormones that affect mental activities are discussed in the next chapter. For now, we concentrate on the steroidal hormones, mainly those involved in the reproductive cycle.

The Steroids: Gallstones, Hardened Arteries, and Good Medicine

Chemists have spent many years determining the structures of hormones. As indicated in Table 22.2, some hormones have complicated protein structures. Others are, by comparison, rather simple. The steroids, for example, all have the same skeletal, four-ring structure.

Not all steroids have hormonal activity. Cholesterol is a common component of all animal tissues. The brain is about 10% cholesterol, but cholesterol's function there is not known. Cholesterol is a major component of certain types of gallstones. It also is found in deposits in hardened arteries (Chapter 18).

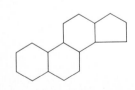

Steroid skeletal structure

Table 22.2 Some Human Hormones and Their Physiological Effects

Name	Gland and Tissue	Chemical Nature	Effect
Various releasing and inhibitory factors	Hypothalamus	Peptide	Triggers or inhibits release of pituitary hormones
Human growth hormone (HGH)	Pituitary, anterior lobe	Protein	Controls the general body growth; controls bone growth
Thyroid-stimulating hormone (TSH)	Pituitary, anterior lobe	Protein	Stimulates growth of the thyroid gland and production of thyroxin
Adrenal cortex-stimulating hormone (ACTH)	Pituitary, anterior lobe	Protein	Stimulates growth of the adrenal cortex and production of cortical hormones
Follicle-stimulating hormone (FSH)	Pituitary, anterior lobe	Protein	Stimulates growth of follicles in ovaries of females, sperm cells in testes of males
Luteinizing hormone (LH)	Pituitary, anterior lobe	Protein	Controls production and release of estrogens and progesterone from ovaries, testosterone from testes
Prolactin	Pituitary, anterior lobe	Protein	Maintains the production of estrogens and progesterone, stimulates the formation of milk
Vasopressin	Pituitary, posterior lobe	Protein	Stimulates contractions of smooth muscle; regulates water uptake by the kidneys
Oxytocin	Pituitary, posterior lobe	Protein	Stimulates contraction of the smooth muscle of the uterus; stimulates secretion of milk
Parathyroid	Parathyroid	Protein	Controls the metabolism of phosphorus and calcium
Thyroxine	Thyroid	Amino acid derivative	Increases rate of cellular metabolism
Insulin	Pancreas, beta cells	Protein	Increases cell usage of glucose; increases glycogen storage
Glucagon	Pancreas, alpha cells	Protein	Stimulates conversion of liver glycogen to glucose
Cortisol	Adrenal gland, cortex	Steroid	Stimulates conversion of proteins to carbohydrates
Aldosterone	Adrenal gland, cortex	Steroid	Regulates salt metabolism; stimulates kidneys to retain Na^+ and excrete K^+
Epinephrine (adrenaline)	Adrenal gland, medulla	Amino acid derivative	Stimulates a variety of mechanisms to prepare the body for emergency action including the conversion of glycogen to glucose
Norepinephrine (noradrenaline)	Adrenal gland, medulla	Amino acid derivative	Stimulates sympathetic nervous system; constricts blood vessels, stimulates other glands
Estradiol	Ovary, follicle	Steroid	Stimulates female sex characteristics; regulates changes during menstrual cycle
Progesterone	Ovary, corpus luteum	Steroid	Regulates menstrual cycle; maintains pregnancy
Testosterone	Testis	Steroid	Stimulates and maintains male sex characteristics

Another steroid of some interest is the adrenal hormone cortisone. Applied topically or injected, cortisone acts in the body to reduce inflammation. It was once widely used in the treatment of arthritis, but relief was transitory and repeated application caused serious site effects.

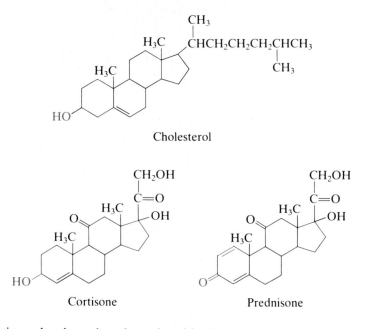

Cholesterol

Cortisone   Prednisone

Cortisone has been largely replaced by the related compound prednisone. Because prednisone is effective in much smaller doses, its side effects are greatly reduced.

Some Sexy Steroids

Closely related in structure to cholesterol, cortisone, and prednisone are the sex hormones. It is interesting to note that male sex hormones differ only slightly in structure from female sex hormones. In fact, the female hormone progesterone can be converted by a simple biochemical reaction into the male hormone **testosterone**. The physiological actions of these structurally similar compounds are markedly different.

Male sex hormones, called **androgens**, are secreted by the testes. These hormones are responsible for the development of the sex organs and for secondary sexual characteristics such as voice and hair distribution. The most important male hormone is testosterone.

There are two important groups of female sex hormones. The **estrogens** are produced mainly in the ovaries. They control female sexual functions such as the menstrual cycle, the development of breasts, and other secondary sexual characteristics. Two important estrogens are estradiol and estrone. Another female sex hormone is progesterone, which prepares the uterus for pregnancy and prevents the further release of eggs from the ovaries during pregnancy.

Sex hormones—both natural and synthetic—sometimes are used therapeutically. For example, a woman who has had her ovaries removed may be given female hormones to compensate for those no longer being produced by her ovaries. Some of the earliest compounds employed in cancer chemotherapy were sex hormones. The male hormone testosterone was used to treat carcinoma of the breast in females; estrogens, female sex hormones, were given to males to treat carcinoma of the prostate.

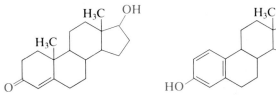

Testosterone

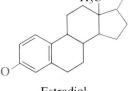

Estradiol

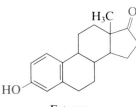

Estrone

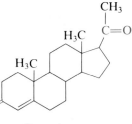

Progesterone

Chemistry and Social Revolution: The Pill

When administered by injection, progesterone serves as an effective birth-control drug. This knowledge led to attempts by chemists to design a contraceptive that would be effective when taken orally.

The structure of progesterone was determined in 1934 by Adolf Butenandt. Just four years later, Hans Inhoffen synthesized the first oral contraceptive, ethisterone. Ethisterone had to be taken in large doses to be effective, and it was never widely used as an oral contraceptive.

The next breakthrough came in 1951, when Carl Djerassi synthesized 19-norprogesterone, progesterone with one of its methyl groups

Figure 22.11 Carl Djerassi, professor of chemistry at Stanford University and president of Zoecon Corporation, Palo Alto, California.

Figure 22.12 Frank Colton, chemist at G. D. Searle Co., synthesized norethynodrel, a progestin used in Enovid, the first birth control pill.

missing. This compound was four to eight times as effective as progesterone as a birth-control agent. Like progesterone itself, it had to be given by injection, an undesirable property. Djerassi then put it all together. Removal of a methyl group made the drug more effective. The ethynyl group (—C≡CH) allowed oral administration. He then synthesized 17α-ethynyl-19-nortestosterone, mercifully known by the trade name Norlutin. This drug proved effective when it was taken in small doses. Djerassi's work was published in 1954, and a patent was issued to him in 1956. At about the same time, Frank Colton (Figure 22.12) synthesized norethynodrel. Note that the two substances differ only in the position of a double bond (Figure 22.13). Patents were issued to Colton's employer, G. D. Searle, in 1954 and 1955. Searle produced the first approved contraceptive, Enovid, in 1960. The pill contained 9.85-mg norethynodrel and 150-μg mestranol. Norethynodrel, norethindrone, and related compounds are called **progestins** because they mimic the action of progesterone. Mestranol is a synthetic estrogen, added to regulate the menstrual cycle. The progestin acts by establishing a state of false pregnancy (Figure 22.14). A woman does not ovulate when she is pregnant (or in the state of false pregnancy established by the progestin). Since the woman does not ovulate, she cannot conceive.

The ultimate effects of prolonged tampering with the reproductive biochemistry of the human female remain to be seen. In use in the United States since 1960 by millions of women, the pill appears to be relatively safe. There are undesirable side effects in some women. Among these are hypertension, acne, and abnormal bleeding. The pill also causes blood clotting in some women, but so does pregnancy. Such

Figure 22.13 Some synthetic steroids.

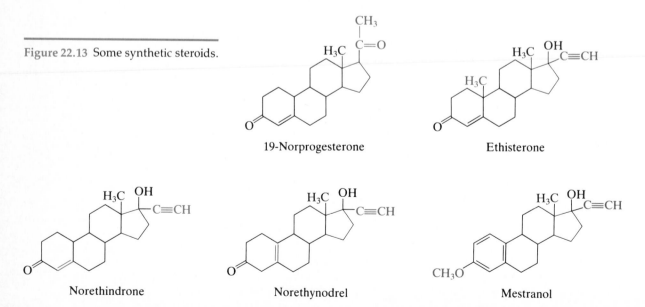

19-Norprogesterone   Ethisterone

Norethindrone   Norethynodrel   Mestranol

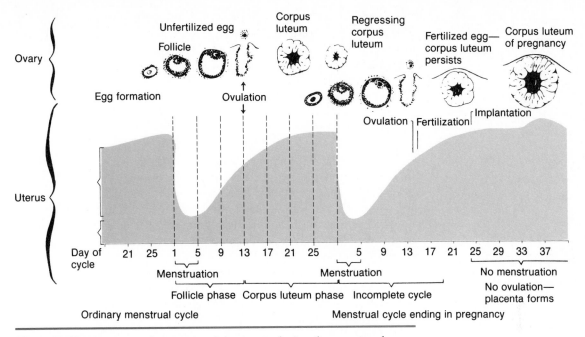

Figure 22.14 Changes in the ovary and the uterus during the menstrual cycle. Pregnancy—or the pseudopregnancy caused by the birth-control pill—prevents ovulation. [Adapted with permission from Philip D. Sparks et al., *Student Study Guide for the Biological Sciences*, 3rd ed. (Minneapolis, Burgess, 1973).]

blood clots can clog a blood vessel and cause death by stroke or coronary heart attack. The death rate associated with the use of the pill is about 3 in 100 000. This is only 10% of that associated with child-birth—about 30 in 100 000.

For the general population, the pill is probably as safe as aspirin. Women who have blood with an abnormal tendency to clot should not take the pill, nor should those who experience any serious side effects. Recently, the FDA has advised women over 40, particularly those who smoke, to use another method of contraception. For these women, the risk associated with using the pill is greater than the risk associated with having a baby.

Most of the side effects of the pill are associated with the estrogen component. In view of that, the amount of estrogen has been reduced from 150 μg in early formulations to about 3 μg today. Also, a minipill containing only a small amount of progestin and no estrogen has been marketed. It supposedly avoids many of the problems associated with the combination pill. The FDA now requires a warning about the possible harmful effects of estrogens on all drugs containing them.

Unless there are some unforseen developments, some form of chemical contraceptive seems to be with us to stay. The pill has already brought revolutionary social changes. The impact of science on society is perhaps nowhere more evident.

DES and the Morning After

In 1973, the FDA approved the use of diethylstilbestrol (DES) to induce abortion. The action was surprising to some, because DES had been banned earlier in 1973 as an additive in animal feeds. Cattle fed DES-treated feeds had achieved marketable weights much more rapidly. The DES ban had a major impact on the cattle industry and was not undertaken lightly. The ban came only after a relation between DES and cancer was established. DES had been administered to women with a history of miscarriages in an attempt to prevent spontaneous abortion. No adverse effects were noted in those women. Fifteen years later, however, an abnormally high incidence of vaginal cancer was noticed among their daughters. This delayed effect emphasizes one of the problems involved in evaluating the possible harmful effects of any chemical.

The FDA had no choice about banning the use of DES in animal feeds. The Delaney clause of the Food and Drug Act of 1958 states clearly that any agent that causes cancer cannot be used in food. Despite efforts to prevent it, DES residues continued to appear in meat from animals that ate feeds containing this additive. More recently, DES has been shown to affect the sons of mothers who took the drug while pregnant. They are more likely than the general population to be sterile and to have poorly developed sex organs.

How could the FDA ban DES from animal feeds and approve its use as an abortifacient? The answer lies in the way the law is written and in the way the chemical is used. The Delaney clause *requires* the banning of any food additive that causes cancer in humans or laboratory animals. DES had been shown to cause vaginal cancer. Therefore, it had to be banned as a food additive. There is no such requirement for drugs; DES could be approved for use in inducing abortions. It also was pointed out that, if the drug were effective, there would be no offspring to have cancer.

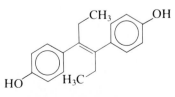

DES

DES is not a steroid. The shape of the molecule shows some similarity to the structure of estradiol (page 503). Thus, DES is a synthetic female hormone. It was not significantly effective in maintaining pregnancy, and, in order to induce abortion, a woman must take large doses. Unpleasant side effects are noted frequently.

The control of human reproduction is an area subject to great controversy—social, moral, political, and legal. Much more research is needed not only to design new drugs for birth control but also to increase our understanding of the biochemistry and physiology of human reproduction.

Prostaglandins: The Latest Miracle Drugs

Perhaps no compounds since birth-control pills have stimulated as much activity in pharmaceutical companies as have the group of compounds called **prostaglandins**. Over 6000 papers are published each year on these compounds. Biochemically, prostaglandins are derived from arachidonic acid, a fatty acid with 20 carbon atoms (Figure 22.15). There are six primary prostaglandins, and many others have been identified. These compounds are widely distributed throughout the

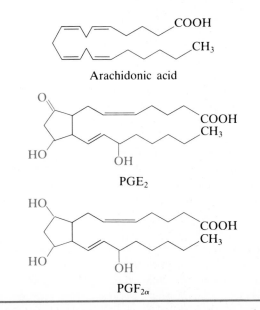

Arachidonic acid

PGE$_2$

PGF$_{2\alpha}$

Figure 22.15 Prostaglandins are derived from arachidonic acid, an unsaturated carboxylic acid with 20 carbon atoms.

body at extremely low concentrations. They are among the most potent biological chemicals, and extremely small doses can elicit marked changes.

The prostaglandins act as mediators of hormone action. They regulate such things as smooth-muscle activity, blood flow, and secretion of various substances. This range of physiological activity led to the synthesis of hundreds of prostaglandins and their analogs. Two such derivatives are now in use in the United States to induce labor. Others have been employed clinically to lower or increase blood pressure, to inhibit stomach secretions, to relieve nasal congestion, to provide relief from asthma, and to prevent the formation of the blood clots associated with heart attacks and strokes.

$PGF_{2\alpha}$ is used in cattle breeding. A prize cow is treated with the hormone gonadotropin (see Table 22.2) and with $PGF_{2\alpha}$ to induce the release of many ova. The ova are then fertilized with sperm from a champion bull. The developing embryos are implanted in less valuable cows. This enables a farmer to get several calves a year from one outstanding cow.

Over 30 prostaglandins and related compounds are now under investigation as possible drugs. They hold great promise for a variety of purposes.

A Pill for Males?

Why do females have to bear the responsibility for contraception? Why not a pill for males? Recent studies have shown that many men—including a majority of the younger ones—are willing to share the risks and the responsibility of contraception. Nevertheless, there are biological reasons for females to bear the burden: women get pregnant when contraception fails, and, in females, contraception has only to interfere with one monthly event—ovulation. On the other hand, males produce sperm continuously.

Actually, a good deal of research has gone into male contraceptives. Estrogens would work, but they would bring about the development of female characteristics in men, including a complete loss of interest in sexual relations with women. Gossypol, a pigment found in cottonseed, has been used in China as a contraceptive for males.

A drug called danazol has also been tested and found safe and effective. Danazol is testosterone enanthate, administered along with testosterone itself. It suppresses sperm production, and the effect is reversed when the drug is withdrawn. Like the pill for women, danazol often causes weight gain in men. The main problem, however, is that the drug is much too expensive for widespread use.

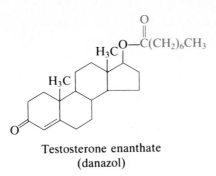

Testosterone enanthate
(danazol)

Perhaps the best bet for a male contraceptive is some sort of drug that doesn't stop sperm formation, but does block growth or transport of the sperm.

The availability of vasectomies—simple surgical procedures that block the emission of sperm—has lessened the demand for a male contraceptive. A major drawback to vasectomy, however, is that it is often irreversible. A safe, cheap effective male contraceptive is still a goal of the drug industry.

Chemicals Against Cancer

If chemists can design molecules to relieve headaches, cure infectious diseases, and prevent conception, why can't they do something about cancer? They have done something, but much remains to be done. Treatment with drugs, radiation, and surgery has led to high rates of cures for some forms of cancer (e.g., skin cancer). For others, such as lung cancer, the rate is still near zero. Over 30 different chemical substances are used widely in the treatment of cancer. We can examine only a few representative ones here. First, let's look at a group called antimetabolites.

In cancer chemotherapy, **antimetabolites** are usually compounds that inhibit the synthesis of nucleic acids. Rapidly dividing cells, characteristics of cancer, require large quantities of DNA. The antimetabolites block DNA synthesis and the increase of the number of cancer cells. Because cancer cells are undergoing rapid growth and cell division, they generally are affected to a greater extent than normal cells. Eventually, though, the normal cells are affected to such a degree that treatment must be discontinued.

The most widely used anticancer drug is Cisplatin, a platinum-containing compound. Cisplatin binds to DNA and blocks its replication.

Two other prominent antimetabolites are 5-fluorouracil and its deoxyribose derivative, 5-fluorodeoxyuridine. In the body, both of

Cisplatin

these can be incorporated into a phosphate–sugar–base unit called a **nucleotide**. The fluorine-containing nucleotide inhibits the formation of a thymine-containing nucleotide required for DNA synthesis. Thus, both compounds slow the division of cancer cells. These compounds have been employed against a variety of cancer, especially those of the breast and the digestive tract.

Another common antimetabolite is 6-mercaptopurine. This compound can substitute for adenine in a nucleotide. The pseudonucleotide then inhibits the synthesis of nucleotides incorporating adenine and guanine. Hence, DNA synthesis and cell division are slowed. 6-Mercaptopurine has been used in the treatment of leukemia.

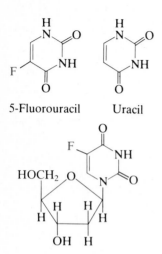

5-Fluorouracil   Uracil

5-Fluorodeoxyuridine

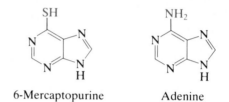

6-Mercaptopurine   Adenine

Another antimetabolite, methotrexate, acts in a somewhat different manner. Note the similarity between its structure and that of folic acid.

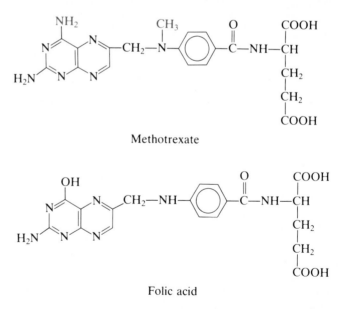

Methotrexate

Folic acid

Like the pseudofolic acid formed from sulfanilamide, methotrexate competes successfully with folic acid for an enzyme but cannot perform the growth-enhancing function of folic acid. Again, cell division is slowed and cancer growth retarded. Methotrexate is used frequently against leukemia.

Alkylating Agents: Turning War Gases on Cancer

Alkylating agents are highly reactive compounds that can transfer alkyl groups to compounds of biological importance. These foreign alkyl groups then block the usual action of the biological molecules. A variety of alkylating agents are used against cancer. Typical among these are the nitrogen mustards, compounds that arose out of chemical warfare research.

The original mustard "gas" was a sulfur-containing blister agent used in chemical warfare during World War I. The compound is a liquid, but it vaporizes readily. Contact with either the liquid or the vapor causes blisters that are painful and slow to heal. Inhalation of the vapor can cause death. The action of the "gas" is insidious. The effects do not appear for several hours after exposure. It is easily detected, though, by its garlic or horseradish odor. Mustard gas is denoted by the military symbol H.

The development of new agents for chemical and biological warfare (CBW) continued after World War I. Attempts were made to increase the effectiveness of blister agents by varying their molecular architecture. The nitrogen mustards (symbol HN) were developed about 1935. Though not quite as effective overall as mustard gas, the nitrogen mustards produce greater eye damage and don't have an obvious odor. Structurally, the nitrogen mustards are chlorinated amines.

$$Cl-CH_2CH_2-S-CH_2CH_2-Cl$$

Mustard gas

(H)

$$CH_3CH_2-N \begin{array}{l} CH_2CH_2-Cl \\ CH_2CH_2-Cl \end{array}$$

HN_1

$$CH_3-N \begin{array}{l} CH_2CH_2-Cl \\ CH_2CH_2-Cl \end{array}$$

HN_2

$$Cl-CH_2CH_2-N \begin{array}{l} CH_2CH_2-Cl \\ CH_2CH_2-Cl \end{array}$$

HN_3

> Chemical agents were used extensively during World War I. Over 30 such substances were employed, killing 91 000 and wounding 1.2 million (many of them for life). Fritz Haber supervised the release of chlorine gas in the first attack by the Germans (Chapter 17). Adolf Hitler was among those wounded when the British retaliated with phosgene a few days later. Haber considered gas warfare to be "a higher form of killing." Fortunately, his views have not become widely accepted. The use of chemical warfare agents was largely avoided during World War II. They have been used in smaller wars, most recently by Iraq in its war with Iran.

To our pleasant surprise, the nitrogen mustards have been found to be effective in the treatment of certain types of cancer. We have seen in earlier chapters how, on occasion, a chemical designed for one (beneficial) purpose was later found to have undesirable side effects. For example, we saw how DDT, once thought to be the perfect insecticide, came under attack for threatening entire species of higher animals. With the nitrogen mustards, we see just the opposite. These molecules, designed for war, have been used most successfully against a formerly fatal type of skin cancer. Complete remission is often obtained by bathing patients in a solution of nitrogen mustards.

Through this example, we see that knowledge gained through science is neither good nor evil. How we use that knowledge does involve morality. The knowledge can be used either for our benefit or for our destruction. In this respect, science is not different from the social sciences or the humanities.

The nitrogen mustard of choice for cancer therapy nowadays is a compound called cyclophosphamide (Cytoxan). It is of interest to note that cyclophosphamide can defleece sheep chemically, saving the high cost of shearing. After treatment with the drug, the wool can be removed almost as easily as taking off an overcoat.

It should be noted that alkylating agents are two-edged swords. They can cause cancer as well as cure it. For example, the nitrogen mustard HN_2 causes lung, mammary, and liver tumors when injected into mice; yet, it can be used with some success in the management of certain human tumors. There is still of lot of mystery—and seeming contradiction—in available data on the causes and cures of cancer.

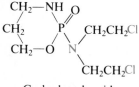

Cyclophosphamide

Miscellaneous Anticancer Agents

There is a bewildering variety of anticancer agents that defy ready classification. Alkaloids from vinca plants have been shown effective against leukemia and Hodgkin's disease. Actinomycin, a mixture of complex compounds obtained from the molds *Streptomyces antibioticus* and *Streptomyces parvus*, is used against Hodgkin's disease and other types of cancer. It is quite effective but extremely toxic. Actinomycin acts by binding to the double helix of DNA, thus blocking the replication of RNA on the DNA template. Protein synthesis is inhibited.

Sex hormones can be used against cancers of the reproductive system. For example, the female hormones estradiol (a natural hormone) and DES (a synthetic hormone) can be used against cancer of the prostrate gland. Conversely, male hormones such as testosterone can be used against breast cancer. Such treatment often brings about a temporary cessation—or even a regression—in the growth of cancer cells.

The food additive butylated hydroxy toluene (BHT) has been shown to be anticarcinogenic in tests involving laboratory animals. Some people involved in cancer research speculate that the use of this additive as a preservative in foods may account for the declining rate of stomach cancer in the United States.

Similarly, there is some evidence that vitamin A may confer a resistance to some cancers. For example, persons suffering from vitamin A deficiencies exhibit a higher incidence of lung cancer. Vitamin C also may have an anticancer function. It has been shown to inhibit the formation of nitrosamines (Chapter 19) under conditions similar to those in the human stomach.

Chemotherapy is only a part of the treatment of cancer. Surgical removal of tumors and radiation treatment remain major weapons in the war on cancer. Modern management of cancers can involve surgery, radiation, and one or more anticancer drugs. Indeed, a combination of drugs is often considerably more effective than any one alone. It is unlikely that a single agent will be found to cure all cancers. Steady progress is being made, however. Rates of cure should improve as research progresses. Perhaps a greater hope lies in the prevention of cancer. Much active research is underway on the mechanisms of carcinogenesis. Once we find the causes, rapid progress toward cures should follow.

Chemotherapy doesn't have all the answers to all human ailments—and it probably never will. Certainly, though, it has helped to give us longer, healthier lives. Let's hear it for chemistry!

Problems

1. What is an analgesic?

2. What is an antipyretic?

3. What is the chemical name for aspirin?

4. What is methyl salicylate? How is it used?

5. How can one brand of aspirin differ from another? In what ways must two brands of aspirin be the same?

6. What is "buffered" aspirin? Are such products truly buffered?

7. How does an "extra strength" pain reliever differ from a regular one?

8. How does aspirin suppress inflammation?

9. How does aspirin reduce the chance of heart attack or stroke?

10. What is acetaminophen?

11. In what ways do aspirin and acetaminophen act similarly?

12. Why is aspirin chosen over acetaminophen for treatment of arthritis?

13. Why is acetaminophen chosen over aspirin for the relief of pain associated with surgical procedures?

14. What is ibuprofen? To what class of compounds does it belong?

15. What are some of the ingredients in combination pain relievers? Do these combinations offer any advantages over plain aspirin?

16. List three effective cough suppressants.

17. List two effective nasal decongestants.

18. List three effective antihistamines.

19. Discuss Linus Pauling's contention that vitamin C prevents colds.

20. What is a placebo?

21. Which cold remedies cure colds? Which shorten the duration of colds?

22. What kinds of diseases were the principal causes of death in 1900?

23. What are sulfa drugs? How do they work?

24. What are antibiotics?

25. How can a drug such as penicillin kill bacterial cells without killing human cells?

26. What structural feature is shared by tetracyclines?

27. Tetracylines are broad spectrum antibiotics. What does that mean?

28. How do tetracyclines work against bacteria?

29. What is meant by "a resistant strain of bacteria"?

30. List some antiviral drugs.

31. What is a hormone?

32. What structural feature is shared by all steroids?

33. What is cortisone?

34. What is an androgen?

35. What is an estrogen?

36. How do birth-control pills work?

37. What structural feature makes a birth control steroid effective orally?

38. What is DES? What problems have been associated with its use?

39. What are prostaglandins?

40. List some possible future uses of prostaglandins as drugs.

41. Which compounds are classified as steroids?
    a. tristearin
    b. cholesterol
    c. testosterone
    d. prostaglandins

42. List some possible contraceptives for males. What problems might be associated with their use?

43. List two major classes of anticancer drugs.

44. How does Cisplatin work as an anticancer agent?

45. What are alkylating agents? How do they act against cancer?

46. What natural substance does 6-mercaptopurine resemble? How does 6-mercaptopurine act against cancer?

47. How does 5-fluorouracil act against cancer?

48. What natural substance does methotrexate resemble? How does methotrexate act against cancer?

49. What is actinomycin? How does it act against cancer?

50. Cancer cells multiply exponentially. (Exponential growth is discussed in Chapter 17.) A certain type of mouse cancer cell doubles in number in 0.5 day. About 1 billion cancer cells kill the mouse. Fourteen days after the injection of one viable cancer cell, there are over 268 000 000 cancer cells. How many more days before the mouse dies?

51. Examine the labels of at least five combination pain relievers. Make a list of the ingredients in each. Look up the properties (medical uses, dosages, side effects, toxicities) of each in a reference work such as *The Merck Index*.

52. Examine the labels of three over-the-counter sleeping pills (such as Nitol, Sominex, and Sleep-eze). Make a list of the ingredients in each. Look up the properties (medical uses, dosages, side effects, toxicities) of the ingredients in a reference work such as *The Merck Index*.

53. Examine the labels of at least three over-the-counter antitension formulations (such as Cope and Compoz). Make a list of the ingredients in each. Look up the properties (medical uses, dosages, side effects, toxicities) of the ingredients in a reference work such as *The Merck Index*.

54. Do a cost analysis of five brands of plain aspirin. Calculate the cost per gram of each.

55. Compare the cost per gram of an arthritis pain formulation to that of plain aspirin.

56. Examine three liquid cold remedies. What is the alcohol content of each? What other drugs are present? What is the function of each drug?

57. Examine five cold tablets or cold capsules. What are their active ingredients and what are the functions of these ingredients?

58. Discuss some of the problems involved in proving a drug safe or proving it harmful. Which is easier? Why?

References and Readings

1. Abraham, E. P. "The Beta-Lactam Antibiotics." *Scientific American*, June 1981, pp. 76–86. About penicillins and cephalosporins.

2. Angier, Natalie. "Bugs That Won't Die." *Discover*, August 1982, pp. 33–35.

3. Chain, Ernst. "Penicillin—The Crucial Experiment." *ChemTech*, August 1980, pp. 474–481.

4. Committee on Chemistry and Public Affairs. *Chemistry in Medicine: The Legacy and the Responsibillity*. Washington, DC: American Chemical Society, 1977.

5. "Cough Remedies: Which Ones Work Best?" *Consumer Reports*, February 1983, pp. 59–61.

6. Dolin, Raphael. "Antiviral Chemotherapy and Che-

moprophylaxis." *Science*, 15 March 1985, pp. 1296–1303.

7. Douglas, Kenneth T. "Anticancer Drugs: DNA as a Target." *Chemistry and Industry*, 15 October 1984, pp. 738–742.

8. Hales, Dianne, and Robert E. Hales. "Testosterone: The Bonding Hormone." *American Health*, November–December 1982, pp. 37–44.

9. Hecht, Annabel. "More Yesses, No's and Maybes for OTC Drugs." *FDA Consumer*, April 1985, pp. 16–19.

10. Hecht, Annabel. "Sulfa: Yesterday's Hero Is Still Taking Bows." *FDA Consumer*, October 1984, pp. 8–11.

11. "Is Bayer Better?" *Consumer Reports*, July 1982, pp. 347–349.

12. Julien, Robert M. *A Primer of Drug Action*, 4th ed. San Francisco: W. H. Freeman, 1985.

13. Liska, Ken. *Drugs and the Human Body*, 2nd ed. New York: Macmillan, 1986.

14. Maugh, Thomas H., II. "Male 'Pill' Blocks Sperm Enzyme." *Science*, 17 April 1981, p. 314. About gossypol.

15. Nelson, Norman A., Robert C. Kelly, and Roy A. Johnson. "Prostaglandins and the Arachidonic Acid Cascade." *Chemical and Engineering News*, 16 August 1982, pp. 30–44.

16. "The New Pain Relievers." *Consumer Reports*, November 1984, pp. 636–638. Discusses Advil and Nuprin.

17. Pauling, Linus. *Vitamin C, the Common Cold, and the Flu*. San Francisco: W. H. Freeman, 1976.

18. "Pills That Compete with Aspirin." *Consumer Reports*, August 1982, pp. 395–399.

19. Saxena, Anil. "Chemistry of Birth Control Pills." *Journal of Chemical Education*, December 1984, pp. 1075–1076.

20. Windholz, Martha (Ed.). *The Merck Index*, 10th ed. Rahway, NJ: Merck and Co., 1983.

23

Drugs

Chemistry and the Human Mind

In the preceding chapter, we discussed drugs that affect our bodies—drugs that relieve minor pain and cold symptoms, drugs that cure infectious diseases, drugs that prevent pregnancy, and drugs that cure or slow the growth of cancer. In this chapter, we consider **psychotropic drugs**—those that affect the mind.

There is no clear distinction between drugs that affect the mind and those that affect the body. Most of the drugs we take probably affect our minds as well as our bodies. Aspirin, sex hormones, and even antibiotics may have some influence on our mental state as well as our physical well-being. On the other hand, the narcotic drugs relieve pain at the same time they depress mental activities. It is becoming increasingly clear that the mind and the body interact closely; they are not separate as people once believed. It can still be useful, however, to distinguish between those drugs that act primarily on the body and those that affect primarily the mind.

Probably the first drugs to be used by primitive peoples influenced the mind. Alcohol, marijuana, opium, cocaine, peyote, and other plant materials have been used for thousands of years. Generally, only one or two of these were used in any one society. Today, however, thousands of drugs that affect the mind are available to us. Some still come from plants, but the great majority are synthetic, brought to you by the creative genius of the chemists of the world.

Drug misuse (for example, using penicillin, which has no effect on viruses, to treat a viral infection) is all too common. **Drug abuse** (using a drug for its intoxicating effect) is even more widespread. Worldwide, people spend more money for illegal drugs than for food. Annual revenues for the drug dealers amount to $500 billion.

Drugs are chemical substances. A knowledge of their chemistry can help us to understand their use and abuse.

Drugs can be mixtures, but often a given drug is a single chemical substance with fixed properties. The drugs often act differently in different people, however. While the drug may be simple, we are complex. Each of us is an enormously complex, unique set of chemicals. It is in interacting with our unique systems that drugs produce these different results.

Figure 23.1 We can change our moods with chemicals. Only a few drugs—such as caffeine and nicotine—are available in vending machines, although many others are easily available (from legal and illegal sources).

Drugs: Up, Down, or Sideways

The drugs that affect the mind can be divided into three classes. The **stimulants** increase alertness, speed up mental processes, and generally elevate the mood. This group, which includes the amphetamines and related compounds, sometimes is referred to as the "uppers." Another category—which includes alcohol, barbiturates, and minor tranquilizers—is the **depressants**. These often are called "downers." They reduce the level of consciousness and the intensity of reactions to environmental stimuli. In general, depressants dull emotional responses. (We discussed ethyl alcohol, the most widely abused drug of all, in some detail in Chapter 11.) The third category—variously called **hallucinogenics**, **psychotomimetics**, and **psychedelics** (because they induce hallucinations, psychoses, and colorful visions)—are popularly called the "mindbenders." These drugs don't so much depress or stimulate the mind as they alter qualitatively the way we perceive things. Notable among the mindbenders are LSD, mescaline, and marijuana. Let's examine some of the major drugs in each category.

Anesthetics: Under and Out

Anesthetics are another type of depressant. A **general anesthetic** acts on the brain to produce unconsciousness as well as insensitivity to pain. A **local anesthetic** renders one part of the body insensitive to pain, yet leaves the patient conscious.

Diethyl ether was the first general anesthetic. It was introduced into surgical practice in 1846 by a Boston dentist, William Morton. Inhalation of ether vapor produces unconsciousness by depressing the activity of the central nervous system. Ether is relatively safe. There is a fairly wide gap between the effective level of anesthesia and the lethal dose. The disadvantages are its high flammability and its side effect, nausea.

Nitrous oxide, or laughing gas (N_2O), was tried by Morton without success before he tried ether. Nitrous oxide was discovered by Joseph Priestley in 1772. Its narcotic effect was noted, and it soon came to be used widely at laughing gas parties among the nobility (Figure 23.2). Nitrous oxide, mixed with oxygen, finds some use in modern anesthesia. It is quick acting but not very potent. Concentrations of 50% or greater must be used to be effective. When nitrous oxide is mixed with ordinary air instead of oxygen, not enough oxygen gets into the patient's blood and permanent brain damage can result.

Chloroform ($CHCl_3$) was introduced as a general anesthetic in 1847. Its use quickly became popular after Queen Victoria gave birth to her eighth child while anesthetized by chloroform in 1853. Chloroform was used widely for years. It is nonflammable and produces effective anesthesia, but it has a number of serious drawbacks. For one, it has a narrow safety margin; the effective dose is close to the lethal dose. It also causes liver damage, and it must be protected from oxygen during storage to prevent the formation of deadly phosgene gas.

The most potent anesthetic gas, cyclopropane (Chapter 10), was first used at the University of Wisconsin General Hospital, Madison, in 1934. Small amounts rapidly produce insensitively to pain without rendering the patient unconscious. The great drawback is that cyclo-

Figure 23.2 An 1802 caricature by James Gillray of a lecture demonstrating the effect of ether vapor and nitrous oxide. [Courtesy of the Bureau of Narcotics and Dangerous Drugs, Washington, DC.]

propane forms explosive mixtures with air throughout its effective range of concentrations. Special equipment and an experienced anesthetist are required for its use.

Modern anesthetics include fluorine-containing compounds such as halothane, enflurane, and methoxyflurane (Figure 23.3). These compounds are nonflammable and relatively safe for the patient. Their safety, particularly that of halothane, for operating-room personnel, however, has been questioned. For example, female workers suffer a higher rate of miscarriages than the general population.

Modern surgical practice has moved away from the use of a single anesthetic. Generally, a patient is given an intravenous anesthetic such as thiopental (page 522) to produce unconsciousness. The gaseous anesthetic then is administered to provide insensitivity to pain and to keep the patient unconscious. A relaxant, such as curare*, also may be employed. Curare and related compounds produce profound relaxation; thus, only light anesthesia is required. This practice avoids the hazards of deep anesthesia.

Nearly all gaseous and volatile-liquid organic compounds exhibit anesthetic action. (Methane [Chapter 10] is an exception; it appears to be physiologically inert. Nitrous oxide is the only common inorganic anesthetic.) It is the anesthetic action of organic solvents that leads to their abuse. Sniffing glue solvents, gasoline, aerosol propellants, and other inhalants is perhaps the deadliest form of drug abuse. The dose required for intoxification is not far from that which will stop the heart. And it is difficult to measure a dose from a plastic or paper bag—the usual method of "glue sniffing." Also, as with nitrous oxide, sublethal doses can cause permanent brain damage by cutting down the oxygen supply to the brain.

The potency of an anesthetic is related to its solubility in fat. General anesthetics seem to work by dissolving in the fatlike membranes of nerve cells. This changes the permeability of the membranes, and the conductivity of the neurons is depressed.

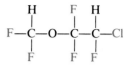
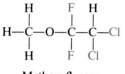
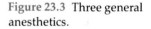

Halothane

Enflurane

Methoxyflurane

Figure 23.3 Three general anesthetics.

Local Anesthetics

For dental work and minor surgery, it is usually desirable to deaden the pain in one part of the body only. The first local anesthetic to be used successfully was cocaine. This drug was first isolated in 1860 from the leaves of the coca plant (Figure 23.4). Its structure was determined

Figure 23.4 Coca leaves and illicit forms of cocaine. [Courtesy of the Bureau of Narcotics and Dangerous Drugs, Washington, DC.]

---

* Curare is the arrow poison used by South American Indian tribes. Large doses of curare kill by causing a complete relaxation of all muscles. Death occurs because of respiratory failure.

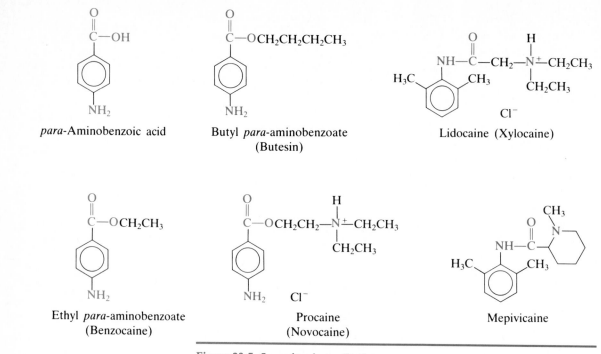

Figure 23.5 Some local anesthetics.

by Richard Willstätter in 1898. Even before Willstätter's work, there were attempts to develop synthetic compounds with similar properties. Cocaine is a powerful stimulant. Its abuse is discussed later in this chapter.

Certain esters of *para*-aminobenzoic acid act as local anesthetics (Figure 23.5). The ethyl and butyl esters are used to relieve the pain of burns and open wounds. These are applied as ointments, usually in the form of picrate salts.

More powerful in their anesthetic action are a series of derivatives with a second nitrogen atom in the alkyl group of the ester. Perhaps the best known of these is procaine (Novocaine), first synthesized by Alfred Einhorn in 1905. Procaine can be injected as a local anesthetic, or it can be injected into the spinal column to deaden the entire lower portion of the body. Local anesthetics work by blocking nerve impulses to the brain. When the block involves the spinal cord, messages of pain from the lower parts of the body are prevented from reaching the brain.

The local anesthetic of choice nowadays is often lidocaine or mepivicaine. Each compound is highly effective and yet has a fairly low toxicity (Table 23.1). Note that lidocaine and mepivicaine are not derivatives of *para*-aminobenzoic acid, but they do share some structural features with the compounds that are.

Table 23.1 Toxicities of Various Drugs*

	Drug	LD$_{50}$ (in milligrams per kilogram of body weight)	Method of Administration	Experimental Animal
Local anesthetics	Lidocaine	292	Oral	Mice
	Procaine	88	Intravenous	Mice
	Cocaine	17.5	Intravenous	Rats
Barbiturates	Barbital	600	Oral	Mice
	Pentobarbital	118	Oral	Rats
	Secobarbital	125	Oral	Rats
	Phenobarbital	600	Oral	Rats
	Amobarbital	575	Oral	Rabbits
	Thiopental	149	Intraperitoneal	Mice
Narcotics	Morphine	500	Subcutaneous	Mice
	Heroin	150	Subcutaneous	Rabbits
	Meperidine	170	Oral	Rats
Stimulants	Caffeine	355	Oral	Rats
	Nicotine	230	Oral	Mice
		0.3	Intravenous	Mice
	Amphetamine	55	Oral	Rats
	Methamphetamine	70	Intraperitoneal	Mice
	Mescaline	500	Intraperitoneal	Mice

* Comparisons of toxicities in different animals—and extrapolation to humans—are at best crude approximations. The method of administration can have a profound effect on the observed toxicity.
SOURCE: Martha Windholz (Ed.). *The Merck Index*, 10th ed. Rahway, NJ: Merck and Co., 1983.

The Barbiturates: Sedation, Sleep, and Suicide

As a family of related compounds, the barbiturates display a wide variety of properties. They can be employed to produce mild sedation, deep sleep, and even death.

Barbituric acid was first synthesized in 1864 by Adolph von Baeyer, a young student of August Kekulé (Chapter 10). He made it from urea, which occurs in urine, and malonic acid, which occurs in apples. The term **barbiturates**, according to Willstätter, came about because, at the time of the discovery, von Baeyer was infatuated with a girl named Barbara. The word comes from *Barbara* and *urea*.

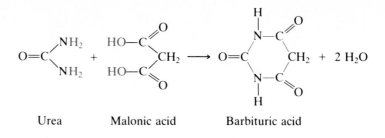

Urea   Malonic acid   Barbituric acid

The medicinal value of the barbiturates was discovered in 1903 by Joseph von Mering. A derivative called barbital (Figure 23.6) was found to be useful in putting dogs to sleep. Several thousand barbiturates have been synthesized through the years, but only a few have found widespread use in medicine. Pentobarbital (Nembutal) is employed as a short-acting hypnotic drug. Before the discovery of the modern tranquilizers, pentobarbital was used widely to calm anxiety and other disorders of psychic origin.

Phenobarbital (Luminal) is a long-acting drug. It, too, is an hypnotic and can be used as a sedative. Phenobarbital is employed widely as an anticonvulsant for epileptics and brain-damaged people. The action of amobarbital (Amytal) is intermediate in duration. Thiopental* (Pentothal), a compound that differs from pentobarbital only in that an oxygen atom on the ring has been replaced by a sulfur atom, is used widely in anesthesia (Figure 23.6).

The barbiturates were once used in small doses as sedatives. The dosage for sedation was generally a few milligrams. In larger dosages (about 100 mg), the barbiturates induce sleep. They were once the sleeping pills so widely used—and abused—by middle-class, often middle-aged, people. The lethal dose is in the vicinity of 1500 mg (1.5 g). Barbiturates are the drug of choice for many suicides. News reports list the cause of death as "an overdose of sleeping pills." There is also potential for accidental overdose. After a couple of tablets, the person becomes groggy. If the person is unable to remember whether he or she took the sleeping pills or not, he or she may take more pills.

The barbiturates are especially dangerous when ingested along with ethyl alcohol. This combination produces an effect much more drastic than just the sum of the effects of two depressants. The effect of the barbiturate is enhanced by factors of up to 200 when taken with, or drinking, alcoholic beverages. This effect of one drug in enhancing the action of another is called a synergistic effect.

---

* Thiopental has been investigated as a possible truth drug. It does seem to help psychiatric patients recall traumatic experiences. It also helps uncommunicative individuals talk more freely. It does not, however, prevent one from withholding the truth or even from lying. No true truth drug exists.

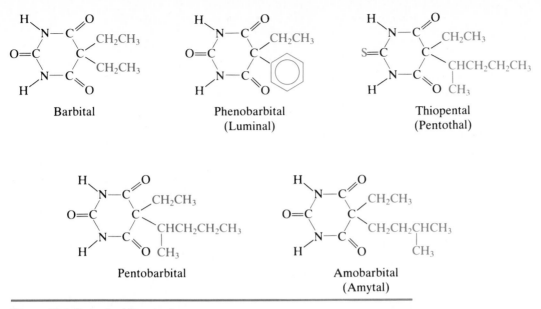

Figure 23.6 Some barbiturate drugs.

Synergism probably has led to many deaths. According to news reports, an autopsy showed that a well-known Hollywood gossip columnist ingested only about 200 mg of barbiturate and 2 oz of alcohol. Either alone would have produced only mild sedation; the combination killed her. Synergistic effects are not limited to alcohol–barbiturate combinations. You should never take two drugs at the same time without competent medical supervision.

The barbiturates are strongly addictive. Habitual use leads to the development of a tolerance to the drugs and ever-larger doses are required to get the same degree of intoxication. Barbiturates are legally available by prescription only, but they are a part of the illegal drug scene also. They are known as "downers" because of their depressant, sleep-inducing effects. They are also called "goof balls," "yellow jackets" (pentobarbital), "red devils" (secobarbital), "blue heavens" (amobarbital), and other terms generally descriptive of the capsules' color.

The side effects of barbiturates are similar to those of alcohol. Abuse leads to hangovers, drowsiness, dizziness, and headaches. Withdrawal symptoms are often severe, accompanied by convulsions and delirium. In fact, some medical authorities now say that withdrawal from barbiturates is more dangerous—that is, more likely to cause death—than withdrawal from heroin.

Barbiturates are cyclic amides (Chapter 11). Notice, however, that the barbiturate ring resembles that of thymine, one of the bases found in nucleic acids. Recent evidence indicates that barbiturates may act by

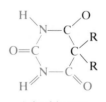

A barbiturate

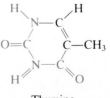

Thymine

substituting for thymine (or cytosine or uracil) in nucleic acids, thus interfering with protein synthesis.

Dissociative Anesthetics: Ketamine and PCP

Ketamine

Phencyclidine (PCP)

Ketamine, like thiopental an intravenous anesthetic, is called a **dissociative anesthetic**: it induces hallucinations similar to those reported by people who have had near-death experiences. They seem to remember observing their rescuers from a vantage point above it all, or moving through a dark tunnel toward a bright light. Unlike thiopental, ketamine seems to affect associative pathways before it hits the brain stem.

Little is known of the action of ketamine at the molecular level. If it acts by fitting receptors in the body, we can assume that our bodies produce their own chemicals that fit those receptors. These compounds may be synthesized or released only in extreme circumstances—such as in near-death experiences. Is it possible that we are on the threshold of the discovery of the chemistry of "life after death"?

Closely related to ketamine is phencyclidine (PCP), known on the street as "angel dust." PCP is soluble in fat and has no appreciable water solubility. It is stored in fatty tissue and released when the fat is metabolized; this accounts for the "flashbacks" commonly experienced by users.

PCP has become an important part of the illegal drug scene. It is cheap and easily prepared. It was tested and found to be too dangerous for human use, but found use as an animal tranquilizer. Many users experience bad "trips" with PCP. About 1 in 1000 develops a severe form of schizophrenia. Laboratory tests show that PCP depresses the immune system. This could lead to increased risk of infection. Despite these well-known problems, every few years a new crop of young people appears on the scene to be victimized by this hog tranquilizer.

Figure 23.7 Opium poppy and derivatives: crude and smoking opium, codeine, heroin, and morphine. [Courtesy of the Bureau of Narcotics and Dangerous Drugs, Washington, DC.]

The Opium Alkaloids: Narcotics

Narcotics are drugs that produce narcosis (stupor or general anesthesia) and relief of pain (analgesia). Many drugs produce these effects, but in the United States only those that are also *addictive* are legally classified as narcotics. Their use is regulated by federal law.

Opium is the dried, resinous juice of the unripe seeds of the oriental

Figure 23.8 Trade card advertising Mrs. Winslow's soothing syrup. [Courtesy of the Bureau of Narcotics and Dangerous Drugs, Washington, DC.]

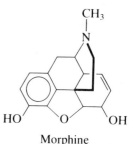

Morphine

poppy (*Papaver somniferum*). It is a complex mixture of some 20 nitrogen-containing organic bases (alkaloids), sugars, resins, waxes, and water. The principal alkaloid, morphine, makes up about 10% of the weight of raw opium. Raw opium was used in many of the patent medicines of the nineteenth century. Ayer's Cherry Pectoral, Jayne's Expectorant, Pierce's Golden Medical Discovery, and Mrs. Winslow's Soothing Syrup were but a few of the many such products.

Morphine was first isolated in 1805 by Frederich Sertürner, a German pharmacist. With the invention of the hypodermic syringe in the 1850s, a new method of administration became available. Injection of morphine directly into the bloodstream was more effective for the relief of pain, but it also seriously escalated the problem of addiction. Morphine was used widely during the American Civil War (1860–1865). It was effective for relief of pain caused by battle wounds. One side effect of morphine use is constipation. Noting this, soldiers came to use morphine as a treatment for that other common malady of men on the battlefront—dysentery. During their wartime service, over a hundred thousand soldiers became addicted to morphine. The affliction was so common among veterans that it came to be known as "soldier's disease."

Morphine and other narcotics were placed under the federal government's control by the Harrison Act of 1914. Morphine is still used by prescription for relief of severe pain. It also induces lethargy, drowsiness, confusion, euphoria, chronic constipation, and depression of the respiratory system. It is strongly addictive if administered in amounts

Figure 23.9 An old-fashioned hypodermic syringe. [Courtesy of the Bureau of Narcotics and Dangerous Drugs, Washington, DC.]

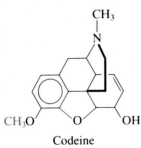

Codeine
(Methylmorphine)

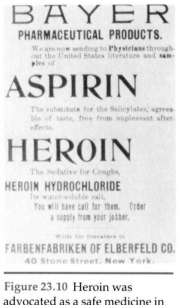

Figure 23.10 Heroin was advocated as a safe medicine in 1900. It was widely marketed as a sedative for coughs. [Courtesy of the Bureau of Narcotics and Dangerous Drugs, Washington, DC.]

greater than the prescribed doses or for a period longer than the prescribed time.

Slight changes in the molecular architecture of morphine produce altered physiological properties. Replacement of one of the —OH groups by an OCH_3 group produces codeine. Actually, codeine occurs in opium to an extent of about 0.5%. It is usually synthesized, however, by methylating the more abundant morphine molecules. About 55 000 kg of codeine is produced each year in the United States, enough for 16 doses of 15 mg for every person in the country.

Codeine is similar to morphine in its physiological action, except that it is less potent and has less tendency to induce sleep. It is also thought to be less addictive. In amounts of less than 2.2 mg/mL, codeine is exempt from the stringent narcotics regulations and is used in cough syrups.

In the laboratory, reaction of morphine with acetic anhydride (acetic anhydride is derived from acetic acid by removal of water) produces heroin. This morphine derivative was first prepared by chemists at the Bayer Company of Germany in 1874. It received little attention until 1890, when it was proposed as an antidote for morphine addiction. Shortly thereafter, Bayer was widely advertising heroin as a sedative for coughs, often in the same ads as aspirin! It soon was found, however, that heroin induced addiction more quickly than morphine and that heroin addiction was harder to cure.

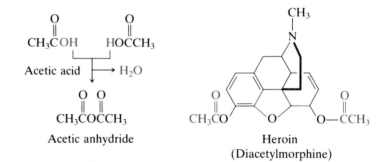

Acetic acid   Acetic anhydride   Heroin
                                       (Diacetylmorphine)

The physiological action of heroin is similar to that of morphine, except that heroin seems to produce a stronger feeling of euphoria for a longer period of time. Heroin is not legal in the United States, even by prescription. It has, however, been advocated for use in relief of pain in terminal cancer patients and has been so used in Britain.

Addiction probably has three components: emotional dependence, physical dependence, and tolerance. Psychological dependence is evident in the uncontrollable desire for the drug. Physical dependence is shown by acute withdrawal symptoms such as convulsions. The body cells are conditioned to the drug and cannot function normally without it. The difference in physical and psychological dependence may be only one of degree, not kind. Both are likely to be biochemical in origin. Tolerance for the drug is evidenced by the increasing dosages required to produce in the addict the same degree of narcosis and analgesia.

Deaths from heroin usually are attributed to overdoses, but the situation is not altogether clear. The problem seems to be a matter of quality control. As an illustration, the office of the chief medical examiner for New York City analyzed 132 samples of drugs, supposedly heroin, that had been confiscated on the streets. Twelve contained no heroin at all. The remaining 120 varied from 1 to 77% heroin. A user could think he or she was getting a dose of 1 unit, when he or she was actually getting 77 times as much—a catastrophic overdose.

Figure 23.11 Street heroin. [Courtesy of the Bureau of Narcotics and Dangerous Drugs, Washington, DC.]

Synthetic Narcotics: Analgesia and Addiction

Much research has gone into developing a drug that would be as affective as morphine for the relief of pain but that would not be addictive. Perhaps the best known of the synthetic narcotics is meperidine (Demerol). Meperidine is somewhat less effective than morphine, but it has the advantage that it does not cause nausea. Repeated use, unfortunately, does lead to addiction.

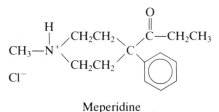

Meperidine
(Demerol)

Another synthetic narcotic is methadone. This drug has been widely used to treat heroin addiction. Like heroin, methadone is highly addictive. However, when taken orally, it does not induce the sleepy stupor characteristic of heroin intoxication. Unlike a heroin addict, a person on methadone maintenance usually is able to hold a productive job. Methadone is available free in clinics. If an addict who has been taking methadone reverts to heroin, the methadone in his or her system effectively blocks the euphoric rush normally given by heroin, and so reduces the addict's temptation to use heroin.

Methadone maintenance is not a perfect answer. Perhaps it is not even a good one. When injected into the body, methadone gives an effect similar to that of heroin, and methadone has been diverted for

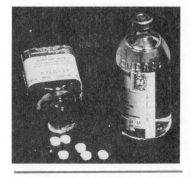

Figure 23.12 Forms of meperidine, a synthetic narcotic. [Courtesy of the Bureau of Narcotics and Dangerous Drugs, Washington, DC.]

illegal use in this manner. And an addict on methadone is still an addict. All the problems of tolerance (and cross-tolerance with heroin and morphine) still exist.

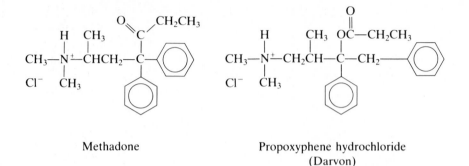

Methadone   Propoxyphene hydrochloride
                                            (Darvon)

Still another synthetic narcotic is propoxyphene hydrochloride (Darvon). This drug has been used widely as a prescription pain reliever. Independent studies have shown that Darvon is no more effective than aspirin for the relief of pain. Structurally, Darvon is quite similar to meperidine and methadone. Recently, it also has been shown to be addictive.

Chemists have synthesized thousands of morphine analogs. Only a few have shown significant analgesic activity. Most are addictive. Morphine acts by binding to receptors in the brain. Those molecules that have morphinelike action are called **agonists**. Morphine **antagonists** are drugs that block the action of morphine, most likely by blocking the receptors. Some molecules have both agonist and antagonist activity. These show great promise as analgesics. An example is pentazocine (Talwin). It is less addictive than morphine and yet it is effective for relief of pain. There is some hope that an effective analgesic could be developed that is not addictive, but to date the two effects seem inseparable.

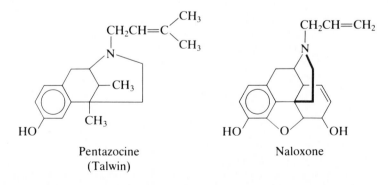

Pentazocine   Naloxone
(Talwin)

Pure antagonists such as naloxone are of value in treating opiate addicts. Overdosed addicts can be brought back from death's door by an injection with naloxone. Long-acting antagonists can block the action of heroin for as much as a month, thus aiding an addict in overcoming his or her addiction.

A Natural High: The Brain's Own Opiates

Morphine acts by fitting specific receptor sites in the brain. These morphine receptors were first demonstrated in 1973 by Solomon Synder and Candace Pert at Johns Hopkins University School of Medicine.

Why should the human brain have receptors for a plant-derived drug like morphine? There seemed to be no good reason, so several investigators started a search for morphinelike substances produced by the human body. Not one, but several such substances soon were found. Each was a short peptide chain composed of amino acid units. Those with five amino acid units are called **enkephalins**. There are two enkephalins, and they differ only in the amino acid at the end of the chain. *Leu*-enkephalin has the sequence *Tyr-Gly-Gly-Phe-Leu* and *Met*-enkephalin is *Try-Gly-Gly-Phe-Met*. The substances with chains of 30 amino acids are called **endorphins**.

Some of the enkephalins have been synthesized and shown to be potent pain relievers. Their use in medicine is quite limited, however, because, after being injected, they are rapidly broken down by the enzymes that hydrolyze proteins. It is hoped, though, that analogs more resistant to hydrolysis can be employed as morphine substitutes for the relief of pain. Unfortunately, both the natural enkephalins and the analogs, like morphine, seem to be addictive.

It appears that enkephalins and endorphins are released as a response to pain deep in the body. Bruce Pomeranz of the University of Toronto has collected evidence that indicates that acupuncture anesthetizes by stimulating the release of the brain "opiates." The long needles stimulate deep sensory nerves that cause the release of the peptides that then block the pain signals.

Endorphin and enkephalin release also has been used to explain other phenomena once thought to be largely psychological. A soldier, wounded in battle, feels no pain until the skirmish is over. His body has secreted its own painkiller. The production of these compounds during strenuous athletic activity is discussed in the next chapter.

We shall return shortly to the chemistry of drugs. Before we do so, however, let's examine some chemistry of the nervous system and explore how nerve cells work.

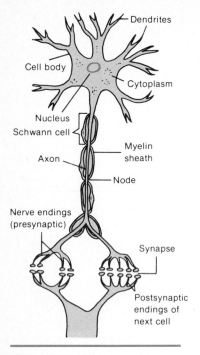

Figure 23.13 A human nerve cell.

Some Chemistry of the Nervous System

The nervous system is made up of billions of **neurons** (nerve cells) with 10^{15} connections between them. The brain operates with a power output of about 25 W and has capacity for about 10 trillion bits of information. Nerve cells vary a great deal in shape and size. One type is shown in Figure 23.13. The essential parts of each cell are the cell body, the axon, and the dendrites. We discuss here only those nerves that make up the involuntary (autonomic) nervous system. These nerves carry messages between the organs and glands that act involuntarily (such as the heart, the digestive organs, and the lungs) and the brain and spinal column.

Although the axons on a given nerve cell may be up to 60 cm long, there is no continuous pathway from an organ to the central nervous system. Messages must be transmitted across tiny, fluid-filled gaps, or

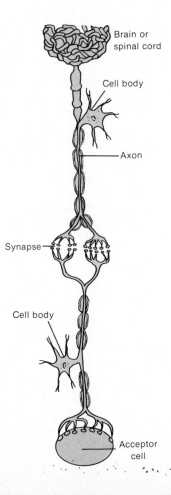

Figure 23.14 Schematic diagram of the pathway by which messages are transmitted to and from an acceptor cell in a gland or an organ to the central nervous system.

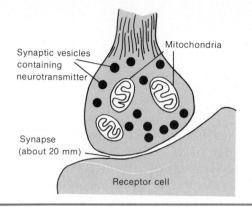

Figure 23.15 Schematic diagram of a synapse. When an electrical signal reaches the nerve ending, neurotransmitter molecules are released from the vesicles. They then migrate across the narrow gap (synapse) and move to the receptor cell, where they fit specific sites.

synapses (Figure 23.14). When an electrical signal from the brain reaches the end of an axon, specific chemicals (called **neurotransmitters**) that carry the impulse across the synapse to the next cell are liberated. There are perhaps a few dozen neurotransmitters. Each has a specific function. Messages are carried to other nerve cells, to muscles, and to the endocrine glands (such as the adrenal glands). Each neurotransmitter fits a specific receptor site on the receptor cell (Figure 23.15). Many drugs (and some poisons) act by mimicking the action of the neurotransmitter. Others act by blocking the receptor and preventing the neurotransmitter from acting on it. Several of the neurotransmitters are amines (Chapter 11), as are some of the drugs that affect the chemistry of our brains.

Brain Amines: Depression and Mania

We all have our ups and downs in life. These moods probably result from multiple causes, but it appears likely that a variety of chemical compounds formed in the brain are involved. Before we consider these ups and downs, though, let's take a look at epinephrine, an amine formed in the adrenal glands.

Commonly called adrenaline, epinephrine is secreted by the adrenal glands. A tiny amount of epinephrine causes a great increase in blood pressure. When a person is under stress or is frightened, the flow of adrenaline prepares the body for fight or flight. Because culturally

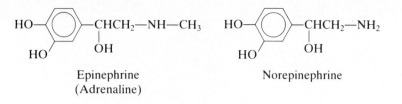

Epinephrine
(Adrenaline)

Norepinephrine

imposed inhibitions prevent fighting or fleeing in most modern situations, the adrenaline-induced supercharge is not used. This sort of frustration has been implicated in some forms of mental illness.

One widely held theory of a biochemical basis for mental illness involves a relative of ephinephrine. Norepinephrine (NE) is a neurotransmitter formed in the brain. When NE is formed in excess, the person is in an elated, perhaps hyperactive, state. In large excess, NE could induce a manic state. A deficiency of NE, on the other hand, could cause depression.

Drugs that block the action of NE could also lead to depression, while those that enhance or mimic its action act as stimulants. Several such drugs are discussed in the sections that follow.

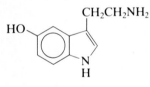

Serotonin

Another neurotransmitter, serotonin, also seems to play a role in mental illness. Serotonin is involved in sleep, sensory perception, and the regulation of body temperature. Its exact role in mental illness is not clear. A metabolite of serotonin, 5-hydroxyindoleacetic acid (5-HIAA), is found in unusually *low* levels in the spinal fluid of suicide victims. This indicates that abnormal serotonin metabolism may play a role in depression.

Richard Wurtman of the Massachusetts Institute of Technology has found a relationship between diet and serotonin levels in the brain. Serotonin is produced in the body from the amino acid tryptophan (Figure 23.16). The synthesis involves several steps, each catalyzed by an enzyme. Wurtman found that diets high in carbohydrates lead to high levels of serotonin. Lots of protein lowers the serotonin concentration. That may seem strange, for protein has lots of tryptophan and carbohydrates have little. But, Wurtman says, protein is only 1% tryptophan. In the presence of all those other amino acids, little tryptophan reaches the brain. With a carbohydrate meal, the hormone insulin lowers the level of the other amino acids in the blood, allowing relatively high levels of tryptophan to reach the brain.

Norepinephrine also is synthesized in the body from an amino acid. It is derived from tyrosine. The synthesis is complex and proceeds through several intermediates (Figure 23.17). Each step is catalyzed by one or more enzymes. The intermediate compounds also have physiological activity: dopa has been used successfully in the treatment of Parkinson's disease, and dopamine has been employed to treat low blood pressure. Since tyrosine is also a component of our diets, it may well be that our mental state depends to a fair degree on our diet.

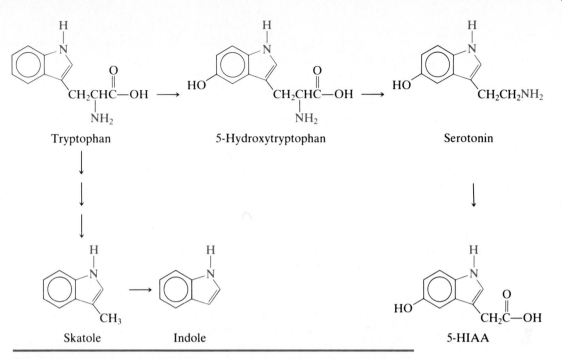

Figure 23.16 Serotonin is produced in the brain from tryptophan. A variety of other interesting compounds are also produced from tryptophan. One route leads (through several intermediates) to skatole, the principal odiferous ingredient in human feces, and to indole. Skatole and indole are used in perfumes.

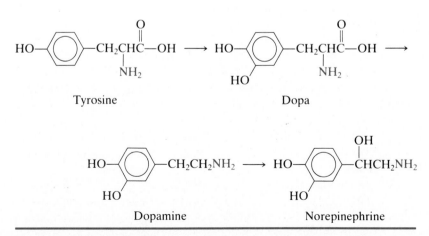

Figure 23.17 The biosynthesis of norepinephrine from tyrosine.

It has been estimated that nearly 1 out of every 10 people in the United States suffers from mental illness. Over half the patients in hospitals are there because of mental problems. When the biochemistry of the brain is more fully understood, mental illness may be cured (or at least alleviated) by administration of drugs. In subsequent sections of this chapter, we see just how far we have already come in learning to control our moods with drugs. As is true for so many things, the potential for good that such compounds represent is matched by a potential for abuse.

Love: A Chemical Connection

Phenylethylamine

Phenylacetic acid

The notion that love might be chemical in origin is an unsettling one. However, it is possible that the emotions that trigger romantic relationships are governed in part by a chemical called phenylethylamine (PEA). PEA functions as a neurotransmitter in the human brain. It appears to create excited, alert feelings and moods. Increased levels of PEA produce a "high" feeling identical to that people describe as "being in love." Not surprisingly, the chemical structure of PEA resembles that of norepinephrine.

How much PEA does it take to get back that old feeling? Levels of PEA in the brain can be estimated by measuring levels of its metabolite, phenylacetic acid, in the urine. Low levels of urinary phenylacetic acid correlate with depression. This has prompted researchers to investigate factors that increase PEA levels in the brain. There are no food sources of PEA, but protein-rich foods contain phenylalanine (Chapter 18), an amino acid precursor of PEA. Perhaps a steak dinner is a way to his or her heart, after all.

Stimulant Drugs: Amphetamines

Among the more widely known stimulant drugs are a variety of synthetic amines related to phenylethylamine (Figure 23.18). Note the simularity of these molecules to those of epinephrine and norepinephrine; all are derived from the basic phenylethylamine structure. The amphetamines probably act as stimulants by mimicking the natural brain amines.

Amphetamine and methamphetamine have been widely abused. Amphetamine has been extensively used for weight reduction. It has also been employed for treating mild depression and narcolepsy, a rare form of sleeping sickness. Amphetamine induces excitability, restlessness, tremors, insomnia, dilated pupils, increased pulse rate and blood pressure, hallucinations, and psychoses. It is no longer recommended for weight reduction. It was found that, generally, any weight loss was only temporary. The greatest problem, however, was the diversion of vast quantities of amphetamines into the illegal drug market. Amphetamines are inexpensive. Armed forces personnel, truck drivers, and college students have been among the heavy users.

Methamphetamine has a more pronounced psychological effect than amphetamine. Generally, the "speed" that abusers inject into their veins is methamphetamine. Such injections, at least initially, are said to give the abuser a euphoric rush. Shooting methamphetamine is quite dangerous, though, because the drug is rather toxic (see Table 23.1).

Another amphetamine derivative, phenylpropanolamine, is widely used as an over-the-counter appetite suppressant. Like its relatives, this compound is a stimulant. Studies show that it is at best marginally effective as a diet aid, and it poses a threat to people with hypertension. Nevertheless, sales of phenylpropanolamine are one billion tablets at $150 million dollars each year.

One controversial use of amphetamines has been their employment in the treatment of hyperactivity in children. The drug of choice is often methylphenidate (Ritalin). Although it is a stimulant, the drug seems to calm kids who otherwise can't sit still. This use has been criticized as "leading to drug abuse" and as "solving the teacher's problem, not the kid's."

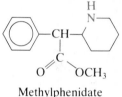

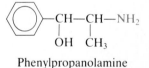

Figure 23.18 Amphetamine and related compounds.

Dextroamphetamine (Dexedrine) is another stimulant drug that has been abused widely. It is structurally related to amphetamine in a very subtle way. Actually, amphetamine is not a single compound but a mixture of two isomers. These isomers have the same atoms and same groups of atoms, but the relative spatial orientations of these constituent atoms and groups are different in the two isomers. One isomer is the mirror image of the other.

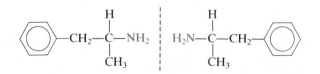

These isomers are not superimposable but are related to one another in much the same way that your right hand is related to your left. You can't fit a right-hand glove on your left hand or vice versa. These mirror-image isomers fit enzymes differently; thus they have different effects. The **dextro** (right-handed) isomer is a stronger stimulant than the **levo** (left-handed). Dexedrine is the trade name for the pure dextro isomer. Benzedrine is the trade name for a mixture of the two isomers in equal amounts. Dexedrine is two to four times as active as Benzedrine.

Mirror-image isomerism is quite common in organic chemicals of biological importance. Usually only one of the two isomers occurs in nature; thus, for many purposes, it is not necessary to emphasize the isomerism. However, any molecule that has four different groups attached to a single carbon atom can exist as mirror-image isomers. Lactic acid is a familiar example.

$$CH_3-\underset{\underset{OH}{|}}{\overset{\overset{H}{|}}{C}}-\overset{\overset{O}{||}}{C}-OH \quad \quad HO-\overset{\overset{O}{||}}{C}-\underset{\underset{OH}{|}}{\overset{\overset{H}{|}}{C}}-CH_3$$

The dextro isomer occurs in blood and muscle tissue, where it is formed by the oxidation of glucose. Levo lactic acid occurs in sour milk.

Caffeine: Coffee, Tea, or Cola

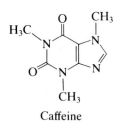
Caffeine

The beverages coffee and tea and some soft drinks (Chapter 18) contain the mild stimulant caffeine. The effective dose of caffeine is about 200 mg, corresponding to about two cups of strong coffee or tea. Caffeine is also available in tablet form as a stay-awake or keep-alert type of drug. The best known brands are probably No-Doz and Vivarin. No-Doz contains about 100 mg of caffeine per tablet; each Vivarin tablet has 200 mg.

Is caffeine addictive? The "morning grouch" syndrome indicates that it may be mildly so. There is also evidence that caffeine may be involved in chromosome damage. To be safe, people in their childbearing years should avoid large quantities of caffeine. Overall, the hazards of caffeine ingestion seem to be slight.

Going Up in Smoke: Nicotine Addiction

Nicotine

Another common stimulant is nicotine. This drug is taken by smoking or chewing tobacco. Nicotine is highly toxic to animals see (Table 23.1). It is especially deadly when injected; the lethal dose for a human is estimated to be about 50 mg. Nicotine is used widely in agriculture as a contact insecticide. Nicotine seems to have a rather transient effect as a stimulant. This initial response is followed by depression.

Is nicotine addictive? Casual observation of a person trying to quit smoking seems to indicate that it is. There is also evidence of the development of tolerance. It is difficult, however, to separate all the social factors involved in smoking from the physiological effects.

Cocaine: The Snow Sniffers

Cocaine, mentioned earlier as a local anesthetic, also serves as a powerful stimulant. Cocaine is obtained from the leaves of a shrub that grows almost exclusively on the eastern slopes of the Andes Mountains. Many of the Indians living in and around the area of cultivation chew the leaves—mixed with lime and ashes—for their stimulant effect. Cocaine usually arrives in the United States as glistening white crystals (hence the slang name "snow"). Some cocaine is legally imported for legitimate human and veterinary medical purposes. A far greater quantity is smuggled in for the illegal drug market.

Cocaine acts by preventing norepinephrine from being taken back up after it is released by nerve cells. High levels of NE develop, causing nerve cells to fire wildly. The brain becomes like an overloaded telephone switchboard. Use of cocaine increases stamina and reduces fatigue, but the effect is short-lived. Depletion of NE occurs in less than an hour, leading to depression and a craving for more cocaine. Once quite expensive and limited to use mainly by the wealthy, cocaine has become available in cheaper and more potent forms. Hundreds, including several well-known athletes, have died from cocaine overdose.

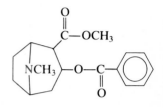

Cocaine

Antianxiety Agents

The hectic pace of life in the modern world has driven people to seek rest and relaxation in chemicals. Ethyl alcohol is undoubtedly the most widely used tranquilizer. The drink before dinner—to "unwind" from the tensions of the day—is very much a part of the American way of life. Many people, however, seek their relief in other chemical forms.

Several over-the-counter drugs—Cope, Vanquish, and Compoz among others—claim to be able to help us cope with or vanquish our problems, or at least to compose ourselves in the face of minor adversity. Such products usually contain a little aspirin plus an antihistamine (Chapter 22). The latter has a side effect of making one drowsy. These products have come under attack by consumer groups for being worthless at best—and perhaps even dangerous.

Another group of drugs, available only by prescription, is widely employed to calm nervous tension. Several of these drugs are carbamates (Figure 23.19). Simple derivatives, such as ethyl carbamate, act as mild soporifics (sleep-inducing agents). The best-known tranquilizer in

Figure 23.19 Some carbamates.

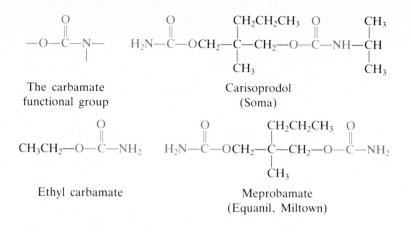

The carbamate
functional group

Carisoprodol
(Soma)

Ethyl carbamate

Meprobamate
(Equanil, Miltown)

this group is meprobamate (Equanil or Miltown). Proceeds from the sale of meprobamate amount to many millions of dollars per year. Another carbamate, carisoprodol (Soma), is employed as a muscle relaxant.

Another class of widely used antianxiety drugs is the benzodiazepines, compounds that feature seven-membered heterocyclic rings (Figure 23.20). Of these, perhaps the best known are diazepam (Valium) and chlorodiazepoxide (Librium). For many years, Valium was the most prescribed drug in the United States. A related drug, flurazepam (Dalmane), is widely used to treat insomnia. It has replaced the barbiturates as the sleeping pill of choice.

The benzodiazepine derivatives and the carbamates were formerly called "minor tranquilizers." They are still widely used for the treatment of anxiety. Some studies have shown them to be remarkably effective. Others have found the drugs to be little better than a placebo—a pill that looks like the drug tablet but which contains no active ingredient. The actual effect apparently depends upon the expectation of the patient. To the extent that these antianxiety agents really work, they do so simply by making people feel better by making

Figure 23.20 Some
benzodiazepine drugs.

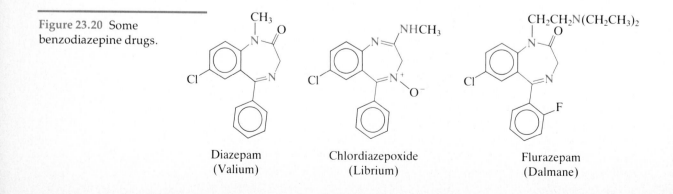

Diazepam
(Valium)

Chlordiazepoxide
(Librium)

Flurazepam
(Dalmane)

them feel dull and insensitive. They do not solve any of the underlying problems that cause anxiety.

Like most other mind-altering drugs, Valium acts by fitting specific receptors. Presumably our bodies produce Valiumlike compounds that fit these receptors. To date, no such compound has been found. Rather, scientists have found compounds, called beta-carbolines, that act on the brain's anxiety receptors to produce *terror.* There is yet so much to learn about the chemistry of the brain.

Anyway, what price tranquillity? After 20 years of use, it was finally found that Valium is addictive. People trying to go off it after prolonged use go into painful withdrawal.

Antipsychotic Agents

For centuries, the people of India used the snakeroot plant, *Rauwolfia serpentina*, to treat a variety of ailments including fever, snakebite, and other poisonings, and—most importantly—to treat maniacal forms of mental illness. Western scientists became interested in the plant near the middle of the twentieth century—after disdaining such remedies as quackery for many generations.

In 1952, rauwolfia was introduced into American medical practice as a hypertensive (blood-pressure-reducing) agent by Robert Wilkins of Massachusetts General Hospital. In the same year, Emil Schlittler of Switzerland isolated an active alkaloid, which he named reserpine, that has an impressive (intimidating?) structure.

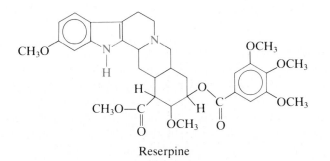

Reserpine

Rauwolfia was found not only to reduce blood pressure but also to bring about sedation. The latter finding attracted the interest of psychiatrists, who found reserpine so effective that by 1953 it had replaced electro-shock therapy for 90% of psychotic patients.

Also in 1952, chlorpromazine (Thorazine), which had been used in France as an antihistamine, was tried as a tranquilizer on psychotic patients in the United States. It was found to be extremely effective

Figure 23.21 Three tranquilizers and a psychic energizer.

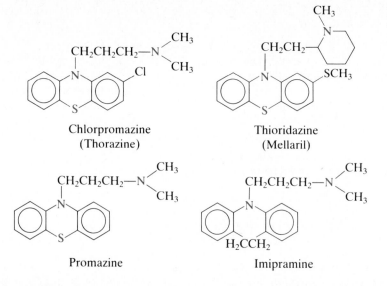

Chlorpromazine
(Thorazine)

Thioridazine
(Mellaril)

Promazine

Imipramine

against the symptoms of schizophrenia. Many compounds related to chlorpromazine have been synthesized. Several of these have been found to have interesting physiological properties. Promazine itself is a tranquilizer, but not as potent a one as chlorpromazine. Thioridazine (Mellaril) is a potent tranquilizer, reputed to be without some of the undersirable side effects of chlorpromazine.

It is worth noting that imipramine, which differs from promazine only in that the sulfur atom is replaced by a —CH_2CH_2— group, is not a tranquilizer at all. It is a psychic energizer. This indicates that slight structural changes sometimes can result in profound changes in properties and that we have a long way to go in understanding why drugs act as they do.

The antipsychotic drugs—reserpine and the promazine derivatives—have been one of the real triumphs of chemical research. They have served to greatly reduce the number of patients confined to mental hospitals by controlling the symptoms of schizophrenia to the extent that 95% of all schizophrenics no longer need hospitalization. These drugs are not cures. We can only hope that continued research will ascertain the causes of schizophrenia. At that time perhaps a real cure—or, better yet, a preventative—can be found.

The "Mindbenders": LSD

The third major class of drugs is popularly called the "mindbenders," because they qualitatively change the way we perceive things. Probably the most powerful of these drugs is LSD (from the German *lysergsaure diethylamid*). The physiological properties of this compound were discovered quite accidentally by Albert Hofmann in 1943. Hofmann, a chemist at the Sandoz Laboratories in Switzerland, unintentionally ingested some LSD. He later took 250 μg, which he considered a small dose, to verify that LSD had caused the symptoms he had experienced. Hofmann had a very rough time for the next few hours, exhibiting such symptoms as visual disturbance and schizophrenic behavior.

Lysergic acid diethylamide
(LSD)

Lysergic acid is obtained from ergot, a fungus that grows on rye. It is converted to the diethylamide by treatment with thionyl chloride (SOCl$_2$), followed by diethylamine. Note that a part of the LSD structure (color) resembles that of serotonin. LSD seems to act by fitting serotonin receptors; it is a serotonin agonist.

LSD is a potent drug, as indicated by the small amount required for a person to experience its fantastic effects. The usual dosage is probably about 10 to 100 μg. No wonder Hofmann had a bad time with 250 μg! To give you an idea of how small 10 μg is, let's compare that amount of LSD to the amount of aspirin in one tablet—one aspirin tablet contains about 300 000 μg of aspirin.

Is LSD a dangerous drug? A few facts are known, but most are disputed. In 1967, Maimon Cohen of the State University of New York at Buffalo reported that LSD damages chromosomes, especially those of the leucocytes (white blood cells). The report received wide publicity. Fear of damage to germ cells—and the subsequent birth of deformed babies—caused a decline in the use of LSD. Additional studies produced mixed results. Some seemed to confirm Cohen's findings. For example, in hamsters, LSD administered to pregnant females caused gross fetal deformities. Other studies, however, seemed to exonerate LSD as a cause of chromosomal damage. The question still has not been resolved.

Marijuana: Some Chemistry of Cannabis

Many complete books have been written about marijuana, yet all that is known for certain about the drug would fill only a few pages. Let's look at some chemistry of the weed and at some of the ways chemists are involved with the marijuana problem.

Tetrahydrocannabinol
(THC)

Figure 23.22 The marijuana plant. [Courtesy of Carolina Biological Supply Company.]

The weed *Cannabis sativa* (Figure 23.22) has long been useful. The stems yield tough fibers for making ropes. *Cannabis* has been used as a drug in tribal religious rituals. Marijuana also has a long history as a medicine, particularly in India. In the United States, marijuana is second only to alcohol as an intoxicant.

The term **marijuana** refers to a preparation made by gathering the leaves, flowers, seeds, and small stems of the plant (Figure 23.23). These are generally dried and smoked. They contain a variety of chemical substances, many of them still unidentified. The principal active ingredient, however, has been identified as tetrahydrocannabinol (THC). Actually, there are several active cannabinoids in marijuana. Only one is shown here (margin). Raphael Mechoulam of the Hebrew University was the first to isolate the active ingredient (1949). The compound was first synthesized in 1967.

Marijuana plants, as they grow in nature, vary considerably in THC content. Most of the marijuana sold in North America has a THC content of about 1%. That native to the United States has a low THC content, usually about 0.1%. Potency depends on the genetic variety of plant, not to any significant extent on the climate or the soil where it is grown.

More potent preparations sometimes are made from marijuana. By selecting only the flowering tops and tender top leaves, you get a stronger product called *ganja*. (The ordinary marijuana is called *bhang* in India.) Jamaican *ganja* has a THC content of between 4 and 8%. Indian *ganja* is generally somewhat less potent. By collecting only the resinous secretions of the flowering parts, you get a product called hashish, or "hash" (known as *charas* in India). Hash has a THC content of between 5 and 12%. Liquid hash and hash oil are probably solvent extracts of marijuana.

The effects of marijuana are difficult to measure, partly because of the variable THC content of different preparations. A variety of standard potency is now grown and supplied for controlled clinical studies. With this standard product, some of the effects of marijuana can be measured in reproducible experiments. Smoking *Cannabis* increases the pulse rate, distorts the sense of time, and impairs some complex motor functions. These effects can be measured easily. Other results also have been noted widely, if less quantitatively. Marijuana smoking sometimes induces a euphoric sense of lightness—a floating sensation. Sometimes it causes a feeling of anxiety. Often, the user has an impression of brilliance, although studies have shown no mind-expanding effects. Users sometimes experience hallucinations, although these are much less frequent than with LSD. Marijuana seems to heighten one's enjoyment of food, with users relishing beans as much as they normally would enjoy steak.

The long-term effects of marijuana use are more difficult to evaluate. Some people claim that smoking marijuana leads to the use of harder drugs. There is little objective evidence that this is the case. In fact, studies have shown that more heroin addicts started out on alcohol than on marijuana.

There is some evidence—both direct and indirect—that marijuana causes brain damage. In September 1971, it was announced that marijuana caused brain lesions in rats. Researchers in Great Britain associated laziness, passivity, and mental sluggishness with brain damage in young men who had been heavy smokers of marijuana. But there have been millions of users, and, even if the observed brain damage is due to marijuana, it is less extensive than that caused by alcohol.

Some people also have claimed that excessive use of marijuana leads to psychoses. It has long been known to induce short-term psychotic episodes in those already predisposed and in others who take excessive amounts. Long-term psychoses, however, occur at the same rate among regular marijuana users as among the general population.

One of the more interesting reports has come from two surgeons at the Harvard Medical School. Menelaos Aliapoulios and John Harman claim to have treated 13 young men for gynecomastia—enlarged breasts. All were heavy marijuana users. Their breasts also discharged a white, milky liquid. The painful swelling receded in three of the men after they stopped smoking *Cannabis*, but three others needed surgery. The two doctors are convinced that marijuana contains a "feminizing ingredient." There is a slight structural similarity of THC to the female hormones.

Figure 23.23 Retail forms of marijuana. [Courtesy of the Bureau of Narcotics and Dangerous Drugs, Washington, DC.]

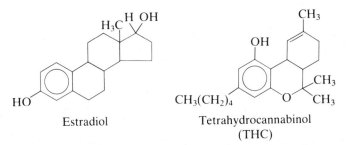

Estradiol

Tetrahydrocannabinol
(THC)

Figure 23.24 Marijuana dealers were early converts to the metric system. Shown here is a kilo brick of marijuana. [Courtesy of the Bureau of Narcotics and Dangerous Drugs, Washington, DC.]

Some studies have shown that THC bonds to estrogen receptors. Despite its apparent feminizing properties, THC in both high and low doses causes an initial rise in testosterone levels in men. With high doses, however, the slight increase is followed by a rapid fall to below-normal testosterone levels.

What do chemists have to do with all this? Well, they have isolated the active components and synthesized them. They can monitor the THC content of marijuana. They can monitor THC in the bloodstream and identify the products of its breakdown. They cannot, however, tell how it changes the body chemistry or what its long-term effects are.

Perhaps the most significant harm from marijuana comes through its impairment of complex motor functions—such as those used in driving an automobile. In addition to dodging drunks, we have to look out for potheads. Incidentally, chemists have developed a THC detector similar to the one used to test blood alcohol. Let's hope the road is cleared of all intoxicated people, no matter what they are intoxicated with.

Unlike alcohol, THC persists in the bloodstream for several days, presumably because it is soluble in fats. The products of its breakdown remain in the blood for as long as 8 days. The persistence of these chemicals inside the body indicates that some of a given dose may still be active in the body at the time another dose is taken. This might account for the fact that an experienced pot smoker can get high on a dose that doesn't affect a novice.

Marijuana may well have some legitimate medical uses. It reduces pressure in the eyes of people who have glaucoma. If not treated, the buildup of pressure eventually causes blindness. Also, marijuana seems to relieve the nausea that afflicts cancer patients undergoing radiation treatment and chemotherapy.

Chemistry has the potential to reveal a great deal about the effects of marijuana. What we may learn may be unimportant, or it may be too late. It took us 300 years to find out that cigarette smoking can cause cancer and even longer to realize that alcohol is responsible for cirrhosis of the liver. We may be up against a third such toxic substance. It does appear that, if the decision to legalize marijuana is made any time soon, it will most likely have to be made on something other than a scientific basis.

Drugs and Deception: Chemistry and Quality Control

One recurring problem on the illegal drug scene is that drugs are not always what they are supposed to be. The buyers simply have to trust the sellers—and their own experience—about the identity and quality of the products they buy. Even with marijuana, a readily identified weed, there are problems of quality control. The nonpotent variety often is harvested and used to dilute the potent product to increase profits. The nonpotent variety also has been found to be laced with other drugs to produce some kind of physiological effect.

Deception with other drugs is even more common. There have been numerous reports of synthetic THC on the illegal-drug scene. Little if any has been confirmed by crime laboratories. Most "THC" has been shown to be either LSD or PCP.

Many fraudulent products frequent the illegal drug scene. The scare about chromosomal damage from LSD led to increased interest in mescaline, a drug derived from the peyote cactus. (Note that mescaline is another phenylethylamine.) Mescaline is scarce and expensive; yet a product called mescaline became readily available on the street in some

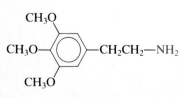

Mescaline

areas. Lab analyses proved that the product was really PCP. In fact, crime labs generally have found nearly two thirds of all drugs (other than marijuana) brought in for analysis to be something other than what the dealers said they were.

Use of illegal drugs is costly to society. From 50 to 80% of all personal injuries and accidents in the workplace are drug related. Drug users are absent from work 2.5 times as often as nonusers and 5 times as likely to file claims for compensation. In short, drug users cause a 25% loss of income to their employers.

Chemistry can provide drugs of enormous benefit to society. It can provide drugs that present society with serious problems. But it can't solve the drug problem. However, chemistry can provide information on which intelligent choices can be based. The choices, though, are up to you—as an individual and as a member of society. One thing seems fairly certain, though. No recreational drug has ever made anyone a better person or any society a better society. The best we can hope for is that little harm will be done.

Problems

1. What are psychotropic drugs?
2. What is the difference in drug misuse and drug abuse? Give an example of each.
3. List the three classes of drugs that affect the mind.
4. What are the effects of a stimulant drug?
5. What are the effects of a depressant drug?
6. What are the effects of a hallucinogenic drug?
7. What is a general anesthetic?
8. Which of the anesthetics in Figure 23.3 are ethers?
9. What is a local anesthetic?
10. Which of these anesthetics are dangerous because of flammability?
    a. diethyl ether   b. halothane
    c. chloroform   d. cyclopropane
11. For each of the following anesthetics, describe a disadvantage associated with its use. Do not include flammability.
    a. nitrous oxide b. halothane c. diethyl ether
12. What application does curare have in modern anesthesiology?
13. How do local anesthetics work? What is spinal anesthesia?
14. Refer to Table 23.1. Which is more toxic, procaine or cocaine? Can you use the data in Table 23.1 to compare the toxicities of lidocaine and cocaine?

15. Many local anesthetics are esters of what acid?
16. What is the basic structure common to all barbiturate molecules? How is the basic structure modified to change the properties of individual barbiturate drugs?
17. Describe a synergistic reaction involving drugs.
18. Name two dissociative anesthetics.
19. How do dissociative anesthetics work?
20. What are narcotics?
21. What is opium?
22. How does codeine differ from morphine in structure? In physiological effects?
23. How does heroin differ from morphine in chemical structure? In its action as a drug?
24. What are the three components of addiction? Can they be separated easily?
25. How does methadone maintenance work? Do you think it is a good idea?
26. How are endorphins and enkephalins related to each of the following?

    a. the anesthetic effect of acupuncture
    b. the absence of pain in a badly wounded soldier

27. What is an agonist?
28. What is an antagonist?

29. How does naloxone act in the treatment of a heroin overdose?

30. Define or identify each of the following.
    a. neuron   b. synapse   c. neurotransmitter

31. Which two naturally occurring amines are thought to play major roles in the biochemistry of mental health? What are their proposed roles?

32. Which amino acids serve as precursors for the amines of Problem 31?

33. How may our mental state be related to our diet?

34. From what basic structure are the amphetamines derived?

35. How do amphetamines exert a stimulant effect?

36. How does cocaine exert a stimulant effect?

37. Is caffeine addictive? Is nicotine?

38. What type of ingredient is found in over-the-counter drugs that purport to treat anxiety?

39. List several antianxiety agents.

40. List several antipsychotic agents.

41. How does imipramine differ from promazine in chemical structure? In its action as a drug?

42. Are tranquilizers a cure for schizophrenia?

43. To what class of compounds does LSD belong?

44. What are some of the problems involved in the clinical evaluation of LSD?

45. What are some of the problems in the clinical evaluation of marijuana?

46. Is marijuana a narcotic legally? Clinically?

47. What is the role of the chemist in the marijuana controversy?

48. How might marijuana have a feminizing effect on males?

49. Why is tetrahydrocannabinol retained in the body for several days?

50. If the minimum lethal dose (MLD) of amphetamine is 5 mg/kg, what would be the MLD for a 70-kg person? Can toxicity studies on animals always be extrapolated to humans?

51. Following are the generic names of several of the most widely prescribed drugs. Use *The Merck Index* or a similar reference to determine (if possible) the chemical structure, medical use, toxicity, and side effects of each.

    a. fursemide   b. methyldopa
    c. hydrochlorothiazide   d. digoxin
    e. ibuprofen   f. naproxen
    g. propanolol   h. indomethacin

52. When administered intravenously to rats, the LD_{50} of procaine is 50 mg/kg and that of nicotine is 1.0 mg/kg of body weight. Which drug is more toxic?

References and Readings

1. Baum, Rudy M. "New Variety of Street Drugs Poses Growing Problems." *Chemical and Engineering News*, 9 September 1985, pp. 7–16. Discusses designer drugs.

2. Bennett, William. "The Cigarette Century." *Science 80*, September–October 1980, pp. 36–43.

3. Davis, Audrey B. "The Development of Anesthesia." *American Scientist*, September–October 1982, pp. 522–528.

4. Gilman, A. G., et al., *Goodman and Gilman's The Pharmacological Basis of Therapeutics*, 7th ed. New York: Macmillan, 1985.

5. Hill, John W., and Susan M. Jones. "Consumer Applications of Chemical Principles: Drugs." *Journal of Chemical Education*, April 1985, pp. 328–331.

6. Julien, Robert M. *A Primer of Drug Action*, 4th ed. San Francisco: W. H. Freeman, 1985.

7. Krassner, Michael B. "Brain Chemistry," *Chemical and Engineering News*, 29 August 1983, pp. 22–23.

8. Langone, John. "Acupuncture: New Respect for an Ancient Remedy." *Discover*, August 1984, pp. 70–73.

9. Liska, Ken. *Drugs and the Human Body*, 2nd ed. New York: Macmillan, 1986.

10. *Physicians' Desk Reference*, 36th ed. Oradell, NJ: Medical Economics Company, 1982.

11. Pines, Maya. "Suicide Signals." *Science 83*, October 1983, pp. 55–58.

12. Van Dyck, Craig, and Robert Byck. "Cocaine." *Scientific American*, March 1982, pp. 128–141.

13. Windholz, Martha (Ed.). *The Merck Index*, 10th ed. Rahway, NJ: Merch and Co., 1983.

Sports

The Chemical Connection

We have spent years developing technology to reduce the amount of physical labor involved in routine jobs. Now we are pursuing the challenging task of being thin and physically fit despite the lazy lifestyle we worked so hard to achieve. Modern labor-saving devices give us plenty of leisure time to use for athletic activities. Some of us participate directly, and even the less ambitious are often involved indirectly by viewing athletic performances at a stadium or on television.

Sports performances have been improved dramatically in recent years by research that has enabled us to understand more fully the action of muscles. Chemical research also has provided new materials for sports, from protective plastic helmets to polymeric playing fields. Sports also have been changed by the ready availability of drugs—both legal and illegal—that affect the way athletes perform.

Some Chemistry of Muscles

Studies have shown that frequent exercise prolongs life and lowers the incidence of disease. Humans have about 600 muscles each. Exercise can make muscles stronger, more flexible, and more efficient in

Figure 24.1 Sports have been transformed in recent years. Biochemistry has brought a better understanding of the athlete's body and the way it works. Chemistry has provided drugs that keep athletes healthier (but increase the potential for abuse) and materials that make sports safer and improve performance.

their use of oxygen. Strong muscles can do more work than weak ones. That is good, because the heart is an organ comprised mainly of muscle. A strong heart is a healthy heart. With regular exercise, resting pulse and blood pressure usually decline. After several months of an effective exercise program, pulse and blood pressure remain lower even *during* exercise. The net result, called the **training effect**, is that a person who exercises regularly is able to do more physical work with less strain.

People who expand their capacity to do more physical work under less strain often begin to think of doing more—faster and with more agility and accuracy—of whatever they do. These people become athletes. Exercise is an art, but it is increasingly also a science—a science in which chemistry plays a vital role.

When cells metabolize glucose or fatty acids, only a part of the chemical energy in those substances is converted to heat. Some of it is stored in another chemical compound called adenosine triphosphate (ATP).

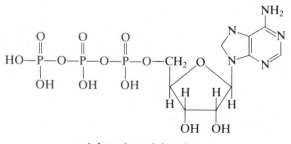

Adenosine triphosphate

The stimulation of muscle causes it to contract; that contraction is work and requires energy. The immediate source of energy for muscle contraction is ATP. The energy stored in this molecule powers the physical movement of muscle tissue. Two proteins, actin and myosin, play important roles in this process. Together actin and myosin form a loose complex called **actomyosin**, the contractile protein of which muscles are made (Figure 24.2). When ATP is added to isolated actomyosin, the protein fibers contract. It seems likely that the same process occurs *in vivo*, that is, in muscle in living animals. Not only does myosin serve as part of the structural complex in muscles, it also acts as an enzyme for the removal of a phosphate group from ATP. Thus, it is directly involved in liberating the energy required for the contraction.

In the resting person, muscle activity (including that of the heart muscle) accounts for only about 15 to 30% of the energy requirements of the body. Other activities, such as cell repair or the transmission of nerve impulses or even the maintenance of body temperature, account for the remaining energy needs. During intense physical activity, the energy requirements of muscle may be more than 200 times the resting level.

The ATP in muscle tissue is sufficient for activities lasting at most a few seconds. Fortunately muscles store a more extensive energy supply in the form of glycogen (Chapter 18). This starch is a storage form of dietary carbohydrate that has been ingested, digested to glucose, and absorbed. This blood glucose is stored as glycogen when not needed immediately for energy.

When muscle contraction begins, the glycogen is converted by muscle cells in a series of steps to pyruvic acid.

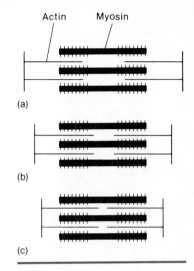

Figure 24.2 Diagram of actomyosin complex in muscle. (a) Extended muscle. (b) Resting muscle. (c) Partially contracted muscle.

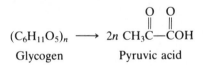

$$(C_6H_{11}O_5)_n \longrightarrow 2n\ CH_3\overset{\overset{O}{\|}}{C}-\overset{\overset{O}{\|}}{C}OH$$

Glycogen Pyruvic acid

Then, if sufficient oxygen and other factors are readily available, the pyruvic acid is converted in a series of steps to carbon dioxide (CO_2) and water (H_2O).

$$2\ CH_3COCOOH + 5\ O_2 \longrightarrow 6\ CO_2 + 4\ H_2O$$

Muscle contractions that occur under these circumstances—that is, in the presence of oxygen—constitute **aerobic exercise**.

If sufficient oxygen is not available, pyruvic acid is reduced (Chapter 8) to form lactic acid.

$$CH_3COCOOH + [2H] \longrightarrow CH_3CHOHCOOH$$

This is **anaerobic exercise**. If it persists, an excess of lactic acid results. The buildup of lactic acid causes a weaker response of muscle cells to

Figure 24.3 The race is over, but the metabolism that fueled the effort continues. [Photo by Nancy Crochiere.]

the stimuli that originally causes them to contract; hydronium ions (H_3O^+) from lactic acid are embedded in the muscle cell tissue. Athletes recognize this weakness as fatigue and sometimes as pain. At this point they are tempted to quit, and if they do, the **oxygen debt** is repaid (Figure 24.3). After exercise ends, the cells' demand for oxygen decreases, making more oxygen available to oxidize lactic acid that results from anaerobic metabolism back to pyruvic acid. This acid is then converted to CO_2, H_2O, and energy. Athletes usually emphasize one type of training (anaerobic or aerobic) over the other. For example, an athlete training for a 60-m dash will do mainly anaerobic work, but one planning to run a 10-km race will do mainly aerobic training.

The aerobic oxidation that occurs during aerobic exercise requires oxygen; thus, the athlete sometimes gulps air during or after his or her event. Sprinting and weight lifting are largely anaerobic activities; a marathon run is largely aerobic. During a marathon, an athlete must set a pace to run for hours. His or her muscle cells depend on slow, steady aerobic conversion of carbohydrates to energy. During anaerobic activities, however, the muscle cells use almost no oxygen. Rather, they need the quick energy provided by anaerobic metabolism.

After glycogen stores are depleted, muscle cells can switch over to fat metabolism. Fats are the main source of energy for sustained activity of low or moderate intensity, such as the last part of a marathon run (42 km).

Muscle Fibers: Fast and Slow Twitch

Another element in athletic performance is the quality and type of the muscle fiber involved. It stands to reason that the quality of the "machinery" might play a role in how the work is done. For example, there is more than one way to get snow off a driveway. One person may remove it in 10 minutes with a snowblower; another may spend 50 minutes shoveling. Muscles are tools with which we accomplish work. They are classified according to the speed and effort required to get work done.

The two classes of muscle fibers are **fast-twitch fibers** (those stronger and larger and most suited for anaerobic activity) and **slow-twitch fibers** (those best for aerobic work). Table 24.1 lists some characteristics of these two types of muscle fibers. The Type I fibers described in Table 24.1 are called on during activity of light or moderate intensity. The respiratory capacity of these fibers is high, which means they can provide much energy via aerobic pathways. Notice that for Type I fibers myoglobin levels are also high. Myoglobin is the heme-containing protein in muscle that transports oxygen (as hemoglobin does in the blood). Aerobic oxidation requires oxygen, and this muscle tissue is

Table 24.1   A Comparison of Types of Muscle Fiber

	Type I	Type IIB*
Category	Slow twitch	Fast twitch
Color	Red	White
Respiratory capacity	High	Low
Myoglobin level	High	Low
Catalytic activity of actomyosin	Low	High
Capacity for glycogen use	Low	High

* A Type IIA fiber exists that resembles Type I in some respects and Type IIB in others. We will discuss only the two types described in the table.

geared to supply high levels of oxygen. The capacity of Type I muscle fibers for use of glycogen is low. This tissue is not geared to anaerobic generation of energy and does not require the hydrolysis of glycogen. The catalytic activity of the actomyosin complex is low. Remember that actomyosin is not only the structural unit in muscle that actually undergoes contraction; it is also responsible for catalyzing the hydrolysis of ATP to provide energy for the contraction. Low catalytic activity means that the energy is parceled out more slowly, which is not good if you want to lift 200 kg but is perfect for a jog of 15 km.

The Type IIB fibers described in Table 24.1 have characteristics just the opposite of Type I fibers. Low respiratory capacity and low myoglobin levels argue against aerobic oxidation. A high capacity for glycogen and high catalytic activity of actomyosin allow this tissue to generate ATP rapidly and also to hydrolyze that ATP rapidly in intense muscle activity. Thus, this type of muscle tissue gives you the capacity to do short bursts of vigorous work. We say *bursts* because this type of muscle tissue fatigues relatively quickly. A period of recovery in which lactic acid is cleared from the muscle is required between brief periods of activity.

Endurance exercise training, such as long distance running, increases the myoglobin levels in skeletal muscles, providing for faster oxygen transport (Figure 24.4). These changes can be observed shortly after training begins, that is, within a week or two. Muscle changes resulting from endurance training do not necessarily include a significant increase in the size of the muscle, in contrast to the effect of strength exercises such as weight lifting (Figure 24.5).

Muscle-fiber type seems to be inherited. Research shows that world-class marathon runners may possess up to 80 to 90% slow-twitch fibers, compared with championship sprinters who may have up to 70% fast-twitch muscle fiber. Some exceptions have been noted, however.

Thus, factors such as training and body composition are important in athletics. Other factors are equally important, including nutrition, fluid and electrolyte balance, and drug use or misuse.

Figure 24.4 The Boston Marathon, a test of the respiratory capacity of muscle. [Photo courtesy of Stock, Boston; © Arthur Grace.]

Figure 24.5 Strength exercises build muscle mass, but do not increase respiratory capacity of muscles. [Photo courtesy of Stock, Boston; © Gale Zucker.]

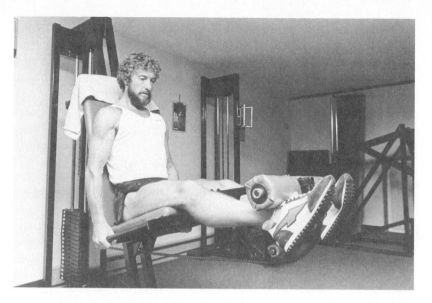

Nutrition and the Athlete

Nutrition is a subject that has many professed experts. In a sense, we all are experts based on our years as consumers of nutrients. The athlete may be interested in nutrition for other reasons, primarily because it provides fuel for the sport. Nutrients such as water, carbohydrates, proteins, fats, vitamins, and minerals can make or break the athlete when it comes to performance. The wrestler employs starvation and dehydration to achieve a certain weight class. Other athletes worry that they don't get enough vitamins or minerals; thus they take twice what they need in the form of supplements. Athletes have even been known to eat whole bowls of vitamin pills for breakfast, completely overlooking the fact that they need Calories for physical activity, and vitamins do not supply them. If a person consumes a reasonably balanced diet, vitamin supplements are a waste of money. Athletes would be wise to consume a balanced diet that meets the **Recommended Daily Allowance** (RDA) for their age, height, and weight. The RDA would be adequate in all respects except for Calories.

Athletes need more Calories because they expend more than the average sedentary individual. The budget-conscious athlete would probably choose to get these extra Calories in the form of carbohydrates. Foods rich in carbohydrates, especially the starches, are the cheapest source of Calories, as well as being the preferred source of energy for the healthy body. Fat- and protein-rich foods are also good sources of Calories, but they are more expensive; protein metabolism also produces more toxic wastes that tax the liver and kidneys. The pregame steak dinner consumed by football players in the past was based on a myth that protein builds muscles; if athletes want more muscle, says the myth, they should eat more protein. This is just not true. Although athletes do need the RDA for protein (based on grams per kilogram of body weight), they do not need an excess. Protein consumed in amounts greater than needed for synthesis and repair of tissue will only make the athlete fatter (due to excessive Calorie intake) and not more muscular.

Muscles are built through exercise, not through consumption of excess protein. When a muscle contracts against a resistance, an amino acid called creatine is released. Creatine stimulates the production of the protein myosin, thus building more muscle tissue. If the exercise stops, the muscle begins to shrink after about two days. After about two months without exercise, muscle built through the exercise program is almost completely gone. (The muscle does *not* turn to fat, as some athletes believe. Former athletes often get fat, though, because they continue to take in the same number of Calories and expend fewer.)

$$HN{=}\underset{\underset{\displaystyle CH_3}{|}}{\overset{\overset{\displaystyle NH_2}{|}}{C}}{-}N{-}CH_2{-}\overset{\overset{\displaystyle O}{\|}}{C}{-}OH$$

Creatine

Diet and Exercise

In American culture today, thin is in and fat is not. Many former nonathletes have joined the ranks of joggers, walkers, tennis players, body-builders, and bikers in pursuit of the bulgeless body. About a million people per week are treated in weight-loss clubs and clinics. Countless others treat themselves with crash diets, dietary supplements, diet pills, and other programs. Diet books populate the best seller lists. Many of these plans are simply fads. They are aimed more at increasing their creator's wealth than in decreasing a person's weight. Many are grossly unbalanced when compared with the RDA. An unbalanced diet, especially over an extended period of time, can cause a variety of nutritional deficiencies, a decrease in resistance to disease, and a decline in general health.

Minerals such as iron, calcium, and potassium also tend to be deficient in many crash diets. A deficiency of these minerals may seriously interrupt the smooth function of nerve impulse transmission to muscle. Such interruption obviously impairs athletic performance. Impulse transmission to vital organs may also be impaired under situations of severe restriction. Several deaths from cardiac arrest have resulted from variations of the "liquid protein diet."

Weight loss or gain is based on the law of conservation of energy (Chapter 13). If we take in more Calories than we use up, the excess Calories are stored as fat. If we take in fewer Calories than we need for our activities, our bodies burn some of the stored fat to make up for the deficit. One pound of adipose (fatty) tissue contains about 3500 Calories. This tissue requires 200 miles of blood capillaries to serve its

Figure 24.6 Percent of body fat can be determined by weighing a person completely submerged in water. The calculation must include a correction for the volume of air in the lungs. [Photo courtesy of The National Institutes of Health.]

cells. Excess fat therefore puts extra strain on the heart; it has to work harder to supply blood to the extra tissue.

How much fat is enough? The male body requires about 3% essential body fat; the average female 10 to 12%. It is difficult to measure percent body fat accurately. Skinfold calipers are quite inaccurate; they measure water retention as well as fat. Dunk tanks for measuring body density (Figure 24.6) are better, but results vary with the amount of air in the lungs. Fat is less dense (0.903 g/mL) than the water (1.000 g/mL) that makes up most of the mass of our bodies. The higher the proportion of body fat in a person, the more buoyant that person is in water.

It is possible to lose weight through dieting. If you reduce your intake by 100 Cal per day and keep your activity constant, you will burn off a pound of adipose tissue (3500 Cal) in 35 days. Unfortunately, people are seldom so patient, and they resort to more stringent diets. To achieve their goals more rapidly, they exclude certain foods and reduce the amounts of others. Such diets are harmful. Diets with fewer than 1200 Cal per day are likely to be deficient in necessary nutrients, particularly in B vitamins and iron. Further, dieting slows down metabolism. Weight loss through dieting is quickly regained when the dieter resumes old eating habits.

According to one theory, hunger is regulated by the hypothalamus. When the hypothalamus senses that the level of fatty acids circulating in the bloodstream is low, it triggers the hunger mechanism. According to the **set-point theory**, each of us has a unique level at which the hypothalamus acts. Some of us must have a higher level of body fat than others to avoid constant hunger. This is consistent with the fact that obesity seems to be inherited (Chapter 18). It is also rather grim news for those of us who would like to lose weight; we can do so only by being constantly hungry. There is some good news, however. It appears that our set point can be lowered through exercise.

People who do not increase their food intake when they begin an exercise program will lose weight. And contrary to myth, exercise (up to one hour a day) will not cause an increase in appetite. Most of the weight loss from exercise is due to the increase in metabolic rate during the activity, but the increased metabolic rate continues for several hours after completion of the exercise. Exercise helps us maintain both fitness and thinness.

Crash Diets: Quick = Quack

Most quick weight-loss diets depend on factors other than fat metabolism to hook a prospective customer. Many contain a **diuretic**, such as caffeine, to increase the output of urine. Weight loss is water loss; weight is regained when the body is rehydrated.

Other such diets depend on depleting the body's stores of glycogen. When carbohydrates are eliminated from the diet, the body draws on its glycogen reserves, depleting them in about 24 hours. Recall (Chapter 18) that glycogen is a polymer of glucose. Glycogen molecules have lots of hydroxyl groups. These OH groups can form hydrogen bonds to water molecules. Each pound of glycogen carries with it about three pounds of water held to it by these hydrogen bonds. Depleting the pound or so of glycogen results in a weight loss of about four pounds (1 lb glycogen + 3 lb water). No fat is lost, and the weight is quickly regained when the dieter resumes eating carbohydrates.

If your normal energy expenditure is 2400 Cal per day, the most fat you could lose by total fasting for a day would be 0.69 lb (2400 Cal/day divided by 3500 Cal/lb adipose tissue). That assumes your body would burn nothing but fat. It won't. Recall (Chapter 18) that the brian runs on glucose, and if that glucose isn't supplied in the diet, it is obtained from protein. Any weight loss brought about by restricting carbohydrate intake is accompanied by a loss of muscle mass as well as fat.

Any weight-loss program that promises a loss of more than a pound or two a week is likely to be dangerous quackery.

The most sensible approach to weight loss is to adhere to a balanced low-Calorie diet that meets the RDA for essential nutrients, and to engage in a reasonable, consistent, individualized exercise program. Hence, the principles of weight loss are met by decreasing intake and increasing output. All people, athletes included, should be aware of the possible risks in dieting. Vitamin deficiency often develops slowly over months or years. This problem can be corrected or avoided by careful nutritional planning or by taking vitamin supplements (although most nutritionists will argue that food is the best source of the body's nutritional needs). Some nutritionists suggest vitamin supplements for those who refuse (or are not able) to eat a balanced diet. These people should take vitamin supplements under the direction of a physician or registered dietitian.

Carbohydrate Loading and Blood Doping

Athletes, it seems, will try almost anything to gain a competitive edge. Those engaged in endurance activities, such as long distance running, often try a technique called **carbohydrate loading**. Muscles run on glycogen as the preferred fuel, but the human body can store only a limited amount, perhaps about 500 g in a 70-kg person.

In carbohydrate loading, glycogen stores are first depleted by limiting intake of carbohydrates and training vigorously. Then, a few days before competition, the athlete cuts back on training and eats a diet high in carbohydrates. The presumption is that, under this regimen, the body will store more glycogen than usual. The benefits of carbohydrate loading, even for top athletes, are questionable. For casual athletes, the technique is probably ridiculous.

A more radical attempt to improve athletic performance is called **blood doping**. A quantity of blood is withdrawn from the athlete, then stored for 6 weeks while the person's body manufactures replacement blood. Then, a few days before competition, the red blood cells from the stored blood are returned to the athlete. This increased supply of red cells enables the athlete's cardiovascular system to transport oxygen more efficiently. In controlled studies, blood-doped runners were able to improve their times by 9 seconds in an 8-km treadmill race, compared with a group that took a saline solution. These runners were already well trained. Blood doping will not convert a mediocre runner into an outstanding one. Debate about ethics of this procedure continues.

Body Fluids: Electrolytes

There is yet another aspect of chemistry and nutrition that athletes should be aware of: the balance between fluid and electrolyte intake. **Electrolytes** are substances that conduct electricity when dissolved in water. Sodium ions (Na^+), potassium ions (K^+), and chloride ions (Cl^-) are the major electrolytes essential for proper cellular function. Water is also an essential nutrient, a fact obvious to anyone who has been deprived of it. Modern Western people seem to prefer to meet the body's need for water by consuming soda pop, coffee, beer, fruit juice, and other beverages. Many of these—such as beer, caffeinated soda pop, and coffee—actually contain drugs that impair the body's use of the water in the beverages. The alcohol in beer and other alcoholic drinks promotes water loss by blocking the action of the **antidiuretic hormone** (ADH). The caffeine in colas and other soda pops and in coffee and tea has a diuretic effect on the kidneys; it promotes urine formation and consequent water loss.

Based on available information, the best way to replace water loss as it occurs through sweat, tears, respiration, and urination is to drink water (Figure 24.7). Unfortunately, thirst is often a *delayed* response to water loss, and it may be masked by such symptoms as exhaustion, confusion, headache, and nausea (the symptoms are a result of dehydration). Body sweat contains 99% water, some sodium ions (Na^+) and chloride ions (Cl^-), minute amounts of calcium (Ca^{2+}) ions and potassium (K^+) ions, urea, lactic acid, and body oils. The normal American diet probably contains too much sodium chloride (NaCl). Thus, it makes the most sense to replace the water component of lost sweat with pure water itself.

Figure 24.7 The best replacement for fluids lost during exercise is plain water. Electrolyte replacement fluids do not help, and salt tablets will likely do more harm than good.

Commercial "thirst quenchers" have been increasing in popularity for both serious and weekend athletes. These drinks, designed to replace the salts and water your body loses during long, sweat-provoking periods of exercise, have recently become the target off some sports-health specialists for being too concentrated—a hazard that could lead to diarrhea. At best, the thirst quenchers are worthless, another waste of the consumer's money. Thus they join the ranks of their relatively nonnutritive counterparts, sugared and artificially sweetened soft drinks.

How do you know if you are drinking enough water and are in a proper state of hydration? Just monitor a source of water loss. The urine is a good indicator of hydration. When a person is dehydrated, their urine is cloudy and yellow because the kidneys are trying to conserve water. The body water shifts to conserve the shrinking blood volume and to prevent shock. As dehydration worsens, muscles tire and cramp. Dizziness and fainting may follow, and brain cells may shrink, resulting in mental disturbances. The heat regulatory system may also fail, causing **heat stroke**. Without prompt medical attention, a victim of heat stroke may die.

Drugs and the Athlete

We have discussed muscle chemistry and metabolism, aerobic and anaerobic exercise, and nutrition and fluid balance. All of this relates to normal internal processes and how they adapt through training to do more work in the form of physical activities. It appears unlikely that a well-fed, well-trained athlete can safely improve his or her performance with magic, superstition, or drugs. Many athletes, however, are not satisfied with the idea that hard work and proper nutrition are the answers to improved performance. They turn instead to dangerous drugs—to stimulants such as amphetamines and cocaine, and to "muscle builders" like anabolic steroids.

Athletes also use drugs to remedy the effects of performance, particularly pain or soreness, and for injury. These drugs, called **restoratives**, include analgesics (painkillers), such as aspirin and acetaminophen (Chapter 22), and antiinflammatory drugs, such as aspirin and cortisone. Cortisone derivatives are often injected to reduce swelling in damaged joints and tissues. Relief is often transitory, however, and side effects are often severe. Prolonged use of these drugs can cause fluid retention, hypertension, ulcers, disturbance of the sex hormone balance, and other problems. Another substance, one that can be bought over the counter and applied externally, is methyl salicylate, or oil of wintergreen. This substance causes a mild burning sensation when applied to the skin, thus serving as a counterirritant for sore muscles. An aspirin derivative, it also acts as an analgesic.

Figure 24.8 Many athletes believe that anabolic steroids help speed the development of muscle mass. These drugs are dangerous, however, and they have devastating side effects.

Many strenuous athletic performances depend on well-developed muscles. Men generally have larger muscles than women because muscle mass is related to the male hormone testosterone. This hormone and some of its semisynthetic derivatives are taken by athletes in an attempt to build muscle mass more rapidly (Figure 24.8). Used in this manner, these hormones are called **anabolic steroids**. It is thought that these chemicals aid in the building (anabolism) of body proteins and thus of muscle tissue.

There are no good controlled studies that demonstrate the effectiveness of anabolic steroids. They do seem to work—at least for some people—but the side effects are many. In males, side effects include testicular atrophy and loss of function, impotence, acne, liver damage, edema (swelling), elevated cholesterol levels, and growth of breasts. Liver cancer is now showing up at an alarming rate in athletes who began using steroids in the 1960s.

Anabolic steroids, derived from male hormones, make women more masculine. In addition to more muscles, they develop baldness, extra body hair, a deepened voice, and menstrual irregularities. What price will an athlete pay for improved performance?

Despite the widespread use of steroids and other drugs, the 1980s will be remembered as the decade of cocaine. A multipurpose drug, cocaine was first used in the late 1800s as a local anesthetic to deaden pain during dental work or minor surgery. It is also a powerful stimulant, the reason for its current popularity among athletes. Like amphetamines, cocaine stimulates the central nervous system, increasing alertness, respiration, blood pressure, muscle tension, and heart rate. Both stimulants mask symptoms of fatigue and give the athlete a sense of increased stamina. The death from cocaine intoxication of several thousand people, including prominent professional athletes, brought the problem to public attention. Other athletes have been arrested, suspended, and banned for cocaine use.

The use of drugs to increase athletic performance is not only illegal and physically dangerous, but it can also be psychologically damaging. Stimulant drugs give the user a false sense of confidence by affecting the person's ego, giving the delusion of invincibility. Such "greatness" often ends as the competition begins, however, because the stimulant effect is short-lived (it lasts for only 30 minutes to an hour for cocaine). What usually follows is exhaustion and extreme depression (Figure 24.9). More stimulants are needed to combat these "down" feelings, and the user then runs the risk of addiction or overdose.

Because drug use for the enhancement or improvement of performance is illegal, chemists have another role to play in sports: screening blood and urine samples for illegal drugs. Using sophisticated instruments, chemists can detect minute amounts of illegal drugs. Drug testing has been used at the Olympic games since 1968. It is rapidly

becoming standard practice for athletes in college and professional sports.

Drugs are used at great risk to the athlete and with little evidence of benefit in most cases. Use of illegal drugs violates the spirit of fair athletic competition. Let us work toward a world in which events and medals are won by drug-free athletes who are highly motivated, well nourished, and well trained.

Figure 24.9 Cocaine is a powerful stimulant that gives athletes a sense of increased stamina. The effect is short-lived, however, and users soon experience extreme fatigue.

Sports and the Brain

Some people are discouraged that there are no chemical substances that magically improve performance the way hard training does. These people should find reassurance in the fact that the body manufactures pain relievers of its own called endorphins and enkephalins (Chapter 23).

Research has shown that there is an increased level of endorphins in the blood of athletes after vigorous activity. Deep sensory nerves stimulated by exercise trigger the release of endorphins to block the pain message. Exercise such as distance running, which is extended over a long period of time, subjects the athlete to many of the same symptoms experienced by opiate users. Runners get a euphoric high during or after a hard run. Unfortunately, both the natural endorphins and their analogs, like morphine, seem to be addictive. Athletes, especially runners, tend to suffer from withdrawal; they feel bad when they don't get the vigorous exercise of a long, hard run. Of course this may be positive if the preventive health predictions about daily exercise are true. Being in good physical condition, taking part in daily vigorous exercise, and experiencing a natural, cheap, legal high is certainly far safer than resorting to illegal narcotics. You may get hooked on exercise, but that seems far better than getting hooked on narcotic opiates.

Figure 24.10 Imaginative use of synthetic fibers and other materials has made it possible for athletes to perform in relative comfort even in inclement weather.

Chemistry of Sports Materials

The athlete's world is shaped by chemicals. The ability to perform is a reflection of physiology and biochemistry within the athlete. Chemicals taken internally may hinder or enhance performance to some degree. Deep sensory nerves, stimulated by rigorous exercise, trigger the release of the brain's own painkillers, the endorphins, which may block pain and provide a euphoric high—a reward for strenuous endeavor and reinforcement for future challenge.

There is yet another chemical dimension to the sports world: the chemicals involved in providing athletic clothing and equipment (Table 24.2). Athletic clothing gets its shape and protective qualities from chemicals. Swimsuits, ski pants, and elastic supports stretch because of synthetic fibers. Joggers run in thunderstorms and windstorms in Gore-Tex suits without getting cold and wet. Gore-Tex is a thin, membranous material made by stretching fibers of polytetrafluoroethylene (Teflon). The material has billions of tiny holes that are too small to pass drops of water but readily pass water vapor. Raindrops falling on the outside are held together by surface tension and run off rather than penetrating the fabric and wetting the wearer. Water vapor from the sweating skin can pass from the warm area inside the suit where vapor concentration is high to the cooler area outside the suit where the vapor concentration is lower. The runner stays relatively dry inside the suit (Figure 24.10).

The jogger is not the only athlete pampered by equipment made of materials from the chemical industry. Football, hockey, and baseball players are protected by plastic helmets. Protective pads of synthetic foamed rubbers help these players avoid injuries to the torso and legs.

Table 24.2 Some Sports and Recreational Items Made From Petroleum

Golf balls	Whistles	Tote bags	Hockey pucks
Wet suits	Motorcycle helmets	Darts	Ice buckets
Parachutes	Dune-buggy bodies	Tents	Fishing nets
Card tables	Checkers	Stadium cushions	Hiking boots
Golf-cart bodies	Chess boards	Finger paints	Frisbees
Warm-up suits	Shorts	Foul-weather gear	Fishing boots
Ping-pong paddles	Tennis shoes	Foot pads	Diving masks
Rafts	Paddles	Visors	Guitar picks
Uniforms	Decoys	Swimming pool liners	Beach balls
Phonographs	Volley balls	Vinyl tops for cars	Sunglasses
Racks	Sleeping bags	Ice chests	Dog leashes
Track shoes	Sports-car bodies	Life jackets	Dice
Dominoes	Tennis balls	Audio tape	Pole-vaulting poles
Windbreakers	Reclining chairs	Model planes	Aquariums
Sails	Insect repellent		

It's a Grand Slam Home Run, Folks!!

Figure 24.11 Baseball in a domed stadium on an artificial surface is quite different from a game played outside on grass.

Brightly dyed nylon uniforms with synthetic colors add to the glamour of amateur and professional teams. Sports events are played on artificial turf, a carpet of nylon or polypropylene with an under-pad of synthetic foamed polymers. Stadiums are protected from the weather by Teflon covers reinforced with glass fibers and held aloft by air pressure. Baseball or football on artificial turf in air-conditioned, enclosed stadiums is quite different from games played outside on grass (Figure 24.11). Balls bounce differently and players wear shoes especially designed to cope with the artificial surface.

The dramatic effect of new materials is perhaps best illustrated in pole vaulting. Between 1940 and 1960, the world-record height for this event increased only 23.5 cm. After the development of plastic poles reinforced with glass fibers, however, the record rapidly rose another 100 cm. Other sports have been affected similarly, if less dramatically (see references at the end of the chapter for examples).

From the soles of sports shoes, to the wax used on cross-country skis, to tennis sweaters that stretch yet retain their shape, athletes are immersed in a world of chemicals. Even the gum they chew contains polyvinyl acetate resin and synthetic flavors and colors. Athletes (and nonathletes) that chew tobacco chew on leaves modified by natural and synthetic flavors, moisturizers, and other chemical additives. From within, athletes can strive to achieve goals and records by monitoring their body chemistry. They will be comfortable and protected by materials designed exclusively for them. With almost half the population exercising on a regular basis, and many of the rest involved at least in watching athletic events, chemistry will continue to play an active role in our activity-oriented culture.

Problems

1. List three ways that chemistry has had an impact upon sports.
2. How many muscles do humans have?
3. Describe the training effect.
4. What is the *immediate* source of energy for muscle contraction?
5. What two proteins make up the actomyosin complex?
6. What are the two functions of the actomyosin protein complex?
7. What is aerobic exercise?
8. What is anaerobic exercise?
9. Which type of metabolism (aerobic or anaerobic) is primarily responsible for providing energy for intense bursts of vigorous activity?
10. Which type of metabolism (aerobic or anaerobic) is primarily responsible for providing energy for prolonged low levels of activity?
11. What is meant by *oxygen debt*?
12. Identify Type I and Type IIB muscle fibers as
    a. fast twitch or slow twitch
    b. suited to aerobic oxidation or to anaerobic use of glycogen.
13. Explain why high levels of myoglobin are appropriate for muscle tissue that is geared to aerobic oxidation.
14. Why does the high catalytic activity of actomyosin in Type IIB fibers suggest that these are the muscle fibers engaged in brief, intense physical activity?
15. Why can the muscle tissue that utilizes anaerobic glycogen metabolism for its primary source of energy be called on only for *brief* periods of intense activity?
16. Which type of muscle fiber is most affected by endurance training exercises? What changes occur in the muscle tissue?
17. Birds use large, well-developed breast muscles for flying. Pheasants can fly 80 km/hr, but only for short distances. Great blue herons can fly only about 35 km/hr, but can cruise great distances. What kind of fibers would each have in its breast muscles?
18. How do the nutritional needs of an athlete differ from those of a sedentary individual? How is that extra need best met?
19. Muscle is protein. Does an athlete need extra protein (above the RDA) to build muscles? What is the only way to build muscles?
20. Describe the biochemical process by which muscles are built.
21. List two ways to determine percent body fat. Describe a limitation of each method.
22. List some problems with low-Calorie diets.
23. How many Calories of energy are stored in a pound of adipose tissue?
24. Why does excess body fat put a strain on the heart?
25. Describe the set-point theory of body weight.
26. Why does a diet that restricts carbohydrate intake lead to loss of muscle mass as well as fat?
27. List two ways in which fad diets lead to a "quick weight loss." Why is this weight rapidly regained?
28. How much glycogen can the average human body store?
29. What is carbohydrate loading? How is it presumed to aid an athlete in an endurance event?
30. What is blood doping? How does it work?
31. List the three major electrolytes essential for proper cellular function.
32. What is a diuretic?
33. Describe the function of antidiuretic hormone (ADH).
34. What fluid is best for replacing water lost during exercise?
35. Why are beer and cola drinks not recommended for fluid replacement?
36. Is thirst a good indicator of dehydration? Are you always thirsty when dehydrated?
37. How is the appearance of urine related to dehydration?
38. What is heat stroke?
39. What are restorative drugs?
40. How is cortisone used in sports medicine? What are some of the side effects of its use?
41. What are anabolic steroids?
42. List some side effects of the use of anabolic steroids in males.
43. What are the effects of anabolic steroids on females?
44. Does cocaine enhance athletic performance? Explain fully.

45. What happens when the effect of cocaine wears off?

46. What is the role of chemists in control of drug use by athletes?

47. Give a biochemical explanation for the "runners high."

48. Give a biochemical explanation of addiction to long distance running.

49. How does Gore-Tex clothing work? What is it made of?

50. List as many synthetic materials as you can that are used in each of the following sports.

   a. baseball
   c. football
   e. golf
   g. swimming
   i. skiing
   k. jogging

   b. basketball
   d. ice hockey
   f. tennis
   h. sailing
   j. track and field

51. A giant double-decker hamburger provides 1200 Cal of energy. How long would you have to walk to burn off that energy if 1 hour of walking uses 300 Cal?

52. One kilogram of fat tissue stores about 7700 kcal of energy. An average person burns about 40 kcal walking 1 km. If that person walks 5 km a day, how much fat will be burned in 1 year?

53. How long would you have to run to burn off the 110

Cal in one glass of beer if 1 hour of running burns of 1100 Cal?

54. How far would you have to run to burn off 5 kg of fat if your running burns off 100 kcal/km? (Assume that there are 7700 kcal in 1 kg of fat.)

55. Fat tissue has a density of about 0.9 g/mL, lean tissue a density of about 1.1 g/mL. Calculate the density of a person with a body volume of 80 L who weighs 85 kg. Is the person fat or lean?

56. The RDA for protein is about 0.8 g per kilogram of body weight. How much protein is required each day by a 125-kg football player?

57. How much protein is required each day by a 50-kg gymnast (See Problem 56)?

58. A 70-kg man can store about 2000 kcal as glycogen. How far can the man run on this stored starch if he expends 100 kcal/km while running?

59. A 70-kg man can store about 100 000 kcal of energy as fat. How far can the man run on this stored fat if he expends 100 kcal/km while running?

60. An athlete can run a 400-m race in 45 seconds. Her maximum oxygen intake is 4 L/minute, yet working muscles at maximum exertion require about 0.2 L of O_2 per minute for each kilogram of body weight. If an athlete weighs 50 kg, what oxygen debt will she incur?

References and Readings

1. Bent, Henry A. "Energy and Exercise." *Journal of Chemical Education*, July 1978, pp.456–458; August 1978, pp. 526–528; September 1978, pp. 586–587; October 1978, pp. 659–660; November 1978, pp. 726–727; December 1978, pp. 796–797.

2. Bishop, Marvin. "Food In/Energy Out." *Chem-Tech*, August 1983, pp. 494–496.

3. "Boats to Badminton: Chemicals Build Athletic Products." *ChemEcology*, April 1982, pp. 1–13.

4. Chase, Anthony. "Cross-Country Skiing: The Hows of Wax." *Science 82*, March 1982, pp. 90–91.

5. Gurin, Joel. "What's Your Natural Weight?" *American Health*, May 1984, pp. 43–47.

6. Hill, John W. "Weight-Loss Diets and the Law of Conservation of Energy." *Journal of Chemical Education*, December 1981, p. 996.

7. "It's Shoetime." *Discover*, June 1985, p. 9. Describes running shoes with built-in computers.

8. Layman, Donald K. (Ed). *Nutrition and Aerobic Exercise*. Washington, DC: American Chemical Society, 1986.

9. Lineback, David R. "Nutrition (Diet) and Exercise." *Journal of Chemical Education*, June 1984, pp. 536–539.

10. Looney, Douglas S. "A Test with Nothing But Tough Questions." *Sports Illustrated*, 9 August 1982, pp. 24–29. About testing athletes for drug use.

11. Rapaport, Roger. "The Blade Runners." *Science 82*, November 1982, pp. 96–97. Talks about artificial turf.

12. Sprague, Ken. *The Athlete's Body*. Los Angeles: J. P. Tarcher, 1981.

13. Zuckerman, Sam. "Food for Sport." *Nutrition Action*, July–August 1984, pp. 6–11.

14. Zurer, Pamela S. "Drugs in Sports." *Chemical and Engineering News*, 30 April 1984, pp. 69–78.

25

Chemical Toxicology

Hemlock, Anyone?

When the Greek philosopher Socrates was accused of corrupting the youth of Athens in 399 B.C., he was given the choice of exile or death. He chose death and implemented his decision by drinking a cup of hemlock (Figure 25.1).

Poisons from plant and animal sources were well known in the ancient world. Snake and insect venoms and plant alkaloids were used. Today, many primitive tribes still use a variety of poisons in hunting and warfare. Curare (Chapter 23), used by certain South American tribes, is one of the more notorious examples.

Although poisonous substances have been known and used for centuries, it is only within the last 150 years that scientists have learned the nature of the chemicals that are the active components of the various poisons. Socrates' hemlock probably was prepared from the fully grown but unripened fruit of *Conium maculatum* (poison hemlock). Usually, the fruit is dried carefully and then brewed into a "tea." Hemlock contains several alkaloids, but the principal one is coniine. This drug causes nausea, weakness, paralysis, and—as in the case of Socrates—death.

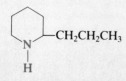

Coniine

564

Figure 25.1 Jacques Louis David's painting *The Death of Socrates* (1787) shows Socrates drinking the cup of hemlock to carry out the death sentence decreed by the rulers of Athens. [Courtesy of the Metropolitan Museum of Art, Wolfe Fund, 1931. Catherine Lorillard Wolfe Collection. (31.45)]

Poisons have always been with us, but our knowledge of them is greater than ever before. Industrial accidents, such as that at Bhopal, India, have made the public acutely aware of problems with toxic substances. At Bhopal, the accidental release of methyl isocyanate (CH_3—N=C=O), an intermediate in the synthesis of carbamate insecticides (Chapter 17), killed more than 2000 people and injured countless others. People are also concerned about long-term exposure to toxic substances in the air, in their drinking water, and in their food. Chemists can detect exceedingly tiny quantities of such substances. It is still quite difficult, though, to determine the effect of these trace amounts of toxic materials on human health.

A study of the response of living organisms to drugs is called **pharmacology**. As we have seen (Chapters 22 and 23), organisms respond in a variety of ways to the many chemical substances we call drugs. In this chapter, we deal with those substances that are poisonous or otherwise injurious. **Toxicology** is the branch of pharmacology that deals with the effects of poisons, their identification or detection, and the development and use of antidotes.

All Things Are Poisons

What is a poison? Perhaps a better question would be, How much is a poison? Substances may be harmless—or even necessary nutrients—in one amount, and injurious—or even deadly—in another. Even common substances such as salt and sugar can be poisonous when eaten in abnormally large amounts. Too much sugar—candy or sweets—can give a child a stomachache. Too much salt—sodium chloride—can

Figure 25.2 *Conium maculatum* (poison hemlock).

induce vomiting. There have even been cases of fatal poisoning when salt was accidentally substituted for lactose (milk sugar) in formulas for infants. Some substances are obviously more toxic than others, however. It would take a massive dose of salt to kill the average healthy adult, whereas only a few micrograms of some of the nerve poisons (page 576) can be fatal. Toxicity depends on the chemical nature of the substance.

People also respond differently to the same chemical. To cite an extreme case, a few grams of sugar would cause no acute symptoms in a normal person but would be dangerous to a diabetic. Excessive amounts of salt would be especially serious to a person with edema (swelling due to excessive amounts of fluid in the tissues).

Still another complicating factor is that chemicals behave differently when administered in different ways. Nicotine is more than 50 times as toxic when applied intravenously as when taken orally. Good, fresh water is delightful when taken orally, but even water can be deadly when inhaled in sufficient quantity. Further complications arise from the fact that even closely related animal species react differently to a given chemical. Even individuals within a species may react to different degrees.

In our discussion here, we cover only a few of the many toxic substances. These are organized into groups with similar activities. Those more likely to be encountered in everyday life are given priority.

Corrosive Poisons: A Closer Look

In Chapter 7, we examined the corrosive effect of strong acids and bases on human tissue. These chemicals indiscriminately destroy living cells. Corrosive chemicals, in lesser concentrations, also exert a more subtle effect.

Both acids and bases, even in dilute solutions, break down the protein molecules in living cells. These reactions involve the breaking of the amide (peptide) linkages in the molecules.

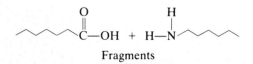

Because the reaction involves water, the process is called **hydrolysis**.

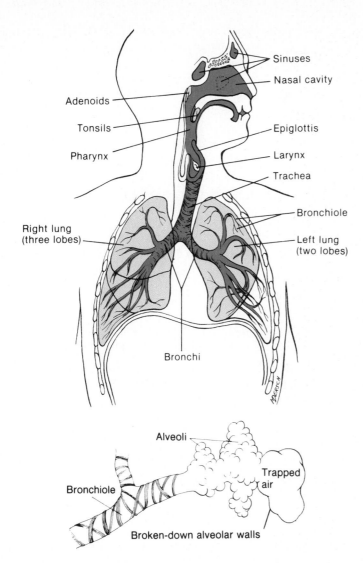

Figure 25.3 The respiratory system, showing the route of air through the nose, pharynx (throat), larynx (voice box), and trachea (windpipe) into the bronchi and bronchioles (bronchial tubes), and ending in the alveoli (air sacs).

Figure 25.4 Pulmonary emphysema. The loss of elasticity and the deterioration of the alveolar walls deter the exhalation of carbon dioxide.

Generally, the fragments are not able to carry out the functions of the original protein. In cases of severe exposure, the fragmentation continues until the tissue has been completely destroyed.

Acids in the lungs are particularly destructive. In Chapter 15, we saw how sulfuric acid is formed when sulfur-containing coal is burned. Acids are also formed when plastics and other wastes are burned. The damage these pollutants do can be explained as the breakdown of lung tissue by the acids (Figures 25.3 and 25.4).

Other air pollutants also damage living cells. Ozone, peroxyacetyl nitrates (PAN), and the other oxidizing components of photochemical smog probably do their main damage through the deactivation of enzymes. The active sites of enzymes often incorporate the sulfur-containing amino acids cysteine and methionine. Cysteine is readily oxidized by ozone to cysteic acid.

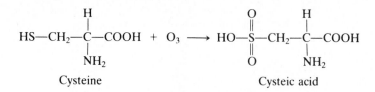

Cysteine — Cysteic acid

Methionine is oxidized to methionine sulfoxide.

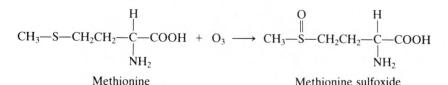

Methionine — Methionine sulfoxide

Still another amino acid, tryptophan, is known to react with ozone. Tryptophan, which does not contain sulfur, undergoes a ring-opening oxidation.

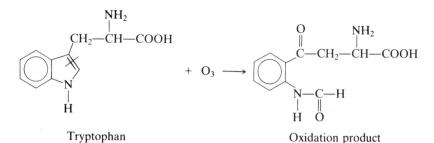

Tryptophan — Oxidation product

Most likely, oxidizing agents can break bonds in many of the chemical substances in a cell. Such powerful agents as ozone are more likely to make an indiscriminate attack than to react in a highly specific way.

Blood Agents

Certain chemical substances serve to block the transport of oxygen in the bloodstream or to prevent the oxidation of metabolites by oxygen in the cells. All act on the iron atoms in complex protein molecules. Probably the best known of these blood agents is carbon monoxide. The action of this lethal gas was described in Chapter 15. Recall that carbon monoxide binds tightly to the iron atom in hemoglobin, blocking the transport of oxygen.

Nitrates, which occur in dangerous amounts in the groundwater in some agricultural areas (Chapter 16), also serve to diminish the ability of hemoglobin to carry oxygen. Nitrates are reduced by microorganisms in the digestive tract to nitrites.

$$NO_3^- \longrightarrow NO_2^-$$

Nitrate ion   Nitrite ion

The nitrite ions oxidize the iron atoms in hemoglobin from Fe^{2+} to Fe^{3+}. The resulting compound, called **methemoglobin**, is incapable of carrying oxygen. The resulting oxygen-deficiency disease is called methemoglobinemia. In infants, this disease is called the blue baby syndrome.

Hemoglobin is bright red and is responsible for the red color of the blood. Methemoglobin is brown. During cooking, red meat turns brown due to the oxidation of hemoglobin to methemoglobin. Dried blood stains turn brown for the same reason.

Cyanide: Agent of Death

Cyanide is one of the most notorious poisons in both fact and fiction. It acts almost instantaneously and only a minute amount constitutes a lethal dose. The average fatal dose is only 50 or 60 mg. Cyanide is used as gaseous hydrogen cyanide ($H—C{\equiv}N$) and as solid salts that contain the cyanide ion ($C{\equiv}N^-$). Hydrogen cyanide is used (with great care by specially trained experts) to exterminate insects and rodents in the holds of ships, in warehouses, in railway cars, and on citrus and other fruit trees. Sodium cyanide (NaCN) is used to extract gold and silver from ores. It also is used in electroplating baths. Hydrogen cyanide is generated easily enough from the sodium salt by treatment with an acid.

$$NaCN + H_2SO_4 \longrightarrow HCN + NaHSO_4$$

Unlike carbon monoxide, cyanide does not react with hemoglobin. Instead, it blocks the oxidation of glucose inside the cell by forming a stable complex with the oxidation enzymes. The enzymes, called cytochrome oxidases, contain iron and copper atoms. They normally act by providing electrons for the reduction of oxygen in the cell. Cyanide ties up those mobile electrons, rendering them unavailable for the reduction process. Thus, cyanide brings an abrupt end to cellular respiration, causing death in a matter of minutes.

Any antidote for cyanide poisoning must be administered quickly. The treatment of choice nowadays involves sodium thiosulfate (the "hypo" used in developing photographic film). The sulfur atom is transferred from the thiosulfate ion to the cyanide ion, converting the latter to relatively innocuous thiocyanate.

$$CN^- + S_2O_3^{2-} \longrightarrow SCN^- + SO_3^{2-}$$

Cyanide   Thiosulfate   Thiocyanate   Sulfite

Unfortunately, few victims of cyanide poisoning survive long enough to be treated.

Some researchers speculate that life on Earth may have developed from molecules formed by the polymerization of hydrogen cyanide. This deadly gas is readily formed when an electric discharge is passed through a mixture of gases designed to simulate the Earth's early atmosphere. Many of the amino acids and the organic bases necessary to form DNA and RNA can be derived by the polymerization of HCN with the rearrangement of only a few atoms. Glycine, the simplest amino acid, can be derived from aminomalononitrile by hydrolysis, decarboxylation, and a second hydrolysis.

$$3\ H-C{\equiv}N \longrightarrow H_2N-\underset{\underset{C{\equiv}N}{|}}{CH}-C{\equiv}N \xrightarrow{+H_2O}$$

Aminomalononitrile

$$H_2N-\underset{\underset{COOH}{|}}{CH}-C{\equiv}N \xrightarrow{-CO_2} H_2N-CH_2-C{\equiv}N \xrightarrow{+H_2O}$$

$$H_2N-CH_2-COOH$$

Glycine

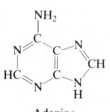

Adenine
(HCN)$_5$

Simple inspection of the formula for adenine shows it to be a pentamer of HCN. Would it not indeed be ironic if life arose from compounds made from such a deadly poison?

Make Your Own Poison: Fluoroacetic Acid

CH$_3$COOH   F—CH$_2$COOH

Acetic acid   Fluoroacetic acid

Although the body generally acts to detoxify poisons, there are notable exceptions in which the body converts an essentially harmless chemical into a deadly poison. Fluoroacetic acid is one such compound.

The body cells use acetic acid to produce citric acid. The citric acid is then broken down in a series of steps, most of which release energy. When fluoroacetic is ingested, it is incorporated into fluorocitric acid. The latter effectively blocks the citric-acid cycle by tying up the enzyme that acts on citric acid. Thus, the energy-producing mechanism of the cell is shut off and death rapidly ensues.

Sodium fluoroacetate (Compound 1080) is widely used to poison rats and predatory animals. It is not selective, and thus is highly dangerous to humans, pets, and other desirable animals. Sodium fluoroacetate was once widely used by ranchers to poison coyotes, eagles, and other animals suspected of preying on sheep and cattle. Such use drove the eagles nearly to extinction. As a result, the poisoning of predators with sodium fluoroacetate and other deadly chemicals was banned on federal land.

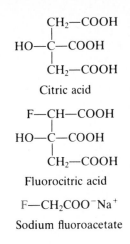

$$CH_2-COOH$$
$$HO-\overset{|}{\underset{|}{C}}-COOH$$
$$CH_2-COOH$$

Citric acid

$$F-CH-COOH$$
$$HO-\overset{|}{\underset{|}{C}}-COOH$$
$$CH_2-COOH$$

Fluorocitric acid

$$F-CH_2COO^-Na^+$$

Sodium fluoroacetate

Heavy Metal Poisons

People have long used a variety of metals in industry and agriculture and around the home. Most metals and their compounds show some toxicity when ingested in large amounts. Even the essential mineral nutrients (Chapter 18) can be toxic when taken in excessive amounts. Quite often, too little of a metal (deficiency) can be as dangerous as too much (toxicity). For example, the average adult requires 10 to 18 mg of iron every day. If less is taken in, the person suffers from an anemia. Yet an overdose can cause vomiting, diarrhea, shock, coma, and even death. As few as 10 to 15 tablets containing 5 grains (324 mg) each of iron (as $FeSO_4$) have been fatal to children.

It is not known exactly how iron poisoning works. The heavy metals—those near the bottom of the periodic table—exert their action primarily by inactivating enzymes. It is well known in inorganic chemistry that heavy metal ions react with hydrogen sulfide (H_2S) to form insoluble sulfides.

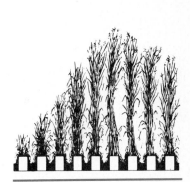

Figure 25.5 The effect of copper on the height of oat seedlings. From left to right, the quantities of copper present are 0, 3, 6, 10, 20, 100, 500, 2000, and 3000 $\mu g/L$. Plants on the left show the effect of a deficiency; those on the right exhibit copper toxicity. Plants in the middle have an optimum amount of copper. [Redrawn from C. S. Piper, *Journal of Agricultural Science*, 32 (1942), 143.]

$$Pb^{2+} \quad + \quad H_2S \longrightarrow \quad PbS \quad + \quad 2\,H^+$$

Lead ion   Lead sulfide
(in solution)   (insoluble)

$$Hg^{2+} \quad + \quad H_2S \longrightarrow \quad HgS \quad + \quad 2\,H^+$$

Mercury ion   Mercury sulfide
(in solution)   (insoluble)

Most enzymes have amino acids with sulfhydryl (—SH) groups at or near the active sites. Heavy metal ions tie up these groups, rendering the enzymes inactive.

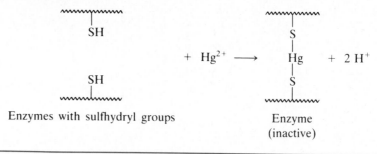

Enzymes with sulfhydryl groups

Enzyme
(inactive)

Arsenic is not a metal, but it has some metallic properties. In commercial poisons, arsenic usually is found as arsenate (AsO_4^{3-}) or arsenite (AsO_3^{3-}) ions. These also render enzymes inactive by tying up sulfhydryl groups.

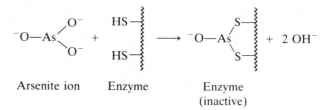

Arsenite ion   Enzyme   Enzyme
(inactive)

Organic compounds containing arsenic are well known. One such compound, arsphenamine, was the first antibacterial. It once was used widely in the treatment of syphilis.

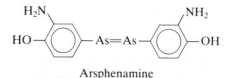

Arsphenamine

Another arsenic compound was developed as a blister agent for use in chemical warfare. This agent was first synthesized by (and named for) W. Lee Lewis. The United States started large-scale production of Lewisite in 1918. Fortunately, World War I came to an end before this gas could be employed.

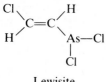

Lewisite

Quicksilver → Slow Death

Mercury (Hg) is a most unusual metal. It is the only common metal that occurs as a liquid at room temperature. This bright, silvery, dense liquid—formerly known as quicksilver—has long held the fascination of people. Children sometimes play with the mercury from a broken thermometer, often with tragic results. Mercury vapor is hazardous. An open container or a few droplets spilled on the floor can put enough mercury vapor into the air to exceed the established maximum safe level by a factor of 200.

Mercury presents a hazard to those who work with it (Color Plate S). Dentists use it to make amalgams for filling teeth. Laboratory workers use mercury and its compounds in a variety of ways. Farmers use seeds treated with compounds of mercury. Since mercury is a cumulative poison (it takes the body about 70 days to rid itself of *half* a given dose), chronic poisoning is a real threat to those continually exposed.

Fortunately, there are antidotes available for mercury poisoning. British scientists, searching for an antidote for the arsenic-containing war gas Lewisite, came up with a compound effective for heavy metal poisoning as well. The compound, a derivative of glycerol (Chapter 18), came to be known as BAL (British Anti-Lewisite). BAL acts by chelating (Greek *chela*: "claw") the metal ion. Thus tied up, the mercury cannot attack vital enzymes.

Now for the bad news. The symptoms of mercury poisoning may not show up for several weeks. By the time the symptoms—loss of equilibrium, sight, feeling, hearing—are recognizable, extensive damage has already been done to the brain and the nervous system. Such damage is largely irreversible. The BAL antidote is effective only when a person knows that he or she has been poisoned and seeks treatment right away.

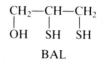

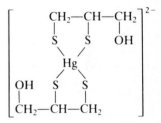

Mercury atom chelated by two BAL molecules

Metallic mercury seems not to be very toxic when ingested (swallowed). Most of it passes through the system unchanged. Indeed, there are numerous reports of mercury being given orally in the eighteenth and nineteenth centuries as a remedy for obstruction of the bowels. Doses varied from a few ounces to a pound or more. Reports from the poison control center of the New York City Department of Health confirm the low toxicity of metallic mercury taken orally. Eighteen incidents without serious effects were noted over a 2-year period.

When inhaled, however, mercury vapor is quite hazardous, particularly when exposure occurs over a long period of time. Such chronic exposure usually involves mining, extraction, or regular occupational use of the metal. The body seems able to convert the inhaled mercury, by some as yet unknown mechanism, to Hg^{2+} ions. All the compounds of mercury, except those that are essentially insoluble in water, are poisonous no matter how they are administered.

Lead in the Environment

Compounds of the element lead (Pb) are widespread in the environment. This reflects the many uses we have for this soft, dense, corrosion-resistant metal and its compounds. Lead (as Pb^{2+}) is present in many foods, generally in concentrations of less than 0.3 ppm. However, condensed milk, sold in cans sealed with lead solder, may contain as much as 0.5 ppm. Lead (again as Pb^{2+}) also gets into our drinking water (up to 0.1 ppm) from lead-sealed pipes. Lead compounds, mainly from automobiles that burn leaded gasoline, even permeate the air we breathe. (This exposure is decreasing as we switch to unleaded gasoline.)

Lead compounds are quite toxic. Metallic lead is generally converted to Pb^{2+} in the body. So, with all that lead, why aren't we dead? The answer lies in the fact that we can excrete about 2 mg of lead per day. Our intake from air, food, and water is generally less than that. If intake exceeds excretion, lead builds up in the body and chronic lead poisoning results.

Lead poisoning is a major problem with children, particularly those in slum areas. Some children develop a craving that leads them to eat unusual things. This syndrome, called **pica**, causes them to eat chips of peeling, lead-based paints. These children probably also pick up lead compounds from the streets, where they are deposited by automobile exhausts. They also may get lead from metal toothpaste tubes, canned milk, and other sources. In all, thousands of children suffer from lead poisoning each year. Such poisoning often leads to mental retardation and neurological disorders through damage to the brain and nervous system.

Lead poisoning usually is treated with a combination of BAL and another chelating agent called EDTA (ethylenediaminetetraacetic acid).

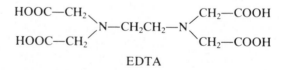

EDTA

The calcium salt of EDTA is administered intravenously. In the body, calcium ions are displaced by lead ions, which the chelate binds more tightly.

$$CaEDTA^{2-} + Pb^{2+} \longrightarrow PbEDTA^{2-} + Ca^{2+}$$

The lead–EDTA complex is excreted.

As with mercury poisoning, the neurological damage done by lead compounds is essentially irreversible. Treatment must be begun early to be effective.

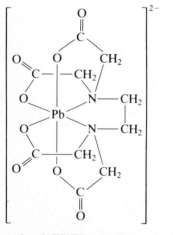

Lead–EDTA complex

Cadmium: The "Ouch-Ouch" Disease

Cadmium (Cd) is used less extensively than lead or mercury, but it too has caused major catastrophes. Cadmium is used widely in alloys, in the electronics industry, in nickel–cadmium rechargeable batteries, and many other applications. Cadmium poisoning leads to loss of calcium ions (Ca^{2+}) from the bones, leaving them brittle and easily broken. It also causes severe abdominal pain, vomiting, diarrhea, and a choking sensation.

The most notable cases of cadmium poisoning occurred along the upper Zintsu River in Japan. The metal (as Cd^{2+}) entered the water in milling wastes from a mine. Downstream the water was used by farm families for drinking, cooking, and other household uses. It was also used to irrigate the rice fields. Soon the farm folk began to suffer from a strange, painful malady that became known as *itai-itai*, the "ouch-ouch" disease.

The Chemistry of the Nervous System

Some poisons—among them the most toxic substances known—act upon the nervous system. Some chemistry of nerve cells was considered in Chapter 23. Recall that signals are shuttled across synapses between cells by chemical substances called neurotransmitters. One such chemical messenger is acetylcholine. It is thought to activate the next cell by fitting a specific receptor and thus changing the permeability of the cell membrane to certain ions.

Once acetylcholine has carried the impulse across the synapse, it is rapidly hydrolyzed to acetic acid and choline. This reaction is catalyzed by an enzyme, cholinesterase.

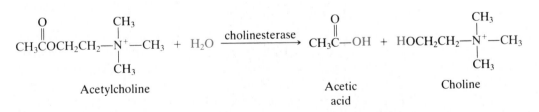

Acetylcholine   Acetic acid   Choline

Choline is relatively inactive. This breakdown enables the receptor cell to receive further impulses. Other enzymes, such as acetylase, convert the acetic acid and choline back to acetylcholine, completing the cycle (Figure 25.6).

It is interesting to note that manic-depressive people are overly sensitive to acetylcholine; they seem to have too many receptors. This may account for their wild swings in mood.

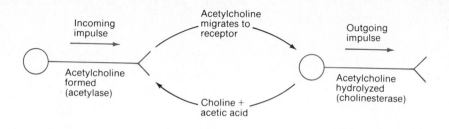

Figure 25.6 The acetylcholine cycle.

Nerve Poisons

Various chemical substances can disrupt the acetylcholine cycle at different points. Botulin, the deadly toxin given off by *Clostridium botulinum* (an anaerobic bacterium) in improperly processed canned food, blocks the synthesis of acetylcholine. With no messenger formed, no messages are carried. Paralysis sets in and death occurs, usually by respiratory failure.

Curare, atropine, and some of the local anesthetics (Chapter 23) act by blocking the receptor sites. In this case, the message is sent but not received. In the case of local anesthetics, this can be good for pain relief in a limited area. These drugs, too, can be fatal in sufficient quantity.

The third category of poisons, called anticholinesterase poisons, blocks the action of cholinesterase (Figure 25.7). The organic phosphorus insecticides (Chapter 17) are well-known nerve poisons. The phosphorus–oxygen linkage is thought to bond tightly to cholinesterase. This blocks the breakdown of acetylcholine, and the acetylcholine builds up and causes the receptor nerves to fire repeatedly. This overstimulates the muscles, glands, and organs. The heart beats wildly and irregularly. The victim goes into convulsions and dies quickly.

While doing research on organic phosphates as possible insecticides during World War II, German scientists discovered some extremely toxic compounds with a frightening potential for use in warfare. The Russians captured a German plant that manufactured a compound called Tabun (designated as agent GA by the United States Army). The Soviets dismantled the factory and moved it to Russia. The United States Army captured some of the stock, however. Presumably, the gas has been produced and stocked by both of the former Allies.

The United States has developed other nerve poisons. One is called Sarin (agent GB). Sarin is four times as toxic as Tabun. It also has the "advantage" of being odorless. Tabun has a fruity odor. Another organophosphate nerve poison is Soman (agent GD). It is moderately persistent, whereas Tabun and Sarin are generally nonpersistent.

Note again the variation in structure from one compound to the next. The approach for developing chemical warfare agents is much the

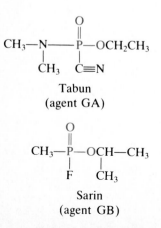

Tabun
(agent GA)

Sarin
(agent GB)

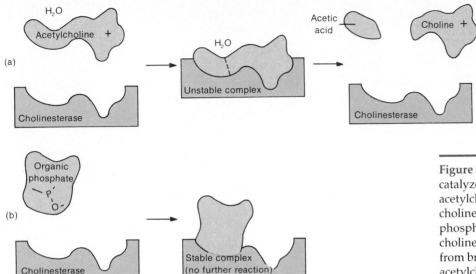

(a)

(b)

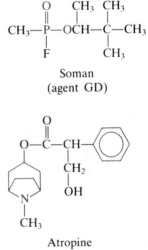

Figure 25.7 (a) Cholinesterase catalyzes the hydrolysis of acetylcholine to acetic acid and choline. (b) An organic phosphate ties up cholinesterase, preventing it from breaking down acetylcholine.

Soman
(agent GD)

Atropine

same as the approach for developing drugs—find one that works and then synthesize and test structural variations.

The nerve gases are among the most toxic synthetic chemicals known. They kill by inhalation or by absorption through the skin. They result in the complete loss of muscular coordination and subsequent death by cessation of breathing. The usual antidote is atropine injection and artificial respiration. Without the antidote, death may occur in 2 to 10 minutes.

The similarity of insecticides such as malathion and parathion to these nerve gases should be recognized. Though somewhat less toxic than the nerve gases, these phosphorus-based insecticides and others like them should be used with extreme caution. Even the relatively safe (to mammals) chlorinated hydrocarbon pesticides act as nerve poisons. Acute DDT poisoning causes tremors, convulsions, and cardiac or respiratory failure. Chronic exposure to DDT leads to the degeneration of the central nervous system. Other chlorinated compounds, such as the PCBs (Chapter 12), act in a similar manner.

Nerve poisons aren't all bad. Even the nerve gases, with their tremendous potential for death and destruction, are not nearly as toxic as the natural toxin botulin. More importantly, the nerve poisons have helped us gain an understanding of the chemistry of the nervous system. It is that knowledge that enables scientists to design antidotes for the nerve poisons. In addition, our increased understanding should contribute to progress along more positive lines—in the control of pain, for example. Even deadly botulin has found use in medicine. To correct crossed eyes, tiny amounts of botulin are injected into the optic muscles, temporarily blocking muscles that are pulling too hard and causing the eyes to cross.

Your Liver: A Detox Tank

The human body can handle a moderate amount of some poisons. The liver is able to detoxify some compounds by oxidation, reduction, or coupling with amino acids or other normal body chemicals.

Perhaps the most common route is oxidation. Ethanol (Chapter 11) is detoxified by oxidation to acetaldehyde which in turn is oxidized to acetic acid, a normal constituent of cells. The acetic acid is then oxidized to carbon dioxide and water.

$$CH_3CH_2OH \longrightarrow \underset{H}{\overset{\displaystyle O}{\underset{\|}{CH_3C}}}\!-\!H \longrightarrow \underset{}{\overset{\displaystyle O}{\underset{\|}{CH_3C}}}\!-\!OH \longrightarrow CO_2 + H_2O$$

Highly toxic nicotine from tobacco is detoxified by oxidation to cotinine.

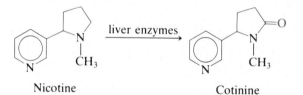

Nicotine   Cotinine

Cotinine is less toxic than nicotine. The added oxygen atom also makes cotinine water soluble, and thus more readily excreted in the urine, than nicotine.

The liver is equipped with a system of enzymes, called P-450, that oxidize fat-insoluble substances that are likely to be retained in the body, into water soluble ones that are readily excreted. It can also conjugate compounds with amino acids. For example, toluene is essentially insoluble in water. The P-450 enzymes oxidize toluene to more soluble benzoic acid. The latter is then coupled with the amino acid glycine to form hippuric acid, which is still more soluble and is readily excreted.

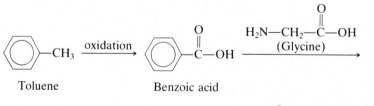

Hippuric acid

It should be pointed out that the liver enzymes just oxidize, reduce, or conjugate. The end product is not always less toxic. For example, methanol is oxidized to a more toxic form, formaldehyde. It is probably the formaldehyde that reacts with the protein in the cells to cause blindness, convulsions, respiratory failure, and death.

$$CH_3OH \xrightarrow{\text{liver enzymes}} H-\overset{\overset{\displaystyle O}{\|}}{C}-H$$

Note that the same enzymes that oxidize the alcohols deactivate the male hormone testosterone. Buildup of these enzymes in a chronic alcoholic leads to a more rapid destruction of testosterone. Thus, we have the mechanism for alcoholic impotence, one of the well-known characteristics of the disease.

> An antidote for methyl alcohol poisoning is ethyl alcohol. Ethyl alcohol is administered intravenously in an attempt to "load up" the liver enzymes with ethyl alcohol and thus to block the oxidation of methyl alcohol until the compound can be excreted.

Enzymes also serve to activate parathion. This sulfur-containing insecticide is itself inactive, but it is transformed biochemically into paraxon, an agent resembling the nerve gases.

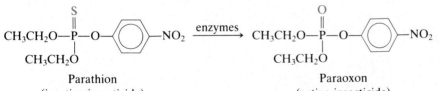

Parathion
(inactive insecticide)

Paraoxon
(active insecticide)

Some drugs also are activated in the liver. Cyclophosphamide, the anticancer drug (Chapter 22), is converted to the active aldophosphamide by liver enzymes.

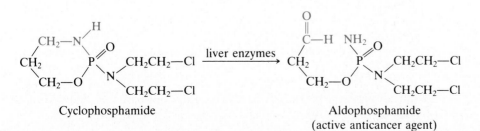

Cyclophosphamide

Aldophosphamide
(active anticancer agent)

And, as we see shortly, some carcinogens are activated in the liver, as well.

Chemical Carcinogens: The Slow Poisons

In earlier chapters, we mentioned carcinogens often. These substances cause the growth of **tumors**. A tumor is an abnormal growth of new tissue. It may be **benign**: benign tumors are characterized by slow growth, they often regress spontaneously, and they do not invade neighboring tissues. **Malignant** tumors often are called **cancers**. They may grow slowly or rapidly, but their growth is generally irreversible. Cancers invade and destroy neighboring tissues. Actually, cancer is not a single disease. It is a catchall term for over a hundred different afflictions. Many are not even closely related to each other.

What causes cancer? The World Health Orgnization estimates that 80 to 90% of the cases are caused by environmental factors, 10 to 20% by genetic factors and (perhaps) viruses. Included most prominently among those "environmental" causes are cigarette smoking (40%), dietary factors (25 to 30%), and occupational exposure (10%). That leaves 10 to 15% that may be caused by environmental *pollutants*. It also should be noted that, contrary to popular opinion, "everything" does not cause cancer. Even among those chemicals that have been suspect, many cannot be shown to be carcinogenic. Only about 30 chemical compounds have been identified as human carcinogens. Another 300 or so have been shown to cause cancer in laboratory animals. Some of these chemicals are widely used.

Not all carcinogens are synthetic chemicals. Some, such as safrole in sassafras and the aflatoxins produced by molds on foods (Chapter 19), occur naturally. Further, a variety of widely different chemical compounds are carcinogenic. We could not attempt to cover all the types of carcinogens here, but we do concentrate on a few major classes.

Some of the more notorious carcinogens are the polycyclic aromatic hydrocarbons (Chapter 10), of which 3,4-benzpyrene is perhaps the best known. Carcinogenic hydrocarbons are formed during the incomplete burning of nearly any organic material. They have been found in charcoal-grilled meats, cigarette smoke, automobile exhausts, coffee, burnt sugar, and many other materials. Not all polycyclic aromatic hyrocarbons are carcinogenic. There are strong correlations between carcinogenicity and certain molecular sizes and shapes. The mechanism of their action is currently under intense investigation. It already appears rather certain that the actual carcinogens are not the hydrocarbons themselves, but the oxidation products formed in the liver.

3,4-Benzpyrene

Another important class of carcinogens is the aromatic amines. Two prominent ones are β-naphthylamine and benzidine. These compounds once were used widely in the dye industry. They were responsible for a high incidence of bladder cancer among the workers whose jobs brought them into prolonged contact with the compounds.

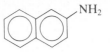

β-Naphthylamine

Benzidine

Several aminoazo dyes have been shown to be carcinogenic. An interesting example is 4-dimethylaminoazobenzene. This compound is also known as "butter yellow." It was used widely as a coloring for butter and oleomargarine before its carcinogenicity became known.

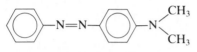

4-Dimethylaminoazobenzene

Not all carcinogens are aromatic. Prominent among the aliphatic (nonaromatic) ones are dimethylnitrosamine (Chapter 19), vinyl chloride (Chapter 12), and ethyl carbamate (Chapter 23). Others include three- and four-membered heterocyclic rings containing nitrogen or oxygen. The epoxides and derivatives of ethyleneimine are examples. Others are cyclic esters called lactones.

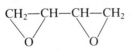

Bis(epoxy)butane

N-Laurylethyleneimine

β-Propiolactone

Keep in mind that this list is not all-inclusive. Rather, its purpose is to give you an idea of the kinds of compounds that have tumor-inducing properties—and the list grows almost daily as the results of research are released.

How do chemicals cause cancer? Their mechanisms of action are probably as varied as their chemical structure. Some carcinogens chemically modify DNA, thus scrambling the code for replication and for the synthesis of proteins. For example, aflatoxin B is known to bind to guanine residues in DNA. Just how this initiates cancer, however, is not known for sure.

There is a genetic component to the development of some forms of cancer. Certain genes, called oncogenes, seem to trigger or sustain the processes that convert normal cells to cancerous ones. Oncogenes are believed to arise from ordinary benign genes, but more research is needed to find out exactly how they develop. It may be that chemical carcinogens, radiation, or other factors act by causing a suppressed gene to be relocated to a place where it is activated. It seems that more than one oncogene must be turned on, perhaps at different stages of the process, before a cancer develops.

Testing for Carcinogens: Three Ways

How do we know that a chemical causes cancer? Obviously, we can't experiment on humans to see what happens. Indeed, there is no way to prove *absolutely* that a chemical does or does not cause cancer in humans. There are, however, three ways to gain evidence against a compound: animal tests, epidemiological studies, and bacterial screening for mutagenesis.

Chemicals suspected for being carcinogens usually are tested on animals. Tests using low dosages on millions of rats would cost too much, so tests usually are done by using large doses on 30 or so rats. An equal number of rats serve as controls. The control group is exposed to the same diet and environment as the experimental group, except that the control group does not get the suspected carcinogen. A higher incidence of cancer in the experimental animals than in the controls indicates that the compound is carcinogenic.

Animal tests are not conclusive. Humans usually are not exposed to comparable doses; there may be a threshold below which a compound is not carcinogenic. Further, human metabolism is different from the metabolisms of their animals. The carcinogen might be active in the rat but not in humans (or vice versa!). Despite the shortcomings of this method, there is good correlation between animal tests and the occurrence of human cancers.

The best evidence that a substance causes cancer in humans comes from epidemiological studies. A population that has a higher than normal rate for a particular kind of cancer is studied for common factors in background. It was this sort of study, for example, that showed that cigarette smoking causes lung cancer, that vinyl chloride causes a rare form of liver cancer, and that asbestos causes cancer of the lining of the pleural cavity (the body cavity that contains the lungs). These studies require sophisticated mathematical analyses. There is always the chance that some other (unknown) factor is involved in the carcinogenesis.

The third way to gain evidence that a substance may be carcinogenic is by use of a screening test (Figure 25.8) developed by Bruce N. Ames of the University of California, Berkeley. The Ames tests use a variant of the bacterium *Salmonella typhimurium* to test for **mutations** (changes in genes). The test is simple and relatively inexpensive, but it is probably the least accurate of the three methods of identifying carcinogens. Some chemicals that are mutagens are not carcinogens. To date, though, 90% of the chemicals that fail the Ames test have been carcinogens as well as mutagens. One shortcoming of the Ames test is that some chemicals, not carcinogenic themselves, are converted to carcinogens by metabolism. These might well pass the Ames test and still cause cancer. A modification of the test uses metabolites from urine or feces in the screening process to overcome this problem.

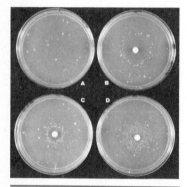

Figure 25.8 The spot test for mutagenic activity. Each petri plate contains, in a thin overlay of agar, the *Salmonella* test strain. Plates C and D also contain a liver microsomal activation system. Mutagens are applied to filter paper discs, which then are placed in the center of each plate. Plate A holds spontaneous revertants; B, the Japanese food additive furylfuramine; C, the mold carcinogen aflatoxin B_1. Mutagen-induced revertants appear in a ring of colonies around each disc. [Reprinted with permission from B. N. Ames, J. McCann, and E. Yamasaki, *Mutation Research*, 31 (1975), 347.]

The Ames test is seldom used alone. If an economically important compound fails the Ames test, the compound is subjected to animal testing. Epidemiological studies also may be done.

Anticarcinogens

There are many natural carcinogens in our food. Why don't we all get cancer? Perhaps there are **anticarcinogens** in food as well. Fiber is thought to protect against colon cancer (Chapter 18). The food additive BHT (Chapter 19) may protect against stomach cancer. Similarly, the natural pigment β-carotene (Chapter 19) has possible anticarcinogenic activity.

Perhaps most notable is the protective value of a diet high in cruciferous vegetables (cabbage, brussels sprouts, broccoli, cauliflower, and kale). This kind of diet has been shown to reduce the incidence of cancer both in animal studies and in studies of human population groups. The exact chemical substances in these vegetables that act as anticarcinogens are not known. It seems quite likely, though, that by eating a balanced diet, including fresh fruit and vegetables, we also balance our carcinogens and anticarcinogens. Please pass the broccoli.

Birth Defects: Teratogens

Another group of toxic chemicals is those that cause birth defects. These substances are called **teratogens**. Perhaps the most notable teratogen is the tranquilizer thalidomide.

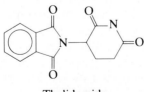

Thalidomide

This drug was linked to a great tragedy during the late 1950s and early 1960s. Thalidomide was considered so safe, based on laboratory studies, that it often was prescribed for pregnant women. In Germany it was available without a prescription. It took several years for the human population to provide evidence that laboratory animals had not. The drug had a disastrous effect on developing human embryos. Women who had taken the drug during the first 12 weeks of pregnancy had babies that suffered from phocomelia, a condition characterized by shortened or absent arms and legs and other physical defects. The drug

Figure 25.9 Frances O. Kelsey, M. D., of the U. S. Food and Drug Administration, refused to approve the tranquilizer drug thalidomide for marketing in the United States. Her action prevented many mothers from facing the tragedy of giving birth to a "thalidomide baby." She is shown here receiving the Distinguished Federal Civilian Service Award from President John F. Kennedy on 7 August 1962. [Courtesy of the *FDA Consumer*, Rockville, MD.]

was used widely in Germany and Great Britain, and these two countries bore the brunt of the tragedy. The United States escaped relatively unscathed because an official of the FDA had believed there was evidence to doubt the drug's safety and had not, therefore, approved it for use in the United States (Figure 25.9).

There may be other chemicals that act as teratogens. More careful testing, however, has helped us to avoid another thalidomide tragedy. Indeed, some now say that our testing procedures are too strict, expensive, and time consuming, and that they keep needed drugs off the market for years.

Hazardous Wastes

Toxic substances, carcinogens, teratogens—the public has become increasingly concerned in recent years about the problems involving hazardous wastes in the environment. Problems with chemical dumps have made household words out of "Love Canal" in New York and "Valley of the Drums" in Kentucky. Although often overblown in the news media, serious problems do exist. Hazardous wastes can cause fires or explosions. They can pollute the air. They can contaminate our food and water. Occasionally they poison by direct contact. As long as we want the products our industries produce, though, we will have to deal with the problems of hazardous wastes (Table 25.1).

The first step in dealing with any problem is to understand what the problem is. A **hazardous waste** is one that can cause or contribute to death or illness or that threatens human health or the environment when improperly managed. For convenience, hazardous wastes are divided into four types: reactive, flammable, toxic, and corrosive.

Reactive wastes tend to react spontaneously or to react vigorously with air or water. They may generate toxic gases, such as hydrogen cyanide (HCN) or hydrogen sulfide (H_2S), or explode when exposed to shock or heat. An example of a reactive waste is sodium metal. Wastes containing sodium caused explosions at Malkin Bank in Great Britain (Figure 25.10). The sodium reacted with water to form hydrogen gas.

$$2 \, Na \; + \; 2 \, H_2O \; \xrightarrow{\text{fast}} \; 2 \, NaOH \; + \; H_2$$

The hydrogen then exploded when ignited in air.

$$2 \, H_2 \; + \; O_2 \; \xrightarrow{\text{explosive}} \; 2 \, H_2O$$

Reactive wastes can usually be deactivated before disposal. Sodium can be treated with isopropyl alcohol, with which it reacts slowly, rather than being dumped without treatment.

Table 25.1 Industrial Products and Their Hazardous-Waste By-products

Product	Associated Waste
Plastics	Organic chlorine compounds
Pesticides	Organic chlorine compounds, organic phosphate compounds
Medicines	Organic solvents and residues, heavy metals (e.g., mercury and zinc)
Paints	Heavy metals, pigments, solvents, organic residues
Oil, gasoline, other petroleum products	Oil, phenols and other organic compounds, heavy metals, ammonia salts, acids, caustics
Metals	Heavy metals, fluorides, cyanides, acid and alkaline cleaners, solvents, pigments, abrasives, plating salts, oils, phenols
Leather	Heavy metals, organic solvents
Textiles	Heavy metals, dyes, organic chlorine compounds, solvents

SOURCE: Adapted from "Everybody's Problem: Hazardous Waste." Washington, DC: U.S. Environmental Protection Agency, 1980.

(a)

(b)

Figure 25.10 (a) A waste dump at Malkins Bank, Cheshire, England, with drums in a lake of chemicals (1970). (b) The same site, cleaned up and restored (1980). [Courtesy of Harwell Laboratory, Oxfordshire, England.]

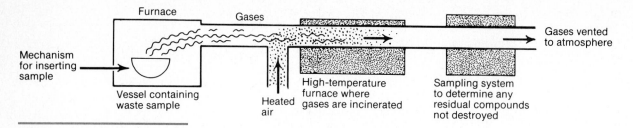

Figure 25.11 A schematic diagram of an incinerator for hazardous wastes. [After drawing from Midwest Research Institute, *MRI Viewpoint*, Summer 1982.]

Flammable wastes are those that burn readily upon ignition, presenting a fire hazard. An example is hexane, a hydrocarbon solvent. In one case, hexane (presumably dumped accidentally by Ralston-Purina) was ignited in the sewers of Louisville, Kentucky.

$$2\ C_6H_{14}\ +\ 19\ O_2\ \longrightarrow\ 12\ CO_2\ +\ 14\ H_2O$$

Explosions ripped up several blocks of streets. Flammable wastes usually can be burned safely in incinerators, rather than discarded.

Toxic wastes are those that contain or release toxic substances in quantities sufficient to pose a hazard to human health or the environment. Most of the toxic substances discussed in this chapter would qualify as toxic wastes if they were improperly dumped in the environment. Some toxic wastes can be incinerated safely (Figure 25.11). PCBs (Chapter 12), for example, are broken down at high temperatures to carbon dioxide, water, and hydrogen chloride. If the hydrogen chloride is removed by scrubbing or is safely diluted and dispersed, this is a satisfactory way to get rid of PCBs. Other toxic wastes cannot be incinerated, however, and must be contained and monitored for years.

Corrosive wastes are those that require special containers because they corrode usual container materials. Acids can't be stored in steel drums because they react with and dissolve the iron.

$$Fe\ +\ 2\ H_3O^+\ \longrightarrow\ \underset{\text{(soluble)}}{Fe^{2+}}\ +\ H_2\ +\ 2\ H_2O$$

Acid wastes can be neutralized (Chapter 7) before disposal, and lime (CaO) is a cheap base for neutralization.

$$2\ H_3O^+\ +\ CaO\ \longrightarrow\ Ca^{2+}\ +\ 3\ H_2O$$

The best way to handle hazardous wastes is not to produce them in the first place. Many industries have modified processes to minimize the amount of wastes. Some wastes can be reprocessed to recover energy or materials. Hydrocarbon solvents such as hexane can be purified and reused or burned as a fuel. Sometimes one industry's waste can be a raw material for another industry. For example, waste nitric acid from the metals industry can be converted to fertilizer (Chapter 17). Finally, if a hazardous waste cannot be used or incinerated or treated to render it less hazardous, it must be stored in a secure landfill. Unfortunately, landfills often leak, contaminating the groundwater. We clean up one toxic waste dump and move the materials to another, playing a rather macabre shell game. The best technology at present for treating organic wastes, including chlorinated compounds, is incineration. At 1260 °C, 99.9999+% destruction is achieved. Research has identified microorganisms that degrade hydrocarbons such as those in gasoline. Other bacteria, when provided proper nutrients, can degrade chlorinated hydrocarbons. Perhaps biodegradation of the wastes will be the way of the future.

In the meantime, hazardous wastes remain on our minds. The issues are emotional, the decisions are political, and the solutions are scientific.

What Price Poisons?

We use so many poisons in and around our homes and work places that accidents are bound to happen. Poison-control centers have been established in several cities to help physicians deal with emergency poisonings. Are our insecticides, drugs, cleansers, and other chemicals worth the price we pay in terms of accidental poisonings? That is for you to decide. Generally, it is the *misuse* of these chemicals that leads to tragedy.

Perhaps it is easy to be negative about chemists and chemistry when you think of such horrors as nerve gases, carcinogens, and teratogens. But keep in mind that many toxic chemicals are of enormous benefit to us and that they can be used safely despite their hazardous nature. The plastics industry was able to control vinyl-chloride emissions once the hazard was known. We still are able to have valuable vinyl plastics, even though the vinyl chloride from which they are made causes cancer.

Increasingly, we will find ourselves having to decide whether the benefits we gain from a chemical substance outweigh the risks we assume by using it. I hope the chemistry you have learned will help you make intelligent decisions. I hope most of all, though, that you will continue to learn chemistry throughout the rest of your life, for chemistry affects nearly everything you do. I wish you success and happiness in your chosen profession. I hope that your knowledge of chemistry enriches your life. May the joy of learning go with you always.

Problems

1. What is pharmacology?
2. What is toxicology?
3. Is sodium chloride (table salt) poisonous? Explain your answer fully.
4. Give an example that shows how the toxicity of a substance depends on the route of administration.
5. Give an example that shows how a substance may be more harmful to one person than another.
6. How do dilute solutions of acids and bases damage living cells?
7. What is hydrolysis?
8. How does ozone damage living cells?
9. What is a blood agent? List two such poisons.
10. What is methemoglobin?
11. How does cyanide exert its toxic effect?
12. How does sodium thiosulfate act as an antidote for cyanide poisoning?
13. How does fluoroacetic acid exert its toxic effect?
14. Iron (as Fe^{2+}) is a necessary nutrient. What are the effects of too little Fe^{2+}? Of too much?
15. How does mercury (as Hg^{2+}) exert its toxic effect?
16. How does BAL act as an antidote for mercury poisoning?
17. List some sources of mercury poisoning.
18. List some sources of lead poisoning.
19. How does EDTA act as an antidote for lead poisoning?
20. How does cadmium (as Cd^{2+}) exert its toxic effect?
21. What is itai-itai disease?
22. What is acetylcholine? Describe its action.
23. How does botulin affect the acetylcholine cycle?
24. How do curare and atropine affect the acetylcholine cycle?
25. What are acetylcholinesterase poisons? How do they act on the acetylcholine cycle?
26. List three nerve poisons developed for use in warfare.
27. How does atropine act as an antidote for poisoning by organic phosphorus compounds?
28. Describe a use of botulin in medicine.
29. What is the P-450 system? What is its function?
30. How does the liver detoxify ethanol?

31. List two ways that the conversion of nicotine to cotinine in the liver lessens the risk of nicotine poisoning.
32. List the two steps in the detoxification of ingested toluene. What is the effect of these steps?
33. Does the P-450 system always detoxify foreign substances?
34. How does ethanol work as an antidote for methanol poisoning?
35. What is a tumor?
36. How are benign and malignant tumors different?
37. What is the single leading cause of cancer?
38. Name two natural carcinogens.
39. How are polycyclic hydrocarbons formed?
40. Name two aromatic amines that are carcinogens.
41. What is butter yellow? How was it used before it was found to be a carcinogen?
42. Name two aliphatic carcinogens.
43. List some of the limitations involved in testing of compounds for carcinogenicity by using laboratory animals.
44. What is an epidemiological study? Can such a study prove absolutely that a compound causes cancer?
45. What is a mutagen?
46. Describe the Ames test for mutagenicity. What are its limitations as a screening test for carcinogens?
47. What is a teratogen?
48. What is a hazardous waste?
49. Define and give an example of a reactive waste.
50. Define and give an example of a flammable waste.
51. What is a corrosive waste?
52. Why can't waste acids be stored in steel drums?
53. What is the best method for disposal of toxic organic wastes?
54. Choose a disposal method for each of the following wastes. Be as specific as possible, and justify your choice.
    a. hydrochloric acid contaminated with iron salts
    b. picric acid (an explosive)
    c. soybean oil contaminated with PCBs
    d. pentane contaminated with residues from penicillin production

55. Should substances be tested for toxicity or carcinogenicity on laboratory animals? On people?

56. Are nerve gases less humane than bullets in warfare?

57. In prowling around an old dump, Murgatroyd B. Muckraker finds a metal container filled with 12 oz of liquid. The label indicates the can contains a carcinogen at a concentration of 8.2 mg/fluid oz. On further investigation, he finds that 20 billion such containers are filled and distributed in the United States each year.
    a. How many milligrams of the carcinogen are in each can?
    b. How many metric tons of the carcinogen are distributed in this manner each year?
    c. What should be done about the problem?

References and Readings

1. Abelson, Philip H. "Treatment of Hazardous Wastes." *Science*, 1 August 1986, p. 509.

2. Cook, Jag D. "Landfill as a Disposal Route for Difficult Wastes." *Chemistry and Industry*, 3 September 1984, pp. 615–619.

3. Duffus, John H. *Environmental Toxicology*. New York: Halsted Press, 1980.

4. "Everybody's Problem: Hazardous Waste." Washington, DC: U.S. Environmental Protection Agency, 1980.

5. Ioannides, Costas, and Denis V. Parke. "The Metabolic Activation and Detoxication of Mutagens and Carcinogens." *Chemistry and Industry*, 1 November 1980, pp. 854–859.

6. Kendall, Ronald J. "Wildlife Toxicology." *Environmental Science and Technology*, August 1982, pp. 448A–453A.

7. Lu, Frank C. *Basic Toxicology*. New York: Hemisphere, 1985.

8. Marshall, Eliot. "Solving Louisville's Friday-the-Thirteenth Explosion." *Science*, 27 March 1981, p. 1405.

9. "Most Dangerous U.S. Waste Dumps Identified." *Chemical and Engineering News*, 3 January 1983, p. 8.

10. Ottoboni, M. Alice. *The Dose Makes the Poison*. Berkeley, Ca: Vincente Books, 1984.

11. Palmark, Mogens. "Future Options for Disposal of Hazardous Wastes." *Chemistry and Industry*, 16 June 1985, pp. 416–422.

12. Reif, Arnold E. "The Causes of Cancer." *American Scientist*, July–August 1981, pp. 437–447.

13. Sills, Thomas W. "Does Beer Drinking Cause Cancer?" *The Science Teacher*, March 1985, pp. 29–31.

14. Stinson, Stephen C. "EPA to Evaluate New Technologies for Cleaning Up Hazardous Waste." *Chemical and Engineering News*, 25 May 1987, pp. 7–12.

15. Wong, John L. "Cancer and Chemicals . . . and Vegetables." *Chem Tech*, July 1986, pp. 436–443.

A

The International System of Measurement

Metric measurement was discussed in some detail in Chapter 1. Further discussion and additional tables are provided here.

The standard unit of length in the International System of Measurement is the **meter**. This distance was once meant to be 0.000 000 1 of the Earth's quadrant, that is, of the distance from the North Pole to the equator measured along a meridian. The quadrant was difficult to measure accurately. Consequently, for many years the meter was defined as the distance between two etched lines on a metal bar (made of a platinum–iridium alloy) kept in the International Bureau of Weights and Measures at Sevres, France. Today, the meter is defined even more precisely as the distance that light travels in a vacuum in 1/299 792 458 of a second.

The primary unit of mass is the **kilogram** (1 kg = 1000 g). It is based on a standard platinum–iridium bar kept at the International

Bureau of Weights and Measures. The **gram** is a more convenient unit for many chemical operations.

The basic SI unit of volume is the cubic meter. The unit more frequently employed in chemistry, however, is the **liter** (1 L = 0.001 m^3). All other SI units of length, mass, and volume are derived from these basic units.

Table A.1 Some SI Prefixes and Their Relationship to the Basic Units

Prefix	Abbreviation	Connotation
Pico-	p	$0.000\,000\,000\,001 \times$ (or $10^{-12} \times$)
Nano-	n	$0.000\,000\,001 \times$ (or $10^{-9} \times$)
Micro-	μ	$0.000\,001 \times$ (or $10^{-6} \times$)
Milli-	m	$0.001 \times$ (or $10^{-3} \times$)
Centi-	c	$0.01 \times$ (or $10^{-2} \times$)
Deci-	d	$0.1 \times$ (or $10^{-1} \times$)
Deka-	da	$10 \times$ (or $10^{1} \times$)
Hecto-	h	$100 \times$ (or $10^{2} \times$)
Kilo-	k	$1\,000 \times$ (or $10^{3} \times$)
Mega-	M	$1\,000\,000 \times$ (or $10^{6} \times$)
Giga-	G	$1\,000\,000\,000 \times$ (or $10^{9} \times$)
Tera-	T	$1\,000\,000\,000\,000 \times$ (or $10^{12} \times$)

Table A.2 Some Metric Units of Length

1 kilometer (km) = 1000 meters (m)
1 meter (m) = 100 centimeters (cm)
1 centimeter (cm) = 10 millimeters (mm)
1 millimeter (mm) = 1000 micrometers (μm)

Table A.3 Some Metric Units of Mass

1 kilogram (kg) = 1000 grams (g)
1 gram (g) = 1000 milligrams (mg)
1 milligram (mg) = 1000 micrograms (μg)

Table A.4 Some Metric Units of Volume

1 liter (L) = 1000 milliliters (mL)
1 milliliter (mL) = 1000 microliters (μL)
1 milliliter (mL) = 1 cubic centimeter (cm^3)

Table A.5 Some Conversions Between Common and Metric Units

Length
1 mile (mi) = 1.61 kilometers (km)
1 yard (yd) = 0.914 meter (m)
1 inch (in.) = 2.54 centimeters (cm)

Mass
1 pound (lb) = 454 grams (g)
1 ounce (oz) = 28.4 grams (g)
1 pound (lb) = 0.454 kilograms (kg)
1 grain (gr) = 0.0648 gram (g)
1 carat (car) = 200 milligrams (mg)

Volume
1 U.S. quart (qt) = 0.946 liter (L)
1 U.S. pint (pt) = 0.473 liter (L)
1 fluid ounce (fl oz) = 29.6 milliliters (mL)
1 gallon (gal) = 3.78 liters (L)

Table A.6   Some Conversion Units for Pressure

1 milliliter of mercury (mm Hg) = 1 torr
        1 atmosphere (atm) = 760 millimeters of mercury (mm Hg)
                          = 760 torr
        1 atmosphere (atm) = 29.9 inches of mercury (in. Hg)
                          = 14.7 pounds per square inch (psi)
                          = 101 kPa

Table A.7   Some Temperature Equivalents*

Phenomenon	Fahrenheit	Celsius
Absolute zero	−459.69 °F	−273.16 °C
Nitrogen boils/liquefies	−320.4 °F	−195.8 °C
Carbon dioxides solidifies/sublimes	−109.3 °F	−78.5 °C
Bitter cold night, northern Minnesota	−40 °F	−40 °C
Cold night, Indiana	0 °F	−18 °C
Water freezes/ice melts	32 °F	0 °C
Pleasant room temperature	72 °F	22 °C
Body temperature	98.6 °F	37.0 °C
Very hot day	100 °F	38 °C
Water boils/steam condenses	212 °F	100 °C
Temperature for baking biscuits	450 °F	232 °C

* Temperature conversions can be made using the following equations:

$$t_F = \left(\frac{9\ °F}{5\ °C}\right) t_C + 32\ °F$$

$$t_C = \frac{5\ °C}{9\ °F}(t_F - 32\ °F)$$

Table A.8   Some Conversion Units for Energy

                    1 calorie (cal) = 4.184 joules (J)
    1 British thermal unit (Btu) = 1053 joules (J)
                                 = 252 calories (cal)
            1 food "Calorie" = 1 kilocalorie (kcal)
                             = 1000 calories (cal)
                             = 4184 joules (J)

B

Exponential Notation

Scientists often use numbers that are so large—or so small—that they boggle the mind. For example, light travels at 300 000 000 m/s. There are 602 300 000 000 000 000 000 000 carbon atoms in 12 g of carbon. On the small side, the diameter of an atom is about 0.000 000 000 1 m. The diameter of an atomic nucleus is about 0.000 000 000 000 001 m.

It is obviously difficult to keep track of the zeros in such quantities. Scientists find it convenient to express such numbers as *powers of ten*. Tables B.1 and B.2 contain partial lists of such numbers.

The speed of light is usually expressed as 3×10^8 (i.e., $3 \times 10 \times 10 \times 10 \times 10 \times 10 \times 10 \times 10 \times 10$) m/s. The mass of an atom of cesium (Cs) is expressed as 2.21×10^{-22} g, that is, as

$$2.21 = \frac{1}{10\,000\,000\,000\,000\,000\,000\,000} \text{ g}$$

Numbers such as 10^6 are called exponential numbers, where 10 is the *base* and 6 is the *exponent*. Numbers in the form 6.02×10^{23} are said to be written in *scientific notation*.

Exponential numbers are often used in calculations. The most common operations are multiplication and division. Two rules must be followed: (1) to *multiply* exponentials, *add* the exponents, and (2) to

A4

Table B.1   Positive Powers of Ten

$10^0 = 1$

$10^1 = 10$

$10^2 = 10 \times 10 = 100$

$10^3 = 10 \times 10 \times 10 = 1000$

$10^4 = 10 \times 10 \times 10 \times 10 = 10\,000$

$10^5 = 10 \times 10 \times 10 \times 10 \times 10 = 100\,000$

$10^6 = 10 \times 10 \times 10 \times 10 \times 10 \times 10 = 1\,000\,000$

$10^{23} = 100\,000\,000\,000\,000\,000\,000\,000$

Table B.2   Negative Powers of Ten

$10^{-1} = 1/10 = 0.1$

$10^{-2} = 1/100 = 0.01$

$10^{-3} = 1/1\,000 = 0.001$

$10^{-4} = 1/10\,000 = 0.000\,1$

$10^{-5} = 1/100\,000 = 0.000\,01$

$10^{-6} = 1/1\,000\,000 = 0.000\,001$

$\vdots$

$10^{-13} = 1/10\,000\,000\,000\,000 = 0.000\,000\,000\,000\,1$

divide exponentials, *subtract* the exponents. These rules can be stated algebraically as

$$(x^a)(x^b) = x^{a+b}$$

$$\frac{x^a}{x^b} = x^{a-b}$$

Some examples follow.

$$(10^6)(10^4) = 10^{6+4} = 10^{10}$$

$$(10^6)(10^{-4}) = 10^{6+(-4)} = 10^{6-4} = 10^2$$

$$(10^{-5})(10^2) = 10^{(-5)+2} = 10^{-5+2} = 10^{-3}$$

$$(10^{-7})(10^{-3}) = 10^{(-7)+(-3)} = 10^{-7-3} = 10^{-10}$$

$$\frac{10^{14}}{10^6} = 10^{14-6} = 10^8$$

$$\frac{10^6}{10^{23}} = 10^{6-23} = 10^{-17}$$

$$\frac{10^{-10}}{10^{-6}} = 10^{(-10)-(-6)} = 10^{-10+6} = 10^{-4}$$

$$\frac{10^3}{10^{-2}} = 10^{3-(-2)} = 10^{3+2} = 10^5$$

$$\frac{10^{-8}}{10^4} = 10^{(-8)-4} = 10^{-12}$$

$$\frac{10^7}{10^7} = 10^{7-7} = 10^0 = 1$$

Problems involving both a coefficient (a numerical part) and an exponential are solved by multiplying (or dividing) coefficients and exponentials separately.

Example B.1   To what is the following expression equivalent?

$$(1.2 \times 10^5)(2.0 \times 10^9)$$

First, multiply the coefficients.

$$1.2 \times 2.0 = 2.4$$

Then multiply the exponentials.

$$10^5 \times 10^9 = 10^{5+9} = 10^{14}$$

The complete answer is

$$2.4 \times 10^{14}$$

Example B.2   To what is the following expression equivalent?

$$\frac{(8.0 \times 10^{11})}{(1.6 \times 10^4)}$$

First, divide the coefficients.

$$\frac{8.0}{1.6} = 5.0$$

Then divide the exponentials.

$$\frac{10^{11}}{10^4} = 10^{11-4} = 10^7$$

The answer is

$$5.0 \times 10^7$$

Example B.3   Give an equivalent for the following expression.

$$\frac{(1.2 \times 10^{14})}{(4.0 \times 10^6)}$$

Before carrying out the division, it is convenient to rewrite the dividend (the numerator) so that the coefficient is larger than that of the divisor (the denominator).

$$1.2 \times 10^{14} = 12 \times 10^{13}$$

Note that the coefficient was made larger by a factor of 10 and the exponential was made smaller by a factor of 10. The quantity as a whole is unchanged. Now divide.

$$\frac{12 \times 10^{13}}{4.0 \times 10^{6}} = 3.0 \times 10^{7}$$

Example B.4   Give an equivalent for the following expression.

$$\frac{(3 \times 10^{7})(8 \times 10^{-3})}{(6 \times 10^{2})(2 \times 10^{-1})}$$

In problems such as this, you can carry out the multiplications specified in the numerator and in the denominator separately and then divide the resulting numbers.

$$(3 \times 10^{7})(8 \times 10^{-3}) = 24 \times 10^{4}$$

$$(6 \times 10^{2})(2 \times 10^{-1}) = 12 \times 10^{1}$$

$$\frac{24 \times 10^{4}}{12 \times 10^{1}} = 2 \times 10^{3}$$

The multiplications and divisions in problems like this can be carried out in any convenient order.

Only one other mathematical function involving exponentials is of importance to us. What happens when you raise an exponential to a power? You just multiply the exponent by the power. To illustrate:

$$(10^{3})^{3} = 10^{9}$$

$$(10^{-2})^{4} = 10^{-8}$$

$$(10^{-5})^{-3} = 10^{15}$$

If the exponential is combined with a coefficient, the two parts of the number are dealt with separately, as in the following example.

$$(2 \times 10^{3})^{2} = 2^{2} \times (10^{3})^{2} = 4 \times 10^{6}$$

For a further discussion of—and more practice with—exponential numbers, see one of the following references.

Goldish, Dorothy M. *Basic Mathematics for Beginning Chemistry*, 3rd ed. New York: Macmillan, 1983. Chapter 3 covers exponential notation.

Loebel, Arnold B. *Chemical Problem Solving by Dimensional Analysis*, 2nd ed. Boston: Houghton Mifflin, 1978. Chapter 1, Section A, deals with exponential notation.

Problems

1. Express each of the following in scientific notation.
   a. 0.00001
   b. 10000000
   c. 0.0034
   d. 0.0000107
   e. 4500000000
   f. 406000
   g. 0.02
   h. 124×10^3

2. Carry out the following operations. Express the numbers in scientific notation.
   a. $(4.5 \times 10^{13})(1.9 \times 10^{-5})$
   b. $(6.2 \times 10^{-5})(4.1 \times 10^{-12})$
   c. $(2.1 \times 10^{-6})^2$

d, $\dfrac{(4.6 \times 10^{-12})}{(2.1 \times 10^3)}$

e. $\dfrac{(9.3 \times 10^9)}{(3.7 \times 10^{-7})}$

f. $\dfrac{(2.1 \times 10^5)}{(9.8 \times 10^7)}$

g. $\dfrac{(4.3 \times 10^{-7})}{(7.6 \times 10^{22})}$

C

Solving Problems by Unit Conversions

Problems in chemistry, such as those in Chapter 1, are often solved by a method that involves converting from one kind of unit to another. This approach is called the **unit-conversion method** or the **factor-label method**. (It is also often called **dimensional analysis**, although the units employed are not always dimensions.) Whatever we call it, the method employs units—such as L, mi/hour, cm/ft, or g/cm^3—as aids in setting up and solving problems. The general approach is to multiply the known quantity (and its units!) by one or more conversion factors so that the answer is obtained in the desired units.

Known quantity and unit × Conversion factor = Answer (in desired unit)

The method is best learned by practice. We urge you to learn it now to save yourself a lot of time and wasted effort later.

Conversions Within a System

Quantities can be expressed in a variety of units. For example, you can buy beverages by the 12-oz can or by the pint, quart, gallon, or liter If you wish to compare prices, you must be able to convert from one unit to another. Such a conversion changes the numbers and units, but it does not change the quantity. Your actual weight, for example, remains unchanged whether it is expressed in pounds, ounces, or kilograms.

You know that multiplying a number by 1 doesn't change its value. Multiplying by a fraction equal to 1 also leaves the value unchanged. A fraction is equal to 1 when the numerator is equal to the denominator. For example, we know that

$$1 \text{ ft} = 12 \text{ in.}$$

Therefore,

$$\frac{1 \text{ ft}}{12 \text{ in}} = 1$$

Similarly,

$$\frac{12 \text{ in}}{1 \text{ ft}} = 1$$

Now, if you want to convert an answer from inches to feet, you can do so by choosing one of the above fractions as a **conversion factor**. which one do you choose? The one that gives you an answer with the right units! Let's illustrate by an example.

Example C.1　My bed is 72 in. long. What is its length in feet?

You know the answer, of course, but let us proceed to show *how* the answer is obtained by using dimensional analysis. We need to multiply 72 in. by one of the above fractions. Which one? The known quantity and unit is 72 in.

$$72 \text{ in.} \times \text{conversion factor} = ? \text{ ft}$$

For the conversion factor, choose the fraction that, when inserted in the equation, cancels the unit *in.* and becomes the unit *ft*.

$$72 \cancel{\text{ in.}} \times \frac{1 \text{ ft}}{12 \cancel{\text{ in.}}} = 6.0 \text{ ft}$$

Just for kicks, let us try the other conversion factor.

$$72 \text{ in.} \times \frac{12 \text{ in.}}{1 \text{ ft}} = \frac{860 \text{ in.}^2}{\text{ft}}$$

Absurd! How can a bed be 860 in.^{2}/ft? You should have no difficulty in choosing between the two possible answers.

One of the advantages of the metric system is the ease of conversion between units. Let us try to demonstrate that with a few examples. Remember that a list of equivalent values is actually a list of conversion factors. Thus, the equality

$$1 \text{ kilogram} = 1000 \text{ grams}$$

can be rearranged into two useful conversion factors:

$$\frac{1 \text{ kilogram}}{1000 \text{ grams}} \quad \text{and} \quad \frac{1000 \text{ grams}}{1 \text{ kilogram}}$$

Example C.2   Convert 0.371 kg to grams.

$$0.371 \text{ kg} \times \frac{1000 \text{ g}}{1 \text{ kg}} = 371 \text{ g}$$

Example C.3   Convert 0.371 lb to ounces.

$$0.371 \text{ lb} \times \frac{16 \text{ oz}}{1 \text{ lb}} = 5.94 \text{ oz}$$

Example C.4   Convert 2429 cm to meters.

$$2429 \text{ cm} \times \frac{1 \text{ m}}{100 \text{ cm}} = 24.29 \text{ cm}$$

Example C.5   Convert 2429 in. to yards.

$$2429 \text{ in.} \times \frac{1 \text{ yd}}{36 \text{ in.}} = 67.67 \text{ yd.}$$

In conversions of customary units, you multiply and divide by factors such as 16 or 36. In metric conversions, you multiply and divide by 100 or 1000 and so on. You need only shift the decimal point when doing metric conversions.

Conversion factors are not usually given in a problem. They may be obtained from listings such as those in Appendix A. However, you would be wise to learn to convert within the metric system without the need of tables. Also, you should remember that this equality

$$1 \text{ centimeter} = 0.01 \text{ meter}$$

is equivalent to

$$100 \text{ centimeters} = 1 \text{ meter}$$

All these fractions

$$\frac{1 \text{ cm}}{0.01 \text{ m}} \qquad \frac{0.01 \text{ m}}{1 \text{ cm}} \qquad \frac{100 \text{ cm}}{1 \text{ m}} \qquad \frac{1 \text{ m}}{100 \text{ cm}}$$

are valid conversion factors.

Example C.6 Knowing that 1 mL = 0.001 L, write four conversion factors relating milliliters and liters.

The first two conversion factors can be formed by arranging the two sides of the equality in the form of a fraction.

$$\frac{1 \text{ mL}}{0.001 \text{ L}} \qquad \frac{0.001 \text{ L}}{1 \text{ mL}}$$

To derive the other two conversion factors, first multiply both sides of the equality by 1000 (in order to obtain the equality in terms of 1 L).

$$1000 \times 1 \text{ mL} = 1000 \times 0.001 \text{ L}$$

$$1000 \text{ mL} = 1 \text{ L}$$

Now just arrange this last equality in fractional form.

$$\frac{1000 \text{ mL}}{1 \text{ L}} \qquad \frac{1 \text{ L}}{1000 \text{ mL}}$$

The conversion factors 1 mL/0.001 L and 1000 mL/1 L would give exactly the same answer if used in a problem. The only difference is convenience. Some people would rather multiply by 1000 than divide by 0.001. In this age of the electronic calculator, perhaps even this difference is no longer significant.

Let us try some more conversions within the metric system.

Example C.7 How many milliliters are there in a 2-L bottle of soda pop?

From memory or from the tables in Appendix A you find that

$$1 \text{ L} = 1000 \text{ mL}$$

$$2\,\cancel{L} \times \frac{1000 \text{ mL}}{1\,\cancel{L}} = 2000 \text{ mL}$$

Notice that we picked the conversion factor that allowed us to cancel liters and obtain an answer in the desired units, milliliters.

Sometimes it is necessary to carry out more than one conversion in a problem.

Example C.8 In the United States, the usual soda pop can holds 360 mL. How many such cans could be filled from one 2.0-L bottle?

The problem tells us that

$$1 \text{ can} = 360 \text{ mL}$$

Using that equivalence, we can calculate the answer.

$$2.0\,\cancel{L} \times \frac{1000 \,\cancel{\text{mL}}}{1\,\cancel{L}} \times \frac{1 \text{ can}}{360 \,\cancel{\text{mL}}} = 5.6 \text{ cans}$$

Example C.9 How many 325-mg aspirin tablets can be made from 875 g of aspirin?

The problem asks us to convert the *given* value of 875 g to tablets. The problem also includes a necessary conversion factor:

$$1 \text{ tablet} = 325 \text{ mg}$$

From memory or the tables, we have another required conversion factor:

$$1 \text{ g} = 1000 \text{ mg}$$

By multiplying the given value by the appropriately arranged conversion factors, we arrive at the answer.

$$875 \cancel{g} \times \frac{1000 \cancel{mg}}{1 \cancel{g}} \times \frac{1 \text{ tablet}}{325 \cancel{mg}} = 2690 \text{ tablets}$$

Conversions Between Systems

To convert from one system of measurement to another, you need a list of equivalents such as that in Appendix A. Let us plunge right in and work some examples.

Example C.10   How many kilograms are there in 33 lb?

$$33 \cancel{lb} \times \frac{1.0 \text{ kg}}{2.2 \cancel{lb}} = 15 \text{ kg}$$

Example C.11   You know that your weight is 142 lb, but the job application form asks for your weight in kilograms. What is it?

From the table we find

$$1.00 \text{ lb} = 0.454 \text{ kg}$$

The solution is simple.

$$142 \cancel{lb} \times \frac{0.454 \text{ kg}}{1.00 \cancel{lb}} = 64.5 \text{ kg}$$

Example C.12   A recipe calls for 750 mL of milk, but your measuring cup is calibrated in fluid ounces. How many ounces of milk will you need?

$$750 \cancel{mL} \times \frac{1.00 \text{ fl oz}}{29.6 \cancel{mL}} = 25.3 \text{ fl oz}$$

Example C.13 How many meters are there in 764 ft (1.00 m = 39.4 in.)?

$$764 \text{ ft} \times \frac{12 \text{ in.}}{1 \text{ ft}} \times \frac{1.00 \text{ m}}{39.4 \text{ in.}} = 233 \text{ m}$$

Example C.14 How would you describe a young man who is 1.6 m tall and weighs 91 kg?

$$1.6 \text{ m} \times \frac{39 \text{ in.}}{1.0 \text{ m}} \times \frac{1 \text{ ft}}{12 \text{ in.}} = 5.2 \text{ ft}$$

$$91 \text{ kg} \times \frac{2.2 \text{ lb}}{1.0 \text{ kg}} = 200 \text{ lb}$$

The young man is 5 ft 2 in. tall and weighs 200 lb. Let us be generous and say that he is well muscled.

It is possible (and frequently necessary) to manipulate units in the denominator as well as in the numerator of a problem. Just remember to use conversion factors in such a way that the unwanted units cancel.

Example C.15 A sprinter runs the 100-m dash in 11 s. What is her speed in kilometers per hour?
   The given speed is 100 m per 11 s.

$$\frac{100 \text{ m}}{11 \text{ s}}$$

The conversion factors that we need are found in the tables or recalled from memory.

$$\frac{100 \text{ m}}{11 \text{ s}} \times \frac{1 \text{ km}}{1000 \text{ m}} \times \frac{60 \text{ s}}{1 \text{ min}} \times \frac{60 \text{ min}}{1 \text{ hr}} = 33 \text{ km/hr}$$

Note that the first conversion factor changes *m* to *km*. It takes two factors to change *s* to *hr*. Note also that we could have first applied the factors that convert *s* to *hr* and then converted *m* to *km*. The answer would have been the same.

Example C.16   If your heart beats at a rate of 72 times per minute, and your lifetime will be 70 years, how many times will your heart beat during your lifetime?

Two equivalences are given in the problem.

$$72 \text{ beats} = 1 \text{ min}$$

$$1 \text{ lifetime} = 70 \text{ yr}$$

Three others that you will need you can recall from memory.

$$1 \text{ yr} = 365 \text{ days}$$

$$1 \text{ day} = 24 \text{ hr}$$

$$1 \text{ hr} = 60 \text{ min}$$

Start now with the factor 72 beats/1 min (the known quantities and units) and apply the conversion factors as needed to get an answer in beats/lifetime (the desired units).

$$\frac{72 \text{ beats}}{1 \text{ min}} \times \frac{60 \text{ min}}{1 \text{ hr}} \times \frac{24 \text{ hr}}{1 \text{ day}} \times \frac{365 \text{ days}}{1 \text{ yr}} \times \frac{70 \text{ yr}}{1 \text{ lifetime}}$$

$$= 2\,600\,000\,000 \text{ beats/lifetime}$$

Significant Figures

Unlike counting, measurement is never exact. You can *count* exactly ten people in a room. If you asked each of those people to *measure* the length of the room to the nearest 0.01 m, however, the values they determine are likely to differ slightly. Table D.1 presents such a set of measurements.

Note that all ten students agree on the first three digits of the measurement; differences occur in the fourth digit. Which values are correct? Actually, all are accurate within the accepted range of uncertainty for this physical measurement. The accuracy of measurement depends on the type of measuring instrument and the skill and care of the person making the measurement. Measured values are usually recorded with the last digit regarded as uncertain. The data in Table D.1 allow us to state that the length of the room is between 14.1 m and 14.2 m, but we are not sure of the fourth digit. The measurements in the table have four *significant figures*, which means that the first three are known with confidence and the fourth conveys an approximate value. **Significant figures** include all digits known with certainty plus one uncertain digit.

In any properly reported measurement, all nonzero digits are significant. The zero presents problems, however, because it can be

Table D.1   A Set of Measurements of the Length of a Room

Student	Length (m)	Student	Length (m)
1	14.14	6	14.14
2	14.15	7	14.17
3	14.17	8	14.17
4	14.14	9	14.16
5	14.16	10	14.17

used in two ways: to position the decimal point or to indicate a measured value. For zeros, follow these rules.

1. A zero between two other digits is always significant.
   Examples: The number 1107 contains four significant figures.
   The number 50.002 contains five significant figures.

2. Zeros to the left of *all* nonzero digits are not significant.
   Examples: The number 0.000 163 has three significant figures.
   The number 0.068 01 has four significant figures.

3. Zeros that are *both* to the right of the decimal point *and* to the right of nonzero digits are significant.
   Examples: The number 0.**2000** has four significant figures.
   The number 0.0**50 120** has five significant figures.
   The number **802.760** has six significant figures.

4. Zeros in numbers such as 40 000 (that is, zeros to the right of *all* nonzero digits in a number that is written without a decimal point) may or may not be significant. Without more information, we simply do not know whether 40 000 was measured to the nearest unit or ten or hundred or thousand or ten-thousand. To avoid this confusion, scientists use exponential notation (Appendix B) for writing numbers. In exponential notation, 40 000 would be recorded as 4×10^4 or 4.0×10^4 or 4.0000×10^4 to indicate one, two, and five significant figures, respectively.

Addition or Subtraction

In addition or subtraction, the result should contain no more digits to the right of the decimal point than the quantity that has the least digits to the right of the decimal point. Align the quantities to be added or subtracted on the decimal point, then perform the operation, assuming blank spaces are zeros. Determine the correct number of digits after the decimal point in the answer and round off to this number. In rounding off, you should increase the last significant figure by one if the following digit is five through nine.

Example D.1 Add the following numbers: 49.146, 72.13, 5.9432.
   Align the numbers on the decimal point and carry out the addition.

$$
\begin{array}{r}
49.146 \\
72.13 \\
\underline{5.9432} \\
127.2192
\end{array}
$$

The quantity with the least digits after the decimal point is 72.13. The answer should have only two digits after the decimal point. Since the third digit after the decimal point is 9, the second digit after the decimal point should be rounded up to 2. Correct answer: 127.22

Example D.2 Add the following numbers: 744, 2.6, 14.812.

$$
\begin{array}{r}
744 \\
2.6 \\
\underline{14.812} \\
761.412
\end{array} \longrightarrow 761
$$

The first quantity has no digits to the right of the decimal point (which is understood to be at the right of the digits). The answer must therefore be rounded so that it, too, contains no digits to the right of the decimal point.

Example D.3 Subtract 9.143 from 71.12496.

$$
\begin{array}{r}
71.124\,96 \\
\underline{-9.143} \\
61.981\,96
\end{array} \longrightarrow 61.982
$$

Since the second quantity has only three digits to the right of the decimal point, so must the answer.

Multiplication and Division

   In multiplication and division, our answers can have no more significant figures than the factor that has the least number of significant figures. In these operations, the *position* of the decimal point makes no difference.

Example D.4   Multiply 10.4 by 3.1416.

$$10.4 \times 3.1416 = 32.672\,64 \longrightarrow 32.7$$

The answer has only three significant figures because the first term has only three.

Example D.5   Divide 5973 by 3.0.

$$\frac{5.973}{3.0} = 1.991 \longrightarrow 2.0$$

The answer has only two significant figures because the divisor has only two.

Exact Values

Some quantities are not measured but defined. A kilometer is defined as 1000 meters: 1 km = 1000 m. Similarly, 1 foot can be defined as 12 inches: 1 ft = 12 in. The "1 km" should not be regarded as containing one significant figure; nor should "12 in." be considered to have two significant figures. In fact, these values can be considered to have an infinite number of significant figures (1.000 000 000 000 000 0...) or, more correctly, to be *exact*. Such defined values are frequently used as conversion factors in problems (Chapter 1). When you are determining the number of significant figures for the answer to a problem, you should ignore such exact values. Use only the measured quantities in the problem to determine the number of significant figures in the answer.

Problems

Perform the indicated operations and give answers with the proper number of significant figures.

1. a. 48.2 m + 3.82 m + 48.4394 m
   b. 151 g + 2.39 g + 0.0124 g
   c. 15.436 mL + 9.1 mL + 105 mL
2. a. 100.53 cm − 46.1 cm
   b. 451 g − 15.46 g
   c. 19.71 L − 10.4 L

3. a. 73 m × 1.340 m × 0.41 m
   b. 0.137 cm × 1.43 cm
   c. 3.146 cm × 5.4 cm

4. a. $\dfrac{5.179 \text{ g}}{4.6 \text{ mL}}$ b. $\dfrac{4561 \text{ g}}{3.1 \text{ mol}}$ c. $\dfrac{40.00 \text{ g}}{3.2 \text{ mL}}$

5. $\dfrac{1.426 \text{ mL} \times 373 \text{ K}}{204 \text{ K}}$

E

Glossary

absolute scale A temperature scale in which the zero point is the coldest temperature possible, or absolute zero.

adsorbed hydrogen A condensed form of hydrogen collected in large volumes on certain metals such as platinum, palladium, and nickel.

acceptor crystal The part of a solar cell that accepts electrons; for example, the boron-doped crystal with a shortage of electrons.

acid A proton donor.

acid anhydride A substance that reacts with water to produce an acid; a nonmetal oxide.

activated sludge method A combination of primary and secondary sewage treatment methods in which some sludge is recycled.

acid rain Rain having a pH less than 5.6.

acidosis Condition that results when the pH of the blood falls below 7.35 and oxygen transport is hindered.

actomyosin The contractile protein of which muscles are made.

addition polymerization A process in which monomers add to one another in such a way that the polymeric product contains all the atoms of the starting monomers.

adequate protein A protein that supplies all the essential amino acids in the quantities needed for the growth and repair of body tissues.

adjunct A sugar or unmalted starch added to malt to increase the production of alcohol without increasing the carbohydrate content.

advanced treatment Sewage treatment designed to remove phosphates, nitrates, and other soluble impurities.

aerated water Water that is sprayed into the air to remove odors and improve its taste.

aerobic process A process that occurs in the presence of oxygen.

aerobic exercise Exercise in which muscle contractions occur in the presence of oxygen.

aerosol Particles of 1 micrometer diameter, or less, dispersed in air.

aflatoxins Compounds produced by molds growing on stored peanuts and grains.

Agent Orange A combination of 2,4-D and 2,4,5,-T used extensively in Vietnam to remove enemy cover and destroy crops that maintained enemy armies.

agonist A molecule that fits and activates a specific receptor.

alkaloid A nitrogen-containing organic compound obtained from plants.

alkalosis A physiological condition in which the pH of the blood rises to life-threatening levels.

alchemy A mystical chemistry that flourished in Europe during the Middle Ages (A.D. 500 to 1500).

alkali metal An element in Group IA of the periodic table.

alkaline earth metal An element in Group IIA of the periodic table.

alkane A hydrocarbon with only single bonds; a saturated hydrocarbon.

alkene A hydrocarbon containing one or more double bonds.

alkylbenzenesulfonate detergent A synthetic, non-biodegradable detergent produced from petroleum products.

alkyne A hydrocarbon containing one or more triple bonds.

allergen A substance that triggers an allergic reaction.

alloy A mixture of two or more elements, at least one of which is a metal; an alloy has metallic properties.

alpha ray A radioactive beam consisting of helium nuclei.

amide An organic compound having the functional group CON in which the carbon is double-bonded to the oxygen and single-bonded to the nitrogen.

amine A compound that contains the elements carbon, hydrogen, and nitrogen; derived from ammonia by replacing one, two, or three of the hydrogens by alkyl group(s).

amino acid An organic compound that contains both an amino group and a carboxylic acid group; amino acids combine to produce proteins.

amino group An NH_2 unit.

amylase A kind of enzyme required for efficient conversion of barley starches and adjuncts into fermentable sugars.

amylopectin A starch with branched chains of glucose units.

amylose A starch with the glucose units joined in a continuous chain.

anabolic steroid A drug that aids in the building (anabolism) of body proteins and thus of muscle tissue.

anaerobic process A process that occurs in the absence of oxygen.

anaerobic exercise Exercise involving muscle contractions without sufficient amounts of oxygen.

analgesic A pain reliever.

androgen A male sex hormone.

anhydrous Without water.

anion A negatively charged ion.

anionic surfactant A surfactant with a hydrocarbon

tail and a water-soluble head that bears a negative charge.

anode An electrode that bears a positive charge.

antagonist A drug that blocks the action of an agonist, by blocking the receptors.

antibiotic A soluble substance, produced by a mold or bacterium, that inhibits growth of other microorganisms.

anticoagulant A substance that inhibits the clotting of blood.

anticodon The sequence of three adjacent nucleotides in a tRNA molecule that is complement to a codon on mRNA.

antidiuretic A water-conserving substance.

antihistamine A substance that relieves the symptoms of allergies: sneezing, itchy eyes, and runny nose.

anti-inflammatory A substance that inhibits inflammation.

antimetabolite A compound that inhibits the synthesis of nucleic acids.

antiperspirant A formulation to stop or retard perspiration.

antipyretic A fever-reducing substance.

antiseptic A compound applied to living tissue to kill or prevent the growth of microorganisms.

applied research Work oriented toward the solution of a particular problem in industry or the environment.

arithmetic growth A process in which a constant amount is added during each growth period.

aromatic compound Any organic compound that contains a benzene ring.

arteriosclerosis Hardening of the arteries.

artificial transmutation A process by which one element is changed into another by an artificial means.

asbestos A group of related fibrous silicates.

asbestosis A severe respiratory disease caused by inhalation of asbestos fibers 5 to 50 micrometers long over a period of 10 to 20 years.

astringent A substance that constricts the opening of the sweat glands, thus reducing the amount of perspiration that escapes.

atmosphere The gaseous mass surrounding the Earth.

atmospheric inversion A warm layer of air above a cool, stagnant, lower layer.

atom Smallest characteristic particle of an element.

atomic mass unit The unit of relative atomic weights.

atomic number The number of protons in the nucleus of an atom of an element.

atomic theory A model that offers a logical explanation for the law of multiple proportions and the law of constant composition by stating that all elements are composed of atoms, all atoms of a given element are identical, but the atoms of one element differ from the atoms of any other element; that atoms of different elements can combine to give compounds and a chemical reaction involves a change not in the atoms themselves, but in the way atoms are combined to form compounds.

autoradiograph A photographic image made by radioactive substances.

background radiation Ever-present radiation from cosmic rays and from natural radioactive isotopes in air, water, soil, and rocks.

barbiturate A depressant anticonvulsant drug.

base A proton acceptor.

basic anhydride A substance that reacts with water to produce a basic solution; a metal oxide.

basic research The search for knowledge for its own sake.

battery A series of electrochemical cells.

beer An alcoholic beverage made from barley and (sometimes) other grains.

benefit Anything that promotes well-being or has a positive effect.

benign tumor A tumor characterized by slow growth; often regresses spontaneously; does not invade neighboring tissues.

beriberi The disease caused by a deficiency of thiamine.

beta ray A beam of electrons emitted from atomic nuclei.

binary compound A compound containing two elements.

binding energy Energy derived from the conversion of mass to energy when neutrons and protons are put together to form nuclei.

biological magnification An increase in concentration of a substance as it moves up the food chain.

biological oxidation demand A measure of the amount of oxygen needed for the degradation of organic material.

biomass The total mass of plants and animals; in energy studies, usually means plant material used as a source of fuel.

bitumen A hydrocarbon mixture obtained from tar sands by heating.

bleach A compound used to remove unwanted color from fabrics, hair, or other materials.

blood doping A method used by athletes to improve their performance. A quantity of blood is withdrawn from the athlete and stored for six weeks while the person's body manufactures replacement blood. Then, a few days before competition, the red blood cells from the stored blood are returned to the athlete.

blood sugar Glucose, a simple sugar circulated in the bloodstream.

breeder reactor A nuclear reactor that is designed to convert nonfissile uranium-238 to fissile plutonium.

broad-spectrum antibiotic An antibiotic effective against a wide variety of microorganisms.

broad-spectrum insecticide An insecticide that kills many kinds of insects.

buffer A compound that consumes either acid or base to keep the pH of a solution essentially constant.

builder Any substance added to a surfactant to increase its detergency.

Calorie 1000 calories (or 1 kcal); used to measure the energy content of foods.

calorie The amount of heat required to raise the temperature of 1 g of water 1 °C.

cancer A malignant tumor; tumor that grows and invades other tissues.

carbohydrate A compound consisting of carbon, hydrogen, and oxygen; a starch or sugar.

carbohydrate loading A method used by athletes to improve the storage of glycogen. Glycogen stores are first depleted by limited carbohydrate intake and vigorous training. Then, a few days before competition, the athlete cuts back on training and eats a diet high in carbohydrates.

carbonate A metal combined with carbon and oxygen.

carbonyl functional group A carbon atom double-bonded to an oxygen atom.

carboxyl group —COOH, the functional group of the organic acids.

carboxylic acid An organic compound that contains the —COOH functional group.

catalyst A substance that increases the rate of a chemical reaction without itself being used up.

catalytic converter A device containing a catalyst for oxidizing carbon monoxide and hydrocarbons to carbon dioxide.

catalytic reforming A process that converts low-octane alkanes to high-octane aromatic compounds.

catenation A process in which carbon atoms join together to form long chains of hundreds or even thousands of atoms.

cathode An electrode that bears a negative charge.

cathode ray A stream of high-speed electrons.

cation A positively charged ion.

cationic surfactant Surfactant with a hydrocarbon tail and a water-soluble head that bears a positive charge.

celluloid Cellulose nitrate, a synthetic material derived from natural cellulose by reaction with nitric acid.

chain reaction A self-sustaining change in which one or more products of one event cause one or more new events.

chemical bond The force of attraction that holds atoms together in compounds.

chemical equation A before-and-after description in which chemical formulas and coefficients represent a chemical reaction.

chemical properties A description of the way in which a substance reacts with another substance to change its composition.

chemistry The study of matter and the changes it undergoes.

chemotherapy The use of chemicals to control or cure diseases.

chillproofing The process in which enzymes are added to beer to act upon and make soluble any solids that might precipitate in cooling.

Chinese-restaurant syndrome A condition thought to be caused by the heavy use of monosodium glutamate in Chinese foods.

chlorofluorocarbon A carbon compound that contains fluorine and chlorine.

chlorophyll A green plant pigment.

chrysotile A magnesium silicate, a form of asbestos.

clinkers Mineral matter left behind after the burning of coal.

cloning Reproducing in identical form.

coal tar The condensed volatile materials given off in the coking process of coal; a source of organic chemicals, particularly aromatic compounds.

codon A sequence of three adjacent nucleotides in mRNA that represents one amino acid.

coke A relatively pure form of carbon left behind after coal is heated to drive off volatile matter.

cologne A diluted perfume.

complete fertilizer A fertilizer containing the three main nutrients: nitrogen, phosphorus, and potassium.

compound A pure substance made up of two or more elements combined in fixed proportions.

condensation The reverse of vaporization; the change from the gaseous state to the liquid state.

condensation polymerization A process in which a portion of the monomer molecule is split out as the polymer is formed and not incorporated in the final polymer.

condensed structural formula A chemical formula that shows the atoms of hydrogen right next to the carbons to which they are attached.

configuration The arrangement of an atom's electrons in space.

congener A by-product of fermentation or any substance other than ethanol that remains in a finished alcoholic beverage.

continuous spectrum A spectrum in which there is a continuous variation from one color to another.

control rod A rod used to absorb neutrons, thus controlling the rate of fission in a nuclear reactor.

copolymer A polymer that is formed by the combination of two or more different monomer units.

copolymerization A process in which a mixture of two monomers forms a product in which the chain contains units of both building blocks.

core A region at the center of the Earth thought to consist largely of iron and nickel.

corrode Eat away by chemical action.

corrosive waste A waste that requires a special container because it corrodes usual container materials.

cosmetic A substance defined in the 1938 U.S. Food, Drug, and Cosmetics Act as "articles intended to be rubbed, poured, sprinkled or sprayed on, introduced into, or otherwise applied to the human body or any part thereof, for cleansing, beautifying, promoting attractiveness or altering the appearance...."

cosmic rays Radiation that comes from outer space and consists of high-energy atomic nuclei.

cough suppressant A drug that provides temporary relief of a cough.

covalent bond A bond formed by a shared pair of electrons.

cracked Decomposed by heat in the absence of air.

cream An emulsion of tiny water droplets in oil.

critical mass The mass of an isotope above which self-sustaining chain reaction can occur.

crust The outer shell of the Earth composed of the lithosphere, the hydrosphere, and the atmosphere.

crystal A solid with plane surfaces at definite angles.

cyclic Ring-containing.

defoliation Premature removal of leaves from plants.

Delaney Amendment A 1958 amendment to the Food and Drug Act that automatically bans any chemical shown to induce cancer in laboratory animals.

density The amount of mass (or weight) per unit volume.

deodorant A formulation of perfumes designed to mask body odor.

deoxyribonucleic acid The type of nucleic acid found primarily in the nuclei of cells.

depressant A drug that slows down both physical and mental activity.

destructive distillation The process of decomposition by heat with volatile substances distilled off.

deuterium An isotope of hydrogen with a proton and a neutron in the nucleus (mass of 2 amu).

dextro isomer A right-handed isomer.

dioxins Chlorinated cyclic compounds once found as contaminants in herbicides.

dipeptide A compound composed of two bonded amino acids.

dipole A molecule that has a positive end and a negative end.

dipole interactions The attractive intermolecular forces that exist among polar covalent substances.

dispersion forces The momentary, usually weak, attractive forces between molecules.

dissociative anesthetic A substance that induces hallucinations similar to those reported by people who have had near-death experiences.

distillation The boiling off of the volatile compounds (such as the alcohol and water), leaving behind the solids and high-boiling compounds.

disulfide unit A covalent linkage through two sulfur atoms.

diuretic A substance that increases the body's output of urine.

donor crystal A crystal with extra electrons; used in solar cells.

double bond The sharing of two pairs of electrons between two atoms.

doubling time The time it takes a population to double in size.

drug abuse Use of a drug for its intoxicating effect.

drug misuse Use of a drug for a purpose other than its intended use.

ductile Able to be drawn into wire.

eigen function A solution to a wave equation that describes the properties of an electron.

elastomer A synthetic polymer with rubberlike properties.

electric current A flow of electrons.

electrochemical cell A device that produces electricity by means of a chemical reaction.

electrodes The carbon rods or metal strips inserted into an electrolytic or electrochemical cell.

electrolysis The process of splitting a compound by means of electricity.

electrolyte A compound that, when melted or taken into solution, conducts an electric current.

electron The unit of negative charge.

electron configuration The arrangement of electrons about a nucleus.

electron dot symbol A representation in which the outer electrons of an atom are indicated by dots.

electronegativity The ability of an atom to attract electron density toward itself when joined to another atom by a chemical bond.

element A fundamental substance in which all atoms have the same number of protons.

emollient An oil or grease used as a skin softener.

emulsion A suspension of submicroscopic particles of fat or oil in water.

end note The portion of perfume that has low volatility; composed of large molecules.

endorphins The naturally occurring peptides that bond to the same receptor sites as the opiate drugs.

endothermic reaction A chemical reaction to which energy must be supplied as heat.

energy The capacity for doing work.

enkephalins Compounds composed of peptide chains of five amino acid units; morphinelike substances produced by the body.

entropy A measure of the randomness of a system.

enzyme A biological catalyst.

epidermis The outer layer of skin.

essential amino acid One of eight amino acids not produced in the body that must be included in the human diet.

ester A compound derived from carboxylic acids and alcohols; the —OH of the acid is replaced by an —OR group.

estrogen A female sex hormone.

eutrophication The excessive growth of plants in a body of water that causes some of the plants to die because of a lack of light. The water becomes choked, depleted of oxygen, and useless for fish or recreational purposes.

excited state That state in which an atom is supplied energy and an electron is moved from a lower to a higher energy level.

exothermic reaction A chemical reaction that releases heat.

expectorant A compound that is supposed to bring mucus up out of the bronchial passages.

experiment (v.) Try out a new idea or activity. (n.) An investigation in which variables are controlled.

eye shadow A product composed of a base of petroleum jelly with fats, oils, and waxes and colored by dyes or made white by zinc oxide or titanium dioxide pigments; used to color eyelids.

facts Pieces of information that remain the same, and are verified by repeated testing.

families of elements Vertical columns of the periodic table; also called groups.

fast-twitch fibers Muscle fibers that are stronger and larger and most suited for anaerobic activity.

fat A compound formed by the reaction of glycerol with three fatty acid units; a triglyceride.

fat-soluble vitamins Nonpolar vitamins with a high proportion of hydrocarbon structural elements; dissolve in the fatty tissue of the body and are stored for future use.

fatty acid A carboxylic acid that contains 4 to 20 or more carbon atoms in a chain.

fermentation The process by which yeast produces alcohol.

fetal alcohol syndrome The syndrome in which babies born to alcoholic mothers are small, deformed, and mentally retarded.

fibrils Bundles of parallel chains of cellulose in the cell walls of plants.

first law of thermodynamics Energy is neither created nor destroyed.

fission The splitting of the nucleus into two large fragments.

fixed nitrogen Nitrogen combined with another element.

flammable waste A substance that burns readily upon ignition, presenting a fire hazard.

flocs Large, porous clumps of sewage.

flotation method A coal-cleaning process that makes use of the different densities of coal and its major impurities.

fluorescence A phenomenon in which, after exposure to sunlight, certain chemicals continue to glow even when taken into a dark room.

fly ash Fine solid particles of soot and dust carried out from burning coal by the draft.

food additive Any substance other than basic foodstuffs that is present in food as a result of some aspect of production, processing, packaging, or storage.

formula mass The sum of the atomic masses of all of the atoms represented in the chemical formula.

fortification The process of adding more alcohol to wine to produce dessert wines with alcohol content greater than that obtainable from fermentation.

free radical A reactive neutral chemical species that contains an unpaired electron.

freezing The reverse of melting; the change from liquid to solid state.

freon A carbon compound containing fluorine, as well as chlorine; Du Pont tradename for chlorofluorocarbon.

fructose A simple sugar found in fruits and honey or made by isomerization of glucose.

fruit sugar Fructose.

fuels Substances that burn readily with the release of significant amounts of energy.

fuel cell A device in which chemical reactions are used to produce electricity directly from fuels and oxygen.

fundamental particles Basic units from which more complicated structures can be fashioned; protons, neutrons, and electrons.

gamma rays Rays similar to X rays that are emitted from radioactive substances; have higher energy and are more penetrating than X rays.

gas The state of matter in which the substance maintains neither shape nor volume.

general anesthetic A depressant that acts on the brain to produce unconsciousness as well as insensitivity to pain.

geometric growth A doubling in size for each growth period.

germination In brewing, the second step in the malting process in which the grain sprouts and proteins and starches are broken down and enzymes are released.

glass transition temperature An important parameter of polymers; above this temperature the polymer is rubbery and tough, while below it the polymer is like glass—hard, stiff, and brittle.

glucose The simple sugar that is circulated in the bloodstream; also called dextrose or blood sugar.

glycogen An animal starch composed of branched chains of glucose units.

GRAS list The list, established by the U.S. Congress in 1958, of additives generally recognized as safe.

greenhouse effect The retention of the sun's heat energy by the Earth as a result of excess carbon dioxide or other substances in the atmosphere; causes an increase in the Earth's atmospheric and surface temperature.

ground state The state of an atom in which all electrons are in the lowest possible energy level.

group A vertical column of the periodic table; a family of elements.

hallucinogen A drug that produces visions and sensations that are not part of reality.

halogen An element in Group VIIA of the periodic table.

hard liquor A distilled alcoholic beverage that tends to be strong tasting and high in ethanol content.

hard water Water containing ions of calcium, magnesium, and iron.

hazardous waste A substance that, when improperly managed, can cause or contribute to death or illness or threatens human health or the environment.

heat A measure of a quantity of energy; of how much energy a sample contains.

heat capacity The amount of heat needed to change the temperature of 1 g of a material by 1 °C.

heat of vaporization The amount of heat involved in the evaporation or condensation of 1 g of a material.

heat stroke A failure of the body's heat regulatory system; unless the victim is treated promptly, the rapid rise of body temperature causes brain damage or death.

herbicide A chemical used to kill weeds.

heterocyclic compound A cyclic compound in which one or more atoms in the ring is not carbon.

high-density polyethylene A polyethylene composed of linear molecules; a strong, rigid plastic.

high-fructose corn syrup A sweetener made by the hydrolysis of starch and the isomerization of a part of the resulting glucose to fructose.

homology The idea that a series of compounds has properties that vary in a regular and predictable manner.

hops Blossoms of the hop plant used to flavor beer and malt liquors.

hormone A chemical messenger that is secreted into the blood by an endocrine gland.

humectant A moistening agent.

hydrocarbon An organic compound that contains only carbon and hydrogen.

hydrogen bond The dipole interaction between a hydrogen atom bonded to F, O, or N in a donor molecule and an F, O, or N atom in a receptor molecule.

hydrolysis The reaction of a substance with water; literally, a splitting by water.

hydronium ion A water molecule to which a hydrogen ion (H^+) has been added; the characteristic ion of an aqueous acid.

hydrosphere The oceans, seas, rivers, and lakes of the Earth.

hydroxide ion OH^-; responsible for the properties of base in water.

hydroxyl group The —OH group.

hyperacidity An excess of acid in the stomach.

hypoallergenic cosmetics Cosmetics claimed to cause fewer allergic reactions than regular products.

hypotheses Guesses that can be tested by experiment.

incidental additive A food additive accidentally introduced during production, processing, or packaging or during storage.

incineration Destruction by burning.

inorganic Mineral; composed of compounds other than those of carbon.

inorganic chemistry The study of the compounds of all elements other than carbon.

intentional additive A substance purposefully put into a food product to perform some specific function.

interionic forces The electrostatic forces between ions.

iodine number The number of grams of iodine that will be consumed by 100 g of fat or oil; an indication of the degree of unsaturation.

ion A charged atom or group of atoms.

ionic bond The chemical bond that results when electrons are transferred from a metal to a nonmetal.

isomerization The conversion of a compound into one or more of its isomers.

isomers Compounds that have the same molecular formula but different structural formulas and properties.

isotopes Atoms that have the same number of protons but different numbers of neutrons.

Joule The SI unit of heat.

keratin The tough, fibrous protein that comprises most of the outermost layer of the epidermis.

kerogen The complex material found in oil shale; has an approximate composition of $(C_6H_8O)_n$, where n is a large number.

ketone bodies Acetoacetic acid, *beta*-hydroxybutyric acid, and acetone.

kilning The act of heating or baking to drive off excessive water.

kilocalorie Measurement of energy content, equal to 1000 calories.

kilogram The SI unit of mass, a quantity slightly greater than two pounds.

kinetic energy The energy of motion.

kinetic-molecular theory A model that uses the motion of molecules to explain the behavior of the three states of matter.

kwashiorkor A disease caused by protein deficiency.

LD$_{50}$ The dosage that would be lethal to 50% of the population of test animals.

lactose The sugar found in milk; composed of two simpler sugars, glucose and galactose.

lactose intolerance The inability of some individuals to break down the sugar lactose; caused by the absence of a necessary enzyme.

lagering A storage process in beer making that aids in the development of flavor.

lakes Colored complexes formed by adhering acid dye compounds to metal ions.

law of conservation of energy The amount of energy within the universe is constant; energy cannot be created or destroyed, only transformed.

law of conservation of mass Matter is neither created nor destroyed during a chemical change.

law of constant composition A compound always contains elements in certain definite proportions, never in any other combination; also called the law of definite proportions.

law of definite proportions A compound always contains elements in certain definite proportions, never in any other combinations; also called the law of constant composition.

law of multiple proportions Elements may combine in more than one proportion to form more than one compound, for example, CO and CO_2.

levo isomer A left-handed isomer.

light beer Beer specially brewed to have fewer Calories through lowered alcohol content or reduction in carbohydrates.

light wine Wine with fewer Calories made by fermenting unripe grapes or by distilling off some of the alcohol from regular wine.

limiting reagent The reactant that is used up first in a reaction, after which the reaction ceases no matter how much remains of other reactants.

line spectrum The pattern of colored lines emitted by each element.

liquid A state of matter in which the substance assumes the shape of its container, flows readily, and maintains a fairly constant volume.

lithosphere The solid portion of the Earth.

liter A measurement of volume equal to a cubic decimeter.

local anesthetic A substance that renders a part of the body insensitive to pain while leaving the patient conscious.

London smog Smog usually from burning coal, consisting mainly of sulfur dioxide, sulfuric acid, ash, soot, smoke, and fog.

Los Angeles smog Smog created by action of sunlight on unburned hydrocarbons and nitrogen oxides, mainly from automobiles; photochemical smog.

lotion An emulsion of submicroscopic fat or oil droplets dispersed in water.

low-density polyethylene A waxy, semirigid, transparent, or translucent plastic that is resistant to chemicals; composed of branched chains of carbon atoms.

LUST Acronym for leaking underground storage tanks.

macromolecule A molecule with a very high molecular mass; a polymer.

malignant tumor Cancer, a tumor characterized by irreversible growth.

malleable Able to be forged and welded.

mantle A region of the Earth thought to be composed mostly of silicates.

marijuana A preparation made from the leaves, flowers, seeds, and small stems of the *Cannabis* plant.

mascara A product composed of a base of soap, oils, fats, and waxes and colored by iron oxide pigments, carbon, chromium oxide, or ultramarine; used to darken eyelashes.

mass A measure of the quantity of matter.

mass number The sum of the numbers of protons and of neutrons in the nucleus of an atom of an element.

matter Stuff of which all materials are made; has mass and occupies space.

mechanism A series of individual steps in a chemical reaction that gives the net overall change.

melanin A brownish black pigment that determines the color of the skin and hair.

melting The process in which a substance changes from the solid to the liquid state.

messenger RNA (mRNA) The type of RNA that contains the codons for a protein; mRNA travels from the nucleus of the cell to a ribosome.

metals The group of elements to the left of the heavy, stepped, diagonal line on the periodic table.

meter The SI unit of length, slightly longer than three feet.

methemoglobin Hemoglobin in which the iron ion has a 3+ charge.

micas Minerals composed of SiO_4 tetrahedra arranged in a two-dimensional, sheetlike array.

micronutrients Substances needed by the body only in tiny amounts.

middle note The portion of perfume intermediate in volatility; responsible for the lingering aroma after most top-note compounds have vaporized.

mineral acids Acids derived from inorganic materials.

mixture Matter with a variable composition.

moderator A substance used to slow down the fission neutrons within a nuclear reactor.

mole The formula weight in grams of an element or compound; or a quantity of chemical substance that contains 6.02×10^{23} units of the substance.

monomer A molecule of relatively low molecular mass. Monomers are combined to make polymers.

monounsaturated fatty acids Fatty acids that contain one double bond per molecule.

mousse A foam or froth; a hair care product composed of holding resins to hold hair in place.

must In wine making: crushed grapes.

mutations Changes in the molecules of heredity in reproductive cells.

narrow-spectrum insecticides Insect-killing chemicals that are directed at one, or a few, insect species.

nasal decongestant A substance that aids in the dilation of the nasal passages for easier breathing.

natural philosophy Philosophical speculation about nature.

neurons Nerve cells.

neurotransmitters Chemicals that carry the impulse across the synapse from one nerve cell to the next.

neutralization The reaction of an acid and a base to produce a salt and water.

neutron A fundamental particle with a mass of approximately 1 amu and no electrical charge.

noble gases Generally unreactive gases that appear on the far right of the periodic table.

nonbonding pairs Pairs of electrons not involved in a bond.

nonionic surfactant A surfactant with a hydrocarbon tail and a polar head whose oxygen atoms attract water molecules and make the head water soluble; bears no ionic charge.

nonmetals The group of elements to the right of the heavy, stepped, diagonal line on the periodic table.

nonpolar covalent bond A covalent bond in which there is an equal sharing of electrons.

note A fraction of a perfume based on differences in volatility (*see also* top note, middle note, *and* end note).

nuclear fusion Combination of two small nuclei to produce one larger nucleus.

nuclear winter A period of dark, cold weather that may be caused by the dust and smoke entering the atmosphere from the explosion of nuclear bombs.

nucleoprotein A combination of a protein with a nucleic acid.

nucleotide A combination of a heterocyclic amine, a pentose sugar, and phosphoric acid; the monomer unit of nucleic acids.

nucleus Concentrated, positively charged matter at the center of an atom; composed of protons and neutrons.

octet rule Atoms seek an arrangement that will surround them with eight electrons in the outermost energy level.

oils Substances formed from glycerol and three unsaturated fatty acids; liquids at room temperature.

oil shale Fossil rock containing kerogen from which oil can be obtained at high cost by distillation.

open dumps Open solid-waste-disposal areas; often occupied by rats, flies, and other pests.

optical brightener A compound that absorbs the invisible ultraviolet component of sunlight and re-emits it as visible light at the blue end of the spectrum.

organic chemistry The study of the compounds of carbon.

orthomolecular medicine Pauling's belief that "having the right molecules in the right amounts in the right place at the right time" is essential to good health.

osteoporosis A disease characterized by a reduction in the quantity of bone.

oxidation Combination of elements and compounds with oxygen; loss of hydrogen; loss of electrons.

oxide A binary compound of oxygen and another element.

oxidized Combined with oxygen; had hydrogen removed; had electrons removed.

oxidizing agent A substance that causes oxidation and is itself reduced.

ozone layer The layer of the stratosphere that contains ozone and shields living creatures on Earth from deadly ultraviolet radiation.

particulate matter A pollutant composed of solid and liquid particles of greater than molecular size.

pathogenic Disease causing.

peptide bond The amide linkage that bonds amino acids in chains of proteins, polypeptides, and peptides.

perfluorocarbon A compound in which all hydrogen atoms have been replaced by fluorine atoms.

periodic table A systematic arrangement of the elements in columns and rows; elements in a given column have similar properties.

periods The horizontal rows of the periodic table.

persistent Does not break down readily in the environment.

pH scale An exponential scale of acidity; below 7, acidic; 7, neutral; above 7, basic.

phaeomelanin A red-brown pigment that colors the hair and skin of redheads.

pharmacology The study of the response of living organisms to drugs.

photochemical smog Smog created by the action of sunlight on unburned hydrocarbons and nitrogen oxides, mainly from automobiles; also called Los Angeles smog.

photoscan A permanent visual record showing the differential uptake of a radioisotope by various tissues.

photosynthesis The chemical process used by green plants to convert solar energy into chemical energy by reducing carbon dioxide.

photovoltaic cell A solar cell; a cell that converts sunlight directly to electrical energy.

physical properties Properties that describe qualities that can be demonstrated without changing the composition of the substance.

pica A disorder in which a person eats dirt, paint chips, and other materials generally regarded as inedible.

placebo A substance that looks and tastes like a real drug but has no active ingredients.

plasma A state of matter similar to a gas but composed of isolated electrons and nuclei rather than discrete whole atoms or molecules.

plasmid A circular piece of DNA that occurs outside the nucleus in bacteria.

plasticizer A chemical substance added to some plastics, such as vinyl, that makes them more flexible and easier to work with.

polar covalent bond A covalent bond in which more than half of the bond's negative charge is concentrated around one of the two atoms.

pollutant A chemical in the wrong place and/or in the wrong concentration.

polyatomic ion A charged particle containing two or more covalently bonded atoms.

polychlorinated biphenyls (PCBs) Compounds derived from the hydrocarbon biphenyl by the replacement of anywhere from 1 to 10 of the hydrogen atoms with chlorine atoms.

polymer A molecule with a large molecular mass; a chain formed of repeating smaller units.

polypeptide A polymer of amino acids; usually of lower molecular mass than protein.

polyunsaturated Containing two or more double bonds.

positive hole Electron-deficient area that maintains a positive charge around an atom, for example, around a boron atom in a silicon matrix.

positron A positively charged particle with the mass of an electron.

potential energy Energy by virtue of position or composition.

pre-emergent A herbicide that is rapidly broken down in the soil and can therefore be used to kill weed plants before the crop seedlings emerge.

primary sewage treatment plant A type of plant with a holding pond intended to remove some of the sewage solids as sludge by settling.

product A substance produced by a chemical reaction and whose formula follows the arrow in a chemical equation.

progestin A compound that mimics the action of progesterone.

proof Twice the percentage of alcohol by volume.

property A quality or trait belonging to a particular form of matter.

prostaglandins Hormonelike compounds derived from arachidonic acid that are involved in increased blood pressure, the contractions of smooth muscle, and other physiological processes.

proton The unit of positive charge in the nucleus of an atom.

proton donor An acid; a substance that gives up an H^+ (proton).

psychedelics Drugs that induce colorful visions.

psychotomimetics Drugs that induce psychoses.

psychotropic drugs Drugs that affect the mind.

pure substance Matter with a definite, or fixed, composition.

quantum A packet of specific size; one photon of energy.

quartz A compound composed of silicon and oxygen.

radioactivity Spontaneous emission of alpha, beta, and gamma rays by the disintegration of the nuclei of atoms.

radioisotopes Radioactive isotopes.

reactant A starting material or original substance in a chemical change; one whose formula precedes the arrow in a chemical equation.

reactive wastes Wastes that tend to react spontaneously or to react vigorously with air or water.

recombinant DNA DNA in an organism that contains genetic material from another organism.

Recommended Daily Allowance (RDA) The recommended level of a nutrient necessary for a balanced diet.

redox reaction A reaction in which oxidation and reduction occur.

reducing agent A substance that causes reduction and is itself oxidized.

reduction A gain of electrons; a loss of oxygen; a gain of hydrogen.

renewable resource A resource that is replenished by natural events.

replication Copying or duplication; the process by which DNA reproduces itself.

resin A polymeric material, usually a sticky solid or semisolid organic material.

respiration "Burning" of glucose by living cells; oxygen is absorbed and carbon dioxide is given off.

restriction endonucleases Enzymes that cleave DNA molecules at specific locations of specific base sequences.

restriction fragment length polymorphisms (RFLP) A pattern of segments of genetic material produced by restriction endonucleases.

reverse osmosis A method of pressure filtration through a semipermeable membrane; water flows from an area of high salt concentration to an area of low salt concentration.

ribonucleic acid (RNA) The form of nucleic acid found mainly in the cytoplasm, but present in all parts of the cell.

risk Any hazard that leads to loss or injury.

risk–benefit analysis An approach that involves the calculation of a desirability quotient by dividing the benefits by the risks.

Rule of 70 A mathematical formula that gives the doubling time for a population growing geometrically; 70 divided by annual rate equals doubling time.

salt bridge An interaction between an acidic side chain on one amino acid residue and a basic side chain on another; the resulting charges serve as an ionic bond between two peptide chains or between two parts of the same chain.

sanitary landfill A method of disposal in which garbage and trash are piled into a trench, compacted, and covered over with dirt.

saponins Natural chemical compounds that produce a soapy lather.

saturated fat A fat composed of a large proportion of saturated fatty acids.

saturated fatty acid A fatty acid that contains no double bonds.

saturated hydrocarbon A compound of carbon and hydrogen with only single bonds.

scientific law A summary of experimental data; often expressed in the form of a mathematical equation.

sebum An oily secretion of the body that protects skin from moisture loss.

second law of thermodynamics The degree of randomness in the universe increases in any spontaneous process.

secondary sewage treatment plant A plant that passes effluent from a primary treatment plant through gravel and sand filters to aerate and to remove additional suspended sewage solids.

segmer The repeating unit within a polymer.

silicates Compounds of metals with silicon and oxygen.

single bond The sharing of one pair of electrons.

skin protection factor The rating of a sunscreen's ability to limit the penetration of ultraviolet radiation.

slag Product of the reaction of limestone with silicate impurities in iron ore.

slow-twitch fibers Muscle fibers that are best suited for aerobic work.

smog The combination of smoke and fog; polluted air.

solar cell A device used for converting sunlight into electricity; a photoelectric cell.

solid A state of matter in which the substance maintains its shape and volume.

solute The substance that is dissolved in another substance to form a solution; usually present in a smaller amount than the solvent.

solution A homogeneous mixture of two or more substances.

solvent The substance that dissolves another substance to form a solution; usually present in a larger amount than the solute.

sparkling wine A wine containing dissolved carbon dioxide.

specific heat The amount of heat required to raise the temperature of 1 g of a substance by 1 °C.

standards of identity The minimum required proportions of ingredients to pass FDA standards for a particular food.

starvation The withholding of nutrition from the body whether voluntary or involuntary.

state of matter A condition or stage in the physical being of matter; a solid, liquid, or gas.

steels Alloys formed by adjusting the carbon content of iron.

steeping Soaking in liquid; the soaking of barley in water during the beer-making process.

stimulant A drug that increases alertness, speeds up mental processes, and generally elevates the mood.

stratosphere The portion of the atmosphere that contains the ozone layer; the next layer above the troposphere.

strong acid An acid that reacts completely; 100% ionized in water; a powerful proton donor.

strong base A powerful proton acceptor; dissociates 100% in water.

structural formula A chemical formula that shows how the atoms of a molecule are arranged, to which other atom(s) they are bonded, and the kinds of bonds.

sucrose Common table sugar derived principally from sugarcane or sugar beets.

sulfides Compounds of metals with sulfur.

sunscreen lotion A type of lotion that promotes tanning of the skin by blocking out the short-wave ultraviolet radiation while allowing the longer-wave ultraviolet radiation to pass through.

surface-active agent Any agent that stabilizes the suspension in water of nonpolar substances such as grease and oil.

synapses The gaps between nerve fibers

synergistic effect An effect much greater than just the sum of the expected effects.

tailings The waste products of ore processing.

tar sands Sands that contain bitumen, a thick hydrocarbon material.

technology The sum total of processes by which humans modify the materials of nature to better satisfy their needs and wants.

temperature A measure of intensity, or how energetic the particles of a sample are.

teratogens Toxic substances that cause birth defects when introduced into the body of the mother.

tertiary treatment The third stage of sewage treatment; designed to remove impurities not affected by primary or secondary treatment.

theories Detailed explanations of the behavior of matter based on experiments; may be revised if new data warrant.

thermal pollution The energy released into the environment that causes undesirable changes in the environment; released during energy conversions for other purposes such as the generation of electricity.

thermoplastic polymers The class of polymers that can be heated and reshaped.

thermosetting polymers The class of polymers that become permanently hard under the influence of heat and pressure; plastics that cannot be softened and remolded.

top note The portion of perfume that vaporizes the quickest; composed of relatively small molecules; responsible for odor when perfume is first applied.

toxicology The division of pharmacology that deals with the effects of poisons on the body, their identification and detection, and remedies against them.

toxic waste A waste that contains or releases poisonous substances in large enough amounts to threaten human health or the environment.

tracers Radioactive isotopes used to trace movement or locate the site of radioactivity in physical, chemical, and biological systems.

training effect The net effect, acquired through repeated exercise, of being able to do more physical work with less strain.

transcription The process by which DNA directs the synthesis of an mRNA molecule during protein synthesis.

transfer RNA (tRNA) A small RNA molecule that contains the anticodon nucleotides; the RNA molecule that bonds to an amino acid.

transition elements Metallic elements situated in the center portion of the periodic table in the B groups.

translation The process by which the information contained in the codon of an mRNA molecule is converted to a protein structure.

transmutation Change of one element into another.

tripeptide A compound composed of three bonded amino acids.

triple bond The sharing of three pairs of electrons between two atoms.

triplet Three-base sequence in the tRNA that determines which amino acid will be attached to an end of tRNA.

tritium A rare radioactive isotope of hydrogen with two neutrons and one proton in the nucleus (a mass of 3 amu).

troposphere The layer of the atmosphere nearest the Earth; harbors almost all living things.

tumor The abnormal growth of new tissues.

univalent atom An atom that forms only one bond.

unsaturated hydrocarbon A hydrocarbon containing a double or a triple bond.

valence The number of covalent bonds that an atom can form.

valence shell Outer shell of electrons of an atom.

valence-shell electron-pair repulsion theory A theory of chemical bonding; states that valence-shell electrons locate themselves as far apart as possible.

vaporization The process in which a substance changes from the liquid to the gaseous (vapor) state.

vitamins Organic compounds that the body cannot produce in the amounts required for good body health.

volt The unit of measurement of electrical potential, or of the tendency of the electrons in a system to flow.

vulcanization The process of making naturally soft rubber harder by reaction with sulfur.

water-soluble vitamins Vitamins with a high proportion of oxygen and nitrogen that are able to form hydrogen bonds with water.

weak acid An acid that reacts only slightly; a poor proton donor with a low percentage ionization in solution.

weak base A base that ionizes to a small degree and produces only a few OH^- ions in solution; a poor proton acceptor.

weight A measure of the force of attraction of the Earth for an object.

work Activity in which one exerts strength or faculties to do or perform something.

X ray Radiation similar to visible light but of much higher energy and much more penetrating.

zwitterion A compound that contains both a positive and negative charge; a dipolar ion.

F

Answers to Selected Problems

Answers are provided for all odd-numbered numerical problems and for other selected odd-numbered problems.

Chapter 1

3. matter: a, b, d, e
19. benefits divided by risks
29. both
33. a. chemical change   b. physical change
39. pure substances: a, b; mixture: c
41. mixture: a; pure substance: b
43. elements: a, b; compound: c
45. element: b; compounds: a, c
49. 1000 mm; 100 cm
51. a. L   b. same size
53. a. 1500 mm   b. 160 mm
55. a. 15 g   b. 86 mg
57. 0.1 mL: 15 mL

59. a. 25 °C   b. 100 °C
61. 1.1 g/cm^3
63. 1.6 g/mL
65. 680 g
67. 11.3 g/cm^3
69. 87.2 mL

Chapter 2

3. the laws of definite proportions and multiple proportions
5. atomistic: a, d, e; continuous: b, c, f
13. 11 g CO_2
15. the law of definite proportions
17. the law of conservation of mass
19. No.
27. 9000 kg H_2O
29. 56 g CO

A35

Chapter 3

11. attract

13. protons and neutrons

19. 32

25. a. 2 b. 11 c. 17 d. 8 e. 12 f. 16

29. $1s^2 2s^2$

31. a. Be b. N c. Al

39. argon

41. No.

45. a. IA b. VIIA c. IIA

47. b

49. b, d, e

51. b, c

53. 2

Chapter 4

5. $_1^1H, \ _1^2H, \ _1^3H$

7. $_{35}^{83}Br$

9. a. $_{31}^{69}Ga$ b. $_{42}^{98}Mo$ c. $_{42}^{99}Mo$ d. $_{43}^{98}Tc$

11. isotopes: b

13. 10.8 amu

19. no change in either

27. a. $_2^4He$ b. $_{-1}^0e$

29. $_{11}^{24}Na$

31. 1500; 750

37. $_{93}^{237}Np \longrightarrow \ _2^4He + \ _{91}^{233}Pa$

39. alpha particles

43. 11 460 years

45. 24.6 years

47. neutrons

49. $_8^{18}O$

51. mass number: 258; atomic number 105

Chapter 5

5. a. Na·   b. ·F̈:   c. ·Ċ·

7. a. ·Äl·   b. K·   c. ·C̈l:

9. a. Mg^{2+} 2:F̈:⁻   b. 2 Na⁺:Ö:²⁻

c. Ca^{2+} 2:C̈l:⁻   d. 2 K⁺:S̈:²⁻

11. a. Mg^{2+} :Ö:²⁻   b. Al^{3+} :N̈:³⁻

13. ionic: a; polar covalent: b; nonpolar covalent: c

15. ionic: a, b; polar covalent: c

17. polar covalent

19. a. :F̈:N̈:F̈:   b. H:C::C:H
        :F̈:   H H

c. H:C:::C:H   d. H:C::Ö:
                        H

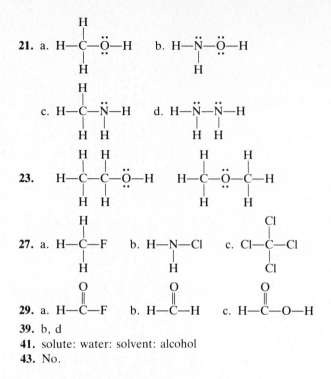

21. a., b., c., d. (Lewis structures)

23. (Lewis structures)

27. a. b. c. (structures)

29. a. b. c. (structures)

39. b, d

41. solute: water: solvent: alcohol

43. No.

Chapter 6

1. a. sodium ion   b. magnesium ion
c. aluminum ion

3. a. iron(III) ion (ferric ion)
b. copper(I) ion (cuprous ion)
c. silver ion

5. a. 3+ b. 2− c. 1+ d. 1−

7. Cations: a and c; anions: b and d

9. a. NH_4^+ b. HCO_3^- c. PO_4^{3-}

11. a. carbonate ion
b. monohydrogen phosphate ion
c. nitrate ion
d. hydroxide ion

13. a. LiF b. CaI_2 c. $AlBr_3$ d. Al_2S_3

15. a. $FeSO_4$ b. $(NH_4)_3PO_4$ c. $Mg_3(PO_4)_2$
d. $CaHPO_4$

17. a. potassium nitrite   b. lithium cyanide
c. ammonium iodide   d. sodium nitrate

19. a. magnesium acetate   b. aluminum acetate
c. ammonium phosphate   d. ammonium oxalate

23. a. N_2O b. OF_2 c. SO_3 d. SCl_4 e. S_4N_4

29. Al = 2; C = 12; H = 18; O = 12

31. balanced: a, d, and e; not balanced: b and c

33. a. $4\,Al + 3\,O_2 \longrightarrow 2\,Al_2O_3$
  b. $2\,C + O_2 \longrightarrow 2\,CO$
  c. $N_2 + O_2 \longrightarrow 2\,NO$
  d. $2\,SO_2 + O_2 \longrightarrow 2\,SO_3$
  e. $2\,NO + O_2 \longrightarrow 2\,NO_2$

35. a. $Cu + 2\,H_2SO_4 \longrightarrow SO_2 + CuSO_4 + 2\,H_2O$
  b. $2\,NH_4Cl + CaO \longrightarrow CaCl_2 + H_2O + 2\,NH_3$

37. a. 80 amu b. 233 amu c. 98 amu d. 123 amu

39. a. 2.0 mol b. 4.0 mol c. 0.0500 mol

41. a. 0.0100 mol b. 0.10 mol c. 0.000100 mol

43. a. 4 g b. 699 g c. 980 g d. 2.46 g

45. a. 4 mol b. 0.6 mol c. 0.684 mol

47. a. 32 g b. 16 g c. 1.6 g

49. a. 4 g b. 44 g

51. a. 6 mol b. 16 mol

53. a. 4.4 g b. 264 g

55. 12 g

57. 110 kg

59. 7.98 g

61. 22 g

63. 23 g

Chapter 7

5. OH^-

7. strong bases: NaOH, KOH; weak base: NH_3

11. H_2SO_4

13. KOH

19. $NaOH + HCl \longrightarrow NaCl + H_2O$

27. acidic: a, c; basic: d; neutral: b

31. 2

33. $1 \times 10^{-10}\ M$

Chapter 8

1. a. CO_2 b. SO_2 c. NO d. H_2O

5. oxidation: a; reduction: b and c

7. *Oxidizing agent*   *Reducing agent*
  a. Cl_2 KBr (actually Br^-)
  b. C_2H_4 H_2
  c. HCl (actually H^+)   Fe

9. *Oxidized   Reduced*
  a. S   N
  b. I   Cr
  c. C   Mn

21. hydrogen peroxide, sodium hypochlorite, chlorine, ozone

Chapter 9

1. crust, mantle, core

3. iron and nickel

5. compounds of metals with silicon and oxygen

7. the chemistry of elements other than carbon

19. iron ore, coal, limestone

23. oxidation

27. iron

31. open dumps, sanitary landfills, incineration

33. 3.75 million t of zinc, 300 000 t of lead, 825 000 t of copper

35. a cube 16 m on each side

37. 2000 kg

39. 120 000 000 kg Al_2O_3, 260 000 000 kg bauxite, 1 100 000 000 kWh of electricity

Chapter 10

3. organic: a, c, d; inorganic: b, e, f

5. a. ethane b. ethylene c. acetylene

7. $CH_3CH_2CH_2CH_3$ $CH_3\overset{\textstyle |}{C}HCH_3$
                                        $|$
                                       CH_3

    butane   isobutane

9. a. same compound b. same compound
  c. isomers d. same compound e. isomers

11.

13. $CH_3CH_2CH_2CH_2CH_3$ $CH_3CHCH_2CH_3$
                                          $|$
                                         CH_3

    CH_3
     $|$
  CH_3CCH_3
     $|$
    CH_3

15. a. homologs b. none of these
  c. same d. isomers

17.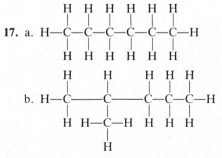

19. aromatic: a, d; not aromatic: b, c

25. $CH_2{=}CH_2 + H_2 \xrightarrow{\text{Ni}} CH_3CH_3$

27. $CH_4 + 2\,O_2 \longrightarrow CO_2 + 2\,H_2O$

29. $2 C_6H_6 + 15 O_2 \longrightarrow 12 CO_2 + 6 H_2O$
130 g CO_2

Chapter 11

1.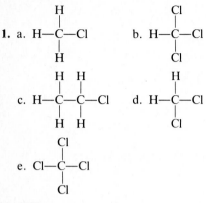
a. H—C(H)(H)—Cl   b. H—C(Cl)(Cl)—Cl

c. H—C(H)(H)—C(H)(H)—Cl   d. H—C(H)(Cl)—Cl

e. Cl—C(Cl)(Cl)—Cl

15. 40%
19. a. propyl   b. methyl
21. as antiseptics and disinfectants

25. —C(=O)—

31. a. CH_3CHCH_3 (OH)   b. CH_3OH c. CH_3CH_2OH
33. sulfuric acid, hydrochloric acid, nitric acid, phosphoric acid

35. a. $CH_3C(=O)$—OH   b. $CH_3CH_2CH_2C(=O)$—OH

c. H—C(=O)—OH

37. a. $CH_3CH_2CH_2C(=O)OCH_2CH_3$

b. $CH_3CH_2CH_2C(=O)OCH_3$

39. ester
41. amino group

43. —C(=O)—N(—)—

45. heterocyclic compounds: a, c
53. ribose: a, b, d; deoxyribose: c, e, f
55. purine: a; pyrimidine: b

57. a. a purine   b. RNA
59. a. guanine; b. thymine; c. cytosine; d. adenine
67. messenger RNA and transfer RNA
69. transfer RNA
71. ...U-A-A-G-C...
73. a. A-A-A   b. G-U-A   c. U-C-G   d. G-G-C

Chapter 12

9. CH_2=CH—Cl
13. A double bond
17. addition
19. ~CH_2CH—CH_2CH—CH_2CH—CH_2CH~ (each with OC(=O)CH_3)

21. ~CH_2CH—CH_2CH—CH_2CH—CH_2CH~ (each with phenyl)

23. CF_2=CF_2
25. CH_2=C(CH_3)—CH=CH_2

29. polybutadiene, polychloroprene (Neoprene), styrene-butadiene rubber (SBR)
31. CH_2=CH—CH=CH_2
35. carbon black, rayon or nylon cord, fiberglass, steel belts
37. titanium dioxide, carbon black, lead compounds, oxides of iron

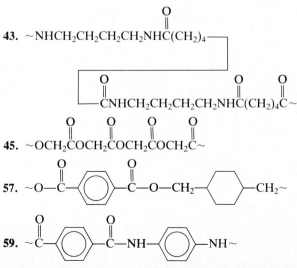

43. ~$NHCH_2CH_2CH_2CH_2NHC(=O)(CH_2)_4$——$C(=O)NHCH_2CH_2CH_2CH_2NHC(=O)(CH_2)_4C$~

45. ~$OCH_2C(=O)OCH_2C(=O)OCH_2C(=O)OCH_2C$~

57. ~O—C(=O)—[benzene]—C(=O)—O—CH_2—[cyclohexane]—CH_2~

59. ~C(=O)—[benzene]—C(=O)—NH—[benzene]—NH~

61.

$$\sim CH_2\underset{\overset{|}{CH_3}}{\overset{\overset{CH_3}{|}}{C}} - CH_2\underset{\overset{|}{CH_3}}{\overset{\overset{CH_3}{|}}{C}} - CH_2CH=CHCH_2 - CH_2\underset{\overset{|}{CH_3}}{\overset{\overset{CH_3}{|}}{C}}\sim$$

Chapter 13

3. nuclear fusion
15. $C + O_2 \longrightarrow CO_2$
17. Soot is unburned carbon.
21. ethylene, hydrogen, methane
23. $CH_4 + 2 O_2 \longrightarrow CO_2 + 2 H_2O$
25. $CH_4 + O_2 \longrightarrow C + 2 H_2O$
29. $C_5H_{12} + 8 O_2 \longrightarrow 5 CO_2 + 6 H_2O$
33. wood
35. petroleum
39. increased
41. 3800 kcal
43. 13 600 kcal
45. 1700 W
47. 36 years
51. the sun
55. 1830 kg

Chapter 14

1. 55%
5. 17%
7. the eastern seaboard and upper midwestern states
11. No.
13. $^{232}_{90}Th + ^{1}_{0}n \longrightarrow ^{233}_{90}Th$
   $^{233}_{90}Th \longrightarrow ^{233}_{91}Pa + ^{0}_{-1}e$
   $^{233}_{91}Pa \longrightarrow ^{233}_{92}U + ^{0}_{-1}e$
17. No.
25. Plasma is a mixture of atomic nuclei and free electrons.
45. $^{30}_{14}Si + ^{1}_{0}n \longrightarrow ^{31}_{14}Si$
   $^{31}_{14}Si \longrightarrow ^{31}_{15}P + ^{0}_{-1}e$

Chapter 15

1. the troposphere
3. 78%, N_2, 21% O_2, 1% Ar
9. $3 O_2 \longrightarrow 2 O_3$
13. No.
17. cool, damp, cloudy
19. $2 SO_2 + O_2 \longrightarrow 2 SO_3$
23. $H_2SO_4 + 2 NH_3 \longrightarrow (NH_4)_2SO_4$
31. to make cement and mineral wool
33. hydrocarbons, nitrogen oxides, ozone, aldehydes, PAN
35. automobiles

39. $2 NO + O_2 \longrightarrow 2 NO_2$
49. operate at a lower temperature; increase the fuel-to-air ratio
53. $CaCO_3 + H_2SO_4 \longrightarrow CaSO_4 + H_2O + CO_2$
67. 8000 μg (or 8 mg)
69. 2.6%

Chapter 16

3. The gasoline will not dissolve in the water. It will float.
9. nearly 98%
11. cholera, typhoid fever, dysentery, hepatitis
13. Na^+, Ca^{2+}, K^+, Mg^{2+}
15. water containing Ca^{2+}, Mg^{2+}, and (sometimes) Fe^{2+}
17. disease-causing bacteria, viruses, etc.
25. acid rain, drainage from abandoned mines
29. add lime (or other basic substances)
37. to kill pathogenic microorganisms
41. organic compounds such as hydrocarbons and chlorinated hydrocarbons
45. $3 Ca(OH)_2 + Al_2(SO_4)_3 \longrightarrow$
   $3 CaSO_4 + 2 Al(OH)_3$
49. usually from fertilizer applied to the soil

Chapter 17

5. 50 weeks; arithmetically
7. in one more month
9. 64 years
11. 12 years
13. carbon, hydrogen, oxygen
27. calcium, magnesium, sulfur
49. 0.5 mg; 0.025 mg
53. 300 mg
57. 420 μg

Chapter 18

1. carbohydrates, fats, proteins
3. a source of energy
5. glucose
17. Fats are esters of glycerol and fatty acids.
19. padding, insulation, a reserve source of energy
23. saturated: d, e
25. a. 18 b. 18 c. 18 d. 16 e. 18
29. about 30% with a maximum of 10% from saturated fats
31. corn oil
33. in every living cell
35. muscle, skin, hair
37. amino groups and carboxylic acid groups
43. red meat, fish, eggs, milk, chicken, cheese...

45. two years
55. proteins
59. Vitamins are organic; minerals are inorganic.
69. water soluble: b, c; fat-soluble: a, d
77. 32 g
83. 83 kcal
85. a. 81 kcal   b. 12 g   c. 15 mL   d. 4.1%

Chapter 19

3. salt, sugar, corn syrup, citric acid, baking soda, vegetable colors, mustard, pepper
5. The additive must be proven safe and effective for its intended use.
17. carboxylic acids and their salts
23. vitamin E
27. cyclamates, saccharin, aspartame
31. antibiotics, PCBs, pesticides, PBBs, DES. . .
39. regular version: 28.3%; low-sodium version: 2.7%
41. 90 T, 121 T, 250 T

Chapter 20

3. a salt of a long-chain carboxylic acid
5. dyes, perfumes, creams, oils, . . .
7. Potassium soaps are softer and produce a finer lather.
11. Hard water contains calcium, magnesium, and iron ions.
29. surfactants and (sometimes) perfumes, colors, oils (to soften skin)
33. surfactants, sodium carbonate, ammonia, solvents, disinfectants, deodorants. . .
37. asphalt tile, wood surfaces, aluminum
45. They are flammable and toxic when swallowed or inhaled.
55. II
57. III

59.
$$CH_2OH$$
$$|$$
$$CH—OH + CH_3(CH_2)_6\overset{O}{\overset{\|}{C}}O^-Na^+ +$$
$$|$$
$$CH_2OH$$

$$CH_3(CH_2)_4\overset{O}{\overset{\|}{C}}O^-Na^+ + CH_3(CH_2)_8\overset{O}{\overset{\|}{C}}O^-Na^+$$

Chapter 21

3. They are claimed to have medical effects, altering body functions or curing ailments such as dandruff.

5. detergent and abrasive
11. Sebum is an oily secretion of the sebaceous glands of the skin.
13. about 10%
15. An oily substance that prevents evaporation of water from the skin.
17. lakes; dyes adhered to metal ions
19. top note (small molecules), middle note, bottom note (large molecules)
21. diluted perfumes
23. to give a cooling effect on the skin
27. aluminum chlorohydrate
29. hydrogen bonds, salt bridges, disulfide linkages, hydrophobic interactions
31. a detergent
33. melanins
35. oxidation
37. diamines
39. an oxidizing agent
41. by coating the hair with a resin
43. I, III, IV
45. I

Chapter 22

3. acetylsalicylic acid
9. Aspirin is an anticoagulant.
11. Both are antipyretics and analgesics.
17. ephedrine, phenylephrine hydrochloride
21. none
35. a female sex hormone
37. the ethynyl group ($—C\equiv C—H$)
41. steroids: b, c
43. antimetabolites, alkylating agents

Chapter 23

1. drugs that affect the mind
3. stimulants, depressants, hallucinogens
15. *para*-aminobenzoic acid
31. norepinephrine and serotonin
35. by mimicking natural brain amines
39. meprobamate, diazepam, chlordiazepoxide
43. amides
49. It is fat soluble.

Chapter 24

5. actin and myosin

9. anaerobic

17. pheasants: Type IIB fibers; great blue heron: Type I fibers

23. about 3500 Cal

31. sodium ions, potassium ions, calcium ions

33. Antidiuretic hormone signals the kidneys to conserve water.

37. Urine is cloudy and yellow when a person is dehydrated.

51. 4 hr

53. 0.1 hr (6 min)

55. 1.06 g/mL (lean)

57. 40 g

59. 1000 km

Chapter 25

1. a study of the response of living organisms to drugs

7. Hydrolysis is a reaction with water; literally, a splitting by water.

21. poisoning by cadmium ions (Cd^{2+})

23. Botulin blocks the synthesis of acetylcholine.

27. Atropine blocks the acetylcholine receptor sites, preventing their continuous stimulation.

31. Cotinine is less toxic and more water-soluble (thus more readily excreted).

37. cigarette smoking

45. a chemical substance that causes mutations

47. a chemical substance that causes birth defects

57. a. 98 mg   b. 2000 t

Index

References to figures have an *f* following the number, and references to tables have a *t* following the number.